9th Brazilian Congress on Metrology (Metrologia 2017)

Quality Assurance in a Changing World

Journal of Physics: Conference Series Volume 975

Fortaleza, Brazil
26 - 29 November 2017

Editor:

Rodrigo P.B. Costa-Felix

ISBN: 978-1-5108-6043-8
ISSN: 1742-6588

Printed from e-media with permission by:

Curran Associates, Inc.
57 Morehouse Lane
Red Hook, NY 12571

Some format issues inherent in the e-media version may also appear in this print version.

Printed by Curran Associates, Inc. (2018)

For permission requests, please contact the Institute of Physics
at the address below.

Institute of Physics
Dirac House, Temple Back
Bristol BS1 6BE UK

Phone: 44 1 17 929 7481
Fax: 44 1 17 920 0979

techtracking@iop.org

Additional copies of this publication are available from:

Curran Associates, Inc.
57 Morehouse Lane
Red Hook, NY 12571 USA
Phone: 845-758-0400
Fax: 845-758-2633
Email: curran@proceedings.com
Web: www.proceedings.com

Table of contents

Volume 975

9th Brazilian Congress on Metrology (Metrologia 2017)

26–29 November 2017, Fortaleza, CE, Brazil

Accepted papers received: 19 February 2018
Published online: 21 March 2018

Preface

Papers

IOP Conf. Series: Journal of Physics: Conf. Series **975** (2018) 011001 doi:10.1088/1742-6596/975/1/011001

PREFACE

METROLOGIA 2017

When choosing the theme "Quality assurance in a changing world" for Metrology 2017, we aimed to bring the importance of metrology and conformity assessment for international processes and decision making. We have not imagined, however, that so many changes and threats would occur in those months that preceded our event.

In face of current uncertainties, we believe that scientific education and a critical knowledge are the most powerful antidotes against the shadows on the horizon. Metrology is a fundamental basis for a strong science, which gives us confidence to be able to move forward. Scientific debate creates an environment of freedom and creativity.

In this changing world, Metrology also changes. The international system of units is about to be revised in a near future. We also deal with new fields of knowledge, which present new challenges. This whole dynamics requires commitment to improving management processes to create more efficient products, which by their turn require more precise measurement techniques.

Metrologia 2017 was held in the face of these new challenges and uncertain future scenarios. The Metrology 2017 was carried out by the Brazilian Society of Metrology (SBM) in partnership with the National Institute of Metrology, Quality and Technology (Inmetro) and the Institute of Radioprotection and Dosimetry (IRD). The meeting aimed to disseminate the culture of metrology and conformity assessment as essential factors for the sustainable development of our society.

Metrology 2017, as in previous years, was composed of simultaneous congresses in the areas of Mechanics, Electrical, Ionizing Radiations, Chemistry and Optics, as well as other Basic Industrial Technology disciplines gathered in the Brazilian Congress of Metrology.

Repeating the initiative held in the last two editions of the Congress, the best papers selected by the Scientific Technical Committee have been published in a special edition of the Journal of Physics Conference Series edited by IOP Publishing. For the present edition, the editors selected the best 78 out of 273 accepted papers to be published.

IOP Conf. Series: Journal of Physics: Conf. Series **975** (2018) 011001 doi:10.1088/1742-6596/975/1/011001

PRESIDENT OF THE CONGRESS

Americo Tristao Bernardes
Federal University of Ouro Preto (UFOP)
atb@iceb.ufop.br

EDITOR-IN-CHIEF

Rodrigo P B Costa-Felix
Brazilian National Institute of Metrology, Quality and Technology
rpfelix@inmetro.gov.br

CHAIR OF THE II BRAZILIAN CONGRESS ON OPTICS METROLOGY (CBMO)

Iakyra Borrakuens Couceiro
Brazilian National Institute of Metrology, Quality and Technology
ibcouceiro@inmetro.gov.br

CHAIR OF THE XII INTERNATIONAL SEMINAR ON ELECTRICAL METROLOGY (SEMETRO)

Gelson Martins da Rocha
Brazilian National Institute of Metrology, Quality and Technology
gmrocha@inmetro.gov.br

CHAIR OF THE III BRAZILIAN CONGRESS ON IONIZING RADIATION (CBMRI)

José Guilherme Pereira Peixoto
Brazilian Institute of Radiation Protection and Dosimetry
guilherm@ird.gov.br

CHAIR OF IV INTERNATIONAL CONGRESS ON MECHANICAL METROLOGY (CIMMEC)

José Carlos Valente de Oliveira
Brazilian National Institute of Metrology, Quality and Technology
jcoliveira@inmetro.gov.br

Peer review statement

All papers published in this volume of *Journal of Physics: Conference Series* have been peer reviewed through processes administered by the proceedings Editors. Reviews were conducted by expert referees to the professional and scientific standards expected of a proceedings journal published by IOP Publishing.

A risk analysis approach applied to field surveillance in utility meters in legal metrology

B A Rodrigues Filho[1], N S Nonato[2] and A D Carvalho[3]

[1] Department of Legal Metrology, National Institute of Metrology, Quality and Technology Inmetro, São Paulo Office, 1922 Santa Cruz St., São Paulo, Brazil
[2] Accreditation Coordenation, Inmetro, 50 Nossa Senhora das Graças Av., Duque de Caxias, RJ, Brazil
[3]Department of Administrative Services, Inmetro, 50 Nossa Senhora das Graças Av., Duque de Caxias, RJ, Brazil

Email: bafilho@inmetro.gov.br

Abstract. Field surveillance represents the level of control in metrological supervision responsible for checking the conformity of measuring instruments in-service. Utility meters represent the majority of measuring instruments produced by notified bodies due to self-verification in Brazil. They play a major role in the economy once electricity, gas and water are the main inputs to industries in their production processes. Then, to optimize the resources allocated to control these devices, the present study applied a risk analysis in order to identify among the 11 manufacturers notified to self-verification, the instruments that demand field surveillance.

1. Introduction

Legal control of measuring instruments and metrological supervision are both the activities responsible for metrological assurance, guarantying the conformity and confidence of measuring instruments under the legal metrology umbrella. Especially the accordance of the measurement errors to the maximum permissible values avoid the economic distortion, which represents the amount of money at risk due to inaccurate measurements [1]. These measurement errors might represent a significant amount of the GDP [2].

Utility meters represent measuring instruments to control public services on utility mains, such as water, electricity and gas. They are also an important part of the economies since they work as main inputs to most of the supply chains in the industry, aggregating their distortion due to measurement inaccuracies, impacting the price of the products impacting the entire economy at all its levels.

While the legal control of measuring instruments comprises type approval and verifications, metrological supervision represents an additional level of control within legal metrology and is responsible for checking the accordance of metrology regulations [3]. Especially in the activities wherein the notified bodies work as substitutes to the metrological authorities, as self-verifications, metrological supervision plays a major role.

The activity is structured in three different and complementary levels [4]:

- Quality surveillance – Aiming at establishing that the quality systems of bodies working in legal metrology complies with regulation;

- Market surveillance – Aiming at a measuring instrument and prepackage before their placement in the market;
- Field surveillance – Aiming at establishing that a measuring instrument in use in the field complies with the statutory requirements.

According to these levels, considering the self-verification as an example, it represents that the notified body needs a structured quality system, demonstrating the capacity to produce measuring instruments in accordance with regulation; then market surveillance checks the accordance of the produced devices; and finally the field surveillance tests the behavior of the instruments in-service. These three stages guarantee the conformity/reliability life cycle of measuring instruments.

However, the utility meters represents the majority of measuring instruments in-service, and the surveillance of each one of them, as well as they replacement would represent great costs to the society. Consequently, the rational usage of the available resources demands for optimization of the budget allocated in legal metrology services.

As the risk analysis is a consolidate tool in quality management and it has been successfully applied to decision-making processes and to predict events wherein risk is involved [5-8] , the present study uses a risk analysis approach to identify the utility meters subjective to field surveillance.

2. Methodology

The present section presents the methodology applied in order to identify the water meters that present greater risk according to the risk analysis approach, as well as the parameters used in the analysis.

2.1. Risk analysis

The risk analysis considers two parameters in order to determine the risk [6]:

- Impact – representing the impact of an unwanted legal interest;
- Probability – Degree to which the unwanted occurrence has happened;

The Likert scale is applied to measure both impact and probability, ranging from 1 – 5, wherein the impact 1 represents the minimal and 5 a significant impact. Probability ranges from very unlikely, 1, to very likely, 5. Then, the risk is defined as (1):

$$Risk = Impact \times Probability \quad (1)$$

We considered as legal interests in order to compute the impact: economic; environmental; and consumer´s protection. Since utility services share similarities, the impact evaluation is a constant for all the manufacturers.

The parameter choose to determine the probability is the non-conformities of measuring instruments during market surveillance.

Finally, matrix displays the risk, showing the higher risk instruments, guiding the field surveillance process more efficiently. Figure 1 shows an example of a risk matrix.

Once the water meters were selected to be tested according to the risk analysis, the field surveillance, the inspection lot sampling was based on the ISO2859-2:2001 [9]. The sampling based on a lot of measuring instruments is also showed in table 1.

The information regarding the location of the water meters identified as greater risk instruments were provided by the Basic Sanitation Utility Company of the State of São Paulo – SABESP.

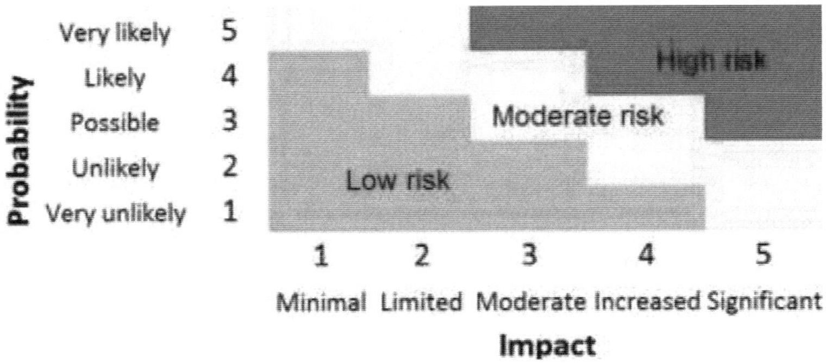

Figure 1. Risk matrix. Adapted from [6].

Table 1. Inspection lot sampling.

Lot	Sample size	Non-conformities	
		Accept	Reject
50 – 1,200	50	1	2
1,201 – 3,200	80	3	4
3,201 – 10,000	125	5	6
10,001 – 35,000	200	10	11

2.2. Case test

We carried out the method for water meters for cold potable water, hereinafter called water meters, considering the 11 manufacturers in the country and the city of São Paulo (estimated population of 12,038,175 in 2016 , representing the biggest metropolitan area in Brazil [10], consequently representing the biggest density of water meters.

To conduct the risk analysis, similar devices must be considered due to sampling procedures. We considered the residential type, class C, nominal flow rate Q_n = 1.5 m³/h, produced from 2013 to 2014.

For confidential purposes, the manufacturers are represented from A to K. The parameters applied for the risk analysis are shown in table 2.

Table 2. Risk analysis parameters.

Manufacturer	Probability	Impact	Risk
A	1	2	2
B	1	2	2
C	1	2	2
D	1	2	2
E	1	2	2
F	1	2	2
G	4	2	8
H	5	2	10
I	1	2	2
J	2	2	4
K	1	2	2

3. Results

The risk matrix, based on the parameters displayed in table 2 is shown in figure 2. All the manufacturers are in the limited impact column once this parameter is a constant. Thus, probability, in this analysis, is the parameter that distinguishes the manufacturers more likely to be surveyed.

According to the risk analysis, nine manufacturers are classified as low risk and two as moderate risk. H followed by G presented more non-conformities elevating their probability, resulting in a higher risk when compared to the others. In addition, no manufacturers were classified in the high-risk region.

After the identification of the manufacturer, we applied a sampling procedure in order to identify the instruments to be collected for metrological tests. Based on the total of water meters installed in the city of São Paulo, the selected sample is represented by 80 instruments. The Google Maps tool was used to display the samples. Figure 3 shows the geographic location of the selected water meters.

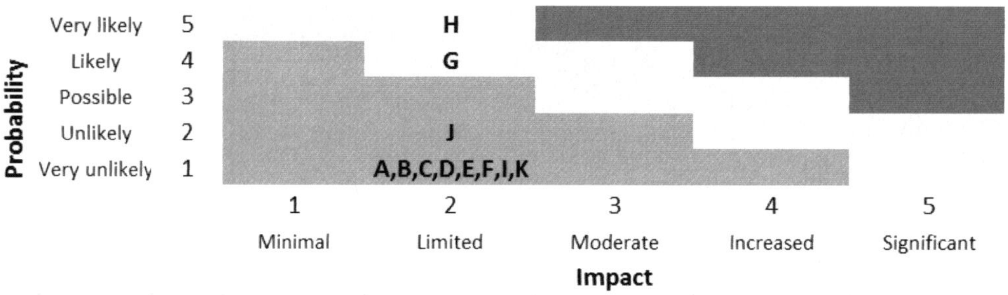

Figure 2. Risk matrix for the case test.

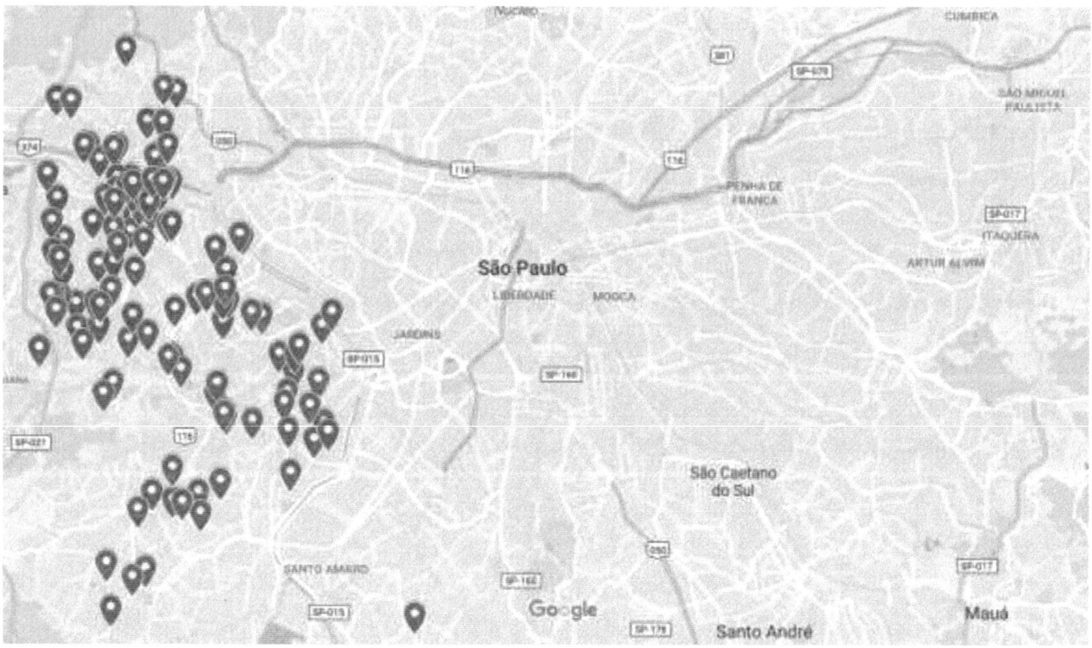

Figure 3. The geographic location of the higher risk water meter, representing the manufacture H

4. Conclusions

The utility services represent a major importance to both the economy as an input to supply chains in basically every industry, and the society since water, energy and gas are vital goods to people´s everyday consumption.

This study presented a risk analysis based method in order to optimize the field surveillance, using water meters as a case test, optimizing the resources in legal metrology in areas that present a higher risk to both economy and society.

The methodology allowed identifying the higher risk instruments (manufacturer) for a specific class of meters, as well as to locate them geographically.

Finally, this study was limited to the identification of the instruments. Future studies may conduct the tests in order to verify the conformity of the selected devices and nonconformity rate.

Acknowledgments
The authors acknowledge the National Institute of Metrology, Quality and Technology – Inmetro, and the Basic Sanitation Utility Company of the State of São Paulo – SABESP, which enabled us to carry out the present study.

5. References

[1] Stiefel S W Management assistance for weights and measures progress, measuring inaccuracy's economic distortion 1973 (The Bureau) Report of the National Conference on Weights and Measures
[2] Filho B A R and Gonçalves R F 2016 Measuring the Economic Impact of Metrological Frauds in Trade Metrology Using an Input-Output Model *Advances in Production Management Systems Initiatives for a Sustainable World.* (Springer, Cham) pp 624–32 doi: 10.1007/978-3-319-51133-7_74
[3] OIML 2013 *International vocabulary of terms in legal metrology.* 1st ed. BIPM (Bureau International des Poids et Mesures) p 57
[4] OIML 2004 *Principles of metrological supervision* (Bureau International des Poids etMesures). Available from: https://www.oiml.org/en/files/pdf_d/d009e04.pdf
[5] ISO.9001:2015 *Quality management systems – Requirements* (International Organizations of Standardization)
[6] WELMEC 2011 *Risk Assessment Guide for Market Surveillance: Weigh and Measuring Instruments* (WELMEC European cooperation in legal metrology) Available from: http://www.welmec.org/fileadmin/user_files/publications/WELMEC_05.03_Risk_Assessment_Guide_issue1.pdf
[7] Ribeiro A S Costa A C Candeias P Sousa J A, Martins L L, Martins A C F, et al 2016 Assessment of the Metrological Performance of Seismic Tables for a QMS Recognition. *J Phys Conf Ser* vol 772 ed. 1 doi: 10.1088/1742-6596/772/1/012006
[8] Emanuel K 2017 A fast intensity simulator for tropical cyclone risk analysis. *Nat Hazards* May vol 6 pp1–18
[9] ISO 2859 2:1985 – *Sampling procedures for inspection by attributes*
[10] IBGE 2013 *Atlas do Censo Demográfico* 2010 (Instituto Brasileiro de Geografia e Estatística: Rio de Janeiro) p 153

An overview of the dynamic calibration of piezoelectric pressure transducers

F R F Theodoro[1], M L C C Reis[2] and C d' Souto[2]

[1] Aeronautics Institute of Technology, ITA. São José dos Campos, SP, Brazil;
[2] Institute of Aeronautics and Space, IAE. São José dos Campos, SP, Brazil.

E-mail: flavio.faciolla@gmail.com

Abstract: Dynamic calibration is a research area that is still under development and is of great interest to aerospace and automotive industries. This study discusses some concepts regarding dynamic measurements of pressure quantities and presents an overview of dynamic calibration of pressure transducers. Studies conducted by the Institute of Aeronautics and Space focusing on research regarding piezoelectric pressure transducer calibration in shock tube are presented. We employed the Guide to the Expression of Uncertainty and a Monte Carlo Method in the methodology. The results show that both device and methodology employed are adequate to calibrate the piezoelectric sensor.

1. Introduction

Since the 1960's researchers and members of the metrology community have realised that steady state pressure sensors were unable to capture the pressure variations that occurred in dynamic phenomena [1]. This has stimulated interest in sensors capable of measuring non-stationary pressure variations and how to calibrate such types of sensors.

A strong motivation for the interest in dynamic calibration is its application in the aerospace and military industries, where the reliability of measurements is very important. Schweppe et al (1963) pointed out that programs related to missile and space vehicles required reliable measurements of pressure signals due to the fact that the values change rapidly [2].

According to Eichstadt (2012) [3], in the automotive industry there is also an important demand for sensors to measure rapid and significant variations in the pressure magnitude. Hjelmgren (2002) [4] cited other areas, such as medicine and robotics, that also need to perform dynamic pressure measurements.

In 1972, in order to standardize the dynamic calibration process of pressure sensors, the American Society of Mechanical Engineers, ASME, published a Guide to Calibrating Pressure Transducers. This document was considered as the state of the art in dynamic calibration of pressure transducers by the Instrumentation Systems and Automation Society, ISA. The ISA reissued the document in 2002 [5]. This document contains information regarding the properties of pressure transducers in the frequency and time domains, mathematical modelling used in dynamic calibration, devices used for calibration, as well as care to be taken with the electronics of the calibration system. This document does not address any methods of estimating uncertainties associated with pressure transducer measurements.

Currently, the dynamic calibration of pressure sensors mobilizes research centres such as the National Physical Laboratory, NPL, in England [6], the Physikalisch-Technischen Bundesanstalt, PTB, in Germany [7], and large corporations such as Rolls-Royce, Volkswagen and Porsche, which are concerned with improving the dynamic pressure measurement processes.

In this article some concepts regarding dynamic measurements of pressure quantities and issues related to the calibration of pressure transducers under dynamic conditions are discussed. The article is the result of studies carried out by the Institute of Aeronautics and Space, IAE, located in São José do Campos, Brazil.

2. Dynamic calibration of pressure transducers

The term "dynamic calibration" for a pressure quantity is used to define the process of characterizing a dynamic pressure transducer, determining its properties such as natural frequency, damping, peak time, stabilization time and sensitivity [8].

The procedures to calibrate a sensor to be employed under static pressure conditions are well defined by metrology organizations. In summary, from a calibrated and traceable instrument, we obtain measurements and compare the data with a standard. However, for a dynamic calibration of a piezoelectric pressure transducer we cannot establish a traceable reference standard which covers all the needs that this calibration demands [8]. For example, we can find the sensitivity of the sensor and compare it with the sensitivity of the standard sensor using different devices, but it is difficult to estimate the peak time or the damping rate of the piezoelectric sensor with the required level of uncertainty so that the dynamic transducer can be considered calibrated.

In dynamic calibration a theoretical signal generated by a device is considered as the standard. Based on the characteristics of this device, a known and repetitive signal is generated which is compared with the signal captured by the transducer under calibration [8].

There are two methods of dynamically characterizing a transducer. In both, a specific type of signal is applied to the analyzed transducer and the response is measured. In the frequency domain method a periodic signal is applied. In the time domain method an aperiodic signal is applied, for example, a step signal [9].

Recently, the authors published a method for characterizing piezoelectric sensors that cover frequency ranges from 1 to 60 kHz and a pressure range from 1 to 1000 kPa [10]. In this method, a shock tube was used as the pressure signal generator to calibrate the sensor. Characteristics such as the rise time, peak time, stabilization time, sensibility, damping and natural frequency where determined, for a Kistler 701A® pressure sensor. This sensor is employed in aerospace testing. In this study, we calculated the uncertainties of the main parameters using the Monte Carlo method [11]. We used also the GUM (Guide to the Expression of Uncertainty in Measurement) methodology [12].

The quality of the signal used in the calibration process is a key factor for the success of the dynamic calibration. The Shock Tube used in this study is located at the Henry T. Nagamazu Laboratory of the Aerodynamics and Hypersonic Division, of the Institute of Advanced Studies, IEAv, Department of Aerospace Science and Technology, DCTA, São José dos Campos, Brazil, and is capable of supplying good repeatability conditions.

Figure 1. Shock Tube [10].

Basically, a shock tube is a device that contains two compartments which are separated by a diaphragm. The first compartment is named the "driver tube" where a gas is stored. The second compartment is the "driven tube" where sensors are located. When the diaphragm is broken by pressure difference, a shock wave is generated and is used as an input step signal in the characterization of the transducers. One can see a schematic view of the main parts of the shock tube with the coupled sensors in Figure 2.

Figure 2. Schematic view of shock tube, diaphragm and sensors [10].

3. Results and discussions

The start of the dynamic characterization process of the sensor occurs before a complete rupture of the diaphragm which separates the driver and driven tubes. Thus, firstly we calculate the velocity of the front shock wave which passes by sensors P2 and P2[1] (see Figure 2). Using a theoretical model [13] and based on the value of the wave front velocity, V, with initial pressure P_1, and initial temperature T_1, of the driver tube the value of the pressure step, ΔP, can be estimated.

The dynamic behaviour of the pressure sensor can be represented by a mass-spring system with damping [5]. Thus, the main dynamic characteristics of the sensor, natural frequency, ω_n, and damping, ζ, can be obtained experimentally, based on the values of its pick time, t_p, steady signal, $x\ (t_\infty)$, $x\ (t_{\max 1})$, and $x\ (t_{\max 2})$. Where $x\ (t_{\max 1})$ and $x\ (t_{\max 2})$ are two consecutives peaks above the mean value of the steady signal, $x\ (t_\infty)$ [4, 5].

An estimation of the uncertainties of the evaluated parameters is an important issue in tests using shock tubes. In order to avoid a large number of experiments, the use of a statistical method such as Monte Carlo in the evaluation of uncertainties can be considered. The experience of the authors is that this strategy has proven to be efficient [8, 10, 11].

Assuming a t-*Student* distribution, we performed 10^6 trials for each measurement model to estimate the mean values and uncertainties of the output parameters ΔP, ζ, ω_n, for a 95% level of confidence. Results is shown in Table 1.

Table 1. Parameters for propagation distributions and estimate values to quantities ΔP, ζ and ω_n

Input parameters	Output Parameters	Mean values (Output)	Standard deviation (Output)	Uncertainties lower limits (Output)	Uncertainties upper limits (Output)	Unit (Output)
V, T_1, P_1	ΔP	5.8×10^2	0.5×10^2	4.8×10^2	6.8×10^2	kPa
$x\ (t_\infty)$, $x\ (t_{\max 1})$, $x\ (t_{\max 2})$	ζ	0.58×10^{-1}	0.21×10^{-1}	0	1.1×10^{-1}	dimensionless
t_p	ω_n	290×10^4	3×10^4	25×10^4	36×10^4	rad/s

In Table 1, comparing the uncertainties limits shown in column five and six, it can be seen that uncertainties values of the ΔP quantities are symmetric (line one). Thus, when using a coverage factor k = 2 [9], we have a coverage interval for ΔP value which is similar to the uncertainties limits obtained according to a Monte Carlo Method.

The histogram of Figure 3 represent the resulting PDF for output ΔP, considering the input values of Table 1 applied a theoretical model obtained in the reference [13].

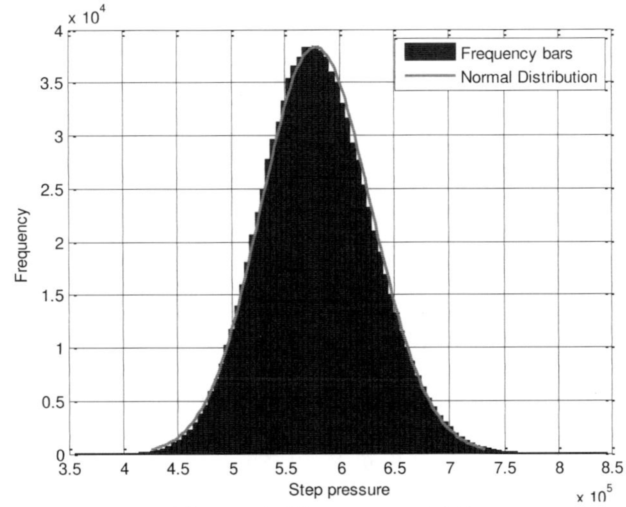

Figure 3. Histogram of ΔP.

In histogram, the red line represents a normal distribution, according to mean values and standard deviation of the quantities. The blue bars represent the results of a Monte Carlo simulation considering the input parameters and model adopted.

4. Conclusion

The aim of this article was to present an overview of the dynamic calibration of pressure transducers and to discuss some of its concepts. The article is the result of studies carried out by the authors in the Institute of Aeronautics and Space (IAE) and Institute of Advanced Studies (IEAv), both part of the Department of Aerospace Science and Technology (DCTA).

A major challenge in dynamic calibration concerns the repeatability of the input pressure signal. In view of this, the entire set of tests must be carefully analyzed: shock tube, data acquisition system, diaphragm construction material, gas used in the shock tube and assumed pressure for the rupture of the diaphragm.

Other precautions such as sensor location, cleaning of the shock tube and the rigidity of its holding can influence the quality of the signal generated, contributing to the uncertainty of the measurements. The results show that the device and a Monte Carlo methodology employed are adequate to calibrate the piezoelectric sensor. In other words, both a Shock tube and a Monte Carlo Method can be used for dynamic calibration of the piezoelectric pressure sensor.

More details and references regarding dynamic calibration can be found in the first author's PhD thesis [10].

References

[1] Ceará C. *Calibration of the pressure transducer* (In Portuguese). Doctoral Tesis. Aeronautics Institute of Technology ITA (1962) In: http://www.bdita.bibl.ita.br/tgsdigitais/lista_resumo.php?num_tg=000560813 Access out 16

[2] Schweppe J L, Eichenberger L C, Muster D F, Michaels E L and Paskusz G F, *Methods for the Dynamic Calibration of Pressure Transducers* NBS Monograph 67 December 12 1963

[3] Eichstädt S. *Analysis of dynamic measurements evaluation of dynamic measurement uncertainty* (2012) Doctoral Tesis Technische Universität Berlin

[4] Hjelmgren J, Dynamic measurement of pressure-A literature survey *SP Rapport 2002:34* In: http://www.diva-portal.org/smash/get/diva2:962216/FULLTEXT01.pdf

[5] ISA–37.16.01-2002. A guide for the dynamic calibration of pressure transducers North Carolina: The Instrumentation Systems and Automation Society

IOP Conf. Series: Journal of Physics: Conf. Series **975** (2018) 012002 doi:10.1088/1742-6596/975/1/012002

[6] National Physical Laboratory *Data analysis and uncertainty evaluation* In: http://www.npl.co.uk/science-technology/mathematics-modelling-and-simulation/research/data-analysis-and-uncertainty-evaluation

[7] Physikalisch Technische Bundesanstalt *Traceable dynamic measurement of mechanical quantities*. In: https://www.ptb.de/emrp/2175.html

[8] Theodoro F R F, Reis M L C C, Souto C d' and Barros E, Measurement uncertainty of a pressure sensor submitted to a step input. 2016 *Measurement Vol.* ***88***. pp. 238-247 doi:10.1016/j.measurement.2016.03.043.

[9] Diniz A C G C, Vianna J N S and Neves F J R, Dynamic calibration of the pressure sensors: method and means (In Portuguese) 2006. *Metrologia (SBM Congress)* September 01-05 Brazil

[10] Theodoro F R F. *Evaluation of the uncertainties to dynamic calibration of pressure sensor submitted to a step input* Doctoral Tesis Aeronautics Institute of Technology ITA (2016). In: http://www.bdita.bibl.ita.br/tesesdigitais/lista_resumo.php?num_tese=71732

[11] BIPM/JCGM 101:2008 Evaluation of Measurement Data – Supplement 1 to the Guide to the expression of uncertainty of measurement - Propagation of distributions using a Monte Carlo method. Joint Committee for Guides in Metrology BIPM

[12] BIPM/JCGM 100:2008 Evaluation of Measurement Data - Guide to the expression of uncertainty of measurement, Joint Committee for Guides in Metrology BIPM

[13] Matthews C, Pennecchi F, Eichstadt S, Malengo A, Esward T, Smith I, Elster C, Knott A, Arrhen A and Lakka A, Mathematical modelling to support traceable dynamic calibration of pressure sensors. 2014 *Metrologia* ***51***. pp. 326–338 doi:10.1088/0026-1394/51/3/326.

Chemical and Biological Mobile Laboratory: infrastructure employed by Brazilian Army in emergency response actions

M Cardozo[1], V G M Oliveira[1], R B Sousa[1] and R L de Paula[1,2]

[1] Instituto de Defesa Química, Biológica, Radiológica e Nuclear – IDQBRN, Centro Tecnológico do Exército – CTEx, Rio de Janeiro, BR.
[2] Programa de Pós-Graduação em Biotecnologia do Instituto Nacional de Metrologia, Qualidade e Tecnologia – PPGBIOTEC/INMETRO, Duque de Caxias, BR.

E-mail: monique@ctex.eb.br, reuel.lopes@eb.mil.br

Abstract. The Brazilian Army specified and acquired a mobile chemical and biological laboratory in order to confirm in a fast and mobile way with more precise analytical techniques the information obtained by the emergency responders field teams. The laboratory was designed for displacement in different scenarios of the national territory. This paper describes the laboratorial structure, the material flow and the deployment of this defense product in the major international events occurred in Brazil from 2011 to 2016, with the objective of providing in situ identification of chemical and biological threats.

1. Introduction

Nowadays a concern about the possible use of Chemical and Biological (CB) agents by non-state actors is increasing in the world. Brazil hosted a series of High Visibility Events (HVE) in recent years, such as FIFA 2014 World Cup and Rio 2016 Olympic Games. For these reasons, Brazil has gained international attention and the probability of terrorist attack might increase. In this context, the Brazilian Army along with other national authorities need to be prepared to act in theatres where there are CB threats.

To operate effectively, the Brazilian Army requires CB agent prevention, detection and identification capabilities. CB agent identification in the field uses a limited number of instruments that are relatively sensitive and easy to operate. However they are very susceptible to environmental interferers, which can generate a large number of false positive responses [1].

To provide a more reliable response an in-depth analysis is required. To do that, samples must be sent to a laboratory with a suitable structure and an improved analytical capability. However, in a continental country as Brazil, a specialized laboratory can be thousands of miles away of a site of attack. Besides that, in the case of a CB threat, the response must be fast to quickly lead the teams in the field to the best possible protective actions, such as area isolation, decontamination and even properly medical care.

In this sense, the Brazilian Army Technological Center (CTEx) initiated, in 2008, a project to specify and develop a mobile chemical and biological laboratory (MCBL). This paper describes the main structure and capability developed and tested by CTEx. It also includes the experience of deployment of MCBL in HVE. All those things demonstrate the potential of using MCBL during operations.

IOP Conf. Series: Journal of Physics: Conf. Series **975** (2018) 012003 doi:10.1088/1742-6596/975/1/012003

2. Laboratorial infrastructure

After all technical specifications made by scientific advisors to guarantee that the MCBL would attend the necessity of an expected defense product, the acquisition of MBCL was made by an international procurement process. The company that achieved the better conditions was Cristanini S.P.A. from Italy. However, part of bodywork was completed in Brazil with in house resources and equipment purchase from other companies such as Agilent Technologies.

2.1. Container characteristics

The main objectives and characteristics (table 1) of the Brazilian Army MBCL project were: to enhance the identification capability with a reduced number of false positives; to have a complete laboratorial structure, to quickly and safely detect and identify the presence of CB agents; and to have mobility to dislocate laboratory structure to any part of Brazilian territory.

The internal walls of shelter are made of stainless steel resistant to decontamination actions. All system is completely airtight and watertight. The structure of MBCL container is composed of three separate areas: utilities; changing room and laboratory.

Table 1. Brazilian Army MCBL characteristics.

Dimensions	Standard Container: 6.0 m x 2.45 m x 2.45 m
Weight	8500 kg
Portability	Transportable by truck, train, ship and plane
	Two (2) people required for field deployment
	Less than 15 minutes for unfolding on the ground
Endurance	Continuous operation for 72 hours without refueling
	17 kVA electric generator
	Self-leveling piston system
	Water tank of 1000 L
	Liquid effluent reservoir
	Refrigerator and air heater

2.2. Utilities area

The shelter is equipped with a self-leveling pistons system that allow the lifting of the whole structure without external aid. The lower part of the container contains clean water tank, waste reservoir and fuel tank with capacity to maintain autonomy of 72 hours without refueling. On the back of the container is all necessary machinery for the general operation of the laboratory, which includes power generator, air-conditioning compressor, equipment of pressurization system, NBC filter and other appliances.

2.3. Changing room and laboratorial area

The changing room is placed right after the entrance door of MBCL. This serves to prevent entry of any contamination into the laboratory environment. There is a shower, which can work with decontamination solution, and a chemical bathroom to provide minimum hygiene conditions in isolated operating situations.

The Laboratorial area works normally with an over pressure system to avoid that any contamination penetrate into the lab. There is also an air filtration system consisting in both particulate (High Efficiency Particulate Airborne – HEPA) and chemical (activated carbon) filters in an NBC filtration system.

Moreover, once the laboratory is closed it remains isolated from the external environment. All samples are manipulated inside a chambers system that includes one for receiving samples, a glove box to manipulate them and a final passage with an autoclave to guarantee safety on the exit of materials. This system operates in an integrated way, with the entrance in each chamber only being allowed when the previous chamber has been completely closed.

In addition, the MBCL has a laminar flow chamber where it is possible to handle biological samples. There is also equipment for conducting necessary CB analyses, refrigerator and freezer for storing samples besides supplies and storage cabinets.

3. Material flow

Sample collection will never be done randomly. It will occur after a suspected use of CB agents (attacks, injured or dead persons/animals, abandoned material in HVE, alarm in a CB detection equipment, etc.). After the suspicion, the primary detection/identification is made with the available means in the following order: to discard the threat of explosion (bomb teams, sniffer dogs and specific detectors); to eliminate the existence of radiological and nuclear risk (use of detectors and identifiers); since only the possibility of a CB threat remains, use available field detectors to guide the collection of representative samples.

3.1. Receiving samples

There is no contact between the team collecting samples and the laboratory personnel. Sample enter in a safe way through a pass box. After that, samples are delivered directly inside the isolator (glove box) of the lab where they are manipulated and prepared for analysis. If possible, samples are divided into four parts: two will be analyzed in MCBL and two will be delivered in CB reference laboratories.

3.2. Initial screening

First step is checking if the sample has characteristics of a biological threat. For unknown suspicious powders, the BioCheck® Screening Test Kit can be used. The kit quickly (5 min) identifies the presence or absence of a biomolecule found in all living materials and many toxins. For some biological targets, it is possible to use immunochromatographic tests intended for the rapid (15 min), in vitro qualitative presumptive identification.

For solids and liquids, the other two technologies that provide relevant information are infrared and Raman. Using portable equipment, it is possible to check for the existence of amino acid content in case of biological threats. In addition, these devices are able to identify some CB agents, even in mixtures, by similarity with spectral data in specific libraries.

3.3. Sample preparation and analysis

For chemical analyses, air and vapor samples can be collected in thermal desorption (TD) tubes where volatile and semi-volatile analytes are captured on the adsorbent. The TD tube can be analyzed by Gas Chromatography (GC) using a suitable TD-GC interface. For identification purposes, Mass Spectrometry (MS) is used [2]. In the same way, solid material can be primary prepared separating a fraction and dissolving the material in a suitable organic solvent. The liquid phase is separated and filtered through a syringe filter to be analyzed directly in a GC-MS system.

For organic liquid samples, a fraction is separated, filtered, diluted and analyzed by GC-MS. Aqueous samples can not be analyzed directly by GC-MS. It is necessary to remove the water or to extract the analytes of interest. One of the most efficient procedures of this type in MCLB conditions is to use solid phase micro extraction (SPME), with the immersion of the fiber directly in the sample for adsorption. In addition to these procedures, others can be required such as extractions under different pH conditions, solvent exchange and derivatizations, depending on the desired targets.

Biohazards are very complex threats. In fact, existent commercial system and instrumentation to be used in mobile applications are limited to certain conditions and types of agents [3]. In MCBL the principal instrument for molecular biology analyses is RAZOR EX System. It works with polymerase chain reaction (PCR) technique to amplify segments of DNA, but even this solution is limited to at least 13 targets.

3.4. Decontamination

In the glove box, the previously prepared and sealed samples as well as all the material used are decontaminated with appropriated solutions. The autoclave can be used as additional measure whenever possible. After decontamination process, samples can be analyzed without present risks.

4. Deployment of NCBL

There are two possible modes that the MCLB can be used, either as a pre-deployed facility used in support of a major event or as a reactive asset following an incident [4]. Fortunately, at the time of writing this paper, Brazilian MCLB has not been required for a large-scale response to an incident. However, the experience in major international events hosted in Brazil indicated the effectiveness of integrating the MCLB in a CBRN response.

4.1. Experience in HVE

The World Military Games in 2011 was the first event where the MCBL was used, when was verified the first difficulties for operation and adaptation of the analytical methodologies to the available means. In 2012, in United Nations Conference on Sustainable Development (Rio+20) there were some suspicions of the use of chemical and biological agents. One of the examples was the case of an abandoned bag in an area of conference containing a white powder inside. The analyses revealed that it was biological yeast and it was discovered later on that the bag belonged to a chef who was working on the event. The MCBL also operates in World Youth Day (2013) and FIFA Confederations Cup (2013).

During the 2014 FIFA World Cup an analysis was conducted to rule out a possibility of chemical threat, more specifically sarin. Conclusions pointed to a kind of cleaning products that have a response similar in portable detection equipment. For Rio 2016 Olympic Games, the Army decided to increase its analytical capacity by acquiring one more MCBL. All of these examples attest that MCBL is a defense solution that meets the objectives listed in its specification as a defense product.

5. Conclusions

The use of MCBL as auxiliary mean in a scenario with CB threats made possible to provide fast and reliable information to guide the actions of first responders and the choices of public decision-makers. The entire structure proved to be capable of carrying out the necessary for the chemical and biological analysis of dangerous agents in environmental samples with safety and comfort, without losing the characteristic of mobility.

Acknowledgements

Authors would like to thank Organisation for the Prohibition of Chemical Weapons – OPCW for Research Grant Support L/ICA/ICB/201062/15 and Centro Tecnológico do Exército – CTEx, for the infrastructure.

References

[1] Hancock J R, Lepage C R J, Chenier C L, Froese D S W and Lukacs M J 2007 *DRDC Suffield.* **216** pp 1-26
[2] Terzic O, Gregg H and de Voogt P 2015 *Trends in Analytical Chemistr.* **65** pp 151-166
[3] Cenciarelli O, Malizia A, Marinelli M, Pietropaoli S, Gallo R, D'Amico F, Bellecci C, Fiorito R, Gucciardino A, Richetta M and Gaudio P 2013 *Defence S&T Tech. Bull.* **6** 1 pp 33-41
[4] Norman K, Roffey P and Royds D 2009 *Australian Journal of Forensic Sciences.* **41** 1 pp 73-80

IOP Conf. Series: Journal of Physics: Conf. Series **975** (2018) 012004 doi:10.1088/1742-6596/975/1/012004

Comparative study between EDXRF and ASTM E572 methods using two-way ANOVA

A Krummenauer, H M Veit and J Zoppas-Ferreira

Programa de Pós-Graduação em Engenharia de Minas, Metalúrgica e de Materiais (PPGE3M), Universidade Federal do Rio Grande do Sul (UFRGS), Av. Bento Gonçalves, 9500, Porto Alegre, RS, Brazil.

alex_krum@yahoo.com.br

Abstract. Comparison with reference method is one of the necessary requirements for the validation of non-standard methods. This comparison was made using the experiment planning technique with two-way ANOVA. In ANOVA, the results obtained using the EDXRF method, to be validated, were compared with the results obtained using the ASTM E572-13 standard test method. Fisher's tests (*F*-test) were used to comparative study between of the elements: molybdenum, niobium, copper, nickel, manganese, chromium and vanadium. All *F*-tests of the elements indicate that the null hypothesis (Ho) has not been rejected. As a result, there is no significant difference between the methods compared. Therefore, according to this study, it is concluded that the EDXRF method was approved in this method comparison requirement.

1. Introduction

In Brazil, the laboratories of RBLE (Brazilian Network of Testing Laboratories), accredited by CGCRE/INMETRO, according to ABNT NBR ISO/IEC 17025:2005 standard [1] must comply, among other things, the following requirements: non-standard methods (item 5.4.4), and validation of methods (item 5.4.5), described in this standard. To assist in this task, INMETRO provides document DOQ-CGCRE-08 - Guidance for validation of analytical methods [2] which guides laboratories in the validation process of a non-standard method, i.e., that is not described in a national or international standard.

The LACOR (Laboratory of Corrosion, Protection and Recycling of Materials), at UFRGS, is accredited by CGCRE/INMETRO (RBLE 1139) and uses the Thermo Scientific Niton XL3t GOLDD+ portable EDXRF analyzer to perform metal determination tests in steel samples, by energy dispersive X-ray fluorescence spectrometry (EDXRF). However, the EDXRF method is a non-standard method and therefore it must be validated. One of the requirements for validation, as determined by item 8.2.6.3.4 of document DOQ-CGCRE-08 [2], is the comparison with a reference method. The purpose of this comparison is to evaluate the accuracy of method to be validated by analyzing samples with concentrations through the entire range in which the method should be validated. Samples were analyzed, with six measurements, by using the two methods separately [2].

There are several techniques that can be used to compare the results obtained by two test methods. In this sense, document DOQ-CGCRE-08 [2] suggests: hypothesis tests and planning of experiment. In this comparison, it was decided to use the two-way factorial experiment (two-way ANOVA) described in the experiment planning technique [3, 4, 5]. In order to obtain the results of the method to be validated, the NITON analyzer was used to analyze the samples 2Q15S1 [6], 2Q15S2 [6], 4Q15S1

[7], and 4Q15S2 [7]. These samples were measured by ASTM E572-13 reference method [8], with the wavelength dispersive X-ray fluorescence spectrometry (WDXRF) method, in the Inter-Laboratory Analysis Programs (ILAP), performed by ASTM International [6, 7], in which LACOR laboratory participated. The term ED "energy dispersive" in EDXRF refers to the technique of detecting X-rays emitted, which is affected by a detector which generates an intensity spectrum as a function of energy [9]. While the term WD "wavelength dispersive" in WDXRF refers to the use of single crystal or synthetic multilayer Bragg diffraction to disperse characteristic X-rays [9]. Figure 1 shows a comparison between the detection techniques used in EDXRF and WDXRF.

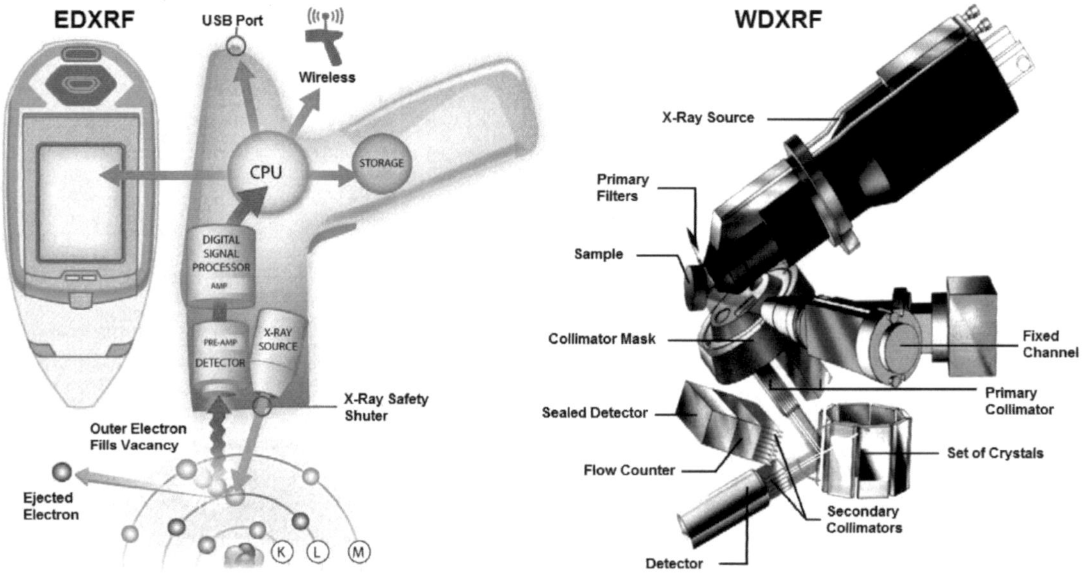

Figure 1. Comparison between the detection techniques used in EDXRF and WDXRF [9, 10].

In this study, the results of the following chemical elements were analysed: molybdenum (Mo), niobium (Nb), copper (Cu), nickel (Ni), manganese (Mn), chromium (Cr), and vanadium (V). Comparison of the results for tungsten (W) and cobalt (Co) was also obtained, but with other methods, outside the scope of this paper. Therefore, these results are not mentioned in this article.

2. Objective

This study performs a comparative study between the non-standard EDXRF method and the ASTM E572 reference method (WDXRF), by using the two-way ANOVA. The objective is to validate the EDXRF method and maintain the accreditation by CGCRE/INMETRO of the metal determination testing by FRX.

3. Materials and methods

In this study, the four samples, used in interlaboratory comparisons [6, 7], were measured with six measurements, under repeatability conditions, by using the NITON analyzer along with the Mobile Test Stand accessory. NITON has been configured with: calibration curve (table 1), general metals analysis from the main menu, two filters – main range (20s) and low range (20s), disable Al, and autoswitch ON [10]. All measurements were taken by using the following procedure quoted in the NITON manual: 1 - preparation of the sample; 2 - clean with isopropyl alcohol and lint-free paper; 3 - perform a system check once every working day; 4 - warm up for ten minutes after start up [10].

Results with six measurements (n = 6), in each sample, were compared with six randomly selected results from other laboratories that participated in interlaboratory comparisons [4, 5]. Two-way ANOVA was used for each chemical element: factor A – concentration – with four levels (a = 4) and factor B – method – with two levels (b = 2).

Table 1. Parameters of the calibration curve and coefficient of determination (R^2).

Chemical element	Intercept (a)	Slope (b)	R^2
Mo	-0.0054	1.0219	0.9999
Nb	0.0033	0.8916	0.9999
Cu	-0.0148	1.1157	0.9993
Ni	0.0021	1.0096	1.0000
Mn	-0.0258	0.9803	0.9997
Cr	-0.0932	0.9758	0.9995
V	0.0019	0.9116	0.9996

3.1. Factorial experiment (two-way ANOVA)

According to Ribeiro and ten Caten [3] and Montgomery [4, 5], the statistical model of two-way ANOVA is equation (1).

$$y_{ijk} = \mu + \tau_i + \beta_j + (\tau\beta)_{ij} + \varepsilon_{ijk} \tag{1}$$

Where μ is the population mean; τ_i is the effect of the i-th level of "concentration"; β_j is the effect of the j-th level of "method"; $\tau\beta_{ij}$ is the effect of the "concentration·method" interaction; ε_{ijk} is the random error, with i = 1, a; j = 1, b and k = 1, n. Assumption: $\varepsilon_{ijk} \rightarrow N(\mu = 0, \sigma)$.

Thus, there are three hypothesis tests to be performed [3, 4, 5]:

- For the factor "concentration": Ho: $\tau_i = 0$ and H1: $\tau_i \neq 0$ for some i, i = 1, a.
- For the factor "method": Ho: $\beta_j = 0$ and H1: $\beta_j \neq 0$ for some j, j = 1, b.
- For the factor "concentration·method": Ho: $\tau\beta_{ij} = 0$ and H1: $\tau\beta_{ij} \neq 0$ for some ij.

For each hypothesis tests, it was computed F tabulated (Ftab) of the Fisher distribution (F-distribution) and compared with the F calculated (Fcalc). If Fcalc > Ftab then the null hypothesis is rejected and therefore the change in levels of the factor analyzed has a significant effect on the mean response [3, 4, 5].

The value of Fcalc is obtained from equation (2) and Ftab is calculated by equation (3).

$$Fcalc = \frac{Variance\ between\ groups}{Variance\ within\ the\ groups} = \frac{MQ(factor)}{MQR} \tag{2}$$

$$Ftab = F_{\alpha;GL\ numerator\ ;GL\ denominator} \tag{3}$$

Where MQ(factor) is mean square or variance, MQR is mean square of error, α is significance level, GLnumerator is degrees of freedom of numerator and GLdenominator is degrees of freedom of denominator [3, 4, 5].

If Fcalc > Ftab, or p-$value$ < 0.05 (α = 5 %), the effect studied is significant for a confidence level of 95 % [3, 4, 5]. According to Ribeiro and ten Caten [3] and Montgomery [4, 5], the formulas for calculations are described below.

$$TC = \frac{(T\ ...)^2}{abn} \tag{4}$$

$$SQA = \sum_{i=1}^{a} \frac{(T_i..)^2}{bn} - TC \tag{5}$$

$$SQB = \sum_{j=1}^{b} \frac{(T_{\cdot j \cdot})^2}{an} - TC \quad (6)$$

$$SQAB = \sum_{i=1}^{a}\sum_{j=1}^{b} \frac{(T_{ij \cdot})^2}{n} - TC - SQA - SQB \quad (7)$$

$$SQR = \sum_{i=1}^{a}\sum_{j=1}^{b}\sum_{k=1}^{n} y_{ijk}^2 - \sum_{i=1}^{a}\sum_{j=1}^{b} \frac{(T_{ij \cdot})^2}{n} \quad (8)$$

$$SQT = \sum_{i=1}^{a}\sum_{j=1}^{b}\sum_{k=1}^{n} y_{ijk}^2 - TC \quad (9)$$

The verification of equation (9) can be done by equation (10).

$$SQT = SQA + SQB + SQAB + SQR \quad (10)$$

Where TC is the correction term, SQ(factor) is the sum of squares of each factor, MQ(factor) is the calculation of the mean square, equation (11), and GDL are the degrees of freedom of each factor. Table 2 presents these calculations.

$$MQ(factor) = \frac{SQ(factor)}{GDL} \quad (11)$$

Table 2. Calculations for two-way ANOVA.

Factor	Sum of squares	GDL	Mean square	Fcalc	Ftab
A	SQA	$(a - 1)$	MQA	MQA/MQR	$F_{\alpha;GLA;GLT}$
B	SQB	$(b - 1)$	MQB	MQB/MQR	$F_{\alpha;GLB;GLT}$
AB	SQAB	$(a - 1)*(b - 1)$	MQAB	$MQAB/MQR$	$F_{\alpha;GLAB;GLT}$
Error	SQR	$ab(n - 1)$	MQR		
Total	SQT	$abn - 1$			

4. Results and discussion

The six results obtained with NITON, in the four samples for Mo, and also the six sample means of these samples from laboratories that participated in ILAP of the ASTM [6, 7], method E572, are presented in table 3. Results of the other chemical elements were obtained by using the same methodology. By using the methods and calculations described in item 2 of this article, with the data from table 3, the following values were obtained for Mo: $a = 4$, $b = 2$, $n = 6$, TC = 53.129788, SQA = 45.504863, SQB = 0.000199, SQAB =0.000051, SQR = 0.003461 e SQT = 45.508574. Table 4 presents the values of two-way ANOVA calculations for Mo.

Table 3. Values obtained with NITON (EDXRF) and ASTM E572 (WDXRF) for Mo.

Concentration	Values measured with NITON (%)			Sample means obtained with E572 (%)		
0.051 % (2Q15S2)	0.049 0.050	0.050 0.051	0.052 0.049	0.058 0.055	0.050 0.054	0.052 0.051
0.102 % (2Q15S1)	0.102 0.101	0.104 0.105	0.101 0.104	0.110 0.108	0.106 0.107	0.110 0.109
2.003 % (4Q15S1)	2.020 2.003	2.002 2.008	2.008 2.011	2.020 2.008	2.003 2.007	2.020 2.001
2.058 % (4Q15S2)	2.044 2.043	2.052 2.046	2.011 2.035	2.046 2.056	2.076 2.035	2.033 2.024

Table 4. ANOVA two-way calculations for Mo.

Factor	Sum of squares	GDL	Mean square	Fcalc	Ftab
Concentration (A)	45.504863	3	15.168288	175298	2.8
Method (B)	0.000199	1	0.000199	2.3	4.1
Interaction (AB)	0.000051	3	0.000017	0.2	2.8
Error	0.003461	40	0.000087		
Total	45.508574	47			

As mentioned previously, the studied effect is significant for Fcalc > Ftab. Table 4 shows that the factor "concentration" was significant and the factor "method" and the interaction were not significant. This result shows that the concentrations had great variation between the samples, but this was expected, since each sample actually had a very different value of concentration of Mo in relation to the others. However, most importantly, the methods do not have significant differences and therefore for Mo the EDXRF method is approved in requirement 8.2.6.3.4 of document DOQ-CGCRE-08 [2], for a confidence level of 95 %. That was the purpose of this study. Figure 2 shows the analysis of the factor method using the *F*-distribution of Fisher.

Figure 3 shows the comparison between the means of measurements made for Mo, with the EDXRF method and the means obtained with the ASTM E572 method (WDXRF) and it can be observed that there is no difference between the values.

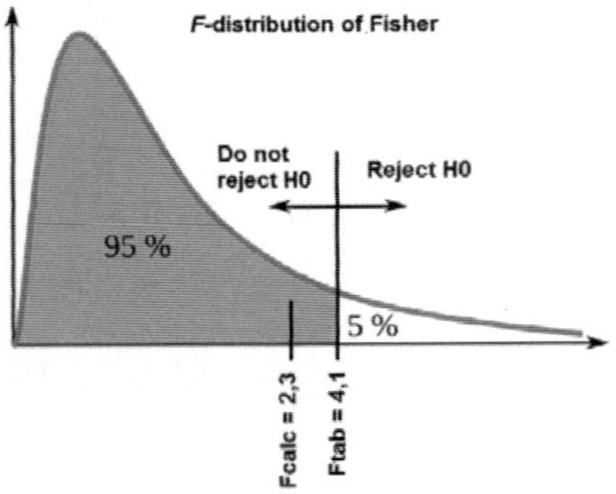

Figure 2. Analysis factor method using *F*-distribution of Fisher.

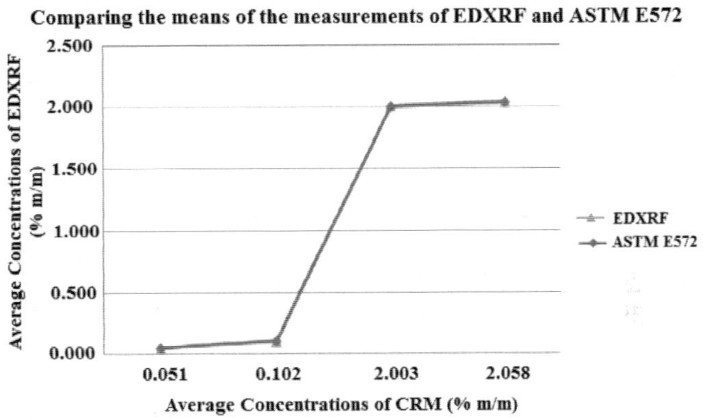

Figure 3. Comparing the sample means of the measurements of EDXRF and ASTM E572.

As for the other elements that followed the same methodology of Mo, table 5 presents a summary of two-way ANOVA calculations of each of the chemical elements.

Analyzing the data in table 5, it is concluded that all chemical elements that have been analyzed and compared with the E572 method [8] are approved for requirement 8.2.6.3.4 of document DOQ-CGCRE-08 [2], for a confidence level of 95 %. There was no significant difference between the methods because no Fcalc was greater than Ftab (see table 5).

Table 5. Summary of two-way ANOVA calculations for the factor "method" of chemical elements.

Chemical element	Sum of squares	GDL	Mean square	Fcalc	Ftab
Mo	0.000199	1	0.000199	2.3	4.1
Nb	0.000006	1	0.000006	4.0	4.1
Cu	0.000284	1	0.000284	4.0	4.1
Ni	0.001349	1	0.001349	1.0	4.1
Mn	0.000524	1	0.000524	1.0	4.1
Cr	0.003485	1	0.003485	1.0	4.1
V	0.000003	1	0.000003	0.2	4.1

5. Conclusion

The comparison of the EDXRF method with the ASTM E572 reference method (WDXRF), using the two-way ANOVA technique, resulted in the approval of the method for measuring Mo, Nb, Cu, Ni, Mn, Cr, and V, in the alloy steels analysis. However, it should be noted that this alone does not characterize the complete validation of the analytical method, since many other requirements have to be approved, as described, for example, in DOQ-CGCRE-08 [2], for final definitive approval of the EDXRF method.

Acknowledgments

The authors thank CAPES for the support with the masters scholarship program; UFRGS, PPGE3M and LACOR by technical and instrumental support; FINEP, for the resources obtained, through the SIBRATEC - Technological Services Provision Networks, for the acquisition, by LACOR, of the portable EDXRF analyzer.

References

[1] ABNT 2005 *NBR ISO/IEC 17025:2005 General Requirements for the Competence of Testing and Calibration Laboratories* (Rio de Janeiro: ABNT)
[2] INMETRO 2016 *DOQ-CGCRE-008 Orientação sobre Validação de Métodos Analíticos* (Rio de Janeiro: INMETRO)
[3] Ribeiro J L D and ten Caten C S 2014 *Projeto de Experimentos* (Porto Alegre: UFRGS)
[4] Montgomery D C 2009 *Introduction to Statistical Quality* (New York: Wiley & Sons)
[5] Montgomery D C 1997 *Design and Analysis of Experiments* (New York: Wiley & Sons)
[6] ASTM International 2015 *Proficiency Test Program: Chemical Analysis of Stainless Steel Second Quarter 2015* (West Conshohocken: ASTM International)
[7] ASTM International 2016 *Proficiency Test Program: Chemical Analysis of Stainless Steel Fourth Quarter 2015* (West Conshohocken: ASTM International)
[8] ASTM International 2016 E572-13 – *Standard Test Method for Analysis of Stainless and Alloy Steels by Wavelength Dispersive X-Ray Fluorescence Spectrometry* (West Conshohocken: ASTM International)
[9] Santos E S 2012 *Determinação de Cobre, Ferro, Níquel e Zinco em Matrizes Etanólicas por Espectrometria de Fluorescência de Raios-X com Energia Dispersiva após Pré-Concentração em Papel Cromatográfico,* Ph.D. dissertation (Salvador: UFBA)
[10] Thermo Fisher Scientific 2010 *Niton XL3 Analyzer User's Guide Revision C*

IOP Conf. Series: Journal of Physics: Conf. Series **975** (2018) 012005 doi:10.1088/1742-6596/975/1/012005

Control Chart on Semi Analytical Weighting

G S Miranda, C C Oliveira, T B S C Silva, T B Stellato, L R Monteiro, J R Marques, M G Faustino, S M V Soares, J C Ulrich, M A F Pires and M E B Cotrim

Instituto de Pesquisas Energéticas e Nucleares, IPEN-CNEN/SP, São Paulo, Brazil
E-mail: gabidsmiranda@gmail.com

Abstract. Semi-analytical balance verification intends to assess the balance performance using graphs that illustrate measurement dispersion, trough time, and to demonstrate measurements were performed in a reliable manner. This study presents internal quality control of a semi-analytical balance (GEHAKA BG400) using control charts. From 2013 to 2016, 2 weight standards were monitored before any balance operation. This work intended to evaluate if any significant difference or bias were presented on weighting procedure over time, to check the generated data reliability. This work also exemplifies how control intervals are established.

1. Introduction

The search for better performance, for results reliability and competitiveness lead laboratories to adopt a Quality Assurance System [1,2]. ISO/IEC 17025 specifies general requirements to calibration and analysis laboratories, including sampling. Laboratories that work in accordance with this standard also follow the requirements of ISO 9001, due to 2005 revision, in order to establish the quality assurance system definitions [3]. Items 5.5 and 5.9 of ISO/IEC 17025 describe data quality assurance from test and calibration, regarding the use of certified reference materials (CRM), interlaboratory programs and/or internal quality control. Also emphasize the requirement of data register in a way that bias can be detected and, when possible, apply statistical analysis to evaluate results [4].

One of the oldest statistical technique to internal quality control is the use of quality control charts. Their applications to data from analytical instruments assure the traceability and reliability of the analytical results [3,4]. ISO 8258/91, revised and altered to ISO 7870/13, presents the basic principles and concepts correlated to Shewhart control charts [2,4]. To achieve results reliability, using a statistical control, is the aim of this study by using quality control charts, to assess the performance of a semi-analytical balance (GEHAKA BG400), in order to identify if any bias is present and to propose corrective actions, if necessary.

2. Materials and methods

The elaboration, deployment and use of quality control charts to the semi-analytical balance assessment (GEHAKA BG400 - Figure 1), in a lab in Centro de Química e Meio Ambiente (CQMA) at IPEN/CNEN-SP, was carried out from June/2013 to 2016, in a total of 766 measurements. The semi-analytical balance specifications are presented in Table 1.

Figure 1. Semi-Analytical Balance - GEHAKA BG400

Table 1. Semi Analytical Balance Specification.

Parameters	Specification
Dimensions	240 x 260 x 55 mm
Weight	5.2 kg
Maximum charge	404 g
Minimum charge	0.001 g
Plate diameter	110 mm

The process to be controlled was the semi-analytical balance response, regarding the variability provided by different analysts, environmental conditions, preventive maintenance and instrument calibration. Two different standard weights, stable and uniform, named "U" and "S" were used. Quality control charts were elaborated using the program Microsoft Excel. Central limits calculations relied on the standard weight total average. Operation limits were established as: Satisfactory (average $\pm$ 1σ); Warning (average $\pm$ 2σ); and Unsatisfactory (average $\pm$ 3σ). The decision criteria adopted meet ISO 8258 – Shewhart Control Charts requirements, as cited below [5]:

a) 1 or more data above superior limit or below inferior limit;
b) 9 consecutive data in the same side of satisfactory zone;
c) 6 consecutive data, all increasing or decreasing;
d) 14 consecutive data alternated between below and above;
e) 2 of 3 consecutive data in unsatisfactory zone and beyond;
f) 4 of 5 consecutive data in warning zone or beyond it;
g) 15 consecutive data in satisfactory zone;
h) 8 consecutive data the warning zone.

3. Results and discussion

Figure 2 presents the histogram and measurement normal distribution performed to standard weights U and S. Figure 3 presents control charts to standard weights U and S, with 383 measurements for each standard weight (from June/2013 to December/2016) using the semi-analytical balance. Warning limits to weight U were between 0.755 g and 0.751g; unsatisfactory limits were between 0.756 g and 0.750 g. To standard weight S, warning limits were between 0.811 g and 0.815 g; unsatisfactory limits were between 0.810 g and 0.816 g. Data over satisfactory limits could arise from measurement errors or typos; failure of unregulated instrument and/or operators failure [4,6]. Table 2 presents the descriptive statistics to standard weights U and S. No value from standard U (n = 383) neither from standard S (n = 383) were out of control.

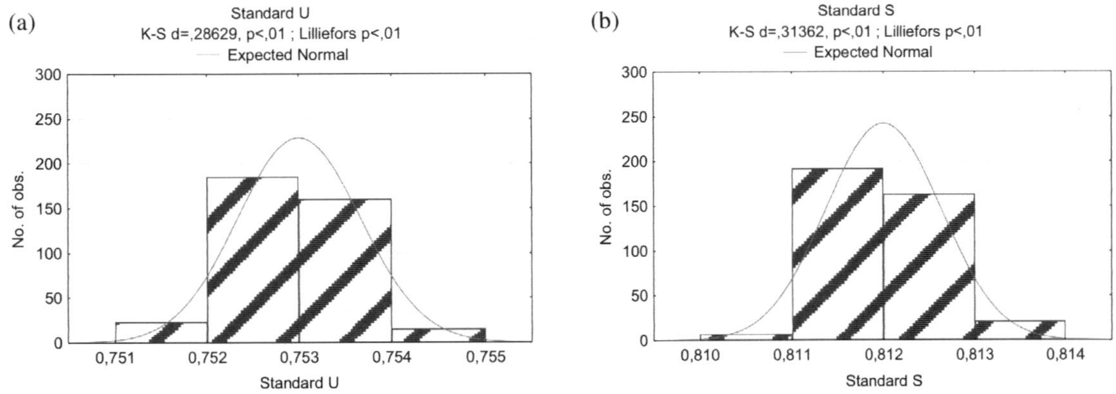

Figure 2: Histogram and expected normal distribution of standard weights (a) U and (b) S.

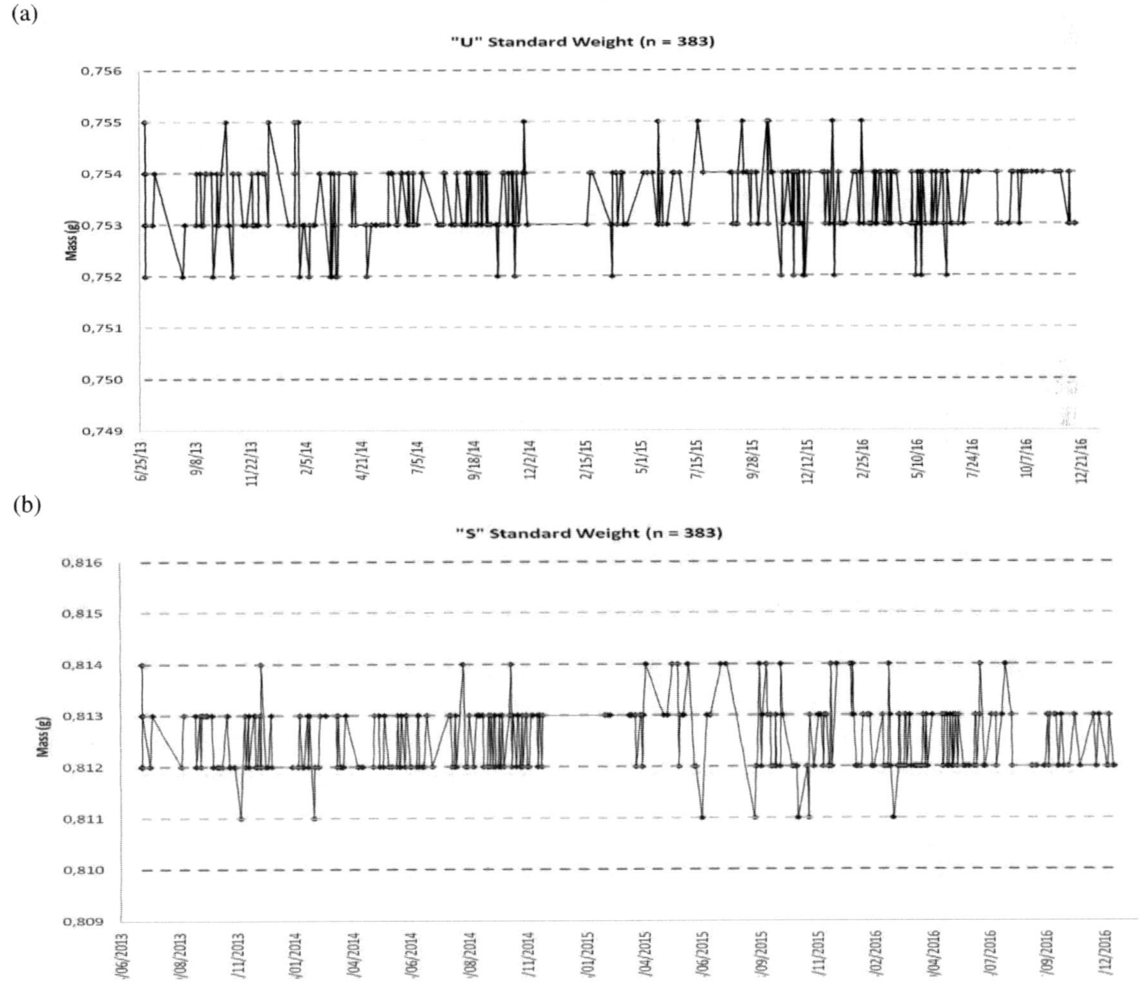

Figure 3. Standard (a) U and (b) S quality control chart.

In order to guarantee the proper functioning and durability of the semi-analytical balance, a calibration is annually performed. Last calibration was carried out in August/2016.

Table 2. Standard U and S descriptive statistics.

Descriptive Statistics	Standard U	Standard S
Average (g)	0.753	0.813
s (g)	0.001	0.001
RSD (%)	0.133	0.123
N	383	383
Median (g)	0.753	0.812
Maximum (g)	0.755	0.814
Minimum (g)	0.752	0.811
Number of warning data	15	7
% of values in the warning limit	3.92	1.83

A periodically performed calibration and the internal quality control are necessary to ensure normal operational conditions to semi-analytical balances, as well as demonstrate its good performance.

4. Conclusion

Quality control chart, used as assessment tool for measurements performed in a semi-analytical balance, could be applied to ensure instrument correct use, its perfect conditions, even after maintenance and calibration process, performed by accredited companies (NBR ISO IEC 17025). As no unsatisfactory data was observed, no corrective actions were necessary. Warning observed values were attributed to measurements performed before equipment stabilization or to laboratory temperature changes.

Acknowledgments
To IPEN-CNEN/SP, especially to the CQMA group by structure; to CNPq, for the financial support (process number 149624 / 2017-4).

References

[1] Souza D R C, Aparecida MS, Ferracini V L, Turuna T B and Bertoni O I R. 2014. Aplicação de cartas controle para avaliar resultados na verificação de balanças. *VII IBEROLAB* 3p.
[2] Turuna T B. 2015. Aplicação de cartas de controle como ferramenta de melhoria frente ás dificuldades operacionais de laboratórios acreditados na ABNT NBR ISSO/IEC 17025. [dissertation]. São Carlos; Universidade de São Paulo. 67 p.
[3] ISO/IEC 17025 2005 General requirements for the competence of testing and calibration laboratories.31p. https://www.iso.org/standard/39883.html
[4] Oliveira C C, Granato D, Caruso M S F, Sakuma A M. 2013. Manual para elaboração de cartas de controle para monitoramento de processos de medição quantitativos. São Paulo; Institute Adolfo Lutz - SES/SP. 76p.
[5] International Organization for Standardization. ISO 8258 - Shewhart control charts. 1991.
[6] Hirata YS. 2002. Gráficos de controle para laboratórios de ensaios. *Biológico.* São Paulo.V64. n.2. p. 183-5.

IOP Conf. Series: Journal of Physics: Conf. Series **975** (2018) 012006 doi:10.1088/1742-6596/975/1/012006

Development of procedures for calibration of meteorological sensors. Case study: calibration of a tipping-bucket rain gauge and data-logger set

M A A Santana[1], P L O Guimarães[1] and L G Lanza[2,3]

[1] Environmental Metrology, Meteorological Instrumentation Laboratory, Center for Weather Forecasting and Climate Research, National Institute for Space Research, Rodovia Presidente Dutra km 39, Cachoeira Paulista, SP, Brazil;
[2] University of Genova, Dept. of Civil, Chemical and Environmental Engineering, 1 Montallegro, 16145 Genova, Italy;
[3] WMO/CIMO Lead Centre "B. Castelli" on Precipitation Intensity, Italy

E-mail: marcio.santana@inpe.br, profmarciosantana@hotmail.com

Abstract. The mechanical assembly of a tipping-bucket rain gauge is often calibrated as a stand-alone instrument. However, the measurement accuracy depends on the associated data-logger as well, especially in case the rainfall intensity is derived from the measurement. This paper reports the calibration of a set comprising the tipping-bucket rain gauge assembly and the associated data-logger. The case study of a specific commercial rain gauge is presented. We conclude that the response time of the data-logger directly affects the measurement of rainfall intensity and that the greatest contribution to the uncertainty budget may arise from the accuracy of the clock of the data-logger.

1. Introduction

Precipitation measurements are very important to the Environmental Sciences. A huge range of applications around the world bases on the observation and investigation of typically measured characteristics of precipitation such as the rainfall amount, intensity and duration, besides the frequency of intense rainfall events [1, 2].

There are instruments that simply inform the status of the rain in that moment, i.e. whether it rains or not, as well as there are those that show the distribution of precipitated drop sizes. The most commonly employed instrument for the measurement of rain amount and intensity is the tipping-bucket rain gauge (TBRG). The TBRG is a device that can send a pulse as an output signal to each nominal amount of rainfall collected in a pivoting two-compartment bucket [3, 4].

Most automatic rain gauges measure a small amount of rainfall over a relatively short time interval, usually less than one minute. Users of rainfall measurements typically require information of accumulated rainfall for longer time intervals, e.g. the hourly, daily, monthly, and even annual total rain depth [4, 5]. Other very important information is the rainfall intensity that is obtained indirectly using a rain gauge and a data acquisition and processing system.

Using a data acquisition system or a data-logger it is possible to record the instant when the various pulses occur, and thus calculate the rainfall intensity (*RI*). Usually, the mechanical device of a tipping-bucket rain gauge is calibrated as a stand-alone instrument. This paper shows the calibration procedure of an instrumentation set comprising a commercial tipping-bucket rain gauge and the associated data-logger, thus making it possible to calculate the uncertainty of the *RI* measurement.

2. Method

Figure 1 shows a classical TBRG. In general, a TBRG opens or closes a contact through a reed-switch sensor, or similar, to each determined volume of precipitated rain, generating a pulse if electrically powered [6]. In order to calculate the accumulated rainfall or rainfall intensity, it is necessary to record the pulses and the instants when they occur in a data acquisition system for suitable processing [7].

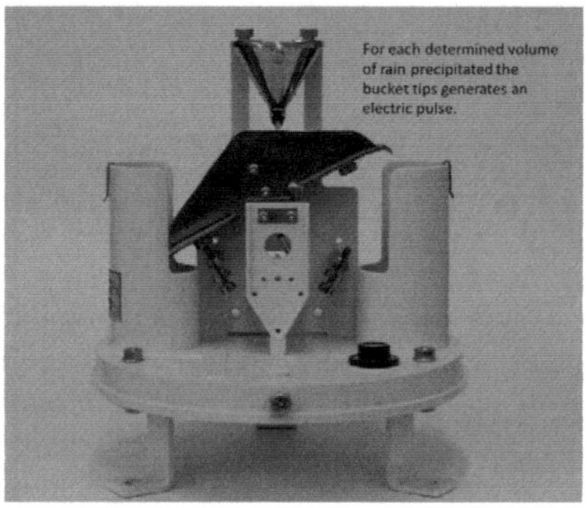

Figure 1. The TBRG and its measuring principle.

Figure 2 shows the set under investigation here, comprising a commercial TBRG, model HD2013, manufactured by Delta Ohm and a commercial data-logger, model CR1000, manufactured by Campbell Scientific as used in a weather station. In the case of TBRGs, at each pulse, the data-logger records the tip and time stamp, i.e. the date and time that pulse occurred. A weather station can also use further sensors to measure other relevant variables. In the case of a rainfall station, the recorded data can be later transmitted using different devices and means of communication. The measured data consist of time stamps of the time of tipping. Depending on the data-logger, it is possible to correct the data and calculate the *RI*, before the transmission.

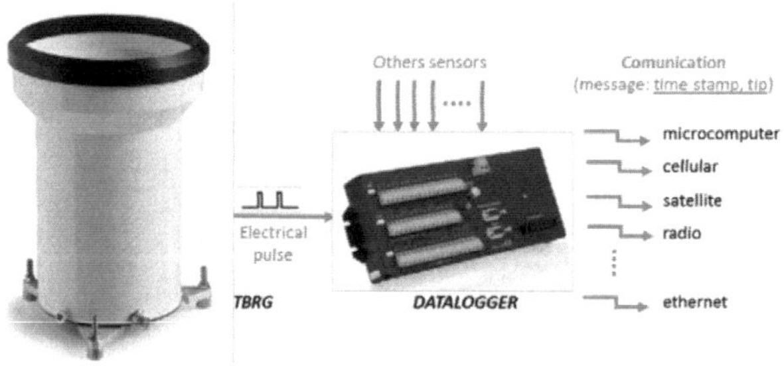

Figure 2. Schematics of a weather station with rainfall record.

The calibration of a TBRG can be performed by the volumetric (input) or gravimetric (output) methods shown in figure 3 [8] and an example of the calibration results is reported in tables 1 and 2. Briefly, the calibration procedure by the input method relies on the indirect comparison to working standards using a peristaltic pump and a data acquisition system, while in the output method the calibration procedure is based on the indirect comparison to the working standards, using a balance and a data acquisition system. A series of fifteen measures are carried out in the verification of the catchment area (here nominally equal to 400 cm^2) and five measurements for each calibration point, and the mean values are reported. The expanded uncertainty of measurement (calibration uncertainty) reported equals the combined standard uncertainty multiplied by the coverage factor "k" for a confidence level of 95.45 %, and was determined according to EURAMET cg-19 - Guidelines on the determination of uncertainty in gravimetric volume calibration [9] and EA-4/02 [10] and ISO / GUM - Guide to the Expression of Uncertainty in Measurement [11].

Figure 3. Calibration methods for a TBRG.

Table 1 shows the results of the calibration for a given *RI* using 100 tips (for *RI* around 15 mm/h) and table 2 shows the *RI* for a given number of tips, for example 50 tips. The calibration certificate also reported the environmental conditions (air temperature, t_{air}, water temperature, t_{water}, air relative humidity, H_R and atmospheric pressure, *P*), the diameter *d* of area of the collector and the measured mean resolution. The measured quantity Q_M is expressed by equation (1), where W_M is the measured mass of water, δ_{water} is the density of water and δ_{air} is the density of air.

$$QM = \frac{W_M \cdot [\delta_{water}(t_{water}) - \delta_{air}(t_{air}, P, H_R)]^{-1}}{\pi\left(\frac{d}{2}\right)^2} \tag{1}$$

Table 1. Calibration of quantity for *RI* = 12.6 to 13.6 mm/h (example).

Number of tips	Nominal quantity Q_N [mm]	Measured quantity Q_M [mm]	Correction *C* [%]	Expanded uncertainty *U* [%]	Coverage factor *k*
1	0.20	0.19	5.3	± 4.0	2.00
2	0.40	0.39	2.6	± 2.2	2.03
5	1.00	0.99	1.0	± 1.5	2.23
10	2.00	2.00	0.00	± 0.41	2.02
20	4.00	4.00	0.00	± 1.0	2.52
50	10.00	9.96	0.40	± 0.85	2.65
100	20.00	20.01	-0.01	± 0.21	2.52

Table 2. Calibration of *RI* versus 50 tips (example).

Calculated *RI* [mm/h]	Nominal quantity Q_N [mm]	Measured quantity Q_M [mm]	Correction *C* [%]	Expanded uncertainty *U* [%]	Coverage factor *k*
26		9.94	0.6	± 1.1	2.65
52		10.05	-0.50	± 0.72	2.65
72	10.00	9.97	0.30	± 0.60	2.52
90		10.04	-0.40	± 0.11	2.13
120		10.04	-0.40	± 0.35	2.52

Resolution (average) = 0.20 mm.

The correction *C* is given by (2), where Q_N is the nominal quantity.

$$C = \frac{[Q_N - Q_M].100}{Q_M} [\%] \quad (2)$$

In this paper, the calibration of the TBRG in conjunction with the data-logger (figure 4B) was performed based on the calibration of the TBRG itself (figure 4(a)). Figure 4(b) shows the RG calibration system obtaining the output signal of TBRG (pulse) and data from the data-logger (tip and timestamp). The RG calibration system consists of a rain simulator, a device for measuring the volume of simulated rainfall and a data acquisition system (data-logger) as well as meters for air temperature, water temperature, air relative humidity, atmospheric pressure and the diameter of the collector. The calibration certificate of the set TBRG/data-logger must also present the results and uncertainties for the rainfall intensity *RI*, including the corrections to be applied, in compliance with the recommendations of the World Meteorological Organization (WMO) [12].

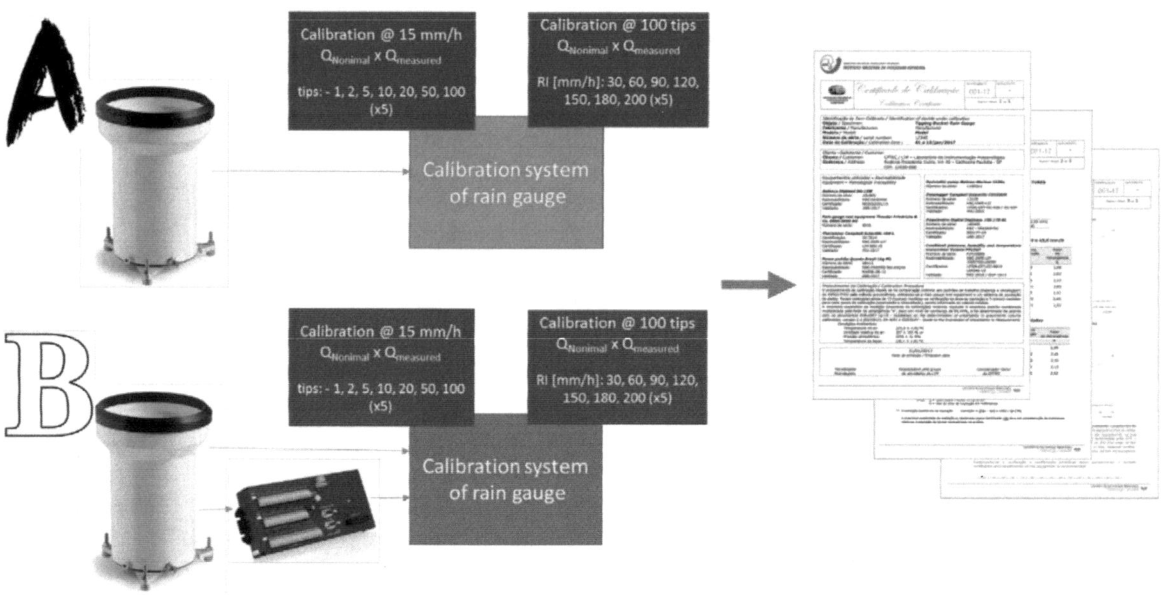

Figure 4. Setups of calibration: (a) TBRG; (b) TBRG + Data-logger.

3. Results

For the calibration of the set comprising the TBRG and the data-logger, the Q_M and *RI* are estimated with their respective uncertainties. The average *RI* in mm/h is expressed by (3), where *Q* is the quantity measured in millimeters and *t* is the time interval, in seconds.

$$\mathrm{RI} = \mathrm{Q.t^{-1}.3600} \tag{3}$$

In equation (4), n is the number of tips and t_n is the time interval between tips.

$$t = \sum t_n . n^{-1} \tag{4}$$

For the measured quantity Q used in equation (5), the uncertainty contributions depend on the method used (volumetric or gravimetric) [8].

$$\mathrm{RI} = \frac{\mathrm{Q}}{\mathrm{t}} + \delta_{\mathrm{RI_{REP}}} \tag{5}$$

For the estimation of the measurement uncertainty of *RI*, the data-logger used as a standard in the calibration system of the rain gauge must have been calibrated in time and frequency in the pulse channel used for the TBRG signal input under calibration. Indeed, physical and electrical factors affecting the crystal oscillator frequency may influence the stability and accuracy of the data-logger [13].

The measurement uncertainty of *RI* is obtained through equation (6), where u_{x1} is the uncertainty of Q, u_{x2} is relative to the calibration certificate of frequency of the data-logger, u_{x3} is due to the specifications of the clock accuracy of the data-logger and u_{x4} is due to the repeatability of the t_n measurements. Equation (7) expresses the relative uncertainty of *RI*.

$$\mathrm{u(RI)} = \left[\sum_{i=1}^{4} \left(\frac{\partial \mathrm{RI}}{\partial \mathrm{x_i}}\right)^2 \mathrm{u^2(x_i)}\right]^{1/2} \tag{6}$$

$$\mathrm{u(RI)(\%)} = \sqrt{\sum_{i=1}^{j} \mathrm{u_R(x_i)^2}} \tag{7}$$

Tables 3 and 4 show the results of calibration (example) where Q_N is the nominal quantity of rain, RI_N is the nominal rain intensity, Q_M is the measured quantity of rainfall, RI_M is the measured rain intensity, C_{QM} and C_{RI} are the corrections, U_{QM} and U_{RI} are the expanded uncertainties and k is the coverage fator for a probability of coverage of approximately 95.45 %.

Table 3. Calibration of quantity for $RI_N \approx 13.00$ mm/h (example).

		TBRG				TBRG with data-logger			
RI_N [mm/h]	Q_N [mm]	Q_M [mm]	C_{QM} [%]	U_{QM} [%]	k	RI_M [mm/h]	C_{RI} [%]	U_{RI} [%]	k
	0.20	0.19	5.3	± 4.0	2.00			± 4.7	2.00
	0.40	0.39	2.6	± 2.2	2.03			± 3.3	2.00
	1.00	0.99	1.0	± 1.5	2.23			± 2.8	2.00
13.00	2.00	2.00	0.00	± 0.41	2.02	12.99	0.1	± 2.5	2.00
	4.00	4.00	0.00	± 1.0	2.52			± 2.6	2.00
	10.00	9.96	0.40	± 0.85	2.65			± 2.5	2.00
	20.00	20.01	-0.01	± 0.21	2.52			± 2.5	2.00

Table 4. Calibration of *RI* versus 50 tips (example).

		TBRG				TBRG with data-logger			
Q_N [mm]	RI_N [mm/h]	Q_M [mm]	C_{QM} [%]	U_{QM} [%]	k	RI_M [mm/h]	C_{RI} [%]	U_{RI} [%]	k
10.00	25.90	9.94	0.60	± 1.1	2.65	25.75	0.58	± 3.6	2.00
	51.43	10.05	-0.50	± 0.72	2.65	51.67	-0.46	± 0.79	2.00
	72.00	9.97	0.30	± 0.60	2.52	71.81	0.26	± 0.66	2.00
	90.00	10.04	-0.40	± 0.11	2.13	90.34	-0.38	± 0.59	2.00
	120.00	10.04	-0.40	± 0.35	2.52	120.49	-0.41	± 0.82	2.00

Resolution (average) = 0.20 mm.

4. Conclusion

The TBRG and data-logger can be calibrated together or separately. When calibration is separately performed, the TBRG should be calibrated in a pluviometry laboratory and the data-logger in a laboratory in the quantity called time and frequency. In the case of set calibration, the calibration must be performed in a pluviometry laboratory, where the rainfall measurand must be simulated and the results for the rainwater amount and the interval of time between tips are indirectly compared to the working standards with metrological traceability evidenced to the SI.

Based on the presented case study, we conclude that the response time of the data-logger directly affects the measurement of rainfall intensity and the greatest contribution of uncertainty may be due to the accuracy of the internal clock of the data-logger. The results showed that the expanded uncertainties for the rainfall intensity *RI* are increased from 0.06 to 2.5 % on the expanded uncertainties of amount of rain *Q*.

This work is the basis for the development of calibration procedures for TBRG and data-logger sets to obtain the amount of rainfall, the calculation of the rainfall intensity and the estimation of its uncertainties and to comply with WMO recommendations.

5. References

[1] Westra S, Fowler H J, Evans J P, Alexander L V, Berg P, Johnson F, Kendon E J, Lenderink G, and Roberts N M. 2014. Future changes to the intensity and frequency of short-duration extreme rainfall, *Rev. Geophysics*, **52**, pp 522–555.

[2] Kidd C and Joe P. Importance, identification and measurement of light precipitation at mid- to high-latitudes. Proc. Joint EUMETSAT Meteorological Satellite Conf. and 15thSatellite Meteorology and Oceanography Conf., Amsterdam, Netherlands, EUMETSAT and Amer. Meteor. Soc., 6 pp, 2007.

[3] Liu X C, Gao T C, and Liu L. 2013. A comparison of rainfall measurements from multiple instruments, *Atmos. Meas. Tech.*, **6**, pp 1585–1595.

[4] Petracca M. Evaluation on accuracy of precipitation data. EUMETSAT H-SAF Visiting Scientist Report, 2011.

[5] Sene K. 2013. Precipitation Measurement. In: Flash Floods. Springer, Dordrecht.

[6] Santana M A A, Guimarães P L O and Lanza L G; Vuerich E.2015. Metrological analysis of a gravimetric calibration system for tipping-bucket rain gauges. *Meteorological Applications (Print)*, v. **22**, p. 879-885.

[7] Stagnaro M, Colli M, Giovanni L L, and Chan P W. 2016. Performance of post-processing algorithms for rainfall intensity using measurements from tipping-bucket rain gauges. *Atmos. Meas. Tech.*, **9**, 5699–5706.

IOP Conf. Series: Journal of Physics: Conf. Series **975** (2018) 012006 doi:10.1088/1742-6596/975/1/012006

[8] Santana M A A, Guimarães P L O and Lanza L G. 2016. Uncertainty contributions in the calibration of rain gauges. MMC 2016, Metrology for Meteorology and Climate conference. Madrid, Spain.

[9] EURAMET. 2011. Guidelines on the Determination of Uncertainty in Gravimetric Volume Calibration. EURAMET cg-19, Version 2.0 (03/2011). Calibration Guide. EURAMET e.V.: Braunschweig, Germany.

[10] European Co-operation for Acreditation (EA). EA-4/02 –Expression of the Uncertainty of Measurement in Calibration. 1999. 79 p. Disponível em: www.european-accreditation.org/n1/doc/ea-4-02.pdf.

[11] JCGM. 2008. Evaluation of measurement data - Guide to the expression of uncertainty in measurement (GUM). BIPM, JCGM 100:2008.

[12] WMO, "Guide to Meteorological Instruments and Methods of Observation", Provisional 2014 Edition, 2014.

[13] Zhou H, Nicholls C, Kunz T and Schwartz H. Frequency accuracy & stability dependencies of crystal oscillators. Carleton University, Ottawa, Ont., Canada. Carleton University, *Systems and Computer Engineering*, Technical Report SCE-08-12, November 2008.

Evaluation of measurement uncertainty for purity of a monoterpenic acid by small-scale coulometry

L C Norte[1], E M de Carvalho[2], M R R Tappin[2] and P P Borges[1]

[1]Inmetro, Electrochemical Laboratory (Label) from Chemical and Thermal Metrology Division(Dimqt), Duque de Caxias, Rio de Janeiro, Brazil

[2]Farmanguinhos, Fiocruz, Rio de Janeiro, Brazil

E-mail: luciana.carreiras@gmail.com

Abstract. Purity of the perylic acid (HPe) which is a monoterpenic acid from natural product (NP) with anti-inflammatory and anticancer properties was analyzed by small-scale coulometry (SSC), due to the low availability of HPe on the pharmaceutic market and its high cost. This work aims to present the evaluation of the measurements uncertainty from the purity of HPe by using SSC. Coulometric mean of purity obtained from 5 replicates resulted in 94.23% ± 0.88% (k = 2.06, for an approximately 95% confidence level). These studies aim in the future to develop the production of certified reference materials from NPs.

1. Introduction

In the quality control of medicines, the pharmaceutical industry uses chromatographic techniques. However, such techniques require a reference material, to be able to quantify it in the plant, which has become a bottleneck in the development of herbal medicines in pharmaceutical market.

However, the markers of the great majority of medicinal plants originate from their secondary metabolism. These substances are synthesized frequently in small quantities.

Brazil is one of the largest producers of oranges in the world and is the largest exporter of orange juice, with about 17.5 million tons produced in the 2016/2017 harvest, according to data from the Ministry of Agriculture, Fisheries and Food. Supply (Mapa). Limonene is a by-product of the orange juice industry. The annual production of this monoterpene is estimated at 50,000 tons, and its price is around US $1–2/kg. It is easily available as waste from the citrus juice industry and is obtained from a renewable source, industries such as pharmaceuticals, chemicals and food have shown great interest in the oxygenated derivatives of limonene, such as HPe.

One of the great obstacles of the natural products (NPs) is to obtain a reference material (RM) compatible with the requirements of the product, according to the new ordinance specific for phytomedicines [5], and considering the high cost of the molecule, the development of a CRM or a RM, will lead to a huge reduction of analysis costs for the molecule studied.

HPe has a low availability, so it is very expensive. Its structural formula is shown in figure 1.

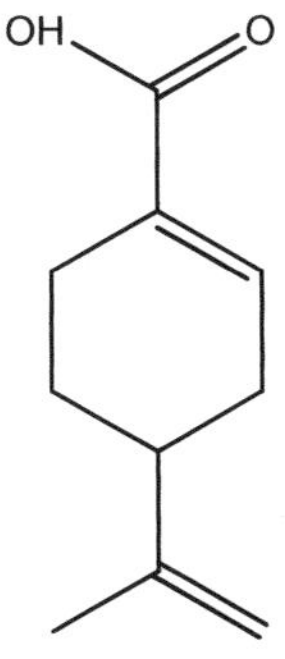

Perylic acid

Figure 1. Perylic acid [1].

Certified reference materials (CRMs) are the most important standards for disseminating measurement units (metrological traceability) to users. The CRM specified property values require metrological traceability with associated measurement uncertainty. These studies aim in the future to develop the production of CRMs from NPs.

An electroanalytical technique explained by Faraday's law, applied in the determination of the amount of substance (mol) of pure chemical compounds or their respective solutions and considered primary, since besides its operation is described and understood in terms of SI units, is a constant current Coulometry, that is widely used to determine the purity of RM and has high accuracy [2].

Microscale chemistry when used requires less reagent consumption and less generation of chemical residues, avoiding environmental pollution The application of the miniaturization of the coulometric technique is due to the need to use less sample, guaranteeing reliability to the chemical measurements, which, as a primary method, provides measurement results with high precision and accuracy, which are crucial for the development of an CRM of secondary metabolism plant [3]. The traceability and measurement uncertainty indicate that the key role of the reliability of chemical measurements is the comparability of results from different times and locations. The comparability of results is obtained mainly by evaluation of the measurement uncertainty[4].

1.1. Objective

This work aims to present the measurement uncertainty for purity of a monoterpenic acid determined by SSC.

2. Experimental

2.1. Materials

All analytical reagent-grade chemicals were used in the development of this work. All the solutions were prepared using deionized water from Elga with an electrolytic conductivity higher than 0.054 µS cm^{-1}. The HPe was purchased from Sigma Aldrich and was certified by the manufacturer to have a 95% purity. Before use, the HPe was dried in an oven at 80 °C for 3 h.

IOP Conf. Series: Journal of Physics: Conf. Series **975** (2018) 012007 doi:10.1088/1742-6596/975/1/012007

The coulometer used was a high-precision coulometry unit from Applied Precision Ltd. It was assembled with a constant current source (model 8011A) containing a high-precision timer with a precision of 1 µs and an indication unit (model 7011A). A gas system for a high-purity argon supply was used for deoxygenation. Parameters of optimization from SSC can be seen at table 1:

Table 1: Optimization from SSC

Optimization from SSC		
Parameters	**Before**	**After**
Coulometric cell (CC) temperature	25°C	35°C
Medium for sample solubilization	Solvent mixtures [a]	MEtOH [b]
Type of sample solubilization	Glass basket	Vacuum
Sample rinsing	KCl 1,0 M	MEtOH
Drying sample before to be used	70°C, 3h	80°C, 3h
Sample mass reduction scale	50 x	100 x
Reduction of supporting electrolyte scale	10 x	10 x

[a]DMSO; ethanol; acetonitrile
[b]Methanol

2.2. The technique

It was in constant-current coulometry carried out manually. An Excel spreadsheet was used to perform all the calculations. The sample of HPe was dried at 80 °C for 3 h and cooled in a desiccator over silica gel pellets. Platinum foil (99.99%) with a thickness of 0.25 mm and an area of 20 mm2 was the generating cathode; a spiral made of silver wire with a purity of 99.99%, from Heraus Vectra, with dimensions of 50 × 55 × 5 mm was the generating anode. The supporting electrolyte consisted of 25 mL of a 1.0 mol dm−1 KCl solution.

Aproximately 10 mL of a similar KCl solution, but with a slightly higher concentration, was introduced in the anode compartment. For the initial titration (endpoint detection at pH ≈ 6.9), approximately 20 µL of 0.01 mol kg−1 HCl solution (nominal value) was added before a 1.0 mA current was passed during a specific time, ending some time after the endpoint was detected. Each HPe sample of approximately 3.0 mg used in the main titration was weighed in a microbalance to which a buoyancy correction was applied to the weighed mass. In the main titration, a 20 mA current was passed until 80% of the sample was titrated. For the final titration (endpoint detection at pH ≈ 6.9), a 1.0 mA current was passed similarly to the initial titration.

The coulometric cell was maintained at 35 °C during the titration. High-purity argon gas (99.999%) was bubbled on both the cathodic and anodic compartments for deaeration or maintained above the cathodic compartment during the main titration. In all the titrations, the supporting electrolyte was stirred with a magnetic stirrer. The Faraday constant, equivalent to 96,485.3289 C mol−1 and the density of the HPe of 1,007 kg m−3 were considered in calculations, as was the molar mass for HPe of 166.217 g mol−1. The endpoint was calculated using the OriginPro8.5.1 program, considering the secondary derivate of the points of the coulometric titration curve.

The uncertainty measurement evaluation was carried out according to the specific literature from uncertainty. The evaluation of the measurement uncertainty was performed with consideration of the physical sources of uncertainty, such as the input quantities of the measurand ν (amount of substance) according to equation 1 and some important chemical sources for the coulometric titration measurement: end-point detection, molar mass, current efficiency, diffusion losses and aerosol losses.

$$\nu = \frac{I_1 \cdot t_1 + I_2 \cdot t_2 + I_3 \cdot t_3}{z \cdot F \cdot m \cdot B} \qquad (1)$$

where ν is the amount of substance; I_1 is the current in A passed in the initial titration; $\mathbf{t}_1$ is the time in s for the initial titration; I_2 is the current in A passed in the main titration; t_2 is the time in s for the main titration; I_3 is the current in A passed in the final titration; t_3 is the time in s for the final titration; z is the charge of the electrons for the electrochemical reaction; F is the Faraday constant; m is the mass sample and B is the buoyancy correction for the mass sample.

For the evaluation of the uncertainty measurement of the purity in terms of mass fraction, w (%), it is necessary to consider the molar mass M of the sample, according to equation. 2.

$$w(\%) = \nu \cdot M \qquad (2)$$

3. Results and Discussion

3.1. The purity of HPe

It was determined after the optimization of the methodology of drying and mass sampling by coulometric titration. Table 2 presents the measurement results obtained in 5 different replicates.

Table 2. Coulometric titration samples

	Coulometric titration		
n[a]	**m (g)**	**Amount content (mol kg^{-1})**	**Mass fraction (%)**
1	0.0034271	5.6303	93.586
2	0.0038015	5.6357	93.675
3	0.0036845	5.6752	94.331
4	0.0035127	5.6897	94.572
5	0.0031659	5.7151	94.994
	Average (%)	**5.6692**	**94.232**
	Standard deviation (%)	**0.036**	**0.6**
	RSD (%)	**0.016**	**0.6**

3.2. Measurement uncertainty evaluation

For comparability of the measurement results at any time and space it is necessary to express the measurement result followed by its uncertainty.

Evaluation of the measurement uncertainty is calculated considering international guides [2,4], which is specifically for chemical analysis. An uncertainty budget of the coulometric titration measurement for HPe in an evaluation considering the relative uncertainty approach is shown in table 3.

Table 3: Uncertainty budget

Uncertainty budget					
Source	**Value**	**Unity**	**Standard uncertainty**	**Unity**	$c_i \cdot u_i \, (v_{HPe})^{-1}$
Repeatability	5.6691954	mol kg $^{-1}$	1.61 x 10^{-2}	mol kg $^{-1}$	0.28%
Weighing	0.000003798	kg	6.5 x 10^{-10}	kg	0.02%
Buoyance correction	1.00102	1	9.0 x 10^{-7}	1	0.00%
Time	275.04	s	5.0 x 10^{-7}	s	0.00%
Current	0.02	A	9.1 x 10^{-10}	A	0.00%
Faraday constant	96,485.3289	C mol $^{-1}$	5.9 x 10^{-4}	C mol $^{-1}$	0.00%
End-point detection	-		5.1 x 10^{-4}	mol kg $^{-1}$	0.01%
Difusion losses	-		3.0 x 10^{-3}	mol kg $^{-1}$	0.05%
Aerosol losses	-		1.5 x 10^{-3}	mol kg $^{-1}$	0.03%
Current efficiency	-		1.5 x 10^{-2}	mol kg $^{-1}$	0.27%
uA	0.28%				
uB	0.28%				
uC	0.40%				

The main sources of uncertainty as can be seen by the Table 3 were the repeatability of the method and the current efficiency. As the method was carried out manually, the precision of the measurements was below than it was expected. On the other hand, the current efficiency was affected, which means that some charge was lost possibly it was used by others secondary electrochemical reactions. The uncertainty budget is also important as a way to indicate the next steps to be followed in order to increase the method performance.

4. Conclusion

The result of the coulometric titration on a small scale was 94.2% ± 0.9% (*k*= 2.06, for an approximately 95% confidence level, according to the specific literature from uncertainty). The purity was also good considering that the sample was added to the glass coulometric cell without any loss and with a vacuum, although the temperature used in the coulometric cell was 35 °C.

The miniaturization of the CC was fundamental for characterizing such substances due to the low availability found in the market, in addition to their high cost. SCC titration results shall be further compared with NMRq results to validate the purity obtained. These studies aim in the future to produce CRMs of NPs.

The future development and use of CRM phytomedicines will contribute to the metrological reliability and traceability of the results of gas chromatographic analysis and other analytical measurements used by chemical laboratories for several other NPs that have a similar matrix to those found in the phytotherapic one studied here.

References

[1] BORGES P P, SILVA Jr W 2014 J Appl Electrochem 44:1411-142 0.

[2] DUETZ W A, BOUWMEESTE H, VAN BEILEN JB, WITHOLT B 2003 Appl Microbiol and Biot 61: 269-277.

[3] GUM 2008Guia para a expressão de incerteza de medição - Avaliação de dados de medição (Rio de Janeiro: Inmetro)

[4] PRATT K W 1994 Anal Chim Acta 289:135-142.

[5] RDC 26/2014 Diário Oficial da União (DOU) 14.05.14 Brasil.

Interaction between the ultrasound velocity and material properties using *ANOVA*

C L F da Rocha [1], D A K Fabricio [2], L C L de Resende[3]

[1] Universidade Federal da Bahia; [2] Instituto Federal de Chapecó, [3] Senai/Cimatec
E-mail: cfanezi@ufba.br

Abstract: By using ultrasound velocity measurements, it is possible to estimate the material anisotropy. Because of material anisotropy, distinct sound velocities are obtained for different measured directions. To qualify the ultrasonic inspection procedure, four conditions of the material, three different directions (with relation of the rolling direction) and two operators were used. After the measurement, the values were analyzed using the *ANOVA* method. Finally, it was possible to verify that there is no interference of the operator, which guarantees a good reproducibility of the inspection procedure, and that there is interference of the direction of inspection, demonstrating the condition of anisotropy of the material.

Keywords: Ultrasound velocity, material anisotropy, *ANOVA*

1. Introduction

Non-destructive testing are techniques used in the industry for inspection of components and equipment, which aims to control the integrity of the material. This type of test does not cause any damage to the component and, beyond that, can be performed in the field, without interfering in the production.

The ultrasound technique is one of the most used nondestructive tests adopted in the industry [1]. This technique is applied, for example, to characterize corrosion process, micro-cracks, intermetallic precipitates that arise on material due to the action of the temperature and the time of exposure in use, porosities, arising from the welding process, between other possibilities.

The basic principle of this technique is the measurement of the ultrasonic velocity of the material, which can be modified depending on the material microstructure, in different conditions [2]. The intensity of the ultrasonic wave decreases with the distance of the source of emission, suffering losses. These losses can be attributed to diffraction, reflection and absorption mechanisms [3].

A measurement has good reproducibility when it follows procedures in such a way that, regardless of the operator, the results can be reproduced.

Thus, this work considers the influence of the rolling direction of the material, the ultrasonic velocity variation and the influence of two operators on the obtained results, using the ANOVA method, a methodology that evaluates the significance of the various factors and interactions [4].

because it is cause loss in mechanical properties and corrosion resistance.

From the point of the inspection, these phases act as barriers to the passage of the sound, thus causing loss in the sound intensity of the material (attenuation), which modifies the value of the ultrasound velocity. The higher the amount of precipitated intermetallic, the greater the attenuation expected.

The test specimens were identified as A (solubilized – without intermetallic), B, C and D, according to the quantity of precipitated intermetallic, in increasing order.

2 . Material and Method

The material selected for inspection in this work is a stainless steel duplex UNS S 31803, in four different specimen conditions: as received from the supplier, and heat treated at 760°C for three different times (10, 30, 60 and 120 minutes), in order to induce the precipitation of intermetallic phases which in practice undesirable in the material because it is cause loss in mechanical properties and corrosion resistence.

From the point of the inspection, these phases act as barriers to the passage of the sound, thus causing loss in the sound intensity of the material (attenuation), which modifies the value of the ultrasound velocity. The higher the amount of precipitated intermetallic, the greater the attenuation expected.

The test specimens were identified as A (solubilized – without intermetallic), B, C and D, according to the quantity of precipitated intermetallic, in increasing order.

2.1. Pulse-Echo Ultrasonic Inspection Technique

The non-destructive ultrasound technique is widely used in industry because it is sensitive to heterogeneities presented by the material, such as cracks and micro-cracks, inclusions, precipitates, porosities, among others, and can characterize them due to the interaction of the sonic beam with the microstructure.

The attenuation is caused by the loss of sound energy when the beam is passing through the material. The intensity of the beam received by the transducer is significantly less than the initial transmitted intensity. According to researchers, losses occur due to interference effects and reflections [6].

For the application of pulse-echo method, basic devices are required: a piezoelectric crystal transducer, which transmits and receives the ultrasonic waves that propagate through the material, and ultrasound equipment capable to transmit the pulse, receiving the echo and amplifying it [7].

This analysis is possible because through the ultrasound velocity (m/s), one can estimate the elastic modulus of the material. By definition, the variation of the elastic modulus indicates the material anisotropy, demonstrating that the rigidity may vary according to the direction of measurement [8]. The higher the ultrasound velocity, the higher the elastic modulus of the material, and thus greater rigidity of the material. In other words, the more significant the increase, the greater the material anisotropy.

2.2. ANOVA

The ANOVA method is the statistical tool used to analyze the results obtained in this study. A complete factorial design was performed, with two replicates for each condition of the material, at a 95% confidence level [9,10].

3 . Results and Discussion

The results obtained by ultrasonic inspection showed that when there is a significant variation of the material condition and when there is a modification of the transducer position at the inspection, there is a variation of the ultrasound velocity. So, the greater the amount of precipitate in the material, the greater the attenuation of the sound, and physical changes in the material occur, as in the elastic modulus (Table 1).

In this way, it can be said that the material experiences change in elastic modulus, allowing to predict that the material undergoes anisotropy. Through the ANOVA method, it was possible to analyze the factors: condition, measurement direction and operator. Table 2 presents the results. Minitab® software was used to perform the analysis, where:

DF: degree of freedom;
SQ: Sum of squares;

MQ: Mean square;
F: Ratio between the variance of the analyzed factor and residual variance;
P-value: Significance level;
F-critical: Tabulated value, obtained from the desired confidence level (95%), the degrees of freedom of the analyzed factor and the residual degrees of freedom;
Significant: A factor or interaction is significant when F > F-critical or when P-value <0.05;

Table 1 – Ultrasound Velocity (m/s) x Operator A and B and differents conditions of heat treatments.

Condition	Direction	Operator A	Operator B
A	0	5889.5	5885.7
	45	2945	2976
	90	5827.3	5822.3
A	0	5891	5893
	45	3081.7	3083.4
	90	5848.9	5849.9
B	0	5850.4	5847.5
	45	5837.6	5846.2
	90	3208	3230
B	0	5863.4	5861.4
	45	3082.1	3080.7
	90	5835.7	5838.2
C	0	5896.6	5892.4
	45	2912.4	2980.2
	90	5811	5813
C	0	5864.9	5870.3
	45	3046.3	3050.2
	90	5898.7	5900.1
D	0	5857	5853
	45	3280.9	3293.2
	90	5858.7	5853.1
D	0	5845.3	5856.7
	45	2995.9	3001
	90	5857.4	5855.6

The interaction between sample condition vs direction was significant for the ultrasound velocity values. This result was expected, since the specimens were submitted to different treatments, which implies a change in the attenuation. In addition, the direction factor was also significant, which was expected, and even extrapolating this result, it is possible to verify a change of its elastic modulus, being the material seen as anisotropic.

The operator factor and the conditions, condition vs operator and direction vs operator were not considered significant.

Table 2 – Results obtained from *ANOVA*.

Source	DF	SQ	MQ	F	P-value	F-critical	Significant (?)
Condition	3	15547	5182	0.01	1.00	3.01	No
Direction	2	56860200	28430100	46.76	0.00	3.40	Yes
Operator	1	453	453	0.00	0.98	4.26	No
Condition vs Direction	6	11224338	1870723	3.08	0.02	2.51	Yes
Direction vs Operator	2	605	302	0.00	1.00	3.40	No
Condition vs Operator	3	178	59	0.00	1.00	3.01	No
Condition vs Direction vs Operator	6	598	100	0.00	1.00	2.51	
Error	24	14593259	608052				
Total	47	82695177					

References

[1] Krautkrämer, J. Ultrasonic Testing of Materials, Springer-Verlag Berlin. Heidelberg

[2] ASM Metals Handbook Volume 17, Nondestructive Evaluation and Quality Control, Materials Park (Ohio), ASM International 1992, 345.

[3] Pandey, D. K.; Pandey, S. Ultrasonics: A Technique of Material Characterization, Acoustic Waves, Don Dissanayake 2010, 414

[4] Rocha, C.L.F; Fabricio, D.A.K, Haag, J.; Strohaecher, T.R. Análise da confiabilidade na determinação das propriedades mecânicas em ensaios de tração, ABM, 2014.

[5] Krueger, S. E. Caracterização microestrutural de ferros fundidos por ultrasom. Tese (Doutorado em Ciência em Engenharia Metalúrgica e de Materiais). Universidade Federal do Rio de Janeiro, 312 p., 2000.

[6] Barella, S. Gruttadauria, A.; Mapelli C.; Mombelli D.; Fanezi, C.L.; Fioletti F.; Formentelli, M.; Guarnieri. M. Effect of heat treatment and of primary austenite grain size on the minimum size of detectable defect on

26NiCrMoV11.5 high strength steel. Advanced Journal of Materials, janeiro, 2014.

[7] CHO, H; LEE, K. Effect of cold working and isothermal aging on the precipitation of sigma phase in 2205 duplex stainless steel. Materials Characterization 75 (2013)29–34.

[8] Garcia, A.; SPIM, J. A.; SANTOS, C. A. Ensaio dos Materiais. Rio de Janeiro: LTC, 2008.

[9] Rocha, C. L. F.; Fabricio, D. A. K.; Costa, V. M.; Reguly, A. Quality assurance of absorbed energy in Charpy impact test. Metrologia 2015.

[10] Fabricio, D. A. K.; Rocha, C.L.F.; Reguly, A.; Strohaecker, R.T. Interlaboratory comparison in the notch measurement of the charpy testing. Tchê Química, ISSN 2179-0302, 2015.

IOP Conf. Series: Journal of Physics: Conf. Series **975** (2018) 012009 doi:10.1088/1742-6596/975/1/012009

Is my bottom-up uncertainty estimation on metal measurement adequate?

J R Marques, M G Faustino, L R Monteiro, J C Ulrich, M A F Pires and M E B Cotrim

Instituto de Pesquisas Energéticas e Nucleares, IPEN–CNEN/SP, São Paulo, Brazil

E-mail: joyce.marques@usp.br

Abstract. Is the estimated uncertainty under GUM recommendation associated with metal measurement adequately estimated? How to evaluate if the measurement uncertainty really covers all uncertainty that is associated with the analytical procedure? Considering that, many laboratories frequently underestimate or less frequently overestimate uncertainties on its results; this paper presents the evaluation of estimated uncertainties on two ICP-OES procedures of seven metal measurements according to GUM approach. Horwitz function and proficiency tests scaled standard uncertainties were used in this evaluation. Our data shows that most elements expanded uncertainties were from two to four times underestimated. Possible causes and corrections are discussed herein.

1. Introduction

Expanded uncertainty (*U*) estimation according to GUM [1] approach is widely applied although time-consuming. Often this procedure is criticized because not all uncertainty sources are covered on its budget [2, 3, 4, 5] Dark uncertainty is a concept defined [3, 5] as the uncertainty not seen at all and therefore unaccounted in the measurement uncertainty estimation budget. So, the emerging question is how to evaluate if the uncertainty estimation was realistic or adequate? Several laboratories follow the procedure and do not evaluate its uncertainty results by other means.

Horwitz [4] treated large amount of analytical data and recommended four approaches to answer this question. These approaches are:

- To run sufficient replicates,
- To reevaluate and revise in details previous uncertainty estimation budget,
- To examine carefully laboratory proficiency test results,
- To compare the uncertainty estimation with Horwitz function results [6].

The present paper aimed to evaluate GUM estimated uncertainty of seven metals measurement by ICP-OES in two different matrixes [7, 8], with different analysts and distinct analytical data. Horwitz approach was used in this work to verify if GUM uncertainty estimation adequately cover all measurement uncertainty.

2. Methods / Experimental

The GUM approach that estimate uncertainty on seven metal measurement (Ba, Cd, Cr, Cu, Fe, Mn and Ni) on wastewater and superficial water by ICP-OES were evaluated. Method 1 and Method 2 were developed and applied to superficial water samples [9] and to wastewater [10] respectively, and details from the uncertainty estimation can be found elsewhere [7, 8]. Despite the fact that both works were developed at CQMA/IPEN, all analytical data were performed separately and therefore are independent. Both methods chosen the same wavelengths for each element measurement.

2.1. Horwitz approach

All four approaches that Horwitz [4] recommends, to verify uncertainty estimation procedure, are discussed in this paper. Some remarks are presented to each recommended step.

- To increase the number of replicates will not change the true value [4], but only increase the confidence on the estimated interval from the true value. In the later uncertainty estimation, this step is not feasible. This must be accounted on the experimental design of the uncertainty estimation procedure.
- Previous studies detailed the uncertainty budget [7, 8], so the uncertainty values estimated with the GUM approach will solely be confronted with complementary data.
- Minimum of two years PT results of Ba (n =8), Cr (n = 12), Cu (n = 4), Fe (n = 12), Mn (n = 6) and Ni (n = 12) helped to evaluate laboratory uncertainty through the Scaled Standard Uncertainty – *SSU* [11] (see equation (1)). The PT provider is EPTIS registered with ISO17043 accreditation [12].
- The Horwitz formula (see equation (2)) was previously extensive tested [4, 5, 13, 14] and herein applied to Method 1 and 2 in order the estimated uncertainty.

$$SSU = \frac{u(y)}{\hat{\sigma}_R} \tag{1}$$

Where: $u(y)$: estimated uncertainty; $\hat{\sigma}_R$: standard uncertainty of the correspondent element PT's

$$RSD = 2\times C^{-0.15} \tag{2}$$

Where: *RSD*: Relative standard deviation in %; *C*: mass to mass fraction, i.e. 1 $mgL^{-1} = 10^{-6}$ g g^{-1}.

3. Results and discussion

Method 1 and 2 summaries of expanded uncertainty (U, k=2) are presented in Table 1 and 2 respectively. With each method concentration range, the relative standard deviation expected by Horwitz function was calculated and ranged between 11 and 16 % to method 1 and between 11 and 25 % to method 2. While the U (k=2) ranged between 3 and 13% to method 1 and 5 to 12% to method 2. Horwitz coverage was calculated as the ratio of each method $U(k=2)$ to expected RSD from Horwitz function. By that way, $U(k=2)$ corresponded roughly to 25 to 80% of all variability expected from Horwitz function in method 1 and to 24 to 60% in method 2, with only one element (Fe) with 106% uncertainty coverage. However, as an empiric method we understand that interpretation needs to be tested.

The *SSU* evaluation, applied as a secondary test, considers that if GUM approach covers most of measurement uncertainty, then *SSU* = 1 or a close value. As presented in Table 1 and 2, most values of method 1 *SSU* ranged from 0.17 to 0.31 with a close to unity value (0.93) only to Cu. Method 2 *SSU* ranged from 0.25 to 0.62. GUM uncertainty and reproducibility studies tested by Walsh et al. [2] and Thompson [5] presented *SSU* < 1, often uncertainty was underestimated by a factor of 2. The underestimation factor in Method 1 and 2 seems to vary from 4 to 2.5 (as *SSU* = 1). The cause of this behavior could be a shorter time in reproducibility studies than ideally required. However, different laboratories, procedures and different analytical techniques seem to face same results [2, 5, 4], as obtained in Method 1 and 2. Which indicates not all variability associated with chemical measurement could be explained with GUM procedure. To solve this issue, a larger coverage factor *k* could be more suitable. Horwitz and *SSU* evaluations were adequate.

Table 1. Method 1 summary of expanded uncertainty estimation (GUM, Horwitz function and *SSU*).

Element	Major Uncertainty component	*U* % (k=2)	Concentration range (mgL^{-1})		*RSD*[a] (%)	Horwitz Coverage %	*SSU*
Ba	Analytical curve	3	0.10	10.00	11	27	0.25
Cd	Analytical curve	5	0.01	1.00	16	38	n.a.[b]
Cr	Standard preparation	4	0.01	1.00	16	25	0.31
Cu	Recovery	13	0.01	1.00	16	81	0.93
Fe	Recovery	5	0.10	10.00	11	44	0.17
Mn	Standard preparation	4	0.01	1.00	11	35	0.27
Ni	Standard preparation	4	0.01	1.00	16	25	0.17

Table 2. Method 2 summary of expanded uncertainty estimation (GUM, Horwitz function and *SSU*).

Element	Major Uncertainty component	*U* % (k=2)	Concentration range (mgL^{-1})		*RSD*[a] (%)	Horwitz Coverage %	*SSU*
Ba	Recovery	5	0.10	10.00	11	44	0.42
Cd	Recovery	6	0.01	0.05	25	24	n.a.[b]
Cr	Recovery	8	0.03	3.00	14	59	0.62
Cu	Recovery	7	0.03	3.00	14	52	0.50
Fe	Recovery	12	0.10	10.00	11	106	0.40
Mn	Recovery	5	0.01	1.00	16	31	0.33
Ni	Recovery	6	0.01	1.00	16	38	0.25

[a] by Horwitz function.
[b] n.a. - not available.

4. Conclusion

Despite worldwide use and largely debated to estimate measurement expanded uncertainty, GUM approach could easily do not cover all uncertainty sources. This work data with detailed uncertainty budget, applied to two different matrixes, seven elements, performed by different analysts with extensive reproducibility and recovery studies, when confronted with much simpler models seems to underestimate uncertainties by a factor from 2 to 4. Therefore, to prevent this effect it is recommended to apply other uncertainty estimation procedures that could guide the analyst and the laboratory find a more suitable uncertainty estimation value, and at the end, if required, to expand the usual coverage factor.

Acknowledgments
Authors thanks to CNPq for the financial support (Processes 134716/2013-2 and 134537/2013-0).

References

[1] Joint Committee for Guides in Metrology 100:2008. Guide to the Expression of Uncertainty in Measurement 1995 with minor corrections Published by the JCGM in the name of the BIPM, IEC, IFCC, ILAC, ISO, IUPAC, IUPAP, and OIML. [Accessed 05 march 2017]. Available: http://www.bipm.org/en/publications/guides/gum.html.
[2] S. Walsh, D. Wegrzynek, A. Venzin and C. Mansoux., "Using reproducibility to test the adequacy of GUM based uncertainty quanti," in Advances in Nuclear Nonproliferation Technology and Policy Conference: Bridging the gaps in nuclear nonproliferation. September 25-30, 2016., Santa Fe, NM., 2016.
[3] Analytical Method Committee. Dark uncertainty - AMC Technical Brief No.53, 2012.

[4] Horwitz W, The certainty of uncertainty, Journal of AOAC International , vol. 86, no. 1, pp. 109-111, 2003.
[5] Thompson M and Ellison S. Dark uncertainty, Accred Qual Assur, vol. 16, pp. 483-487, 2011.
[6] Analytical Method Committee. The amazing Horwitz function- AMC Technical Brief No.17, Royal Society of Chemistry, 2004.
[7] Faustino M G, Marques J R, Monteiro L R, Stellato T B, Soares S M V, Silva T B S C, et al. Uncertainty estimation in the determination of metals in superficial water by ICP-OES, Journal of Physics, no. Conference Series 733/ 012032, 2016.
[8] Marques J R, Soares S M V, Stellato T B, Silva T B S C, Faustino M G, Monteiro L R, et.al. Uncertainty Estimation of Metals and Semimetals Determination in Wastewater by Inductively Coupled Plasma Optical Emission Spectrometry (ICP-OES), Journal of Physics, no. Conference Series 733/ 012023, 2016.
[9] Faustino M G. Desenvolvimento e validação de metodologia para determinação de metais em amostras de água por Espectrometria de Emissão Óptica com Plasma de Argônio (ICP-OES). São Paulo. Tese [Doutorado em Tecnologia nuclear] – Universidade de São Paulo; 2015.
[10] Marques J R, Otimização e validação de metodologia analítica para determinação de metais e semimetais em efluentes líquidos por Espectrometria de Emissão Óptica com Plasma de Argônio (ICP-OES). São Paulo. Tese [Doutorado em Tecnologia nuclear] – Universidade de São Paulo; 2015
[11] International Organization for Standardization. ISO 5725-2 - Accuracy (trueness and precision) of measurement methods and results- Part 5: Alternative methods for the determination of the precision of a standard measurement method. 1994.
[12] Rede Metrológica do Rio Grande do Sul, Programa de Ensaios de Proficiência em Análises Ambientais: Produto: Água Tratada, Água Bruta e Efluente Industrial. 02 Feb 2016. [Accessed 2 jun 2017]. Available: http://www.redemetrologica.com.br/interlaboratoriais/39.
[13] Thompson M, Recent trend interlaboratory precision at ppb and sub ppb concentration in relation to fitness for purpose criteria in proficiencing testing. The analyst, vol. 125, pp. 385-386, 2000.
[14] Korol W, Rubaj J, Bielecka G, Walczynsky S, Zygmunt J and Dobrowolsky R, Criteria for using proficiency test results for estimation of measurement uncertainty: feed analysis example. Accred Qual Assur, vol. 22, pp. 83-89, 2017.

Measurement of ultrasound power using a calorimeter

G Morgado, S Miqueleti and R P B Costa-Felix

Laboratory of Ultrasound, National Institute of Metrology, Quality and Technology (Inmetro), Av. Nossa Sra das Graças 50, Duque de Caxias, RJ, 25250-020, Brazil

E-mail: gisele.morgado@hotmail.com; rpfelix@inmetro.gov.br

Abstract. This paper presents a comparison between the ultrasound power of a 1 MHz therapy equipment on the water using a calorimeter and a radiation force balance. For a range of 5 to 10 W, the results presented a normalized error less than 1, disclosing compatibility of the results from the developed system and the radiation force balance. The calorimetric method might be used as a faster and cheaper means for the verification of the ultrasonic power emitted by an equipment for physiotherapeutic treatment.

1.Introduction

Therapeutic ultrasound equipment typically use frequency from 1 to 3 MHz and intensities ranging from 0.1 to 3.0 W cm^{-2} in continuous and pulsed modes. The equipment comprises a high frequency electric current generator and an amplifier, connected to a piezoelectric ceramic lead zirconate titanate (PZT) which deforms in the presence of an electric field. The radiation produced by ultrasound transducers finds increasing application in several areas of knowledge. In certain cases, it is necessary to know the applied power, especially in the medical field where it is widely used for diagnosis, therapy or surgical procedures [1].

Ultrasound is among the physical resources most used by professional physiotherapists in the treatment of the most diverse disorders of the musculoskeletal system [2][1]. The intensity of the ultrasonic radiation is an essential factor for the success of any therapy, as well as its time of application

The ultrasonic radiation effects on the tissues depend on its applied intensity. The ultrasound power should be calibrated in order to avoid lack of efficiency of the treatment [4]. The equipment must be checked periodically to ensure that the standards are correct [5][6]. According to Repacholi and Benwel [7], this energy depends on the following parameters: acoustic power, time, frequency, and effective radiation area of the transducer [8].

Typically, ultrasound power is measured with a radiation force balance. However, this measurement equipment is expensive and time consuming. Alternative way to assess physiotherapy equipment ultrasonic power would be of interest for equipment in clinical commitment [9].

The work reported in this paper aims to develop a calorimeter [10] for the measurement of the ultrasonic power. Further, we compared the power output from an ultrasound therapy equipment for 1 MHz using as reference value the results from a radiation force balance.

IOP Conf. Series: Journal of Physics: Conf. Series **975** (2018) 012010 doi:10.1088/1742-6596/975/1/012010

2. Methodology

2.1. Heat Flux

For a body to be heated, normally a constant power source is used, that is, a source capable of supplying an amount of heat per unit of time.

We define the heat flux (ϕ) as the quotient between the quantity of heat (Q) and the time interval of exposure (Δt) and is given by the formula:

$$\phi = \frac{Q}{\Delta t} \quad (1)$$

The unit adopted for heat flux in the International System is watt (W), corresponding to joule per second.

2.2 Thermal variation

Given the equation

$$Q = m \cdot c \cdot \Delta\theta \quad (2)$$

in which Q is the quantity of heat (joule), $\Delta\theta$ is the expected thermal variation (°C) when the m mass of water (kg) is heated with a transducer of an ultrasound therapy equipment was obtained by the equation (4):

$$\Delta\theta = \frac{\phi \cdot \Delta t}{m \cdot c} \quad (3)$$

in which, c is the specific heat of water equal to 4,1868 $J \cdot g^{-1} \cdot °C^{-1}$. The expected temperature increase by heating 250 g of water for 5 minutes, for the three nominal power is shown in Table 1.

Table 1. Expected thermal variation for 5 minutes of heating of 250 g of water.

Power [W]	$\Delta\theta$ [°C]
5,0	1,43
6,5	1,86
10,0	2,86

The sample temperature is monitored during the measurement process with the aid of a thermocouple type K and a temperature measuring system (model U1252A, Agilent Technologies, CA). Figure 1 illustrates the experimental setup of the measuring system described above:

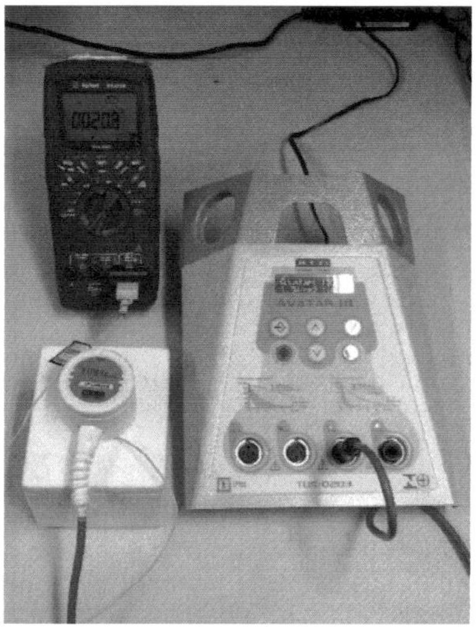

Figure 1. Experimental arrangement used for determining ultrasonic power with the calorimeter.

2.3 Ultrasonic Power

The ultrasonic power estimated from the thermal variation was determined based on four measurements under repeatability conditions, for each power studied. The ultrasonic power is given by the equation:

$$P = \frac{m \cdot c \cdot (\Delta\theta)}{t} \quad (4)$$

2.4 Measurement Uncertainty

The measurement uncertainty is a parameter that provides positive dispersion characteristics of a measured quantity. Here, the measurement uncertainty is calculated according to the Guide of the Expression of Uncertainty in Measurements and the contributions for uncertainty are shown in Figure 2.

Standard uncertainties Type A and Type B were considered [11][12][13]. The standard uncertainty Type A was obtained by calculating the standard deviation of the mean for four repetitions. The standard uncertainty Type B was estimated from the temperature uncertainty acquired from the digital thermometer calibration (0.1 °C), the mass of water (5 g), specific heat of water (0,0010 $J \cdot g^{-1} \cdot {}^{\circ}C^{-1}$), and time exposition (2 seconds).

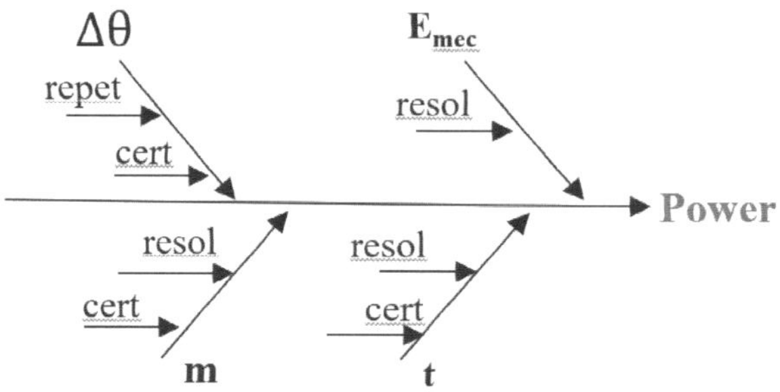

Figure 2. Contributions for uncertainty.

According item 5.1.2 [7], the combined standard uncertainty uc_{power} is the positive square root of the combined variance, and is given by (5):

$$uc_{power} = \sqrt{\sum \left(\frac{\partial P}{\partial xi} \cdot u_{xi} \right)^2} \tag{5}$$

The expanded uncertainty U_{power} was obtained by multiplying the combined standard uncertainty uc_{power} by a coverage factor k = 3.18 based on the t-distribution for $\nu_{eff} = 3$, with coverage probability of 0.95.

3. Results

The measurement results of ultrasonic power obtained with the calorimeter *(P_{cal})* and radiation force balance (P_{bal}), with its respective uncertainties *(U_{cal})* and (U_{bal}) are shown in Table 2.

Table 2. Measurement results of ultrasonic power obtained with the calorimeter, with its respective uncertainties

Nominal Power [W]	Force balance P_{bal} [W]	U_{bal} [W]	Calorimeter P_{cal} [W]	U_{cal} [W]
5.00	4.48	0.32	5.4	1.6
6.50	5.82	0.33	7.07	0.90
10.0	9.18	0.56	10.0	1.1

The results of ultrasonic power obtained with the calorimeter, compared to the ultrasonic power obtained with the force balance of radiation and the nominal power of the equipment are presented in Figure 3.

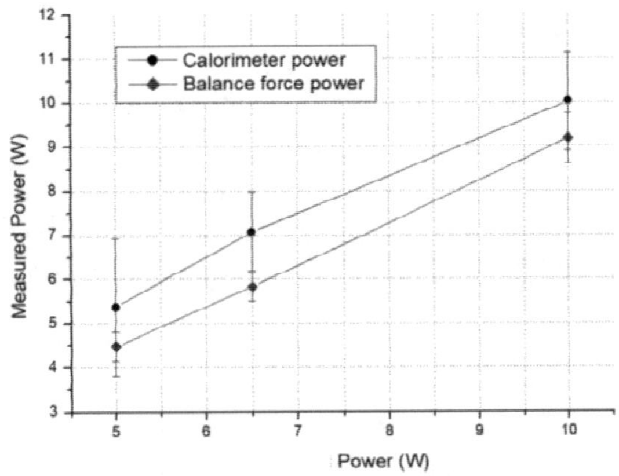

Figure 3. Relation between ultrasonic power obtained with the calorimeter and ultrasonic power obtained with the force balance of radiation.

Normalized error is a statistical evaluation used to compare proficiency testing results in which the uncertainty in the measurement result is included. Typically, it is the first evaluation used to determine conformance or non-conformance (i.e. Pass/Fail) in proficiency testing. The formula used for calculating the normalized error (E_n) is shown below. Values of $E_n < 1$ means that the results can be considered equivalent.

$$E_n = \left| \frac{P_{cal} - P_{bal}}{\sqrt{\left(U_{bal}^2 + U_{cal}^2\right)}} \right| \tag{6}$$

The results of normalized error (E_n) are presented in Table3.

Table 3. Normalized Error (E_n)

Nominal Power [W]	E_n
5.0	0.56
6.5	1.30
10	0.66

4. Conclusion

Ultrasound is the physical resource mostly used by professional physiotherapists in the treatment of the several disorders of the musculoskeletal system. The intensity of the ultrasonic radiation is an essential factor for the success of any therapy, as well as it is the time of application. Here, we compared the results of the ultrasonic power measurements of a physiotherapy equipment obtained from the radiation force balance and a calorimeter. The results show that the expected temperature variation was achieved, but for a better accuracy in measurements, it will be necessary to use a thermometer with lower uncertainty. The largest component of uncertainty (90%) in the result obtained comes from calibration certificate of the thermometer. Larger temperature differences ($\Delta\theta$) would be obtained if a specific heat lower than water were used, for example, castor oil [10], specific heat equal to 1,8 $J \cdot g^{-1} \cdot °C^{-1}$.

The normalized error less than 1, in two powers, attests the compatibility of the results of the measurements of the powers of the ultrasound equipment between the calorimeter and the reference value obtained from the balance of force of radiation.

References

[1] Ferrari, C B. 2008. Avaliação do desempenho de equipamentos de ultrassom terapêutico [dissertação]. São Paulo: *Universidade Metodista de Piracicaba.*
[2] Abrunhosa V M, Soares C P, Possidonio A C B, Alvarenga A V, Costa-Felix R P B and Costa M L P S, Mermelstein C S. 2014 *Ultrasound in Medicine & Biology* v. **40** pp 504-512.
[3] Petrella L I, Maggi L E, Souza R M, Alvarenga A V and Costa-Felix R P B. 2014 *Ultrasonics* **54** pp 1476-1479.
[4] Hildersley K. 1994. *Physioterapy* v. **80**(4) pp 219-23.
[5] Silva C E R, Alvarenga A V and Costa-Felix R P B. 2011 *Journal of Physics. Conference Series* 279 pp 012025.
[6] Maggi L E, Silva C E R, Alvarenga A V and Costa-Felix R P B. 2011 *Journal of Physics. Conference Series* 279 pp 012029.
[7] Repacholi M H and Benwell D A. 1979. *Health Phys.* v. **36** pp 679-86.
[8] Ishikawa N M. 2000. Avaliação de equipamentos de ultra-som para Fisioterapia segundo a norma NBR IEC 1689 da Associação Brasileira de Normas Técnicas [thesis]. Federal University of Rio de Janeiro.
[9] Guirro E C O and Guirro R R. Ultra-som. In: Guirro ECO, Guirro RR. *Fisioterapia dermato-funcional: Fundamentos, recursos*
[10] Delchar T A and Melvin RJ. 1994. A calorimeter for ultrasound total power measurements. *Measurement Science and Technology.*
[11] JCGM, 2008, *Evaluation of Measurement Data – Guide of the Expression of Uncertainty in Measurements* JCGM **100**:2008.
[12] Costa-Felix R P B and Machado J C. 2015 *Measurement* **69** pp 146-154.
[13] Costa-Felix R P B. 2006 *Measurement* **39** pp 169-175.

Acknowledgments
Research reported in this paper has been partially supported by the Carlos Chagas Filho Research Support Foundation (Faperj, grant number E-26/201.563/2014), by the National Council for Scientific and Technological Development (CNPq, grant number 310.392/2014-4) and Pronametro/Inmetro.

IOP Conf. Series: Journal of Physics: Conf. Series **975** (2018) 012011 doi:10.1088/1742-6596/975/1/012011

Measurements in flame retardant textiles and protective clothing using an instrumented

V M Kort-Kamp [1], A M Santos [1], A F Azevedo [1], R M V Lima [1,2], E Bittencourt [2]

[1] National Service of Industrial Training – Technology Center of Chemical and Textile Industry – SENAI CETIQT;
[2] University of Campinas – UNICAMP;
[3]Aeronautics Institute of Technology –ITA

vmkamp@cetiqt.senai.br

Abstract. The flame test manikin system can be used to evaluate the performance of thermal protective clothing under fire simulation conditions. Different weights of thermal protective garments were tested and the total clothed burn injury area decreased as the fabric weight increased. In addition, a comparison of different compositions for the same weight was analyzed too.

1. Introduction

When clothing is exposed to intense heat or flames, polymers and other materials will suffer thermal degradation and combustion, which depend on parameters such as absorption of energy from heat flux (conductive, convective and radiant), molecular composition, formation of products like liquids, char and solid materials. However, the combustion characteristics are too complex, depending on the fabric fiber, cloth structure, moisture trapped in the fibers, nature of finishing and others [1]. In this scenario, flame retardant (FR) finishes are applied in fabrics to reduce clothing flammability and avoid melting or ignition when the garment is exposed to heat.

According to Torvi [2], flame resistance is defined as the property of a material whereby flaming combustion is slowed, terminated or prevented. Textile materials can become flame resistant through chemical treatments or by use of inherently flame-resistant fibers. Chemically treated FR fabrics are produced by applying a flame retardant chemical finish to a fabric or by adding a chemical treatment to the fibers before they are woven or knitted into the fabric. These flame-retardants form a strong bond with the polymer chain, modifying its molecular structure and for this reason, it is difficult to remove these chemicals by washing.

Inherently FR fabrics are made of fibers in which the FR properties are part of the polymer backbone like aramids. The thermal protective performance of the textiles and garments can be evaluated using a variety of standard tests [3-4]. Bench-scale tests such as TPP test, RPP test and full-scale flame manikin test are the advanced methods employed to assess the protective properties of the materials aforementioned. This study proposed to evaluate the performance characteristics of flame resistant clothing and the influence of the fabric weight in the predict burn injury using an instrumented manikin.

2. Methodology

2.1. Teste apparatus
In this work, the tests were performed on an upright flame test manikin developed by Thermetrics (figure 1), which contains a 50th percentile Western male body form. Six groups of propane gas burners distributed evenly around the manikin generate the simulated flash fire expose, with average heat flux of 84 kW/m^2. A total of 134 copper calorimeter sensors, distributed on the manikin body, measure the total heat energy transmitted through protective clothing and this result is used to predict the degree of burn injury to human tissue.

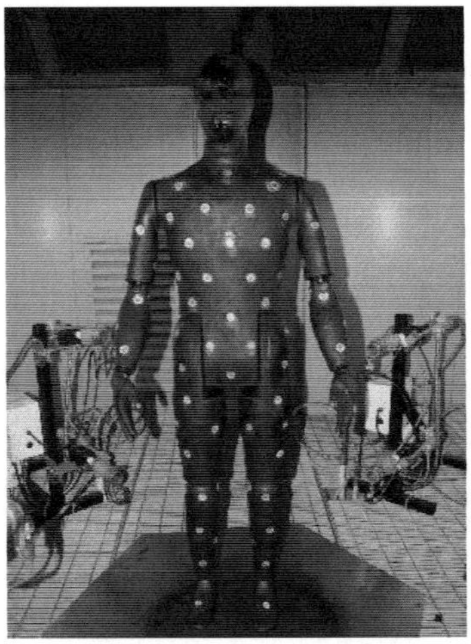
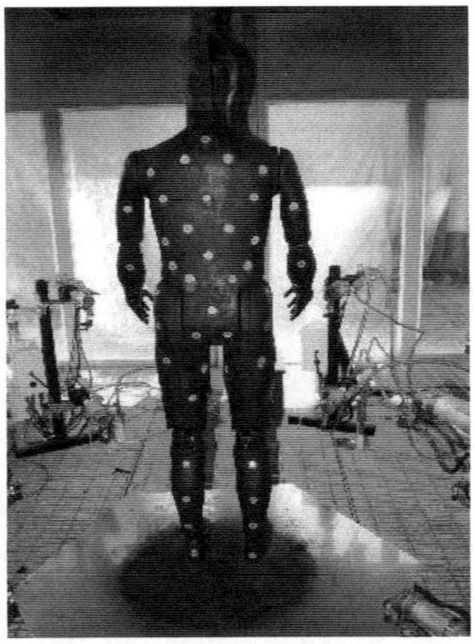

Figure 1. The instrumented manikin of SENAI-CETIQT (front view on left and back view on right).

According to ASTM F1930-15 [5] and ISO 13506-08 [6], the prediction of skin damage related to second-degree and third-degree burn is determined using a skin model. In these tests the head, hands and feet were not included because these parts were not covered with the garment during the flame exposure.

2.2. Materials
In order to evaluate the influence of the weight in the prediction of burn skin damage, table 1 shows the six single layer protective clothing (overalls) tested in combination with t-shirt and underwear. The exposure time was 3 seconds for all garments tested. Three samples of each garment type were measured and the results are given as mean values.

Table 1. Details of the samples.

Garment	Textile Composition	Weight (g/m^2)
I	88% cotton / 12% polyamide	237
II	88% cotton / 12% polyamide	250
III	88% cotton / 12% polyamide	260
IV	88% cotton / 12% polyamide	470
V	100% cotton	260
VI	100% cotton	290

3. Results and discussion

Figure 2 summarizes the results of burn prediction for five different weights and two compositions.

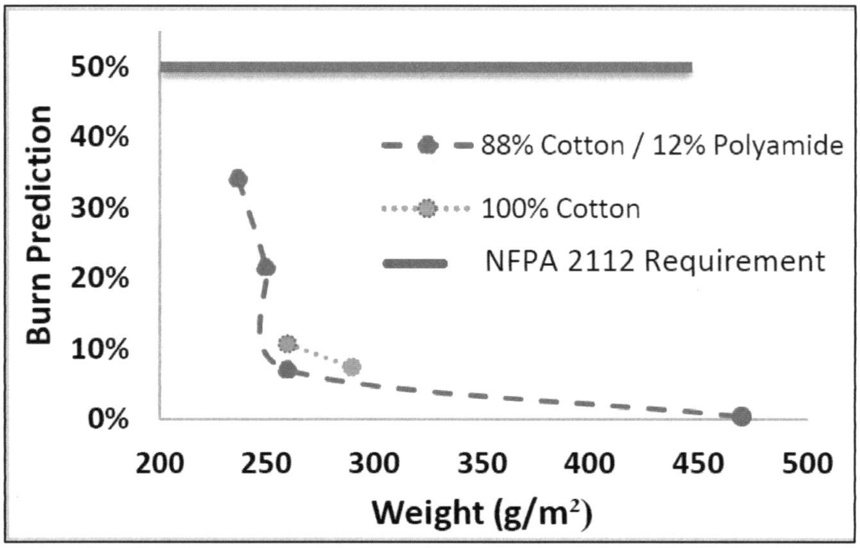

Figure 2. Variation of burn prediction with weight.

In both compositions evaluated, the burn injury decreases as the weight increases. The same behavior can be stated when analyzing the same weight of 260 g/m^2 for different compositions. There is a slightly increase in burn injury for the composition of 100% cotton. It can be explained by the addition of synthetic fibers in the composition of the fabric that may improves the performance of the garment in the test, however there is a limit in this addition that must be obeyed [7].

Kotresh [8] studied the influence of fabric weight and burning rate of different fabrics vertically oriented during burning. He stated that the burning rate decreases as the fabric weight increases. The same pattern could be noted in the study proposed in this article. The results analyzed are the sum of second and third degree burn prediction, which is defined as the total clothed burn injury area.

For both, 88% cotton / 12% polyamide and 100% cotton, it is observed a decrease in the burn prediction as weight increases. Additionally, all the results of burn injury were below the maximum value of 50% stated by NFPA 2112 for flame resistant garments [9]. Figure 3 (a) to (d) shows the variation of burn prediction with time for 88% cotton / 12 % polyamide (a) 237 g/m^2, (b) 470 g/m^2, (c) 260 g/m^2 of weight and for 100% cotton (d) 260 g/m^2.

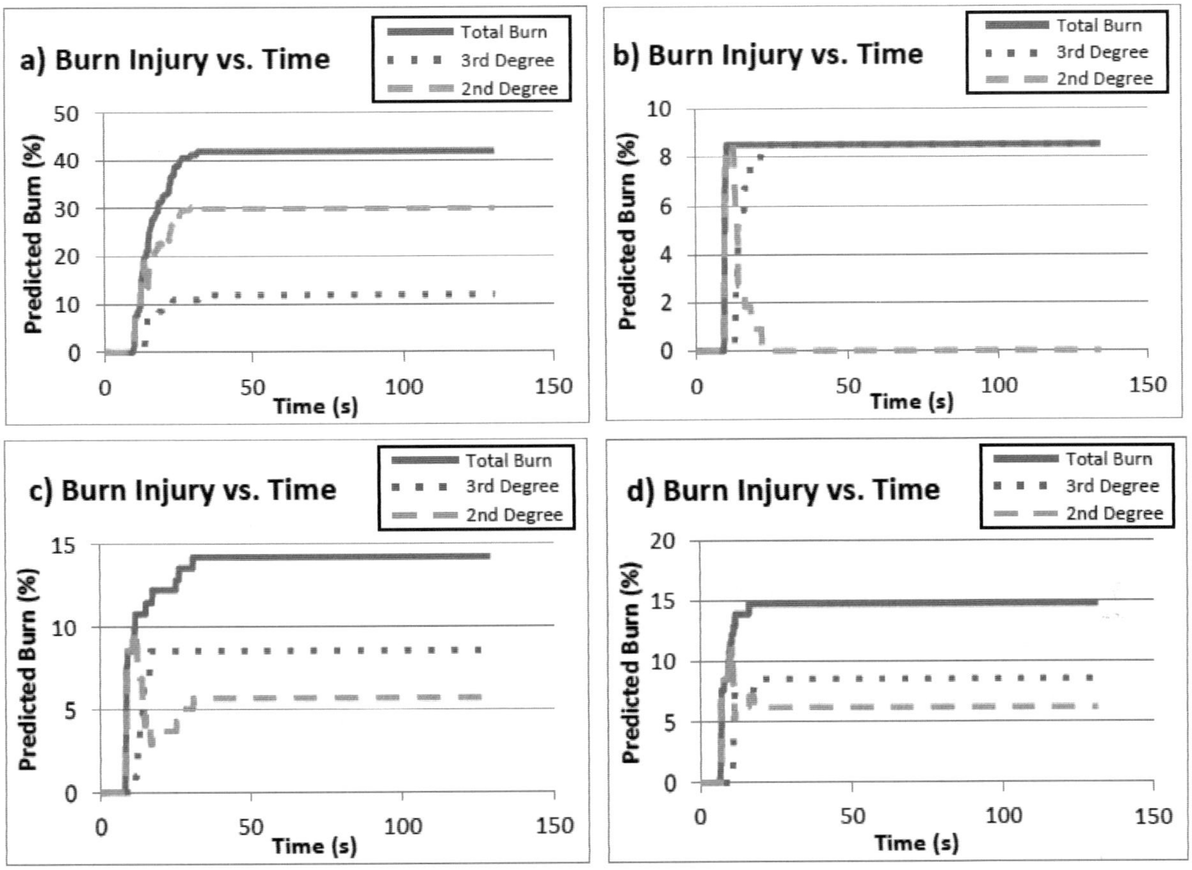

Figure 3. Variation of burn prediction with time for 88% / 12% cotton / polyamide **(a)** 237 g/m², **(b)** 470 g/m², **(c)** 260 g/m² and 100% cotton **(d)** 260 g/m².

4. Conclusion

Significant studies have been done in the field of thermal protective clothing. In general, the performance of thermal protective clothing can only be evaluated using an assessment based on an instrumented manikin due to the most real condition to simulate burn damage in human body.

This study indicates a tendency of decreasing of the total clothed burn injury area with the increasing of the FR fabric weight of protective clothing, tested in an instrumented manikin exposed of simulated flash fire expose, with average heat flux of 84 kW/m².

Nevertheless, regarding protective clothing used in industry, the association of safety and comfort is needed and an increasing of the weight fabric may not be the best option in order to improve the ergonomics for flame protective clothing.

Then, the study was satisfactory in relation to its purpose and it could be observed an inverse relation between fabric weight and predicted burn skin injury.

5. References

[1] Kutz M 2005 Handbook of Environmental Degradation of Materials, *Elsevier*, USA.
[2] Torvi, D A 1997 Heat transfer in thin fibrous materials under high heat flux conditions. Thesis from Department of Mechanical Engineering, University Of Alberta, Edmonton, Alberta.
[3] Torvi D A, Eng P and Threlfall T G 2006 Heat transfer model of flame resistant fabrics during

cooling after exposure to fire. *Fire Technology*. **42**, 27-48.
[4] He J, Wang M and Li J 2015 Determination of thermal protective performance of clothing during bench-scale fire test and flame engulfment test: Evidence from a new index. *J Fire Sci.* **33** (3), 218-231
[5] ASTM F1930-2015. Test method for evaluation of flame resistant clothing for protection against fire. Simulations using an instrumented manikin.
[6] ISO 13506-2008. Protective clothing against heat and flame: test method for complete garments -prediction of burn injury using an instrumented manikin.
[7] Stämpfli R, Bruggmann g and Rossi R M 2005 Comparision of flame spread of textile and burn injury prediction with a manikin. *Fire and Materials*. **29**, 395-406.
[8] Kotresh T M. 1996 Hazard potential of apparel textiles. *Indian Journal of Fiber and Textile Research*. **21** (2),157-160.
[9] NFPA 2112. Standard on Flame-Resistant Garments for Protection of Industrial Personnel Against Flash Fire, 2012.

Method development and validation for simultaneous determination of IEA-R1 reactor's pool water uranium and silicon content by ICP OES

J C Ulrich, S N Guilhen, M E B Cotrim, and M A F Pires

Instituto de Pesquisas Energéticas e Nucleares, IPEN-CNEN/SP - Av. Prof. Lineu Prestes, 2242 - Cidade Universitária - CEP: 05508-000 - São Paulo/SP

E-mail: jculrich@ipen.br

Abstract. IPEN's research reactor, IEA-R1, an open pool type research reactor moderated and cooled by light water. High quality water is a key factor in preventing the corrosion of the spent fuel stored in the pool. Leaching of radionuclides from the corroded fuel cladding may be prevented by an efficient water treatment and purification system. However, as a safety management policy, IPEN has adopted a water chemistry control which periodically monitors the levels of uranium (U) and silicon (Si) in the pool's reactor, since IEA-R1 employs U_3Si_2-Al dispersion fuel. An analytical method was developed and validated for the determination of uranium and silicon by ICP OES. This work describes the validation process, in a context of quality assurance, including the parameters selectivity, linearity, quantification limit, precision and recovery.

1. Introduction

IEA-R1, an open pool type research reactor moderated and cooled by light water, was commissioned in 1957 and is located at the Instituto de Pesquisas Energéticas e Nucleares, IPEN/CNEN-SP, a Brazilian agency managed by the National Nuclear Energy Commission. The reactor's fuel elements use low enriched uranium (LEU) targets, i.e., up to 20% enriched uranium (^{235}U isotope), assembled as fuel plates of U_3Si_2 dispersed in Al [1].

Spent fuel is provisionally stored in the pool until relocation to a final disposal site or until the decay heat is low enough to allow migration to a dry storage facility. Systematic treatment and purification of the pool water is required to its quality assurance, a key factor in preventing corrosion of aluminum clad fuel elements and other structural components in water [2-4].

Also, dispersed impurities in the water may become activated by neutron flux as the water circulates through the reactor core, impairing radiological shielding and, consequently, the attenuation of the radiation emitted by the reaction in the core.

This way, specifying requirements and operational limits for water treatment and purifications systems is essential for the management of water quality in research reactors. Therefore, the Research Reactor Center (CRPq/IPEN) has established sampling procedures and physical-chemical parameters to periodically monitor the water quality. The Chemistry and Environment Center (CQMA/IPEN) provides assistance concerning the water chemistry control, in which the levels of silicon and uranium in the pool's reactor are determined. This way, the detection of either of these elements in the water may indicate a possible leakage of the fuel to the reactor's pool.

However, carbon steels, usually employed in pipelines and valves of the reactor, may also respond for the silicon presence in the water, since carbon steels are essentially alloys of iron and carbon with small additions of elements such as manganese and silicon to provide the desired mechanical properties [5]. Also, silicon in water may also be accounted for its leaching from the concrete.

That's why, more permissible limits for silicon are allowed, given that the amount of silicon determined in the water would only be derived from the fuel leaching process if the equivalent amount of uranium was also determined. When silicon is found in water in the absence of uranium, one can assume that the source is less worrisome and that the water quality is satisfactory, since no corrosion products are being detected in the water. Therefore, uranium is the key parameter.

Naturally, corrosion products are greatly diluted, given that 273 m^3 of water is used to fill the pool. This way, when uranium is detected, immediate attention should be directed to remediate and isolate the fuel element.

An applicable technique for uranium and silicon quantification in water is inductively coupled plasma optical emission spectrometry (ICP OES).

Validation is defined by the International Vocabulary of Metrology (VIM) [6] as "verification, where the specified requirements are adequate for an intended use". Method validation has an important role in analytical laboratories to ensure the quality and to demonstrate the method performance. In this paper required information for method development and validation described by INMETRO document [7].

In this study, the main objective was to develop and validate method which may be applicable for simultaneous analysis of uranium (U) and silicon (Si) in reactor pool water.

2. Experimental

2.1. *Reagents and standards*

Uranium (U) and silicon (Si) single element standard solutions (1000 μg mL^{-1}, Inorganic Ventures, USA) were appropriately diluted to prepare standard calibration solutions ranging from 0.05 to 10 mg L^{-1} in a 1% (v/v) nitric acid (HNO_3, 65%, Merck, Darmstadt, Germany) aqueous solution. Ultrapure water with a resistivity of 18 M□ cm was obtained using a Master system (Gehaka, São Paulo, Brazil).

2.2. *Instrumentation*

A Spectro ARCOS ICP optical emission spectrometer (Spectro Analytical Instruments Co, Kleve, Germany), equipped with an axially viewed plasma, was used for the determination of U and Si in water. Sample introduction system was composed by a cross flow nebulizer and a Scott double-pass spray chamber. Instrumental parameters selected for U and Si determination were: 1400 W RF power,12 L min^{-1} plasma flow, 1 L min^{-1} auxiliary flow, 1 L min^{-1} nebulizer flow, 3 replicates, wavelengths: U = 409,014 nm and Si = 251,612 nm.

2.3. *Method validation*

The first step in method validation was to prepare a written protocol with detailed instructions in the form of analytical procedures and statistical treatment of the obtained data. This protocol assumed that the instrument was previously selected for sensitivity and that the analysts were experienced in analytical methods validation. The second step involved experimental tasks as to prepare and analyze the solutions for each specific parameter defined in the protocol. The following step was the evaluation of the method's performance and to establish the degree of acceptability that is required for each specific parameter. Finally, the appropriate filing of the method's documentation, containing procedures, calculations and the all records, was conducted.

The evaluation of the matrix effect, or selectivity, was investigated comparing two calibration curves obtained from matrix blank solution spiked with U and Si standard solutions and spiked public supply water. Student's t-test was applied for the comparison of the two curves and the (1) was used to calculate experimental t (t_{exp}), which is then compared with the critical (theoretical) t.

$$t_{exp} = \frac{|\overline{X_A} - \overline{X_B}|}{\sqrt{S_{AB}\left(\frac{1}{n_A} + \frac{1}{n_B}\right)}} \quad (1)$$

where X_A, X_B = average of groups A,B; n_A,n_B = number of values each group and S_{AB} = weighted variance.

Linearity was evaluated by the determination of the working range and then, the calibration solutions were prepared and measured. ANOVA statistical test was applied. The total variability of the responses was decomposed into the sum of squares due to regression and the residual (about regression) sum of squares and the later is decomposed into lack-to-fit and pure error sums of square.

The method's precision was evaluated as the coefficients of variation for the recovery tests on the measured concentration of elements in two spiked samples.

Accuracy was estimated through recovery tests using two spiked solutions and the results calculated by (2).

$$Recovery\ (\%) = \left[\frac{(C_1 - C_2)}{C_3}\right] \times 100 \quad (2)$$

where:
C_1 = concentration of the analyte in the fortified sample
C_2 = concentration of the analyte in the unfortified sample
C_3 = concentration of the analyte added to the fortified sample

3. Results and Discussion

3.1. *Selectivity*

As cited in 2.3 section, the t-test was applied to compare the two calibration curves. For U, t_{exp} = 0.04 and $t_{critical}$ = 2.23 and for Si, t_{exp} = 0.39 and $t_{critical}$ = 2.23. The criterion is if $t_{exp} < t_{critical}$ the H_0 is "accept", i.e. both samples (and calibration curves) contain the analytes and exhibit no matrix effect. Therefore, for both analytes, the selectivity is proved.

3.2. Linearity

As cited in 2.3, the calibration experiment was started with the choice of a preliminary linear working range from 0.05 to 10.0 $\mu g\ g^{-1}$ of both analytes. Six calibration solutions (concentrations of 0.05, 0.25, 0.5, 2.5, 5.0 and 10.0 $\mu g\ g^{-1}$) were prepared and analyzed. Three replicates of each concentration were prepared and the calibration curves were measured three times, independently. Table 1 summarizes the analysis of variance. Homogeneity of variances and linearity were verified by a statistical significant test. Table 2 shows the results of linearity and regression efficiency tests for the U and Si.

Table 1. Analysis of variance – uranium (U) and silicon (Si) calibration curves.

Uranium				
Sources of Variability	Sum of Squares (SQ)	Degrees of Freedom	Mean Sum of Squares (MQ)	
Totals	$1.59x10^{11}$	17	$9.35 x10^{9}$	MQT
Regression	$1.59x10^{11}$	1	$1.59x10^{11}$	MQR
Residual	$3.22 x10^{8}$	16	$2.01 x10^{7}$	MQE
Pure Error	$2.98 x10^{8}$	13	$2.29 x10^{7}$	MQEP
Lack-of-Fit	$2.41 x10^{7}$	4	$6.02 x10^{6}$	MQL
Silicon				
Sources of Variability	Sum of Squares (SQ)	Degrees of Freedom	Mean Sum of Squares (MQ)	
Totals	$9.13 x10^{11}$	17	$5.37 x10^{10}$	MQT
Regression	$9.08 x10^{11}$	1	$9.08 x10^{11}$	MQR
Residual	$5.52 x10^{9}$	16	$3.45 x10^{8}$	MQE
Pure Error	$5.27 x10^{9}$	13	$4.06 x10^{8}$	MQEP
Lack-of-Fit	$2.45 x10^{8}$	4	$6.13 x10^{7}$	MQL

Table 2. Results of linearity and regression efficiency tests.

Uranium calibration curve	
Linearity (lack-of-fit) $F_{calculated}$ = MQL/MQEP = 0.26 $F_{0.025;4;12}$ = 4.12 Fcalculated < F0.025;4;12 : Linearity is accepted	Regression validity $F_{calculated}$ = MQR/MQE = 7875.1 $F_{0.025;1;16}$ = 6.12 Fcalculated > F0.025;1;16: Regression validity accepted
Efficiency: R^2 = SQR/SQC = 0.998: $R^2 > 0.95$ regression accepted	
Silicon calibration curve	
Linearity (lack-of-fit) $F_{calculated}$ = MQL/MQEP = 0.15 $F_{0.025;4;12}$ = 4.12 Fcalculated < F0.025;4;12: Linearity is accepted	Regression validity $F_{calculated}$ = MQR/MQE = 2631.4 $F_{0.025;1;16}$ = 6.12 Fcalculated > F0.025;1;16: Regression validity accepted
Efficiency: R^2 = SQR/SQC = 0.994: $R^2 > 0.95$ regression accepted	

It was observed, in table 2, that $R^2 = 0.99$ for both analytes. As $R^2 > 0.95$ was the adopted criterion to accept the regression, the linearity for U and Si was accepted.

A significant MQR/MQE ratio confirms that there is regression. If the ratio MQL/MQEP is higher than the critical level, the linear model appears to be inadequate.

3.3. *Quantification limit*

As mentioned in INMETRO document [7], quantification limit was defined as the first point of calibration curve, i.e. for both analytes the quantification limit is 0.05 $\mu g\ g^{-1}$.

3.4. *Precision*
Coefficients of variation (CV) ranging from 0.02 % to 1.2 % and 0.1 % to 1.6 % (n = 9) for U and Si, respectively. These results demonstrated that the method has good precision.

3.5. *Recovery*
A satisfactory accuracy for the proposed method was achieved, since recovery values, for U, varied from (90.4±0.2) % (5.0 $\mu g\ g^{-1}$) to (100.4±0.8) % (2.5 $\mu g\ g^{-1}$) and for Si the values from (83.6±0.1) % (5.0 $\mu g\ g^{-1}$) to (98.4±0.8) % (2.5 $\mu g\ g^{-1}$).

4. Conclusion
The method development in this work was suitable for the determination of U and Si with good precision and accuracy. The validation process proved the absence of matrix effects and the linearity was suitable since $R^2 > 0.95$ for both analytes, with acceptable regression.

References
[1] Saxena R N and Fulfaro R 2003 IAEA-CN-100/43. Proceedings of an International Conference on Research Reactor Utilization, Safety, Decommissioning, Fuel and Waste Management; organized by the IAEA, Santiago, Nov 10–14
[2] Cegalla M A 1996 6o. Congresso Geral de Energia Nuclear–CGEN, Rio de Janeiro, 27 de outubro a 1 de novembro
[3] Cegalla M A, Neto A J S and Longo E S 1999 7o. Congresso Geral de Energia Nuclear–CGEN, Belo Horizonte, 31 de agosto a 3 de setembro
[4] Cegalla M A and Carvalho M R 2002 International Nuclear Atlantic Conference, 13th National Meeting of Reactor Physics and Thermal Hydraulics, Rio de Janeiro, Aug. 11-16
[5] Haddad R, Rodriguez S, Lanzani L, Quiroga A, Silva E. 2003 IAEA-CN-100/10. Proceedings of an International Conference on Research Reactor Utilization, Safety, Decommissioning, Fuel and Waste Management, organized by the IAEA, Santiago, Nov 10–14
[6] JCGM 2012 International vocabulary of metrology – basic and general concepts and associated terms (VIM). 3rd edition
[7] Instituto Nacional de Metrologia, Qualidade e Tecnologia (INMETRO) - Coordenação Geral de Acreditação (CGCRE). Orientação sobre validação de métodos analíticos. Rev. 04, julho 2011 (DOQ-CGCRE-008)

Metrological traceability of holmium oxide solution

D E F Gonçalves[1], J F S Gomes[2], A P D Alvarenga[2], P P Borges[3] and T O Araujo[3]

[1] Department of Reference Material, Visomes Comercial Metrológica, São Paulo, Brazil
[2] Optical Metrology Division, National Institute of Metrology, Quality and Technology (Inmetro), Xerem, Rio de Janeiro, Brazil
[3] Chemical and Thermal Metrology Division, National Institute of Metrology, Quality and Technology (Inmetro), Xerem, Rio de Janeiro, Brazil

E-mail: debora@visomes.com.br

Abstract. Holmium oxide solution was prepared as a candidate of certified reference material for spectrophotometer wavelength scale calibration. Here is presented the necessary steps for evaluation of the uncertainty and the establishment of metrological traceability for the production of this material. Preliminary results from the first produced batch are shown.

1. Introduction

According to ISO/IEC 17025 [1], a measurement result is reliable when it presents evidence of its traceability, accompanied by an estimative of its measurement uncertainty. In measurements performed with UV/Vis spectrophotometry, these requirements can be fulfilled with the proper use of certified reference materials (CRMs).

1.1. Metrological traceability and CRM

Metrological traceability can be defined as the property of a measurement result related to a reference through an uninterrupted and documented chain of calibrations, contributing to measurement uncertainty [2]. Measurement uncertainty is a parameter associated with each measurand, adequate to the intended use of the measurement results [2]; it indicates data quality and helps interpretation and decisions taking.

CRMs are one of the main bases for attainment of accuracy and reliability of chemical measurements. Its use is a requirement for the process of equipment calibration and method validation. CRM definition in ISO 17034 [3] states that the material must be characterized by a metrologically valid procedure (traceable to the International System of Units or to an internationally accepted scale) and it must be accompanied by a certificate that provides the property value, its associated uncertainty and a statement of metrological traceability.

1.2. Spectrophotometry

The spectrophotometry technique has low operational cost and is easy to use. Spectrophotometer is a fundamental instrument for analysis of several parameters in many areas, and thus, it is very important to verify its performance and also to keep the measurements traceability.

UV/Vis spectrophotometer measures the amount of absorbed light in ultraviolet (UV) and visible (Vis) spectrum by chemical compounds, allowing to quantify its concentration in different samples.

Many analytes have their sensitivity intensely reduced due to a small variation in wavelength, and thus, calibration of the spectrophotometers wavelength scale is essential for measurements accuracy. CRMs are usually employed for periodic verification and calibration of the equipment [4].

Holmium oxide CRM (Ho_2O_3) is the principal substance used as reference for the UV/Vis spectrometers and spectrophotometers wavelength scale calibration due to its stability and characteristics. It presents intrinsic and well resolved sharp peaks from 240 nm to 640 nm, covering the range used by the majority of the laboratories in diverse areas such as petrochemical and pharmacological. There are others CRMs with extended wavelength range into the infrared, but they are compounds into a glass matrix, which results in broader peaks and bands, and are less stable. The development of a CRM for spectrophotometers wavelength accuracy, such as Ho_2O_3, is important to assure reliability of the measurements results made by the laboratories.

1.3. Objective

This work aims to present the methodology for establishment of metrological traceability and for calculating measurement uncertainty of produced Ho_2O_3 CRM used for the wavelength scale calibration of UV/Vis spectrophotometers. This type of CRM will be the first one developed in Brazil.

2. Experimental

2.1. Ho_2O_3 CRM production

The developed CRM was the Ho_2O_3 solution, which is essential to guarantee the reliability of UV/Vis spectrophotometric analysis.

The production was based on the procedure of Weidner et al. [5], with 99.99 % purity Ho_2O_3 and 70% perchloric acid ($HClO_4$). A 4 % Ho_2O_3 solution in 10 % $HClO_4$ was gravimetrically prepared. The acid solution was bottled into 5 mL ampoule of clear glass containing a volume of approximately 4 mL, as shown in figure 1. This procedure was accomplished at Inmetro's Electrochemistry Laboratory.

In order to meet the ISO 17034 requirements [3], the homogeneity and stability of the produced batch must be assessed. These studies are ongoing. The batch characterization will be performed through a comparison between Inmetro's Radiometry and Photometry Laboratory and Visomes Laboratory, and further combination of the results.

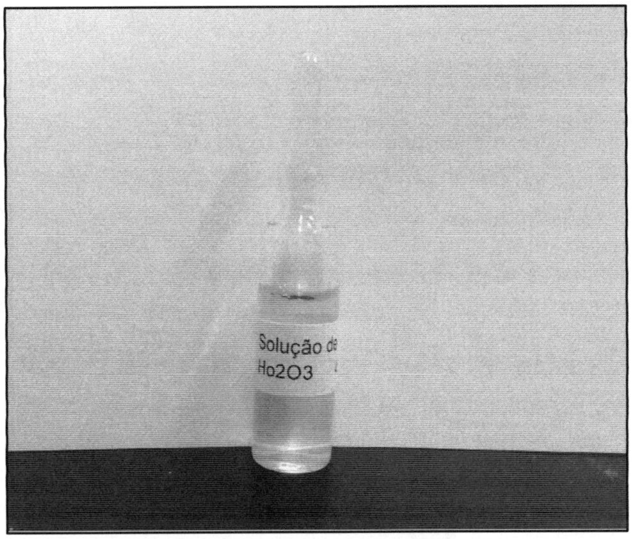

Figure 1. Candidate CRM of Ho_2O_3.

IOP Conf. Series: Journal of Physics: Conf. Series **975** (2018) 012013 doi:10.1088/1742-6596/975/1/012013

2.2. Traceability

Inmetro and Visomes spectrophotometers wavelength scale have been calibrated by using SRM-2034, a standard from the National Institute of Standards and Technology (NIST), in order to guarantee the measurements traceability in the Ho_2O_3 characterization. It was used for both the measurement method validation and the produced CRM batch traceability. The sources of uncertainty considered for the spectrophotometer calibration are showed in figure 2.

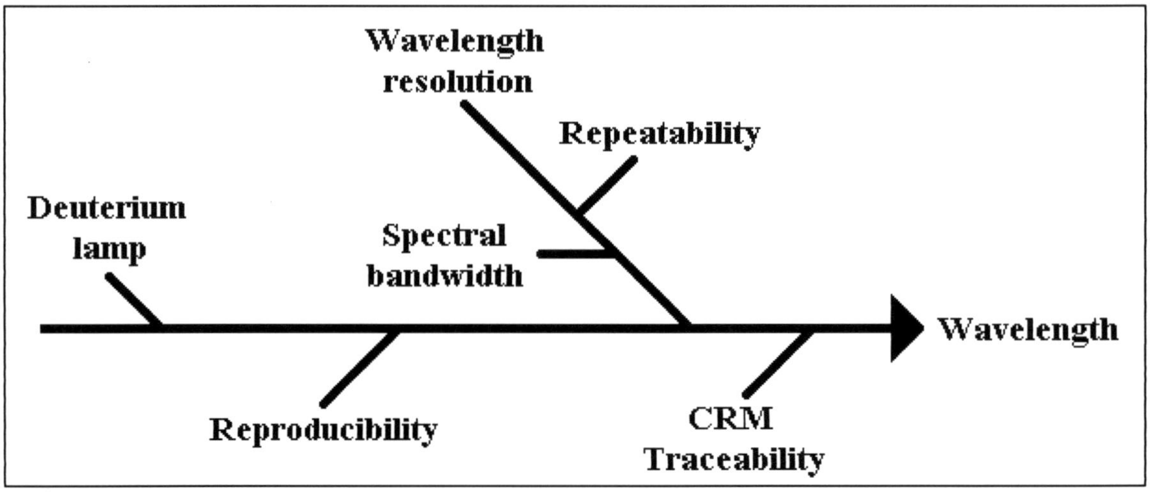

Figure 2. Ishikawa diagram with wavelength uncertainty sources for spectrophotometer calibration.

The SRM-2034 has certified wavelength values of minimum transmittance for 14 bands from 240 nm to 650 nm, for spectral bandwidth of 1 nm. The certified wavelengths are metrologically traceable to the SI derived unit expressed as nanometer.

2.3. Uncertainty evaluation

The uncertainty of the CRM produced will be evaluated according to the ISO GUM [6] and ISO GUIDE 35 [7]. For the CRM certified value, the following uncertainty contributions will be considered:
(a) characterization,
(b) non-homogeneity of the material,
(c) changes in property values during transport (short-term stability),
(d) changes in property values during storage (long-term stability).

3. RESULTS

After bottling the first batch, a sample of the CRM was measured and the spectrum can be observed in figure 3. The spectrophotometer has dual beam, double monochromator, wavelength range from 190 nm to 900 nm. The parameters were: 0.01 nm resolution, 1 nm spectral bandwidth, 20 $nm \cdot min^{-1}$ scanning speed, cycles of 2 measurement and 3 replicates, using a 10 mm optical path quartz cuvette.

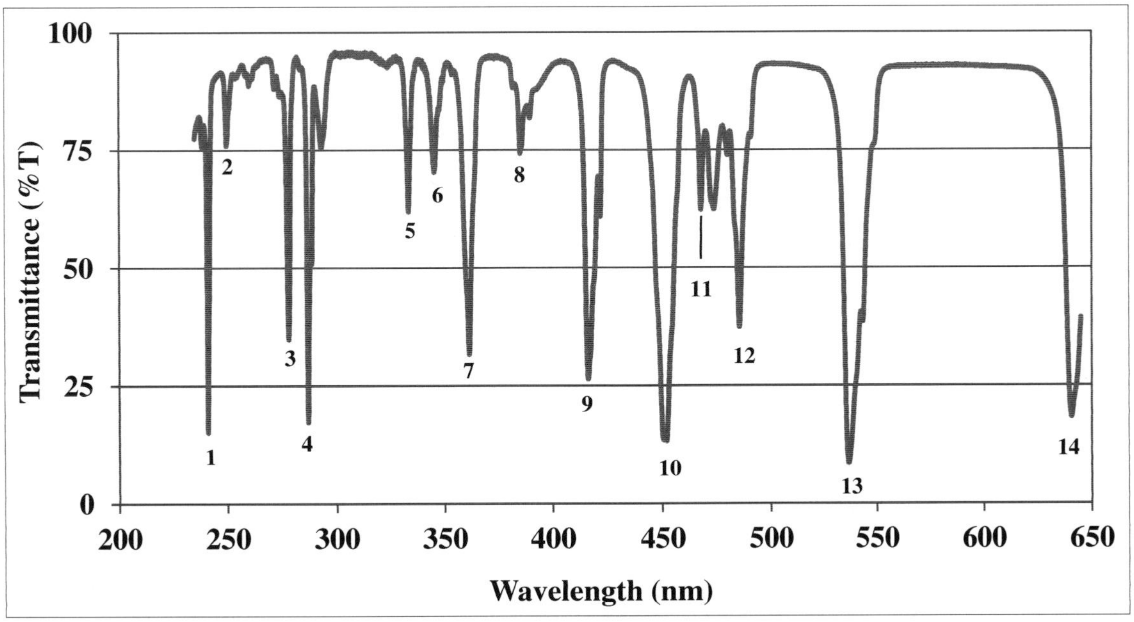

Figure 3. Transmittance spectrum of Ho_2O_3 in perchloric acid and the 14 studied peaks.

Table 1 shows the preliminary results of the measured sample together with the calculated normalized errors (E_n) between the SRM-2034 certified values and the produced Ho_2O_3 CRM candidate measured values. All of the 14 peaks ranging from 240 nm to 650 nm for spectral bandwidth of 1 nm were employed as indicated in figure 3. For all peaks resulted in $E_n < 1$, indicating satisfactory results. According to these preliminary results, one can proceed to the studies for certification of the batch of the candidate CRM of Ho_2O_3.

Table 1. Comparison between SRM-2034 values and measured values from candidate CRM of Ho_2O_3.

Certified value	*U* certificate	Measured value	*U* measured	Normalized error (E_n)
241.12	0.05	241.18	0.09	0.60
249.89	0.05	249.93	0.09	0.40
278.13	0.05	278.17	0.09	0.40
287.22	0.05	287.26	0.09	0.40
333.48	0.05	333.50	0.11	0.17
345.38	0.05	345.43	0.10	0.45
361.25	0.05	361.30	0.09	0.50
385.61	0.04	385.67	0.08	0.66
416.25	0.05	416.32	0.08	0.75
451.45	0.05	451.38	0.10	0.64
467.82	0.04	467.84	0.07	0.26
485.23	0.04	485.30	0.07	0.92
536.56	0.04	536.63	0.07	0.84
640.50	0.04	640.55	0.07	0.59

3.1. Characterization

The uncertainty calculation for the candidate CRM of Ho_2O_3 will be evaluated by each laboratory using its own standards. The following uncertainty sources will be considered:

(a) measurements repeatability,

(b) SRM-2034 traceability,

(c) spectrophotometer calibration uncertainty.

The characterization uncertainty (u_{char}) will be calculated as the square root of the sum of squares of each laboratory uncertainty, according to (1).

$$u_{char} = \sqrt{u_{LAB1}^2 + u_{LAB2}^2} \tag{1}$$

3.2. Non-homogeneity

In order to estimate the degree of non-homogeneity of the produced batch, the data obtained was analyzed using the single-factor analysis of variance (ANOVA), which allows the calculation of the standard deviation between the units of CRM and the standard deviation of the repeatability of the measurement method.

Measurement uncertainty of non-homogeneity (u_h) is a function of the mean square (M) values between the ampoules (M_b) and within the ampoules (M_w). If M_b is greater than the M_w, the uncertainty is calculated by (2). If M_w is greater than M_b, the uncertainty can be obtained by (3).

$$u_h = \sqrt{\frac{M_b - M_w}{n}} \tag{2}$$

$$u_h = \sqrt{\frac{M_w}{n}} \times \sqrt[4]{\frac{2}{df_{Mw}}} \tag{3}$$

where n is the number of repeated measurement on the same ampoule and df_{Mw} is the M_w degree of freedom [7].

3.3. Stability

The stability assessment will be analyzed using the linear regression tool to evaluate data and estimate the uncertainty obtained in the stability studies. For stability studies, it will be checked the absence of a significant trend of the CRM value over the time.

Measurement uncertainties for the short-term (u_{ss}) and long-term (u_{ls}) stabilities are determined by multiplying the standard error (e_{sd}), obtained from the linear regression calculations, by the studied time (t_{stud}) (in the case of the short-term study) or by the estimated time for material validity (t_{estim}) (in the case of the long-term study), as shown in (4) and (5), respectively [7].

$$u_{ss} = e_{sd} \times t_{stud} \tag{4}$$

$$u_{ls} = e_{sd} \times t_{estim} \tag{5}$$

3.4. CRM uncertainty

Uncertainty evaluation of the produced CRM will be finally obtained as described in (6).

$$U = k \times \sqrt{u_{char}^2 + u_h^2 + u_{ss}^2 + u_{ls}^2} \quad (6)$$

where U is the expanded uncertainty of the CRM and k is the coverage factor for an approximately 95% of confidence level [7].

4. Conclusion

In this article it was emphasized the importance of traceability and uncertainty evaluation of the candidate CRM of Ho_2O_3. It was presented all the necessary steps for the metrological certification of this material that is being produced as a future CRM by the first time in Brazil. After certification, this CRM can be used in spectrophotometric wavelength measurements for assuring accuracy and reliability on chemical and photometric measurements. Preliminary results from the first produced batch showed good agreement with the NIST SRM-2034 reference material, therefore the ensuing characterization is under way.

References

[1] ISO/IEC 17025-2017 – *General requirements for the competence of testing and calibration laboratories*
[2] JCGM 200-2012 – *International Vocabulary of Metrology. Basic and general concepts and associated terms*
[3] ISO 17034-2016 – *General requirements for the competence of reference material producers*
[4] Gomes JFS, Guedes MB, Couceiro IB and Grieneisen HP 2008 *Enqualab*
[5] Weidner VR, Mavrodineanu R, Mielenz KD, Velapoldi RA, Eckerle KL and Adams B 1985 *Journal of Research of the National Bureau of Standards* **90** 115
[6] JCGM 100-2008 – *Evaluation of measurement data – Guide to the expression of uncertainty in measurement*
[7] ISO GUIDE 35-2017 – *Reference materials – Guidance for characterization and assessment of homogeneity and stability*

Mycotoxin metrology: Gravimetric production of zearalenone calibration solution

E C P Rego[1], M E Simon[2], Xiuqin Li[3], Xiaomin Li[3], A Daireaux[4], T Choteau[4], S Westwood[4], R D Josephs[4], R I Wielgosz[4], V S Cunha[1]

[1] INMETRO, Chemical and thermal metrology division, Duque de Caxias, Brazil
[2] INTI Argentina, Toxicology and nutrition laboratory, Agro-food center, Buenos Aires, Argentine
[3] NIM China, Division of chemistry metrology and analytical science, Beijing, China
[4] BIPM, Department of chemistry, Paris, France

E-mail: ecrego@inmetro.gov.br

Abstract. Food safety is a major concern for countries developing metrology and quality assurance systems, including the contamination of food and feed by mycotoxins. To improve the mycotoxin analysis and ensure the metrological traceability, CRM of calibration solution should be used. The production of certified mycotoxin solutions is a major challenge due to the limited amount of standard for conducting a proper purity study and due to the cost of standards. The CBKT project was started at BIPM and Inmetro produced gravimetrically one batch of zearelenone in acetronitrile (14.708 ± 0.016 µg/g, k=2) and conducted homogeneity, stability and value assignment studies.

Keywords: zearalenone; mycotoxin; traceability; gravimetric preparation; Certified Reference Material (CRM).

1. Introduction

Zearalenone (ZEN) is a fungal mycotoxin produced by *Fusarium spp.* and present in several types of food but especially in maize and wheat. It is a non-steroid estrogenic compound which can cause changes in reproductive organs and fertility loss and has also been shown to have several other toxic effects [1]. Zearalenone analyses are a matter of health and food safety for many countries around the globe which produce or consume large quantities of corn grains and wheat.

In 2016 the Capability Building and Knowledge Transfer (CBKT) project on Safe Food and Feed in Developing Economies was started at BIPM to allow NMIs (National Metrology Institutes) to work together to strengthen mycotoxin metrology infrastructure; provide knowledge transfer to scientists developing capabilities in this area, including periods as visiting scientists at the BIPM; and enable NMIs to provide mycotoxin calibrant and matrix (C)RMs and PT materials to support mycotoxin testing laboratories within their countries [2]. Inmetro took part in this project and prepared gravimetrically one batch of calibration solution of Zearalenone, and conducted the homogeneity, stability and value assignment studies of this solution, following requirements of ISO 17034 [3] and

ISO Guide 35 [4]. The ZEN stock solution batch used to prepare the calibration solution batch was previously studied by BIPM, that conducted the purity assessment of ZEN powder standard.

2. Methods

2.1. Gravimetric preparation and ampouling

One ZEN calibration solution (named OGP.026A) at about 15 µg/g in acetonitrile was prepared in February of 2017 by dilution of BIPM stock solution OGP.025. The preparation of the solution was carried out using a gravimetric procedure. The final concentration and its uncertainty were calculated. The solution was filled in 10 mL clear glass ampoules with 4 mL of appropriate solution. A batch of 64 ampoules was produced. All the ampoules of OGP.026A were stored at -20°C. The homogeneity, stability and value assignment of the batch was tested by use of UV-Visible Spectrophotometer (Perkin Elmer Lambda 650) and LC-MS/MS system (Applied Q-trap 4000).

2.2. Homogeneity Testing

The first procedure to determine the homogeneity of the main component Zearalenone (ZEN) was based on the UV spectrophotometric measurements at three wavelengths (235 nm, 274 nm and 314 nm).

A second procedure to determine the homogeneity of the impurities of ZEN was based on the LC-MS/MS method. Three impurities were identified: Dehydrozearalenone (dehydroZEN), Zearalanone (ZAN) and Isomers of Zearalenone (isomerZEN) (figure 1). An LC-DAD method was also carried out to determine the main compound concentration (ZEN).

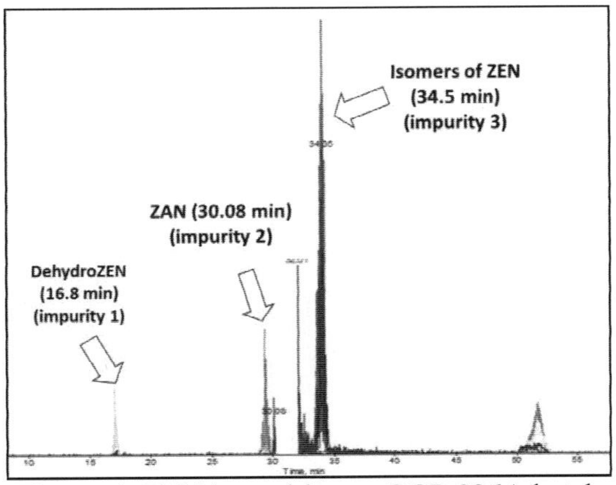

Figure 1. ZEN impurities on OGP.026A batch.

The response data obtained from each method were evaluated by ANOVA to get information concerning the homogeneity of main component and its impurities in the calibration solution.

2.3. Stability Testing

Stability testing is performed to confirm the stability of the material. The study is normally conducted under storage conditions (long-term stability) and under transport conditions (short-term stability).

The stability testing only considers the possible impact of temperature, time and light.

The batch was also submitted to stability testing for ZEN and its related impurities using UV-Visible spectrophotometry and LC-DAD-MS/MS methods.

The stability study of ZEN calibration solutions of ZEN (OGP.026A) followed the isochronous design, under repeatability conditions, with a reference temperature of -20 °C (dark) and study temperatures of 4°C (dark) and 22 °C (with light) during 4 weeks. For each condition/time were used two ampoules.

The response data obtained from each method were tested for trends using linear regression evaluation.

2.4. Value Assignment

Two samples from the OGP.026A batch were analysed directly without dilution. One ampoule of BIPM OGP.024 (ZEN calibration solution) was used as standard for UV measurements and one ampoule of BIPM OGP.025 (ZEN stock solution) was used to prepare standards for LC measurements.

Three independent replicates were analyzed in each method for major compound (ZEN). Two methods were used: UV spectrophotometric method with temperature fixed at 25 °C (Scan and fixed wavelength 235 nm, 274 nm and 314 nm) and LC-DAD method using bracketing quantification.

3. Results and discussion

3.1. Gravimetric preparation and ampouling

The gravimetric value for ZEN in OGP.026A is 14.708 ± 0.016 µg/g (k=2). The sources of uncertainties considered in the calculation were: weighing, buoyancy correction, evaporation correction and ZEN stock solution OGP.025 uncertainty (that includes purity of ZEN powder).

3.2. Homogeneity Testing

No trends for the filling sequence were observed for ZEN at 235nm, 274 nm and 314 nm for UV measurements (figure 2).

No trends for the filling sequence were observed for ZEN and its impurities on LC-MS/MS and LC-DAD methods (figure 3).

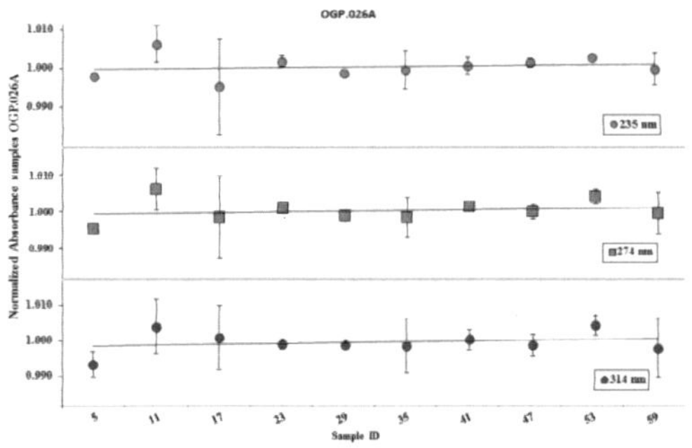

Figure 2. Homogeneity Testing by UV.

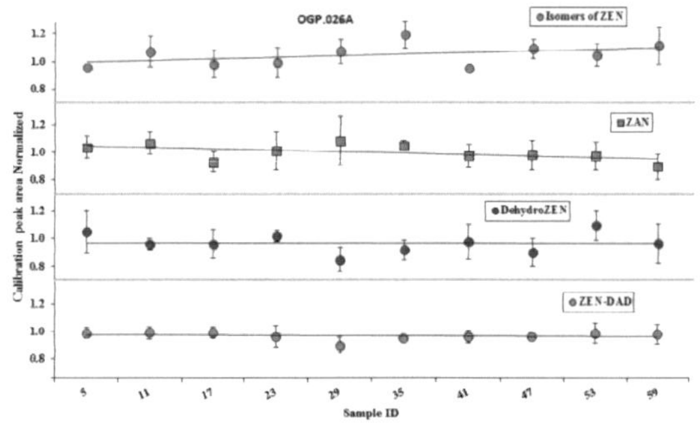

Figure 3. Homogeneity Testing by LC-MS/MS-DAD.

No trends for the analysis sequence were observed. No differences in the within- and between-sample variances could be detected by the F-tests at the 95 % confidence level for Zearalenone (main compound/ZEN-DAD) and its three impurities (dehydroZEN, ZAN and isomerZEN). The material could be regarded as homogeneous. The homogeneity uncertainty for ZEN by UV is 0.16% (274 nm) and 0.27% by LC-DAD. The minimum sample intake is 30 µL for UV and 10 µL for LC-DAD.

3.3. Stability Testing

A significant negative trend at 22°C with light exposure was observed for all wavelengths (235nm, 274 nm and 314 nm) by UV method. The wavelength 274 nm was choosen as it is the most specific wavelenth (figure 4).

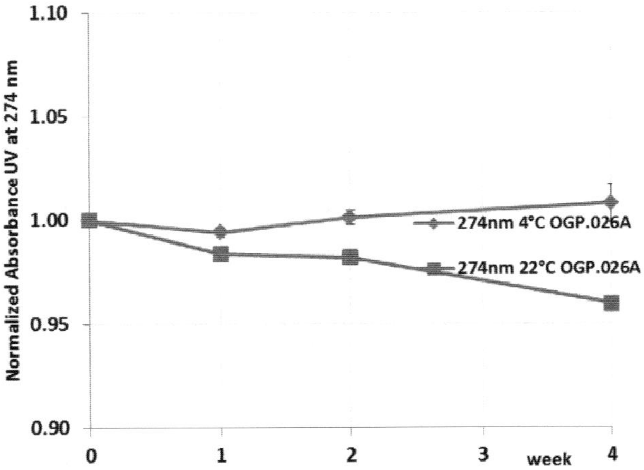

Figure 4. Stability Testing by UV.

A significant decrease of ZEN by LC-DAD and a significant increase of the IsomerZEN impurity were observed at 22 °C with light exposure (figures 5 and 6). No trends could be identified for 4 °C and 22 °C (with light) for the 7'DehydroZEN and ZAN impurities.

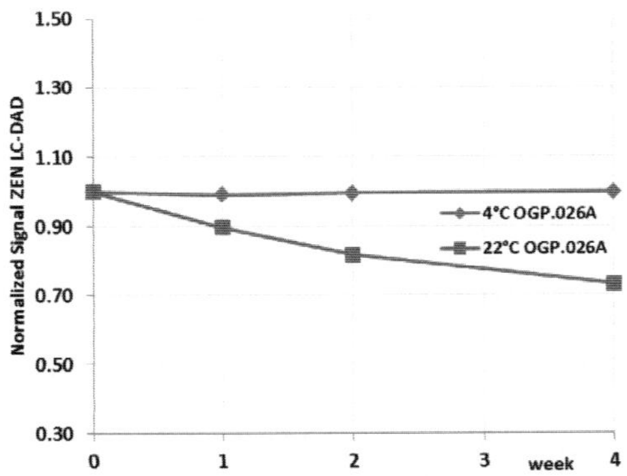

Figure 5. ZEN degradation at 22 °C with light.

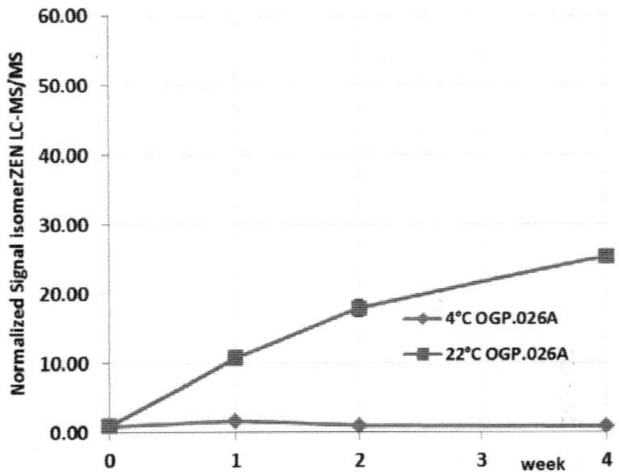

Figure 6. Increase of the IsomerZEN impurity during stability study at 22 °C with light.

It was concluded that 22 °C and light exposure conditions should be avoided during shipment and storage of this material. The batch showed stability at 4 °C without light exposure for ZEN and its impurities.

3.4. Value Assignment

The value assignment for ZEN studied by two methods (UV and LC-DAD), and the analytical values agreed with gravimetric value (table 1).

Table 1. Value assignment of ZEN calibration solution OGP.026A.

	Gravimetric	UV - using standard	UV – using molar absorptivity	LC-DAD
Value (µg/g)	14.708	14.461	15.156	14.645
U (k=2)	0.081*	0.50	1.5	0.13

*Including homogeneity (u_{bb}) from LC-DAD

There are no significant differences between the gravimetrical value and each analytical result, using normalized error ($E_n \leq 1$).

4. Conclusion

One batch of ZEN calibration solution in acetonitrile was successful produced and studied.

The batch OGP.026A was submitted to homogeneity study and showed sufficient homogeneity, with u_{bb} less than 0.27% for ZEN. No evidence of statistically significant trend was observed for filling sequence. The stability study showed that 22 °C and light exposure conditions should be avoided during shipment and storage due a degradation of ZEN in these conditions. The ZEN showed stability at 4 °C without light exposure during 4 weeks (conditions for transportation of the batch). The mass fraction of zearalenone in OGP26A obtained gravimetrically (14.708 ± 0.016 µg/g, k=2) was checked by two methods during value assignment study (UV and LC-DAD) and the analytical values agreed with gravimetric value.

This work was successful and the knowledge acquired at the BIPM will be implemented at Brazil to improve the production of new calibration solutions CRMs and to develop capabilities in mycotoxin metrology.

Acknowledgments

BIPM and NIM China for the training.

PTB for financial support.

References

[1] Gromadszka K, Waskiewicz A, Chelkowski J and Golinski P 2008 *World Mycotoxin Journal* **1** **(2)** 209-20

[2] BIPM website, CBKT project, http://www.bipm.org/en/cbkt/safe-food.html, accessed on June 2017

[3] ISO 17034:2016, General requirements for the competence of reference material producers, 24p.

[4] ISO Guide 35:2006, Reference materials -- General and statistical principles for certification, 64p.

IOP Conf. Series: Journal of Physics: Conf. Series **975** (2018) 012015 doi:10.1088/1742-6596/975/1/012015

New methodology for the analysis of volatile organic compounds (VOCs) in bioethanol by gas chromatography coupled to mass spectrometry

M S G Campos[1], J E S Sarkis[1]

[1]Institute of energy and nuclear research-National (IPEN/CNEN-SP), Av. Professor Lineu Prestes 2242 05508-000 São Paulo, SP, Brazil

E-mail: mauricio_campos@usp.br

Abstract. The present study presents a new analytical methodology for the determination of 11 compounds present in ethanol samples through the gas chromatography coupled to mass spectrometry (GC-MS) technique using a medium polarity chromatography column composed of 6% cyanopropyl-phenyl and 94% dimethyl polysiloxane. The validation parameters were determined according to NBR ISO 17025:2005. The recovery rates of the studied compounds were 100.4% to 114.7%. The limits of quantification are between 2.4 $mg.kg^{-1}$ and 5.8 $mg.kg^{-1}$. The uncertainty of the measurement was estimate in circa of 8%.

1. Introduction

Ethanol is a product of molecular formula CH_3CH_2OH, obtained by fermentation of simple sugar present in biomass [1,2]. It is colorless, volatile, flammable, water-soluble, with characteristic flavor and odor [4]. Ethanol is widely used in automotive industry various other industrial processes for the production cleaning products, paints, perfumes, cosmetics, varnishes, solvents among other products.

In Brazil, ethanol is used as biofuel produced from sugarcane, however, it can also be obtained by other renewable sources as the beetroot (Germany), corn (USA), sorghum saccharine (Africa) and wheat (Europe) can be used [1,2].

However, to be marketed in its various forms, it must comply several specifications regulated by regulatory agencies, laws in force in each country or even by companies that are purchasing the product. Consequently, daily, thousands analysis are performed to guarantee the quality of the product and its safe use. For this reason, it is necessary a continuous improvement of the used analytical, as well as the development of new ones which can improve all analytical process. Based on this premise, this work presents a new methodology for the simultaneous determinations of 11 VOCs in ethanol samples, based on gas chromatography coupled to mass spectrometry (GC-MS) technique.

2. Experimental

2.1. Reagents and standard

All analytical reagents used in this study have purity greater than 99.0%. In the validation of the methodology, were used the reference materials: INMETRO – Cachaça Proficiency Testing – 4th round and NSI Lab Solutions – Lot 052316 – Expiration date: 05/31/2018.

2.2. Instrumentation

A chromatographic Shimadzu system composed of a gas chromatograph, model GC-17A, coupled to a single quadrupole type mass spectrometer, model QP-5050 were used for the measurements. The sample introduction into the chromatograph was performed through an autosampler, model AOC-5000 Shimadzu.

The chromatography column used was a fused silica capillary DB-624 (6% cyanopropyl-phenyl 94% dimethyl polysiloxane), with dimensions of 60 m length x 0.25 mm inner diameter x 1.4 µm of film thickness, J Brand & W Scientific. Helium with 99.999% purity was used as carrier gas. The injection of sample was performed with a 10 µL Hamilton microsyringe. The analytical solutions were prepared gravimetrically, with automatic pipettes 10 µL, 100 µL, 1000 µL, 10 mL volumetric flasks and analytical balance (accuracy of 0.0001 g). The samples were injected, without need for preparation, filtration or dilution.

2.3. Chromatographic conditions and validation parameters
The chromatographic conditions defined for the study were as follows: initial temperature of 35°C held for 10 minutes, followed by an increase of 10°C per minute until 80°C, remaining at this temperature for 5 minutes. Second heating ramp 5°C per minute until it reaches 120°C, remaining at this temperature for 7.5 minutes. Third heating ramp 15°C per minute until 150°C staying at this temperature for 3 minutes. With this temperature program the total chromatographic analysis will be of 40 minutes. The injection temperature was 230°C, and 2µL volume was injected in split mode with a split ratio of 1:25. Carrier gas was helium with 2.0 mL/min flow. The mass spectrometer was operated in electron impact (EI) mode, ion source temperature 230ºC, MS transfer line 230ºC, SIM mode; solvent delay 6.0 min

The validation parameters followed the requirements by NBR ISO 17025:2005: selectivity, linearity, limit of detection, limit of quantification, repeability, precision, accuracy, recovery and uncertainty measurement. In this last parameter, a cause and effect diagram (Ishikawa) was elaborated to identify the sources of uncertainty [8].

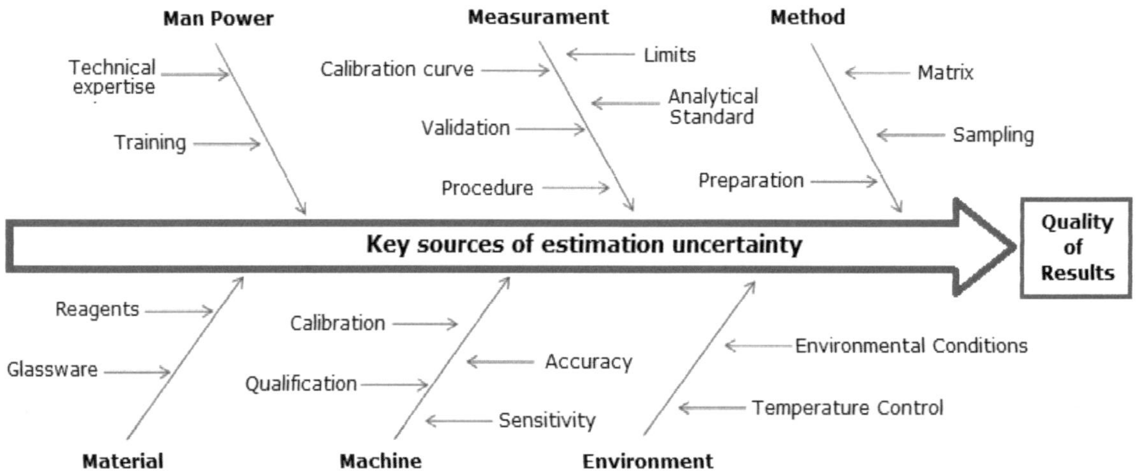

Figure 1: Ishikawa Diagram.

3. Results and discussion

In order to demonstrate capability of the methodology, 6 constituents were selected, 5 of them polar: methanol, 1-propanol, Isobutanol, 1-butanol, 2-butanol and 1 apolar: cyclohexane.

3.1. Limit of detection and quantification
Limit of detection is the lowest concentration of analyte in the sample that can be detected but not necessarily quantified under the conditions established for the assay. Limit of quantification is the

smallest amount of analyte in the sample that can be quantitatively determined with acceptable accuracy and accuracy [3].

Table 1. Limits of detection and limit of quantification in $mg.kg^{-1}$ (ppm).

Limits	**Limit of Detection**	**Limit of Quantification**
Methanol	1.8	5.8
1-propanol	0.8	2.4
Isobutanol	1.6	5.6
1-Butanol	1.1	3.9
2-Butanol	1.0	3.3
Cyclohexane	0.03	0.11

In this study, the LOD and LOQ were calculated by the method based on analytical curve parameters, corresponding to 3 times and 10 times standard deviation of the linear coefficient of the equation by the angular coefficient of the analytical curve, respectively.

3.2. Selectivity
Is the degree to which the method can quantify the analyte in the presence of other analytes, matrices or other potentially interfering material [3].

3.3. Linearity
Linearity refers to the ability of the method to generate results linearly proportional to the analyte concentration in the sample [5]. In this study, linearity ($R^2 > 0.9999$) was obtained in the range from 2.0 $mg.kg^{-1}$ to 300 $mg.kg^{-1}$.

3.4. Precision
Precision is the general term for evaluating the dispersion of results between independent assays, repeated in the same sample, similar samples or standards under defined conditions [3].

3.5. Repeatability
It is the measurement condition in a set of conditions, which includes the same measurement procedure, the same operators, the same measuring system, the same operating conditions and the same place, as well as repeated measurements on the same object or similar objects for a short time [3].

Table 2. Comparison of relative standard deviations at different concentration levels.

Compound	**Level**	**Average Intensity**	**Standard Deviation**	**%RSD**
Methanol	Low	2.9	6980	**9.87**
	Medium	29.4	17126	**2.60**
	High	147.7	31107	**0.96**
1-propanol	Low	3.1	4774	**3.64**
	Medium	31.0	42904	**2.94**
	High	155.6	101018	**1.33**
Isobutanol	Low	3.1	2099	**3.94**
	Medium	31.2	13838	**2.27**
	High	156.7	40063	**1.25**

	Low	3.6	2798	**5.28**
1-Butanol	Medium	36.1	14743	**2.43**
	High	181.1	44661	**1.41**
	Low	3.5	2854	**2.22**
2-Butanol	Medium	14.3	7702	**1.37**
	High	71.2	33091	**1.15**
	Low	0.09	130	**2.39**
Cyclohexane	Medium	1.99	903	**1.41**
	High	10.7	7709	**1.23**

The values shown in Table 02 show that the lower the concentration of the analyte greater the value of the relative default deviation.

3.6. Recovery
Recovery is the proportion of the amount of the substance of interest, present or added in the analytical portion of the analyzed material, which is extracted and quantifiable. It can be estimated as the percentage of the analyte after processing the sample compared to that of a solution containing the analyte in a concentration corresponding to 100%. The reference materials used in recovery, were: INMETRO – Cachaça Proficiency Testing – 4th round and NSI Lab Solutions – Lot 052316 – Expiration date: 05/31/2018. The expected values are shown in table 3.

Table 3. Recovery rate of the studied compounds.

Parameters	Reference Value	Concentration	Recovery (%)
Methanol	82.4	87.9	106.7
1-Propanol	268.0	307.4	114.7
Isobutanol	209.6	210.5	100.4
1-Butanol	35.8	37.9	106.0
2-Butanol	62.1	64.3	103.6
Cyclohexane	3.1	3.2	103.2

3.7. Measurement uncertainty
According to Eurachem/CITAC, the uncertanty is defined as "a parameter associated with the result of a measurement that characterises the dispersion of the values that could reasonably be attributed to the measurand" [7]. So, the uncertainty of an analytical methodology is a parameter characterizing the dispersion of the values attributed to a measured quantity considering all steps of the analytical procedure.

Table 4. Combined and expanded uncertainty in the studied method.

Parâmetros	u^a	U^b	U(%)
Methanol	2.7	5.4	10.1
1-Propanol	15.0	29.9	9.8
Isobutanol	10.6	21.2	9.5
1-Butanol	1.6	3.1	8.5
2-Butanol	2.7	5.4	8.4
Cyclohexane	0.1	0.3	8.3

u^a – Combined uncertainty
U^b – Expanded uncertainty, covarage factor K = 2
U(%) – Relative expanded uncertainty

4. Conclusions

As described previously, the method presented good results of selectivity, linearity, repeatability, recovery and the expanded uncertainties and analytical quality similar to the commonly used methods for this same purpose. In addition, once it uses only one chromatography column for the analysis of compounds with different polarity, it allows a cost savings and enhancement in the laboratory throughput.

References

1. Gnansounou E, Dauriat A 2005 Ethanol fuel from biomass: A review. *Journal of Scientific & Industrial Research* (Switzerland) **64** pp. 809-821
2. Macedo, I C 2007 Situação atual e perspectivas do etanol. *Estudos avançados* **59** pp. 157-165.
3. de Acreditação, C G 2010 Orientação sobre Validação de métodos analíticos. *DOQ-CGCRE-008-Revisão 03*
4. Dias, T P V B 2011 Equilíbrio líquido-vapor de misturas de alcoóis superiores contendo 2-propanol, 2-metil-1-propanol e 3-metil-1-butanol.
5. Dias, J C 2010 Determinação de acetato, nitrato, ferro(II), Ferro(III), e cobre em etanol combustível por cromatografia de íons.
6. Portes N 2010 Análise da qualidade do álcool combustível: uma proposta de aprendizagem. *Instituto Federal de Educação, Ciência e Tecnologia do Triângulo Mineiro – Campus Uberaba.*
7. Ellison, S. L. R. & Williams, A 2015 Eurachem/CITAC guide: Quantifying uncertainty in analytical measurement. *Available from www.eurachem. org.*

IOP Conf. Series: Journal of Physics: Conf. Series **975** (2018) 012016 doi:10.1088/1742-6596/975/1/012016

Participation on official proficiency test of the OPCW: case study of Brazilian Army – IDQBRN

T C Silva[1], **C N Ferreira**[1], **M Cardozo**[1] **and R L de Paula** [1,2]
[1] Instituto de Defesa Química, Biológica, Radiológica e Nuclear – IDQBRN, Centro Tecnológico do Exército – CTEx, Rio de Janeiro, BR.
[2] Programa de Pós-Graduação em Biotecnologia do Instituto Nacional de Metrologia, Qualidade e Tecnologia – PPGBIOTEC/INMETRO, Duque de Caxias, BR.

E-mail: taynara.idqbrn@gmail.com, reuel.lopes@eb.mil.br

Abstract. In recent years, Brazil has been the scene of international events, gaining worldwide repercussion. However, this increases the risk of terrorist attacks using chemical warfare agents. It brings the need to achieve proficiency to do a quick and unequivocal identification of these dangerous compounds. Thinking about it, the Laboratory of Chemical Analyzes (LAQ) of Brazilian Army has participated since the end of 2010 in the interlaboratory tests promoted by the Organisation for the Prohibition of Chemical Weapons (OPCW). In this article, we discuss about the participation of LAQ in the 40th OPCW Proficiency Test, one more step in the way to become a laboratory designated by OPCW and an international reference.

1. Introduction

Chemical agents have been used since earliest times, both in isolated acts and in large scale, as seen in terrorist attacks and in the case of the First World War. The devastating effect of these agents, even if such occurrences are not so frequent, has led to a legislative effort worldwide to ban the use of chemical warfare agents (CWA) under the Chemical Weapons Convention (CWC), which was opened for signature in January 1993 and came into force in 1997 [1].

The threat of chemical weapons use emphasizes the need for specific methods to detect and identify CWAs and their degradation products. The Organisation for the Prohibition of Chemical Weapons (OPCW), as the implementing body of the CWC, has been organizing and conducting Official Proficiency Tests (PT) since 1996 to certify laboratories for the analysis of authentic samples under the provision of the convention.

The manufacture of chemicals related to the convention cannot, however, be completely prohibited as some of them have potential industrial uses. Brazil has the largest industrial park in America Latina. In national territory, several companies work with precursor products of chemical warfare agents and, although the peaceful purposes of using such substances and the strict means of control imposed by the legislation and oversight bodies, there is always the risk of accidents or criminal use of these products. In addition, in recent years, the country has been scene of several events of international repercussion and it has become necessary to ensure the safety of the authorities and civilians involved.

With the objective of quickly and accurately detect agents and take necessary measures of containment, the Brazilian Army seeks to develop research in the CBRN Defense field. The Institute of Chemical, Biological, Radiological and Nuclear Defense, is the agency in Brazilian Army with the mission to plan and execute the research and development of CBRN Defense, providing products,

IOP Conf. Series: Journal of Physics: Conf. Series **975** (2018) 012016 doi:10.1088/1742-6596/975/1/012016

scientific advice and technical support. Among its laboratories is the Chemical Analysis Laboratory (LAQ), assembled with the objective of detecting chemical warfare agents through analytical techniques.

LAQ was the first laboratory on Latin American and Caribbean Group (GRULAC) region to be accredited in accordance with ISO 17025 (since 2015) to conduct assays involving chemical warfare agents. Besides that, LAQ is too the only one in GRULAC to regularly participate in official OPCW proficiency tests reaching the goal of identifying all chemical warfare agents present in submitted samples. These things are requirements for being an OPCW designated laboratory and demonstrate technical capacity, putting LAQ a few steps of becoming an international reference.

2. OPCW Proficiency Tests

These tests are part of a mechanism to ensure that there are around the world laboratories that have proven competence in the analysis of chemicals related to the CWC. To be designated the laboratory need to fulfill some requirements, like maintain a quality system, and successful completed three consecutives tests, what mean that only one spiking chemical can be missed in this period and no one false positive can be presented. The designation proves expertise in off-site analysis of CWC related compounds and makes the laboratory eligible to receive authentic samples. To maintain the designation status, a laboratory must achieve satisfactory results in at least one test offered per calendar year. The tests are organized with the assistance of two laboratories, one preparing the test samples, and other evaluating the test results.

OPCW PT is a qualitative proficiency test in which participating laboratories must determine the presence of any CWC scheduled chemicals, their precursors and/or degradation products spiked at 1–10 ppm level. The samples are received with unknown composition to be analyzed for presence of many possible chemicals in various matrices (such as soil, water, organic waste, wipe, etc.).

The participants receive two sets of three samples each (test sample + control + blank) without indication; each sample must be analyzed in the same way to be identified. Laboratories are not restricted as to sample preparation or analytical methods. Sample preparation procedures often require clean-up, extraction, derivatization and concentration steps. Identifications must be confirmed by at least two different analytical techniques giving consistent results, of which one technique must be spectrometric.

In addition, a laboratory creates its own solvent blank to prove the absence of reported chemicals prior to introduction of sample. Reports must contain names, structures, CAS numbers (if available) of identified chemicals; description of sample preparation methods employed; description of analytical methods employed; analytical data (chromatographic and/or spectrometric) with analytical conditions. There should be an unbroken chain of evidence linking each test sample to each reported chemical in the entire report.

For means of performance scoring, as shown in Table 1, each correct identification is positively scored (+1 point). If the original spiking chemical is no longer present, identification of a degradation product(s) instead of the spiking chemical is positively scored (+1 point). False negative results arising from not finding a spiking chemical or its degradation product is scored negatively (-1 point). Identification of minor constituents of the spiking chemical(s) is considered correct, but does not add to the score. Reporting any false positive result will constitute failure of the proficiency test, even if the identification is not supported by sufficient data.

Table 1. Method of evaluating laboratory performance

Criteria fulfilled	Identification	Scoring	Rating
Yes	Lab identifies all chemicals	Maximum score	A
Yes	Lab identifies all chemicals except one	Maximum score minus two	B
Yes	Lab identifies more than half of the chemicals	Score between zero and maximum minus two	C
Yes	Lab misses more chemicals than it identifies	Negative score	D
No	Lab reports a false positive	No score, failure	F

Besides that, all work need to be done in a short period: participants have only 15 calendar days upon receipt of samples to perform analysis and dispatch the report.

3. LAQ working system

The Brazilian Army Chemical Analysis Laboratory participates in OPCW Proficiency Tests since 2010 using gas chromatography as the main technique used, in the various configurations allowed (different columns, sources and parameters).

The current laboratory staff works as follows: a team for preparation of each sample group, a team to carry out chromatographic tests, a team to analyze the experimental results and a team to prepare the report. All members carry out the report review.

4. Results and discussion

4.1Test scenario and Sample preparation

In the 40th PT the test scenario presented was that the samples were collected one month after an explosion in a residential area on a territory under conflict. It was informed that victims close to the blast area presented symptoms of nerve agent poisoning. During the investigation, the team collected a metal fragment, believed to be a part of an exploded munition with evidence of a residue on its surface; and water present in a crater believed to be associated with the site or explosion.

Upon arrival at the laboratory, the samples were recorded and verified for their integrity; then were divided into aliquots and prepared according to the recommended procedures [2,3]. Each aliquot received a code for its unambiguous identification, and each compound found was identified with the CAS, the name and structure so as not to break the chain of evidence. All preparation steps are described briefly in the following flowchart (Figure 1).

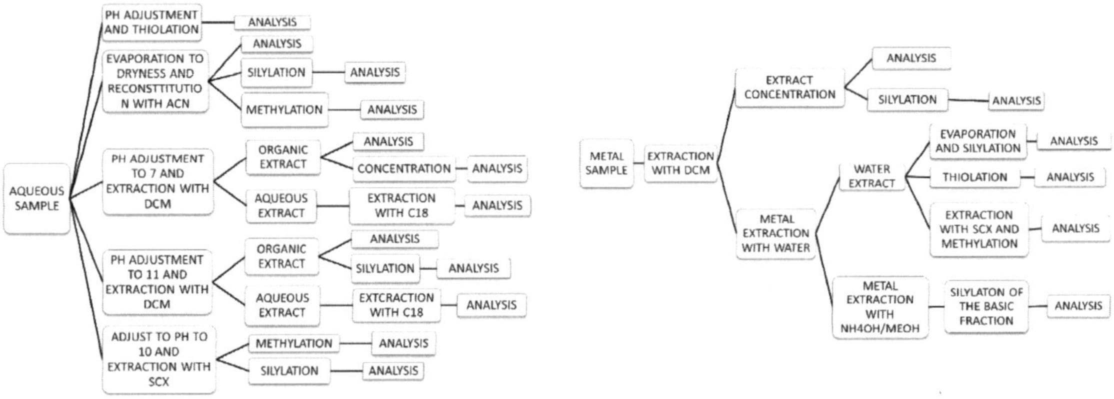

Figure 1. Scheme of Sample preparation

IOP Conf. Series: Journal of Physics: Conf. Series **975** (2018) 012016 doi:10.1088/1742-6596/975/1/012016

4.2 Analysis Evaluation

AMDIS tool and Gas Chromatography (GC) with *flame photometric detector* (FPD) and GC with nitrogen-phosphorous detector (NPD) techniques were used for screening purposes.

Analyzing the aliquot resulting from Dichloromethane (DCM) extraction of metal sample identified with the numeric code 739 was verified the presence of two relevant chemical compounds identified by LAQ as A1 and A2. In addition, screening of the aliquot resulting from silylation of DCM extract of the same metal sample using N,O-Bis(trimethylsilyl) trifluoroacetamide (BSTFA) could reveal the presence of silylated derivative of the same compound A2 and a new one identified as A3. The screening of all aliquots originated from other two metal samples, 740 and 741, such as other aliquots prepared with sample 739 not indicated the presence of any spiking chemicals.

As the next step, the unambiguous identification was achieved by at least two different spectrometric techniques. A2 was identified by the mass spectrum of its trimethylsilyl (TMS) derivative using GC and mass spectrometry (MS) with electron impact ionization (EI). This spectrum showed the base peak at *m/z* 255 and corresponded with the reference spectra from the OPCW Central Analytical Database (OCAD).

The identification of A2 was confirmed by a chromatographic parameter with comparison between retention index (RI) obtained experimentally and the RIs value in the OCAD. The unambiguous identification of A1 e A3 was achieved in a similar way. Both compounds were initially identified by GC-MS-EI, A1 as original compound and A3 as its TMS derivative. The confirmation technique was confirmed by GC-MS with chemical ionization (CI).

Table 2. Spiking Chemicals and relevant analytical data.

Sample code	Chemical reported	Compound structure	GC-MS		
			EI (m/z)	CI (m/z)	RI (m/z)
739 (metal)	3-Quinuclidinyl benzilate A1	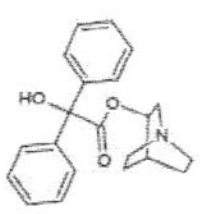	183	183	-
	2,2-Diphenyl-2-hydroxyacetic acid A2	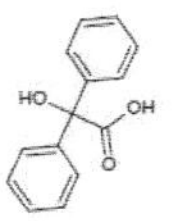	255	-	255
	3-Quinuclidinol A3	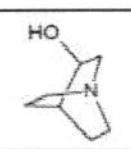	73, 199	200	-
742 (aqueous)	sec-Butyl methylphosphonate D1	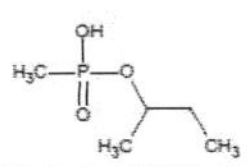	153	-	153
	Butyl methylphosphonate D2	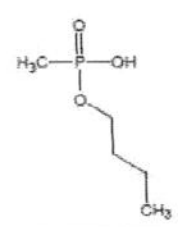	153	-	153
743 (aqueous)	2,2-Diphenyl-2-hydroxyacetic acid E1	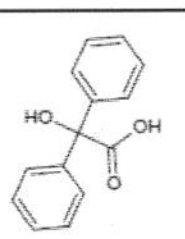	255	-	255
	3-Quinuclidinol E2	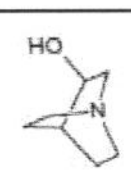	73, 199	200	-

The screening of the aliquots resulting from evaporation to dryness and reconstitution with acetonitrile (ACN) and DCM of water samples identified with the numeric code 742 and 743, followed by silylation using BSTFA could indicate the presence of four relevant chemical compounds identified by LAQ as D1, D2, E1 and E2. The screening of all aliquots originated from the other aqueous sample, 744, such as other aliquots prepared with samples 742 and 743 not indicated the presence of any spiking chemicals.

To report E2 was identified by the mass spectrum of its TMS derivative by GC-MS-EI. This spectrum showed the base peak at *m/z* 73 and the molecular peak at *m/z* 199, according with the reference spectra from the OCAD. Its identification was confirmed by GC-MS-CI technique, which spectrum showed the main characteristic signals for the compound with a strong signal at *m/z* 200 $[M+H]^+$. Chemicals D1, D2 and E1 were identified in a similar way by the mass spectrum of its TMS derivatives using GC-MS-EI technique. For their unambiguous identification, once again, the chromatographic parameter was used with comparison between RIs obtained experimentally and the RIs values in OCAD. Relevant analytical data are summarized in Table 2.

4.3 Report and Results evaluation

At the end of the analysis it was generated a report with all the relevant information of the identified compounds, their structural formula, details of the sample preparation and used analytical procedures, as well a compilation of all pertinent chromatograms and spectral data.

The Proficiency Test reports of all participating laboratories were sent for evaluating laboratory that checks the data presented, verifying if they meet the criteria established. The evaluation was based on the criteria for acceptable performance, defined in specifics PT quality system documentation and in agreement with the procedure for the evaluation of the results of an OPCW PT.

Table 3. Summary of Final Results for the 40th Proficiency Test. Blank space indicates that the chemical was not reported, 'X' indicates that the chemical was identified and its identification was supported by sufficient analytical data; 'O' indicates that insufficient data were presented to support identification; "F" (not for Chem. ID) indicates failure of the test; and "#" indicates trial participation.

	Sample code	**Chemical reported**	**Participant code**																					
			1	2	3	5	6	7	8	9	10	11	12	13	14	15	16	17	18	20	21	22	23	28
Scoring chemicals	739	3-Quinuclidinyl benzilate	X	X	X	X	X	X	X	X	X	X	X	X	X	X	X	X	X	O	O	X	X	X
	742	n-Butyl methylphosphonate	X	X	X	O	X	X	X	X	X	X	X	X	O	X	X		X		X	O	X	X
	742	sec-Butyl methylphosphonate	X	X	X	O	X	X	X	X	X	X	X	X	X	X	X		O		X	X	X	X
	743	2,2-Diphenyl-2-hydroxyacetic acid	X	X	X	X	X	X	X	X	X	X	X	X	X	X	X		X	O	X	X	X	X
	743	Quinuclidin-3-ol	X	X	X	X	X	X	X	X	X	X	X	X	X	X	X		X		O	X	X	X
Non-scoring chemicals	739	2,2-Diphenyl-2-hydroxyacetic acid	X	X	X	X	X	X		X		X	X		X	X	X	X	X			X	X	X
	739	Quinuclidin-3-ol			O			X										O	X			X	X	
	739	Methyl benzilate														X								X
	739	Benzophenone																					X	
	742	Methylphosphonic acid						X		X							X						X	
		Score	A	A	A	B	A	A	A	A	A	A	A	A	B	A	A	#	B	D	C	B	F	A

In the 40th Proficiency Test, all participants submitted their analytical report within the test period. There were sixteen A's, four B's, one C, one D and one F in the test score for the 23 regular participants, including the two A's of the assisting laboratories. No score was assigned to one trial participant in this test, LAQ could detect five spiking chemicals and two non-scoring chemicals, but lost one spiking chemical in the report evaluation, getting score B. In order to have an overview of the results of the participating laboratories OPCW elaborates Table 3. The laboratories that reached performance index A and B have considered results for the designation, requiring in three consecutive proficiency tests at least two results A and one B.

Preliminary results were discussed in a meeting with all participating laboratories at OPCW headquarters. After this, the result was issued, and the laboratories had 30 days to present the follow up report, containing the corrective actions to eliminate the presented errors.

5. Conclusions

GRULAC, as strategic region, needs a laboratory infrastructure that provides chemical analysis relevant to the CWC with a high degree of confidence. Brazilian Army Chemical Analysis Laboratory demonstrated in 40th Proficiency Test its ability to unambiguous identify chemical agents. LAQ will continue to seek knowledge and improvements in its infrastructure, methodologies and equipment to achieve the goal of becoming accredited by OPCW.

Acknowledgements

Authors would like to thank Organisation for the Prohibition of Chemical Weapons – OPCW for Research Grant Support L/ICA/ICB/201062/15 and Centro Tecnológico do Exército – CTEx, for the infrastructure.

References

[1] Dubey V, Velikeloth S, Sliwakowvski M, *et al* 2009 *Accred. Qual. Assur.* **14** pp 431-437

[2] Vanninen, P (editor). *Recommended Operating Procedures for Analysis in the Verification of Chemical Disarmament* 2011 (The Ministry for Foreign Affairs of Finland, Helsinki)

[3] Hooijschuur E, Hulst A, De Jong A, *et al* 2002 *TrAC* **21** pp116-130

Regulation of flow computers for the measurement of biofuels

R O Almeida[1], E A Aguiar Júnior[1], R P B Costa-Felix[1]

1 National Institute of Metrology, Quality and Technology (Inmetro), Av. Nossa Sra. das Graças 50, Duque de Caxias, RJ, Brazil, ZIP 25250-020

E-mail: roalmeida@inmetro.gov.br

Abstract. This article aims to discuss the need to develop a standard or regulation applicable to flow computers in the measurement of biofuels. International standards and recommendations are presented which are possibly adequate to fill this gap and at the end of the article a way is proposed to obtain a single document on the subject.

1. Introduction

With the publication of the Inmetro's Regulation nº 499 of 2015, this Institute established a metrological technical regulation (RTM) of flow computers and conversion devices, whose scope includes fiscal measurement, appropriation, custody transfer, among others of liquid hydrocarbons and natural gas. This new proposal presents several improvements to the legal metrological control of these instruments in Brazil, while also expanding the scope when compared to the previous regulation.

However, these devices also possess the functionality, and are in fact used, to mathematically and continuously convert volumes of biofuels (ethanol and biodiesel) (1), which will not be covered by this new regulation. It is important to note that, in 2014 alone, Brazil produced 3.42 billion liters of biodiesel and 28.8 billion liters of anhydrous and hydrated ethanol (2), a demand for measurement which, due to this lack of legal coverage, is not covered by the services of legal metrology.

What we have here is a real need to regulate the measurement of biofuels with the use of conversion devices. Not only due to the large volume of transported product, as mentioned previously, but also due to the publication of the "Strategic Guidelines for Brazilian Metrology 2013-2017" (3), by the National Council of Metrology, Standardization and Industrial Quality (CONMETRO). This resolution, which guides the actions of the several Brazilian metrology institutions, as well as the application of governmental resources for the effective development of metrology in the country, stipulates that it is up to the legal metrology services to expand their performance in order to incorporate the demand for measurement of, among others, the biofuels market. Considering that legal metrology can cover only measuring instruments that are subject to metrological technical regulations issued by Inmetro (4) and that such regulations are usually based in national and international standards and recommendations, a problem arises.

There are currently no standards or recommendations with technical and metrological requirements applicable to these conversion devices in the measurement of biofuels. There are standards applicable to the measurement of biofuels. We have, for example, OIML R22 (5) and ABNT NBR 5992 (6) dealing with the volume conversion of ethanol and its mixtures with water and ABNT NBR 15512 (7) dealing with the same subject for biodiesel, but none of them mention flow computers. There are also other standards that provide requirements for flow computers, such as OIML R117 (8), WELMEC

10.4 (9), ABNT NBR 16020 (10) and API MPMS 21.2 (11), but most of them are applicable only to measurement systems. Some comments should be made on each of these standards in addition to the regulation currently in vogue for these devices.

2. OIML R22

OIML R22 places great emphasis on the use of tables for the conversion of ethanol volumes, but it is the knowledge of the formula that generated these tables that should be available for use in flow computers. This is due to the fact that the formula will be programmed into the flow computer as a mathematical model for ethanol conversion. Thus, this OIML recommendation could be used, within the context of a metrological technical regulation, as a mathematical model. In contrast to the physical standard, the mathematical model exists only as an equation, but allows the comparison between the value of a measurement obtained by the instrument and the expected value according to the equation.

3. OIML R117

OIML R117 presents technical and metrological requirements applicable to fluid measurement systems, but throughout its contents there are also requirements applicable to flow computers. Maximum permissible errors, test approaches, requirements for physical and electronic sealing, requirements for associated measuring devices, and requirements for the installation of the flow computer are mentioned.

The latter type of requirement is the reason why this recommendation cannot be fully validated as a set of requirements for the measurement of biofuels in flow computers. As you wish to develop the requirements applicable only to conversion devices, its installation requirements in the system cannot be observed. In a deeper analysis, only the physical and electronic sealing requirements, test approaches and maximum permissible errors can be validated in a hypothetical regulation for biofuels measurement in flow computers.

4. WELMEC 10.4

The WELMEC 10.4 guide is presented as an add-on to R117. Initially developed by WELMEC in order to address concerns that arose in the recommendation application. This document acts as a guide for R117's application on conversion devices. In addition to the test approaches presented by the recommendation, it suggests a new approach, which allows for a more flexible application of this guide according to the assembly of the conversion device and its associated measuring devices, as stipulated by the manufacturer.

In addition to this new approach the guide also presents what may be considered a omission in R117 for flow computer essays: a complete detailing of tests for this device with numbers of points to be checked in each test, specific tests to be performed according the device approach and reference values for calculating the maximum permissible errors.

With the presentation of these reference values WELMEC 10.4 can be considered a complete test guide for the deployment of R117 in flow computers. However it addresses use only in complete measuring systems. It is also important to note that the guide does not present any technical requirements, may it be constructive or related to the audit trail.

5. ABNT NBR 16020: 2011

Focusing on ABNT NBR 16020: 2011 requirements, exclusively applicable to flow computers, it can be observed that this standard provides much more robust information than, for instance, WELMEC 10.4. This is due to the fact that the 16020 recommends specific standards for calculating the volume conversion of fluids. Item is absent from WELMEC 10.4, which is complementary to R117.

Also noteworthy is the presence of specific sections on calculation algorithms for the measurement of biofuels. Although R117 and WELMEC 10.4 already recommend a document for conversion of volumes of ethanol (OIML R22), ABNT 16020 goes further and recommends, in addition to a standard for the conversion of ethanol (ABNT NBR 5992), a standard for the measurement of biodiesel (ABNT NBR 15512).

The 16020 also features a table with maximum admissible errors for temperature, pressure, specific mass, volume conversion factor and pulse indication. This point was addressed by R117 and later expanded by WELMEC 10.4. Specific test procedures for each type of approach (first, second and third) have not been mentioned, nor are they presented by the standard. Instead a single procedure for any type of assembly of the flow computer is presented.

6. ABNT NBR 5992: 2008

Like the OIML R22 recommendation, ABNT NBR 5992: 2008 presents a mathematical model for ethanol and its mixtures with water volume calculation. In this case the standard stipulates the whole procedure and laboratory equipment for the tests to determine the specific mass and the alcohol content of the ethanol and water mixture. Part of this process involves using the values determined experimentally in the "Programs of Alcoholic Tables" software, supplied in full with the standard.
In addition to the calculation of the alcoholic concentration and the volume conversion factor, the software is also capable of producing tables where specific mass or alcohol content is related to temperature to provide the conversion factor at 20 ° C. Effectively this program eliminates the need for printed tables (such as those presented by OIML R22) for the temperature conversion of volumes of ethanol.

Thus, the standard can be considered more appropriate to the use in flow computers than the OIML recommendation, since, not only presents all the subsidies for the accomplishment of the tests to obtain the specific mass, but also has an updated mathematical model for insertion into the programming of the conversion device.

7. ABNT NBR 15512: 2014

Although it is indicated by ABNT NBR 16020 as a reference for the biodiesel conversion of its specific mass due to the effect of temperature, ABNT NBR 15512 only mention it briefly. The standard only directs the reader to consult the conversion table used for petroleum in Brazil, in this case the Resolution of the National Petroleum Council No. 6 of June 25, 1970.

Although the scope of this resolution is the conversion of volumes of crude oil, it is also used, as recommended by 15512, for conversion of biodiesel. No further mention of volume conversions is made throughout the document.

8. MPMS API 21.2

API 21.2 focuses on requirements applicable to liquid measurement systems. Although, in this standard, these systems are used for measurement of liquid hydrocarbons, some unique requirements for flow computers can be extracted. Among them we have: Audit trail requirements (especially configuration report, events and alarms);

- Requirements to be observed before the acquisition of the conversion device;
- Requirements for its verification.

The form and content of the audit trail are discussed in detail, including the security measures that the flow computer has in order to protect this record. Despite the technological advancement of flow computers since the publication of the last addition of this standard (2002), this chapter continues to be up to date.

The verification process presented by API 21.2 is not described in detail, is only limited to recommending the calibration of the analog and frequency inputs of the tertiary devices. In addition it points out the need for testing in at least three points along the operational measurement range.

This standard does not clarify maximum permissible errors, types of tests nor even details of the minimum number of points to be collected or provides mathematical models for the measurement of biofuels. However, it provides specifications of measurement standards that should be used to test all components of a liquid metering system, including the its flow computer.

9. Inmetro's Regulation N ° 499, of October 02, 2015

Inmetro's Regulation N°. 499, of October 2nd, 2015, presents the technical metrological regulation applicable to the measurement of oil and gas in flow computers. Of particular interest to the subject of this article is its analysis because:

- Specifically regulate flow computers and their measurement of liquid (although not biofuel);
- Unlike other standards and regulations, it focuses on the conversion device and not on the measurement system;
- It is already structured in the form of a Brazilian regulation.

Additionally, it presents maximum permissible errors, technical sealing requirements and instructions for carrying out checks. However, it does not introduce any details of the tests to be performed (nature of the tests, minimum number of points to be collected and requirements for repeatability) or which mathematical model should be followed for the modeling of fuel fluids. This regulation also possesses electromagnetic, climate and software requirements.

Thus this regulation presents necessary but not sufficient elements for a full regulation for a metrological technical regulation of flow computers for the measurement of liquid biofuels. As an advantage, this regulation was widely discussed with segments of society linked to the fuel sector, which gives them a high degree of robustness.

Finally, the transition rules introduced by this ordinance are highlighted, as they provide clarity to the market of this equipment on how the regulatory body (Inmetro) will treat the equipment in the field. The key point for evaluation of the instruments currently in use is the maximum permissible errors. By evaluating these errors the calculating devices in the field may have their use continued or prohibited after the publication of this regulation.

10. Conclusion

Each of the analyzed documents presents necessary but not sufficient elements to the composition of a full proposal regulation of flow computers for the measurement of biofuels. One possibility for solving this absence of a single text would be to write an amalgam of each of the elements presented by these standards.

The WELMEC's maximum permissible errors and tests, Inmetro's regulation N°. 499/2015 technical requirements, API's field verification instructions and specification of working standards may be used, for instance.

A better analysis of the content of these standards compared to Inmetro's regulation shows that this amalgam can be useful not only for the regulation of conversion devices in the measurement of biofuels, but also in the measurement of any other type of fluid. This fluid limitation whose flow computer regulation currently covers (only oil and natural gas) can be considered one of its greatest weaknesses and should be revised in order to comply with the provisions of the "Strategic Guidelines for Brazilian Metrology 2013-2017".

References

[1] Computers, Dynamic Flow. Dynamic Flow Computers I Flow Meters I Flow Indicators I Prover Metering I Custody Metering I Proving Metering. Dynamic Flow Computers. [Online] Dynamic Flow Computers. [Citado em: 15 de Março de 2016.] http://dynamicflowcomputers.com/wpsite/wp-content/uploads/2012/02/E-Plus-P2DAAA-Data-Sheet.pdf.

[2] ANP - Agência Nacional do Petróleo, Gás Natural e Biocombustíveis. ANUÁRIO ESTATÍSTICO 2015. ANP - Agência Nacional do Petróleo, Gás Natural e Biocombustíveis. [Online] ANP - Agência Nacional do Petróleo, Gás Natural e Biocombustíveis, 23 de julho de 2015. [Citado em: 23 de julho de 2015.] http://www.anp.gov.br/?pg=76798.

[3] Brasil. Resolução Conmetro n° 01. Aprova o documento "Diretrizes Estratégicas para a Metrologia Brasileira 2013-2017". 10 de abril de 2013.

IOP Conf. Series: Journal of Physics: Conf. Series **975** (2018) 012017 doi:10.1088/1742-6596/975/1/012017

[4] Portaria Inmetro nº 585. Cientifica que somente serão passíveis de controle legal os instrumentos de medição que forem objeto de regulamentação técnica metrológica expedida pelo Inmetro. 1 de novembro de 2012.
[5] Organização Internacional de Metrologia Legal. R 22 en: International alcoholometric tables. Internacional Organization of Legal Metrology - English. [Online] [Citado em: 23 de Maio de 2016.] https://www.oiml.org/en/files/pdf_r/r022-e75.pdf.
[6] ABNT - Associação Brasileira de Normas Técnicas. ABNT NBR 5992 - Álcool etílico e suas misturas com água - Determinação da massa esécífica e do teor alcoólico - Método do densímetro de vidro. 2008.
[7] ABNT NBR 15512 - Biodiesel - Armazenamento, transporte, armazenamento e controle de qualidade de biodiesel e/ou óleo BX. 2014.
[8] Organização Internacional de Metrologia Legal. R 117 en: Dynamic measuring systems for liquids other than water. International Organization of Legal Metrology — English. [Online] [Citado em: 14 de julho de 2015.] https://www.oiml.org/en/files/pdf_r/r117-p-e07.pdf.
[9] WELMEC. WELMEC 10.4. Guide for Testing of Electronic Calculators with Conversion Function and Conversion Devices (Measuring Systems for Liquids other than Water). 2006. primeira edição.
[10] ABNT - Associação Brasileira de Normas Técnicas. ABNT NBR 16020 - Medição eletrônica de líquidos — Computadores de vazão. 2011.
[11] American Petroleum Institute. API MPMS Chapter 21.2 - Flow Measurement Using Electronic Metering Systems. 2011.

Social profit in the context of the activities at Fluids Measurement Sector in Legal Metrology Department - Inmetro

L R Cinelli, L G Silva, E A Junior, R O Almeida

Instituto Nacional de Metrologia, Qualidade e Tecnologia - Inmetro
Diretoria de Metrologia Legal - Dimel
Setor de Medição de Fluidos – Seflu
Duque de Caxias - Brasil

E-mail: lrcinelli@inmetro.gov.br

Abstract. This article was prepared in the context of the work of the Fluids Measurement Sector (Seflu) of the Legal Metrology Department of Inmetro (Dimel) in order to try to answer the following question: What is the magnitude of Social Profit generated for brazilian society from the existence of legal control of measuring instruments within the scope of this sector? In this sense, some examples of a case study containing the main measurement instruments related to the evaluation process of models performed at the Seflu are presented.

1. Introduction

The Legal Metrology can be defined as a part of Metrology related to activities resulting from mandatory requirements referring to measurements, measuring unities, instruments and methods of measuring, which are developed by competent organisms. Their main purpose is to protect the consumer treating the measuring unities, methods and measuring instruments according to the mandatory technical and legal requirements. In their essence is an exclusive responsibility of the State, and it consists of a group of activities and technical, legal and administrative procedures established through legal clauses by public authorities aiming at assuring the quality and credibility of measuring results involving the commercial transactions, human health, environmental and citizen safety.

The legal control of measuring instruments (type evaluation, verifications and inspections) makes part of a process, which begins with elaboration/conception/using of this instrument for society. The last ones undergone to examinations and/or essays to assure reliability of the results.

Therefore, can be it is said that the type evaluation intervene in several phases of the process on implementing the legal requirements by defining the minimum characteristics necessary to the construction and performance of the measuring instruments from the technical and metrological requirements.

Moreover, to be more specific, it can still be said that the type evaluation is the act by which Inmetro recognizes that a measuring instrument, measurement materialized or part of a measuring instrument is able to meet the regulatory requirements to be used in order to provide reliable results. Before being launched in the market for trading or usage, all measuring instruments or materialized measuring imported or manufactured in Brazil with an applicable metrological and technical regulation (RTM in portuguese) in effect must have their models approved.

Regarding the initial verification, it can be said that this phase of legal control occurs after the production of instrument in factory and before being traded.

In addition, finally, the subsequent verification is a phase accomplished periodically, in general with a minimum 1-year interval, and always when the instrument is found in use. In this situation, it is also important to clarify that, in the occurrence of repair, maintenance, or modification of the measuring instrument, the instrument must undergo though a new checking (post-repairing checking).

All phases converge to the main objective of assuring credibility to the measuring instrument. It must be reminded that all phases imply metrological services charges (Direct Revenues).

2. Performance in Legal Metrology

The Fluids Measurement Sector (Seflu), within its competence area accomplishes essentially type evaluation, and it works a portfolio with approximately 15 measuring instruments all of which are employed in economic activities and in the achievement /definition of acts in juridical businesses of taxing origin.

Currently, the department is composed of 8 coworkers whose positions involve:

a) Accomplishment of measuring instruments model assessment;

b) Development of metrological legislation and rules implementation studies in reference;

c) The dissemination of knowledge by training in technical courses in metrology in Inmetro and in RBMLQ-I [1]

d) The provision of technical support for activities accomplishment of RBMLQ-I;

e) The representation before regulatory agencies, public agencies, and companies in public services concessionaries, associations and measuring equipment/instruments.

f) Technical support given to the society in general,

g) Technical cooperation with nations from South America for knowledge dissemination and training accomplishment.

[1] *Delegated body*

3. Generated Revenues

Through legal control of measuring instruments under supervision of the Fluids Measurement Sector, it was noted in the year of 2014 the data inserted in the tables 1 and 2.

Table 1. Amount of services undergone to initial and subsequent verifications in the year 2014.

Instrument	Total
Fuel Dispensers	404,528
Water Meters	667,420
Gas Measurement – Diaphragm	5,280
Gas Measurement – Rotating	1,262
Gas Measurement – Turbine	351
Gas Measurement – Coriolis	2
Gas Measurement – Ultrasonic	-
Vessels (20L)	42,424

Source: Inmetro/RBMLQ-I

Table 2. Revenues generated in the year of 2014 for instruments covered in the direct and indirect activities of Seflu.

Type Evaluation	Verifications	Legal Control
US$ 86,236.57	US$ 20,180,175.30	US$ 20,266,411.87

Source: Inmetro/RBMLQ-I

4. Estimation of Social Profit by Legal Control - Cases

4.1. Flow Meters (Oil) – Royalties Collection and Special Participations

According to sector data, it is estimated that in 2012 there were around 1,100 points of fiscal measurement of oil of the biggest oil company in the country alone. In these points, there is at least one (01) flow meter installed which was approved by Type Evaluation by effect of RTM approved by Inmetro Ordinance n° 64/2003 and Joint Resolution ANP/Inmetro n° 1/2013. The Oil production this year was approximately of 2 millions of barrels per day. In addition, in these points there is incidence of Royalties and Special Participations. Rio de Janeiro State (RJ) has received more than US$ 1 billion in this year regarding Union on lending (which collected US$ 10.2 billion in this period).

4.2. Fuel Dispensers – ICMS Collector

In 2014 there were around 2,100 gas stations installed at Santa Catarina State (SC). By regulation every gas station at Santa Catarina State (SC) has at least (01) one dispenser of liquid fuel which passed by Type Evaluation, Initial and Subsequent Verification by effect of RTM approved by Inmetro Ordinance n° 23/1985. In this same year, the State (SC) collected around US$ 1 billion in ICMS[2] referring to the fuel sect (trading). The Union collected in this period around US$ 14.3 billion in ICMS of fuels.

4.3. Cold Water Meters –PIS/CONFINS Collector

In 2013, it is estimated that the amount of water meters installed in the São Paulo State (SP) is close to 7.2 million of instruments. By regulation 1/5 of these instruments are changed yearly. In this context, it can be considered that all models used by the concessionary need to pass by the type approval process through Inmetro, and what all of them need, sometimes along the useful life, pass by checking by force of RTM approved by Inmetro Ordinance n° 246/2000.

In this year, Sabesp[3] presented a revenue around US$ 3.1 billion referred to rendering services related to the provision of water and basic sanitation. Upon which US$ 223,342 were taxed by Union for PIS/CONFINS (Government programs for Social Integration and for Social Insurance and for Financing of Social Insurance) by the Laws nº 10.637/2002 and nº 10.833/2003.

[2] *ICMS is a tax on sales and services and applies to the movement of goods, transportation, communication services and other general supplying of goods.*

[3] *Sabesp is a Brazilian water and waste management company owned by São Paulo state.*

5. Conclusions

Based on the above mentioned, the analysis of social, economic, and financial impact related to activities of public institutions is a question that needs to be studied. The concept of Social Profit can be a decisive management tool to the impact estimation of the performance of the public service in society in general.

However, it is known that the consolidation of this concept is a long process of paradigms breaking, once it brings in itself the deconstruction of a thought that the public service is inefficient, inoperative and excessively bureaucratic. In this way, it is expected that the Seflu contribution can enrich the discussion of activities relevance not only of Inmetro but also of public institutions in Brazil.

References

[1] Vocabulário Internacional de Termos de Metrologia Legal (VIML) - Portaria no. 150, de 29 de março de 2016.
[2] Anuário estatístico brasileiro do petróleo, gás natural e biocombustíveis: 2015.Agência Nacional do Petróleo, Gás Natural e Biocombustíveis. - Rio de Janeiro: ANP, 2008.
[3] Contribuição à Nota Técnica Final - Primeira Revisão Tarifária Da Sabesp - Cálculo do P0, P1 e Fator X. Sabesp. Março 2014.
[4] Carga Tributária no Brasil – 2014 (Análise por Tributo e Bases de Incidência). Ministério da Fazenda. Receita Federal (CETAD). Outubro de 2015.

Speed of sound in biodiesel produced by low power ultrasound

P A Oliveira, R M B Silva, G C Morais, A V Alvarenga and R P B Costa-Felix

Laboratory of Ultrasound (Labus), National Institute of Metrology, Quality and Technology (Inmetro), Av. Nossa Sra. das Graças 50, Duque de Caxias, RJ, Brazil, ZIP 25250-020

passuncaooliveira@gmail.com; rpfelix@inmetro.gov.br

Abstract. The quality control of the biodiesel produced is an important issue to be addressed for every manufacturer or retailer. The speed of sound is a property that has an influence on the quality of the produced fuel. This work presents the evaluation about the speed of sound in biodiesel produced with the aid of low power ultrasound in the frequencies of 1 MHz and 3 MHz. The speed of sound was measured by pulse-echo technique. The ultrasonic frequency used during reaction affects the speed of sound in biodiesel. The larger expanded uncertainty for adjusted curve was 4.9 $m.s^{-1}$.

1. Introduction

The search for alternative energy sources has been motivated due to the limitation and exhaustion prediction of fossil fuels, besides environmental concerns. Thus, the production and use of biodiesel became attractive for being a renewable fuel.

One of the limiting factors for the large-scale commercialization of biodiesel is the relative high cost of production compared to fossil fuels [1]. Hence, improvements in biodiesel production processes are necessary to provide costs reduction. Ultrasound has been used in biodiesel production for this purpose [2][5].

It is known that the quality control of the biodiesel produced is extremely important. In the injection process, fuel quality directly affects engine efficiency. Some properties such as viscosity and density strongly influence the quality of the produced fuel [6]. The speed of sound is way to assess these parameters and consequently the quality of the fuel. Thus, the determination of the speed of sound in biodiesel is relevant [7].

However, few studies available in the literature determine the speed of sound in biodiesel [6][10]. Commonly, the cited authors studied the biodiesel produced from the conventional method (mechanical stirring and heating) and rarely consider the metrological aspects.

Therefore, this paper aims to evaluate the speed of sound for biodiesel produced by ultrasound-assisted through low-power, in the frequencies of 1 MHz and 3 MHz. Throughout the study the calculation of uncertainties of measurements was carried.

2. Materials and methods

2.1. Biodiesel synthesis

The biodiesel was synthesized by the transesterification reaction of soybean oil (Liza®) with methanol (purity 99.8%) using molar ratio of alcohol/oil of 6:1 and potassium hydroxide as a catalyst (1.5%). The reaction was carried out in a mini-reactor using two ultrasound generator devices models 4144 and 4134 (Sonomed, Brasil), operating in frequencies of 1 MHz and 3 MHz with nominal power of 7.2 W. The reaction was performed without the aid of mechanical stirring or heating bath, and its total time was 40 minutes.

2.2. Measuring the speed of sound

The technique described in [11] was used in the measurement of the biodiesel speed of sound. The measurement system is shown in figure 1, and is composed by a thermal bath (a) a cell with biodiesel (b), and an ultrasonic transducer (c) of 1 MHz and 12.7 mm in diameter model A303S (NDT-Panametrics, Olympus Corporation, Japan) [12]. The ultrasonic transducer acted as transmitter and receiver, and it was excited through a wave generator (d) model 33250A (Agilent Technologies, CA, USA). The acquired signals were digitized by an oscilloscope (e) model DSO-X 3012A (Agilent Technologies, CA, USA) and transferred to a computer (f) through a program developed in LabView (National Instruments, TX, USA).

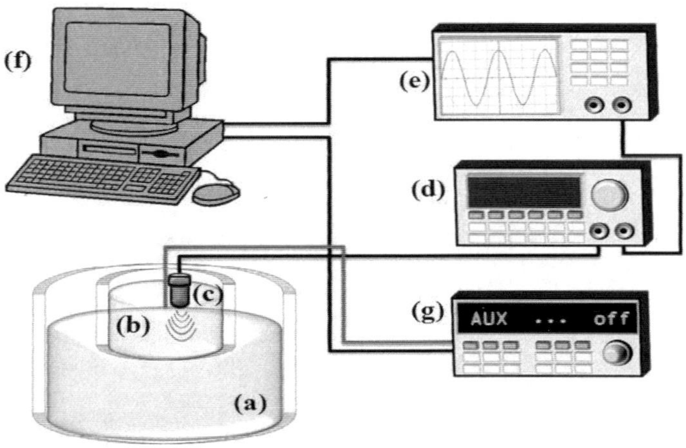

Figure 1. Experimental Setup.

The speed of sound in each temperature, for each repetition, was determined through the measurement of the time that the ultrasonic signal took to cross the sample and return to the same transducer (time of flight). The temperature of bath was set initially at 20 °C and increases up to 50 °C. It was monitored during the measurement using a temperature measuring device (g) model 34970A (Agilent Technologies, CA). The measurements were performed under repeatability conditions, with five measurements for each biodiesel sample [13][14].

According to Oliveira *et al* [11], the speed of sound in vegetable oils as a function of temperature is given by a linear function. Thus, after determining the biodiesel propagation velocity at each point, adjusted curves were established (v_{bio_a}) based on equation (1).

$$v_{bio_a} = (a + b \cdot T) + \delta \tag{1}$$

in which a and b are constants obtained in each repetition (contributing to uncertainty of type A), and T is the temperature in °C. The type B uncertainty, derived from the calibration certificate of the thermometer, was used in the uncertainty budget. Since δ is a term that has been inserted in the equation in order to take into account the uncertainty of each measured velocity point, its value is equal to zero but it has an associated uncertainty value of 0.65 m.s^{-1}.

The expanded uncertainty was calculated according to the Guide of the Expression of Uncertainty in Measurements [15], by a coverage factor k that takes into account a t-distribution with coverage probability of 0.95.

2.3. Normalized error

The results of speed of sound of the produced biodiesel were statistically compared by calculating the normalized error [16], according to equation (2).

$$E_n = \frac{|x_{1\mathrm{MHz}} - x_{3\mathrm{MHz}}|}{\sqrt{{U_{1\mathrm{MHz}}}^2 + {U_{3\mathrm{MHz}}}^2}} \qquad (2)$$

in which E_n is the normalized error; $x_{1\mathrm{MHz}}$ is the speed of sound in biodiesel produced at 1 MHz; $x_{3\mathrm{MHz}}$ is the speed of sound in biodiesel produced at 3 MHz; $U_{1\mathrm{MHz}}$ and $U_{3\mathrm{MHz}}$ are the respective the expanded uncertainties (p = 0.95).

The normalized error tests the compatibility of the results; its evaluation criterion is given by: $|\boldsymbol{E_n}| \leq 1.0$ the biodiesels speeds of sound are metrologically equivalent; $|\boldsymbol{E_n}| > 1.0$ the biodiesel speeds of sound are not statistically equivalent.

3. Results and discussion

The results of speed of sound obtained from the five replicates for soybean biodiesel in 1 MHz and 3 MHz are shown in figures 2 and 3, respectively.

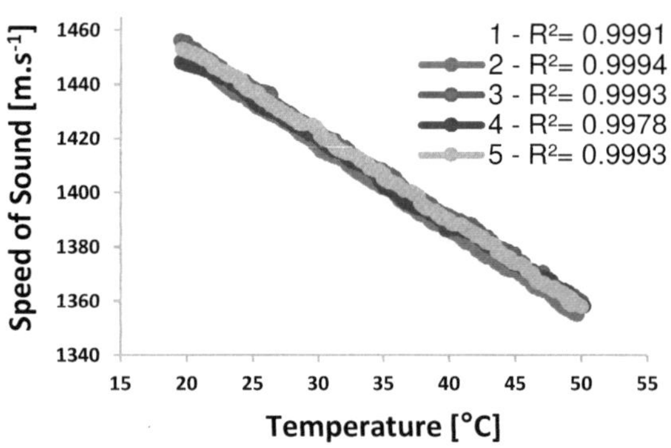

Figure 2. Relationship between speed of sound and temperature for biodiesel produced at 1MHz.

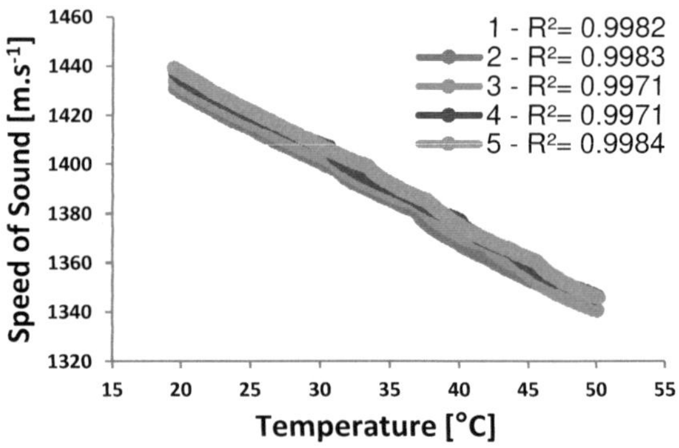

Figure 3. Relationship between speed of sound and temperature for biodiesel produced at 3MHz.

One can observe that the relation between the speed of sound in biodiesel and the temperature tends to be linear for the studied temperature range (figure 2 and 3). Besides, for the all the five repetitions, the coefficients of determination (R^2) are close to 1. This corroborates the linear relation between the variables studied.

Adjusted curves were established based on linear regression data from the five measurements performed for each biodiesel, and the expanded uncertainty was estimated, as shown in tables 1 and 2.

Table 1. Expanded uncertainty(U_{1MHz}) of the adjusted curve for speed of sound for biodiesel produced with 1 MHz ultrasonic frequency.

T [°C]	v_{bio_a} [m.s^{-1}]	k [p=0.95]	U_{1MHz} [m.s^{-1}]
20	1450.8	2.36	4.3
25	1435.1	2.36	4.4
30	1419.5	2.26	4.3
35	1403.8	2.23	4.4
40	1388.2	2.18	4.5
45	1372.5	2.14	4.6
50	1356.9	2.12	4.7

Table 2. Expanded uncertainty (U_{3MHz}) of the adjusted curve for speed of sound for biodiesel produced with 3 MHz ultrasonic frequency.

T [°C]	v_{bio_a} [m.s^{-1}]	k [p=0.95]	U_{3MHz} [m.s^{-1}]
20	1435.0	2.36	4.4
25	1419.9	2.36	4.6
30	1404.7	2.26	4.5
35	1389.6	2.23	4.6
40	1374.5	2.18	4.7
45	1359.4	2.14	4.8
50	1344.2	2.11	4.9

The graphical representation of tables 1 and 2 is shown figure 4.

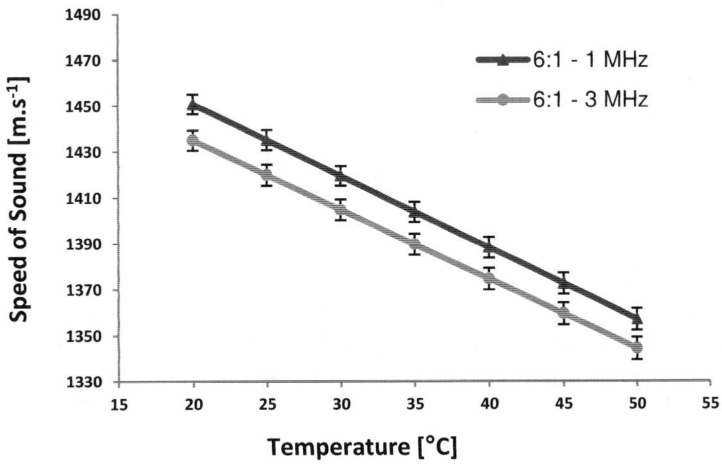

Figure 4. Adjusted curves with expanded uncertainty for each biodiesel studied.

Then, from the data in tables 1 and 2, the normalized error was calculated to determine if the speeds of sound biodiesel samples produced from different frequencies are statistically equivalent. The results are presented in table 3.

Table 3. Normalized error for evaluation between biodiesel produced by different frequencies.

T [°C]	E_n
20	2.56
25	2.40
30	2.38
35	2.23
40	2.10
45	1.98
50	1.86

Analyzing table 3, it is possible to note that speed of sound in biodiesel produced by different ultrasonic frequencies are not statistically equivalent, since the normalized error was greater than 1. The frequency of the ultrasonic waves is one of the factors that lead to the occurrence of the cavitation bubbles. Different frequencies form cavitation bubbles with different diameters, which may influence the emulsification of the reaction and consequently the biodiesel produced.

4. Conclusions

The speed of sound for two biodiesels produced from different frequencies of ultrasonic irradiation was measured, varying the temperature from 20 °C to 50 °C.

For biodiesel studied here, the speed of sound decreases with increasing temperature and this relationship is well described by a linear function, what is not the case for other biological tissues . Thus, the results agree with the behavior of speed of sound of data available in the literature for conventional biodiesel. The correlation coefficient and the low uncertainty disclose the quality of the measurements, reinforcing the importance of metrological aspects for those measurements.

The first set of results show that the ultrasound frequency used during the biodiesel production affects directly the speed of sound in the final product. However, to determine which frequency is most appropriate for the biodiesel production by ultrasound, analysis of the conversion rate of all the reactions performed is necessary, as well a comparison with the properties of the conventional biodiesel.

The difference in the speed of sound in the produced biodiesels is an evidence that they have dissimilar physical chemical properties. As, within this research, biodiesel was produced with the aid of different ultrasound frequencies, one can infer directly that ultrasonic parameters play a key role in the reaction.

References

[1] Glisic S B, Pajnik J M and Orlovic A M 2016 *Applied Energy.* v **70** p 176-185.
[2] Than L T, Okitsu K, Maesa Y and Bandiw H 2014 *Ultrasonics Sonoch.* v **21(2)** p 467-441.
[3] Moholkar V S, Choudhury H A and Srivastava P 2014 *AIChE Journal.* v **60(5)** p 1572-1581.
[4] Ferreira E, Vicente J, Silva R C D, Nele M, Nunes E and Costa-Felix R P B 2014 *Polymer Testing.* **33** pp 16-20.
[5] Oliveira J V, Oliveira D, Treichel H, Rosa C D, Batistella L, Popiolki A S and Trentin C M 2015 *Bioprocess Biosyst. Eng.* v **38** p 437-448.
[6] Lopes A F G, Talavera-Prieto M C, Ferreira A G M, Santos J B, Santos M J and Portugal A T G 2014 *Fuel.* v **116** p 242–254.
[7] Maggi L E, Silva C E R, Alvarenga A V and Costa-Felix R P B 2011 *Journal of Physics. Conference Series.* **279** pp 012029.
[8] Freitas S V D, Paredes M L L, Daridon J, Lima A S and Coutinho J A P 2013 *Fuel.* v **103** p 1018-1022.

[9] Ndiaye H I, Habrioux M, Coutinho J A P, Paredes M L L and Daridon J L 2013 *J Chem Eng Data*. **58** p 1371–1377.
[10] Tat M E and Gerpen J H V 2003 *J Am Oil Chem Soc*. **80** p 1249-1256.
[11] Oliveira P A, Baêsso R M, Morais G C, Alvarenga AV and Costa-Félix R P B 2016 *J. Phys.: Conf. Ser*. **733** pp 012040.
[12] Silva C E R, Alvarenga A V and Costa-Felix R P B 2011 *J. Phys.: Conf. Ser*. **279** pp 012025.
[13] Costa-Felix R P B 2006 *Measurement*. **39** pp 169-175.
[14] Costa-Felix R P B and Machado J C 2015 *Measurement*. **69** pp 146-154.
[15] BIPM, JCGM:100 2008 Paris.
[16] Steele A G and Douglas R J **2006** *Metrologia*. **43(4)** p S235-S243.
[17] Abrunhosa V M, Soares C P, Possidonio A C B, Alvarenga A V, Costa-Felix R P B, Costa M L P S and Mermelstein C S 2014 *Ultrasound in Medicine & Biology*. **40** pp 504-512.
[18] Petrella L I, Maggi L E, Souza R M, Alvarenga A V and Costa-Felix R P B 2014 *Ultrasonics*. **54** pp 1476-1479.

Acknowledgements
Research reported in this paper has been partially supported by the Carlos Chagas Filho Research Support Foundation (Faperj, grant number E-26/201.563/2014), by the National Council for Scientific and Technological Development (CNPq, grant number 310.392/2014-4) and Pronametro/Inmetro.

Synthesis of reference compounds related to Chemical Weapons Convention for verification and drug development purposes – a Brazilian endeavour

S F A Cavalcante[1,2], R L de Paula[1,3], D A S Kitagawa[1], M C Barcellos[1], A B C Simas[2] and J M Granjeiro[3]

[1] Instituto de Defesa Química, Biológica, Radiológica e Nuclear – IDQBRN, Centro Tecnológico do Exército – CTEx, Rio de Janeiro, BR.
[2] Instituto de Pesquisas de Produtos Naturais Walter Mors – IPPN, Centro de Ciências da Saúde, HSS-27, Universidade Federal do Rio de Janeiro, BR.
[3] Programa de Pós-Graduação em Biotecnologia do Instituto Nacional de Metrologia, Qualidade e Tecnologia – PPGBIOTEC/INMETRO, Duque de Caxias, BR.

E-mail: samirfac@yahoo.com.br, samir@ctex.eb.br, reuel.lopes@eb.mil.br

Abstract. This paper deals with challenges that Brazilian Army Organic Synthesis Laboratory has been going through to access reference compounds related to the Chemical Weapons Convention in order to support verification analysis and for research of novel antidotes. Some synthetic procedures to produce the chemicals, as well as Quality Assurance issues and a brief introduction of international agreements banning chemical weapons are also presented.
Keywords. Organisation for the Prohibition of Chemical Weapons, Chemical Weapons Convention, Organic Synthesis, Reference Chemicals, Quality Assurance.

1. Introduction

The use of chemicals as weapons dates from ancient times. Nevertheless, their lethality has been improved over the years. Even though international agreements had already been signed at the end of 19th century, toxic chemicals were deployed in World War I (1914-1918), causing millions of casualties. Examples of chemicals used were gases as chlorine and phosgene, and, especially sulfur mustard, HD (1, figure 1).

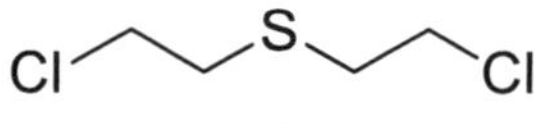

1

Figure 1. Sulfur mustard HD

Between World Wars and after II World War, research on novel pesticides led to a class of very toxic pentavalent phosphorus compounds. They irreversibly inhibit acetylcholinesterase (AChE), enzyme involved in Parasympathetic Neurotransmission, which can lead to life-threatening events. Examples of these compounds are the G-Agents sarin (2), soman (3), tabun (4) and V-Agent VX (5), all known as nerve agents (figure 2) [1-3].

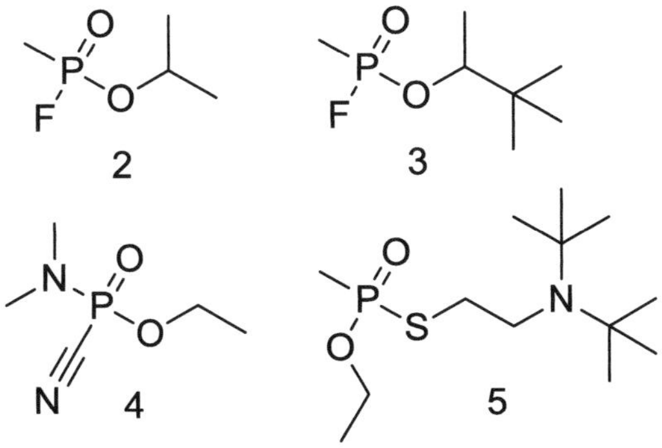

2 3 4 5

Figure 2. G-Agents

Although these organophosphorus compounds were not used in the II World War, their potential as Weapons of Mass Destruction has been demonstrated in different situations, as Iran-Iraq War (1980-1988). After this event, in 1993, the first draft of the Chemical Weapons Convention (CWC) was published. This led to foundation of "Organisation for the Prohibition of Chemical Weapons (OPCW)". The current text of CWC entered into force in 1997 [1,4]. Unfortunately, CWC has not precluded the use of chemical warfare in Japan (1994-1995) and ongoing Syrian Civil War (2013). As of December 2017, there are 192 signatories or State Parties. For its efforts throughout years to eliminate chemical arsenals, OPCW was awarded with Nobel Peace Prize in 2013 [1].

CWC has the "Annex on Chemicals", that list the different chemicals used as weapons (Part A) or precursors (Part B) for their preparation, as well as two toxins (ricin and saxitoxin), which underlines the convergence between Chemistry and Biology. There are three Schedules of Chemicals, divided accordingly to their lethality and possibility of peaceful application. As Brazil is an OPCW State Party, these compounds cannot be produced or used as warfare. They only can be synthesized for peaceful purposes, as testing of protective equipment, detectors, forensics and research on novel medicines. This asserts the need of synthetic capabilities, as these chemicals are not sold by chemical suppliers.

Schedule 1 contains the most toxic substances which have been developed for or used as chemical warfare agents with virtually no peaceful applications. It lists, for example, sulfur mustard (1), G- and V-Agents (2-5). Schedule 2 lists toxic chemicals that could not only be used as weapon, but also as precursors to Schedule 1 chemicals. Nonetheless, they have uses not prohibited by the CWC, as precursors for pesticides and pharmaceuticals, for example. Therefore, they have relevant industrial production. Thiodiglycol (6), precursor to sulfur mustards (figure 3) and used as solvent for inks, is a relevant example of this Schedule.

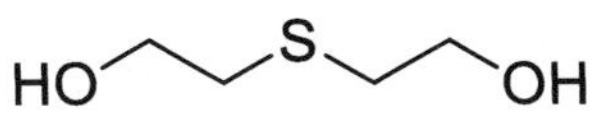

6

Figure 3. Thiodiglycol

Schedule 2 is important for verification purposes, as they can be both precursors and degradation products of Schedule 1 chemicals. For example, thiodiglycol (6) is precursor and hydrolysis degradation product of sulfur mustard (1, scheme 1).

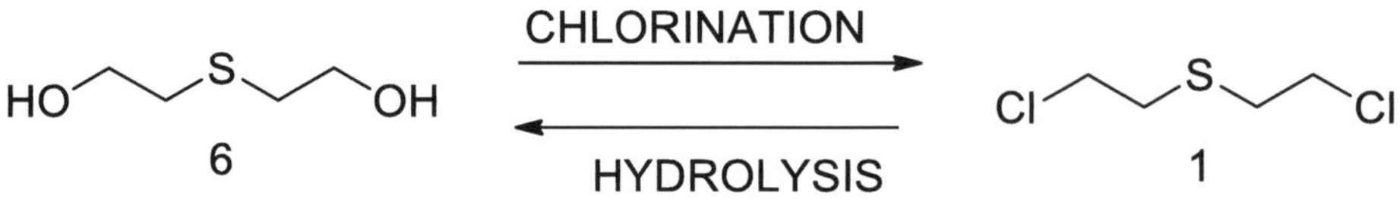

Scheme 1. Relationship between Schedules 1 and 2 chemicals

Schedule 3 refers to chemicals produced in large scale for peaceful purposes; however, they may be used such as warfare agents or precursors for chemicals of the Schedules 1 and 2. Triethyl phosphite (7, figure 4), is a precursor to G- and V-Agents [4], as well flame-retardant materials.

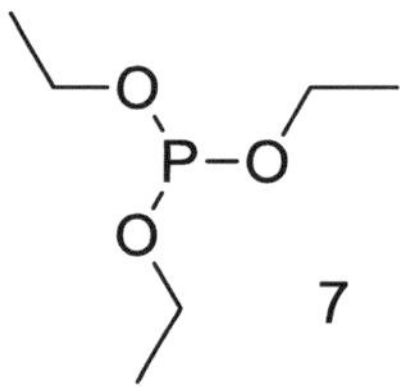

7

Figure 4. Triethyl phosphite

Although Schedule 3 chemicals may be easily purchased in comparison to other Schedules, the production of certified reference materials and reference materials is also restricted. There are few suppliers of reference materials certified in Brazil and also at an international level. Thus, it is extremely important that the laboratories belonging to the Defense Affairs have expertise to synthesize reference materials for CWC chemicals.

2. The LSO experience in synthesis of CWC chemicals – the quest for the reference standards

During the OPCW Proficiency Tests, which occurs twice a year, a forensic exercise to all laboratories which aim to be awarded with "OPCW Designated" status, LAQ has found some challenges to access reference chemicals needed [5]. Even during the ISO 17025 Accreditation process, availability of chemical references becomes a key step, because, as aforementioned, virtually all analytical targets are not sold by chemical suppliers. All Designated Laboratories have in-house procedures to afford them. LSO has been designed to carry off this problem and we have been very successful so far, even facing bureaucratic problems to access key precursors. OPCW has tried to help laboratories to overcome the synthetic limitations through its database, allowing them to take part in Proficiency Tests. This comprehensive database contains different spectroscopic and spectrometric data of chemicals from all Schedules to support laboratories' findings.

For instance, diethyl methylphosphonate (DEMP, 8), a Schedule 2 chemical, key precursor to some G-Agents, has been synthesized from triethyl phosphite (7, Schedule 3 chemical) and iodomethane (9, scheme 2) using Arbuzov-Michaelis Reaction:

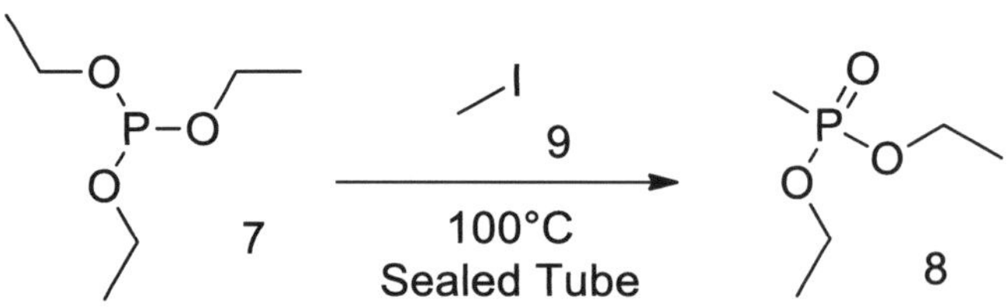

Scheme 2. Synthesis of 8

Recently, after we have sought in the literature for nerve agents surrogates [6], we have successfully prepared O-ethyl methylphosphonochloridate (EMPC, 10), also a Schedule 2, structurally related to G-Agents. From EMPC, we have carried out the synthesis of O-4-nitrophenyl O-ethyl methylphosphonate (NEMP, 11), a VX surrogate, which has been used for testing of our antidote prototypes and evaluation of detectors (scheme 2). NEMP is a potent AChE inhibitor. It is the first time that a scientific work in Brazil use nerve agents' surrogates for in vitro assay of novel antidotes (scheme 3). This work has been submitted to peer-reviewed publications in Medicinal Chemistry.

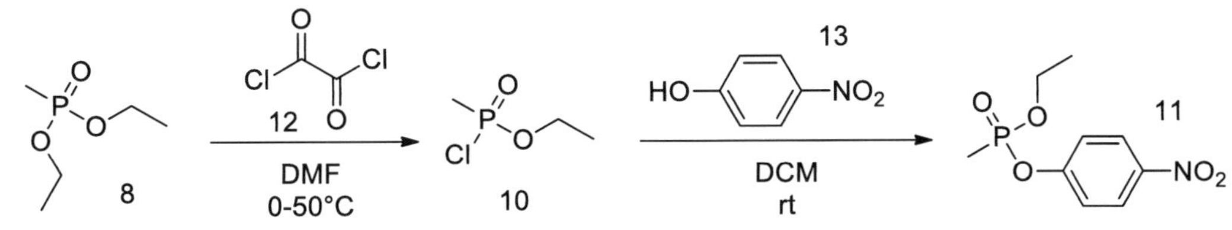

Scheme 3. Synthesis of 11

Recently, in 41st OPCW Proficiency Test, LAQ had to confirm the presence of different degradation products of sesquimustards, some of them very challenging compounds if we consider the availability of precursors. Starting from ethylene oxide (14), LSO synthesized 2-mercaptoethanol (commercial, but no readily available, 15) and 1,2-diiodoethane (16), to afford the synthesis of 1,2-bis(2-hydroxyethylthio)ethane (17) (scheme 4) [6].

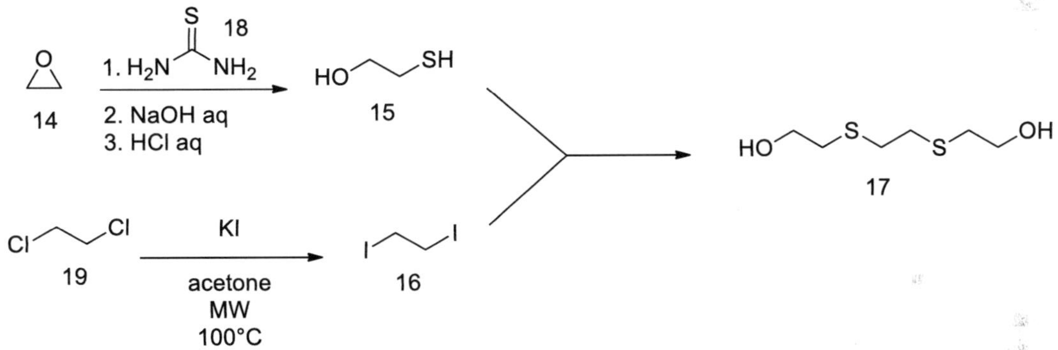

Scheme 4. Synthesis of 17

Based on aforesaid, it is clear that LSO has all required expertise to safely undertake the synthesis CWC related compounds.

3. Perspectives

From a point of view of the Quality Assurance, LSO has striven to develop other synthetic methodologies that can fulfill the need of reference chemicals related to CWC aiming, in a future, to be a supplier of these compounds to other laboratories, complying with all requirements of ISO 17034 to produce chemical references. Naturally, all actions taken must comply with the CWC, as Brazil is a State Party.

Access to CWC Certified Reference Materials (CRM) is a challenge for many laboratories, since they are scarce and require experience in handling hazardous materials. However, their production has several advantages, such as: traceability and measurements reliability, validation of methods, calibration of equipment, quality control and improvement of performance during OPCW Proficiency Tests. Therefore, the production of CRM is one of the priorities for the LSO.

CRM production must comply the requirements of ISO 17034 and Guide 35, i.e., values for homogeneity, stability (short and long terms) characterization should be established for synthesized

IOP Conf. Series: Journal of Physics: Conf. Series **975** (2018) 012020 doi:10.1088/1742-6596/975/1/012020

materials [7,8]. A flowchart to be used by the LSO, according to ISO 17034, for the production of CWC-CRM is given (chart 1).

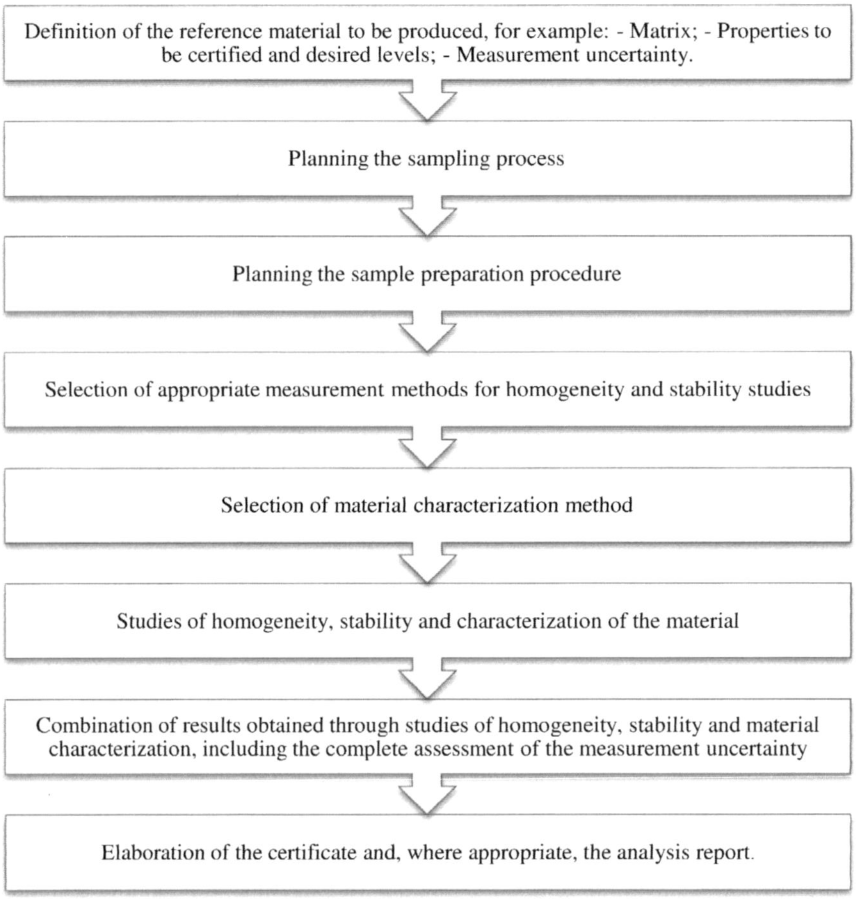

Chart 1. Workflow

The flowchart may be explained using DEMP as example. After synthesis, purification and spectroscopic identification, the reference material should be diluted in methanol at 1000 µg/ml and stored in a suitable vial at -20°C for further studies (homogeneity, stability and characterization). If an initial batch of 100 bottles is produced, 14 of them will be separated for homogeneity studies. This percentage figure is present in the sampling criteria set out in British Standard 5309 [9]. The aliquots required for the stability and characterization studies will be drawn from the separate bottles for the homogeneity study. The choice of bottles will be made with stratified random sampling. The homogeneity of the produced material will be evaluated by analysis of variance (ANOVA), by calculating the standard deviation between the units of a reference material (s_{bb}) and the standard deviation of the repeatability of the measurement method (s_r). The uncertainty regarding homogeneity (u_{hom}) will be the greatest value between the two deviations, s_{bb} or s_r. The calculations referring to the stability study will also be performed with analysis of variance, evaluating regression data. The stability uncertainty (u_{st}) is calculated by multiplying the standard error value (s_{bl}) by the study time [8,10]. The short-term stability (u_{sts}), which evaluates the transport conditions of the CRM, and the long-term stability (u_{lts}), which evaluates the storage conditions of the CRM, will be calculated. The measurement reference temperature in the stability studies shall be 4°C. The short-term stability will be carried out in the period of fifteen days, with temperatures of 25ºC and 50ºC. Long-term stability studies will be carried out over 12 months at temperatures of 4°C and 25°C. The characterization study of the CRM will be carried out using two different analytical techniques, Gas Chromatography coupled to Mass Spectrometry (GC-MS) Nitrogen and Phosphorus Detector (GC/NPD). The methodology of analysis will be carried out according to the established in the "Blue Book", method standardized by the OPCW [11]. The measurement of traceability of the chromatographic and spectrometric analysis procedure will be ensured using certified DEMP methanolic solution, in conjunction with calibration of the laboratory instruments used and sample preparation. The characterization uncertainty may be measured by Interlaboratory comparison if there is a population of laboratories that are equally capable of determining the characteristics of the CRM, providing results that have acceptable accuracy. In the absence of such laboratories the characterization uncertainty can be calculated by the standard deviation of the results of measurement divided by the square root of the number of replicates tested. The final expanded uncertainty (U_{CRM}) will be calculated with the values of the uncertainty of homogeneity (u_{hom}), short term stability (u_{sts}), long term stability (u_{lts}) and characterization (u_{char}), as in equation (1). The coverage factor (k = 2) should be considered for the calculation of the expanded final value.

$$U_{CRM} = k\times\left(u_{hom}^2 + u_{sts}^2 + u_{lts}^2 + u_{char}^2\right)^{1/2} \quad (1)$$

Finally, the certificate of the reference material will be drawn up in accordance with the requirements set out in ISO Guide 31 [12].

Acknowledgements
Authors would like to thank Organisation for the Prohibition of Chemical Weapons – OPCW for Research Grant Support L/ICA/ICB/201062/15 and Centro Tecnológico do Exército – CTEx, for the infrastructure.

References

[1] OPCW. www.opcw.org (accessed in Jul 2017)
[2] Makhaeva G F, Filonenko I V, Yankovskaya V L, Fomicheva S B, Malygin V V 1998 *Neurotoxicology* **19** (4-5) p 623
[3] Rozengart E V, Basova N E, Moraleva S N, Lushchekina S V, Masson P, Varfolomeev S D 2013 *Chemico-Biological Interactions* **203** (1) pp 3-9
[4] OPCW *Chemical Weapons Convention (CWC)* 1997
[5] Dubey V, Velikeloth S, Sliwakowski M, Mallard G 2009 *Accreditation and Quality Assurance* **14** (8) p 431
[6] Meek E C, Chambers H W, Coban A, Funck K E, Pringle R B, Ross M K, Chambers J E 2012 *Tox. Sci.* **126** (2) pp 525-533
[7] *ISO 17034:2016 General requirements for the competence of reference material producers* First edition
[8] *ISO Guide 35:2006 Reference materials - General and statistical principles for certification* Third edition
[9] *BS 5309:1976 Methods for sampling chemical products* (British Standards Institution) Part 1. Introduction and general principles
[10] Miller J C and Miller J N, *Statistics for Analytical Chemistry* Third edition (Ellis Horwood Series in Analytical Chemistry)
[11] Vanninen, P (editor). *Recommended Operating Procedures for Analysis in the Verification of Chemical Disarmament*, 2011 (The Ministry for Foreign Affairs of Finland, Helsinki)
[12] *ISO Guide 31:2015 Reference materials - Contents of certificates, labels and accompanying documentation* Third edition

The influence of temperature on the speed of sound of cortical bone phantom: a metrological view

R M Souza, R P B Costa-Felix and A V Alvarenga

Laboratory of Ultrasound (Labus), Directory of Scientific and Industrial Metrology (Dimci), National Institute of Metrology, Quality and Technology (Inmetro), Brazil, RJ.

E-mail: rebecasouzam@gmail.com; rpfelix@inmetro.gov.br; avalvarenga@inmetro.gov.br

Abstract. This paper presents an experimental study to estimate the group velocity of ultrasound longitudinal propagation in a cortical bone phantom for the temperature range of 19 ºC to 37 ºC, and respective measurement uncertainty estimates. The group velocity was measured using the pulse/echo method. The uncertainty model was proposed based on the BIPM JCGM:100 – Guide of the Expression of Uncertainty in Measurements. The velocity results show a tendency to decrease linearly as the temperature increases, considering the uncertainty estimated.

1. Introduction

The characterization of cortical bone by ultrasound, based on parameters such as attenuation and longitudinal propagation velocity, can reflect the structure and elastic properties of the propagation medium [1][2]. Studying cortical bone properties in vitro has the advantage of working under controlled experimental conditions, while in vivo studies are influenced by the bone geometry and body movements, as well as temperature variations [2]. Ultrasound plays an important role in therapy and other health care applications [3][4].

Bone phantoms are largely used to mimic real bone properties in estimating acoustic parameters (speed of sound, backscattering, and attenuation coefficient), as well as in assessing their heating distribution in thermal phenomena studies [5]. Hence, it is important to know and consider how their acoustic properties vary with different temperatures. This work assesses the group velocity variation of a cortical bone phantom for temperature ranging from 19 ºC to 37 ºC, and respective measurement uncertainty.

2. Materials and methods

2.1. Experimental Setup

For this study, a 10 mm-thick epoxy block with short fibers (Sawbones™, USA) that simulates the mechanical properties of the cortical bone was used. The experimental setup consists on a 5 MHz-nominal frequency, 12.7-mm-diameter single-crystal circular unfocused ultrasonic transducers acting as transmitter and receiver (model V303, Olympus, USA) [6]; an arbitrary waveform generator (model 33250A, Agilent Technologies, USA) to excite the transducer with a sine burst of 3 cycles,

20 V peak-to-peak amplitude at 5 MHz; an oscilloscope (model DSO 5012A, Agilent Technologies, USA) to acquire the echo-signals; a water bath (Fisatom, Brazil) to heat the cortical bone phantom (Sawbones®, USA); a thermometer (ThermoSchneider, Germany) to measure temperature throughout the measurement process. Despite the transducer is typically used in non-destructive testing , its construction characteristics are robust enough to be used in such kind of research.

2.2. *Measurement procedure*

After the system alignment and thermal equilibrium, different times of flight are measured as showed in figure 1. First, the time of flight in the water (t_w), without the sample, is measured. Then, the sample is inserted, and the times of flight on the sample surface (t_1) and on the reflector (t_2), with the sample inserted, are measured. The values of t_w, t_1 and t_2 are measured to six different temperature values between 19 °C and 37 °C. At each step of temperature, four repetitions are carried out.

Between repetitions, the phantom is removed from the system and the alignment procedure is repeated. After the temperature is increased for a next step, a new thermal equilibrium is allowed by at least 30 min.

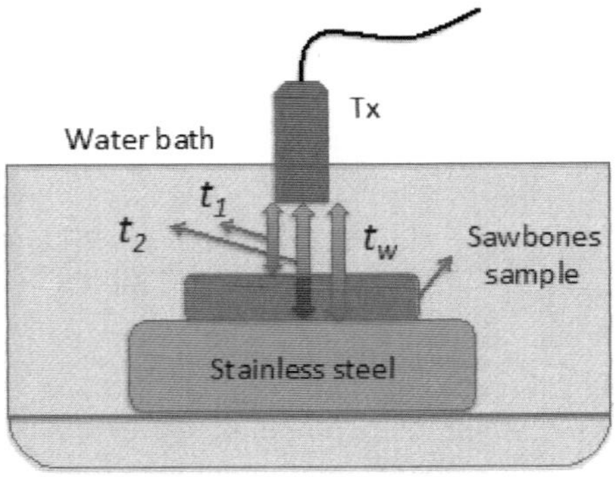

Figure 1. Experimental Setup

2.3. *Group velocity and phantom thickness*

The phantom thickness (x) is determined as:

$$x = v_w.\left(\frac{t_w - t_1}{2}\right) \tag{1}$$

in which $\boldsymbol{v_w}$ is the speed of sound in water as a function of temperature ($\boldsymbol{T}$ in °C) as [9]:

$$v_w = 1405{,}03 + 4{,}624\,.T - 3{,}83.\,10^{-2}.T^2 \tag{2}$$

The group velocity within the phantom is calculated as:

$$v = 2 \cdot x \cdot (t_2 - t_1)^{-1} \tag{3}$$

2.4. *Measurement uncertainty*

The entire procedure is repeated three times and uncertainties are estimated based on the Guide to the expression of uncertainty in measurement [10].

The following sources of uncertainty were considered: an uncertainty component obtained by a Type A evaluation (estimated as the experimental standard deviation of the mean), the standard uncertainty of the temperature obtained from the calibration certificate of the thermometer (μ_T =

0.015 ºC), the uncertainty of the determination of the ultrasound propagation velocity in water (μ_{eq} = 0.18 m s-1) [9] and the uncertainty of the time base (0.04%) obtained from the calibration certificate of the oscilloscope.

Based on the expressions for thickness of the specimen (1) and for the speed of sound in water as a function of temperature (2), the following sensitivity coefficients are determined:

$$c_{v_w} = \frac{\partial x}{\partial_{v_w}} = \frac{t_w - t_1}{2} \quad (4)$$

$$c_{t_w} = \frac{\partial x}{\partial_{t_w}} = \frac{v_w}{2} \quad (5)$$

$$c_{t_1} = \frac{\partial x}{\partial_{t_1}} = -\frac{v_w}{2} \quad (6)$$

$$c_{v_w}^{T} = \frac{\partial v_w}{\partial T} = 4{,}624 - 7{,}66\,.\,10^{-2}.T \quad (7)$$

The combined uncertainty of speed of sound in the water is given by:

$$\mu_{vw}^2 = \mu_{eq}^2 + \left(c_{v_w}^{T}\right)^2.\mu_T^2 \quad (8)$$

The combined uncertainty for x (μ_x) (in meters) is thus expressed as:

$$\mu_x = \sqrt{c^2{}_{v_w}.\mu^2{}_{v_w} + c^2{}_{t_w}.\mu^2{}_{t_w} + c^2{}_{t_1}.\mu^2{}_{t_1}} \quad (9)$$

Based on the model for the group velocity (3), the following sensitivity coefficients are determined:

$$c_x = \frac{\partial v}{\partial x} = 2 \cdot (t_2 - t_1)^{-1} \quad (10)$$

$$c_{t_1} = \frac{\partial v}{\partial t_1} = -2x\,.\,(t_2 - t_1)^{-2} \quad (11)$$

$$c_{t_2} = \frac{\partial v}{\partial t_2} = 2x\,.\,(t_2 - t_1)^{-2} \quad (12)$$

The combined uncertainty of v (u_v) (m/s) is thus expressed as:

$$u_v = \sqrt{c^2{}_x.\mu^2{}_x + c^2{}_{t_1}.\mu^2{}_{t_1} + c^2{}_{t_2}.\mu^2{}_{t_2}} \quad (13)$$

3. Results

The phantom thickness is estimated by the pulse/echo technique as 10.302 mm (U = 0.018 mm; p = 0.95). Tables 1, 2 and 3 present the results of group velocity measured at each temperature, their respective expanded uncertainties (U) and coverage factors (k), from three repetitions of the complete measurement procedure carried out in three different days.

Figure 2 presents the group velocity results from the three repetitions of the measurement procedure all together. One can observe that group velocity tends to decrease linearly as temperature increase in the range of 19 ºC to 37 ºC.

4. Discussions and Conclusion

Many works have been assessing the velocity on bone phantoms and samples, but rarely they mention the temperature observed during the experiments [5][14][15][15]. Moreover, the measurement uncertainties of the published results are never declared.

Litniewski et al. [14] proposed a method for simultaneous assessment of thickness and longitudinal velocity of cortical bone phantoms and samples. The ultrasonic data were obtained at the frequency of 1.0 MHz using a single element focusing transducer. For bone phantoms, the authors mention relative errors for thickness of 5.9%, and 6.8% for the ultrasonic wave velocity.

Table 1. Results of group velocity measured at each temperature, their respective expanded uncertainties (U) and coverage factors (*k*) for the first repetition of the measurement procedure.

Temperature [ºC]	Group Velocity [m/s]	U [m/s]	*k*
19.1	2967	11	2.02
23.0	2948	10	1.97
26.9	2926.1	8.9	1.96
31.6	2909	10	1.96
33.8	2894.6	9.2	1.97
37.2	2875.3	8.8	1.96

Table 2. Results of group velocity measured at each temperature, their respective expanded uncertainties (U) and coverage factors (*k*) for the second repetition of the measurement procedure.

Temperature [ºC]	Group Velocity [m/s]	U [m/s]	*k*
20.9	2958	10	1.96
23.9	2944	10	1.96
27.0	2929	10	1.96
31.0	2909	10	1.96
34.0	2893	10	1.96
37.0	2873	10	1.96

Table 3. Results of group velocity measured at each temperature, their respective expanded uncertainties (U) and coverage factors (*k*) for the third repetition of the measurement procedure.

Temperature [ºC]	Group Velocity [m/s]	U [m/s]	*k*
19.5	2959.6	6.2	1.96
23.0	2948.5	6.1	1.96
26.9	2931.1	5.9	1.96
30.8	2908.7	6.5	2.0
33.9	2891.9	5.9	1.97
37.0	2877.4	5.8	1.97

McCarthy et al. [15] assessed the longitudinal velocity of bone samples from the mid-site of the third metacarpus of 20 horses. The ultrasonic data were obtained at the frequency of 2.25 MHz. The authors observed that over the temperature range of 4 ºC to 42 ºC, velocity varied inversely to temperature. Standard deviation of 3.5% is declared for the velocity measured.

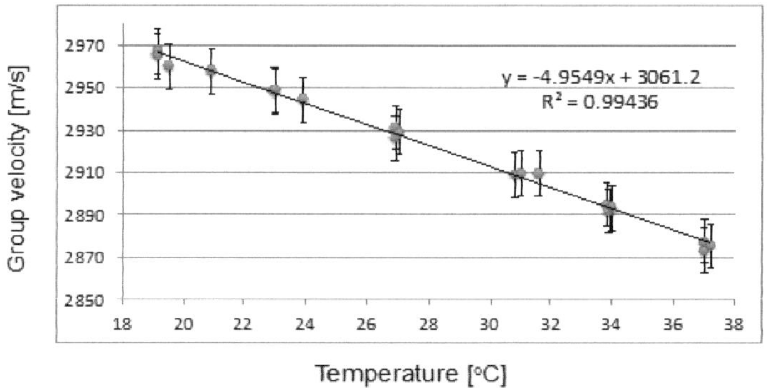

Figure 2. The group velocity results from the three repetitions of the measurement procedure together. Vertical bars indicate the expanded uncertainties.

El-Sariti et al. [15] measured the longitudinal velocity (@ 750kHz) in bovine bone marrow as a function of temperature between 17 °C and 44 °C. The authors pointed out a difference of 74 m s-1 in velocity between 20 °C, temperature normally used in vitro experiments, and 37 °C, temperature in which clinical measurements are carried out.

Sellani et al. [5] assessed the therapeutic ultrasound (TUS) heating distribution on commercial bone phantoms and in vitro femur and tibia human samples using thermographic images. An infrared camera captured images after 5 min of TUS stimulation (@ 1 MHz). The authors pointed out that temperature increased up to 8.2 °C and 9.8 °C for the femur and tibia, respectively. Moreover, they declared minor differences in mean and standard deviation temperatures between bone phantoms and samples. The authors concluded the commercial bone phantoms could be applied on thermal effects of ultrasound research.

In this work, we observe that group velocity on a cortical bone phantom tends to decrease as temperature increase. Moreover, an uncertainty model was presented to help assessing the differences in velocity for different temperatures. Finally, we conclude that different values of group velocity should be considered when temperature varies in simulations or experimental procedures, and this has obvious implications for the research results carried out in bone samples and phantoms.

References

[1] Sasso M, Haïat G, Yamato Y, Naili S and Matsukawa M. 2008 *Journal of Biomechanics* v. **41** pp 347–355.
[2] Laugier P and Haïat G. 2011 *Bone Quantitative Ultrassound*, 1° Ed. Nova York, Ed. Springer
[3] Abrunhosa V M, Soares C P, Possidonio A C B, Alvarenga A V, Costa-Felix R P B, Costa, M L P S and Mermelstein, C S. 2014 *Ultrasound in Medicine & Biology* v. **40** pp 504-512.
[4] Petrella L I, Maggi L E, Souza R M, Alvarenga A V and Costa-Felix R P B. 2014 *Ultrasonics* v. **54** pp 1476-1479.
[5] Sellani G et al. 2016 *Journal of Therapeutic Ultrasound* **4**:13.
[6] Silva C E R, Alvarenga A V and Costa-Felix R P B. 2011 *Journal of Physics. Conference Series* v. **279** pp 012025.
[7] Maggi L E, Silva C E R, Alvarenga A V and Costa-Felix R P B. 2011 *Journal of Physics. Conference Series* v. **279** pp 012029.
[8] Pichardo S, Sin VW and Hynynen V 2011 Phys Med Biol., 56 (1) pp 219–250.
[9] Lubbers J and Graaff R. 1998 *Ultrasound Med Biol.,* 24 (**7**) pp 1065–1068.
[10] JCGM 2008 Evaluation of Measurement Data – Guide of the Expression of Uncertainty in Measurements JCGM 100:2008.
[11] Costa-Felix R P B. 2006 *Measurement* v. **39** pp 169-175.
[12] Costa-Felix R P B and Machado J C. 2015 *Measurement* v. **69** pp 146-154.
[13] AIUM 1988 American Institute of Ultrasound in Medicine, Laurel, MD.
[14] Litniewski J, Tasinkevych Y, Podhajecki J and Falińska K 2015, IUS IEEE *International Ultrasonics Symposium* pp 1-4.
[15] McCarthy RN, Jeffcott LB and McCartney RN. J 1990 *Biomech.*, 23 (**11**):1139-43.
[16] El-Sariti AA, Evans JA and Truscott JG. 2006 *Ultrasound in Med. & Biol.*, 32 (**6**) pp 985–989.

Acknowledgments

Research reported in this paper has been partially supported by the Carlos Chagas Filho Research Support Foundation (Faperj, grant number E-26/201.563/2014), by the National Council for Scientific and Technological Development (CNPq, grant number 310.392/2014-4) and Pronametro/Inmetro.

IOP Conf. Series: Journal of Physics: Conf. Series **975** (2018) 012022 doi:10.1088/1742-6596/975/1/012022

The next GUM and its proposals: a comparison study

J C Damasceno [1,3], P R G Couto [2,3]

[1] Divisão de Metrologia de Materiais – Inmetro, Av. N. Sra. das Graças, 50, Duque de Caxias, RJ, 25250-020, Brazil
[2] Divisão de Metrologia Mecânica – Inmetro, Av. N. Sra. das Graças, 50, Duque de Caxias, RJ, 25250-020, Brazil
[3] Núcleo de Incerteza de Medição e Estatística Aplicada (Nimea) – Inmetro

jcdamasceno@inmetro.gov.br

Abstract. The Guide to the Expression of Uncertainty in Measurement (GUM) is currently under revision. New proposals for its implementation were circulated in the form of a draft document. Two of the main changes are explored in this work using a Brinell hardness model example. Changes in the evaluation of uncertainty for repeated indications and in the construction of coverage intervals are compared with the classic GUM and with Monte Carlo simulation method.

1. Introduction

A few decades ago, no universal guidance for the expression of measurement uncertainty existed. In this way, metrology laboratories in the world could not compare their results with good reliability. Since 1993, when the first edition of GUM (Guide to the Expression of Uncertainty in Measurement) was published, this scenario has changed. The GUM was really well received by metrology laboratories as it harmonized the way measurement uncertainty was evaluated.

The GUM received some minor revisions later in 1995 and again in 2008, when it was made free of charge as a JCGM (Joint Committee for Guides in Metrology) document [1]. GUM supplements [2-5] were also published since then to extend and generalize the applicability of the guide without changing the main document. However, the extension of the GUM concepts in its supplements caused an unavoidable inconsistency with the former GUM guide. The supplements incorporated a number of newer concepts, based on Bayesian analysis and numerical methods that do not coexist well with concepts exposed in the original GUM.

JCGM then proposed a new revision of the GUM and a first draft was circulated among national institutes of metrology in 2012. Remarkable changes were made that could affect the way laboratories deal with the measurement uncertainty results. This revision is still being discussed and some information about it has also been released elsewhere [6].

In this work, we discuss some of these changes and use an example to show their consequences.

2. The GUM and its new proposals

Two main proposals of the new GUM draft are discussed in this work: a correction in the uncertainty for repeated measurement indications and the construction of coverage intervals for the final measurand result.

2.1. Uncertainty for repeated indications

In the classic GUM, repeated indications are treated like having a normal distribution and the uncertainty u_R associated with it is evaluated by:

$$u_R = \frac{s}{\sqrt{n}} \tag{1}$$

where s is the standard deviation of the indicated values and n is the number of repeated indications. This evaluation is not consistent with the GUM supplement 1 [2], where repeated indications are treated as Student's *t*-distributions to account for the lack of degrees of freedom, or the low number of indications. The new proposal for the GUM is to consider repeated indications as *t*-distributions, just like in supplement 1. So its uncertainty would be evaluated as:

$$u_R = \left(\frac{n-1}{n-3}\right)^{1/2} \frac{s}{\sqrt{n}} \tag{2}$$

2.2. Coverage intervals

Once the combined uncertainty is obtained after propagation, the classic GUM uses a *t*-distribution to construct a coverage interval for a designed coverage probability. It uses the Welch–Satterthwaite equation to evaluate the effective degrees of freedom for the final distribution and then calculates a coverage factor k to expand the combined uncertainty u, thus:

$$U = ku \tag{3}$$

where U is the expanded uncertainty.

The draft for the new GUM proposal suggests that the final coverage interval cannot be reliably determined if only an expectation y and a standard deviation $u(y)$ is known, mainly if the final distribution deviates significantly from a normal or *t*-distribution. Thus, they propose distribution-free coverage intervals in the form $y \pm U_p$, with $U_p = k_p u(y)$: (a) If no information is known about the final distribution, then a coverage interval for the measurand Y for coverage probability of at least p is determined using $k_p = 1/(1-p)^{1/2}$. If $p = 0.95$, a coverage interval of $y \pm 4.47u(y)$ is evaluated. (b) If it is known that the distribution is unimodal and symmetric about y, then $k_p = 2/[3(1-p)^{1/2}]$ and the coverage interval $y \pm 2.98u(y)$ would correspond to a coverage probability of at least $p = 0.95$.

3. Case study: Brinell hardness measurement

A Brinell hardness test is used in this work as a case study for the application and discussion of the GUM and its new proposals. In this example, a sphere made from a very hard material is loaded onto the surface of the material sample. This procedure leaves an indentation mark with a diameter that is inversely proportional to the hardness of the material. The Brinell hardness BH is calculated in [kgf/mm^2] as:

$$BH = \frac{2F}{\pi D(D - \sqrt{D^2 - d^2})} \tag{4}$$

where F is the applied load in [kgf], D is the diameter of the sphere in [mm] and d is the diameter of the indentation mark in [mm].

In this particular case study, there are three input quantities, each of them with one source of uncertainty: (a) uncertainty of F is given by a certificate as 2%, with $k = 2$; (b) uncertainty of D is certified as 0.01 mm, with $k = 2$; and (c) uncertainty of d is given by the repeatability of 5 measurements.

In addition to the GUM and its new proposals, supplement 1 was also employed to further compare the results. Monte Carlo numerical simulations were performed using 200,000 iterations.

4. Results and discussion

Table 1 summarizes the values for the input quantities used for the Brinell hardness example.

Table 1. Input quantities for the Brinell hardness example.

Quantity	Value
Load (F)	3000 kgf
Sphere diameter (D)	10 mm
Indenter mark diameter (d)	3 mm

The standard uncertainties for the quantities were calculated as: $u(F) = 0.02 \times 3000/2 = 30$ kgf, for the load; and $u(D) = 0.01/2 = 0.005$ mm, for the sphere diameter. The indenter mark measurement was repeated 5 times and the average was found as 3.00 mm, with a standard deviation of 0.0791 mm. The standard uncertainty for d was calculated using the classic GUM approach (equation 1), as well as using the new GUM proposition (equation 2).

In order to check the impact of the sample size of the repeated indications (n) on the uncertainty propagation using both methods, all calculations were performed using n as 5, 10, 15, 20, 30 and 40. Results for the resulting combined uncertainty are shown on Figure 1.

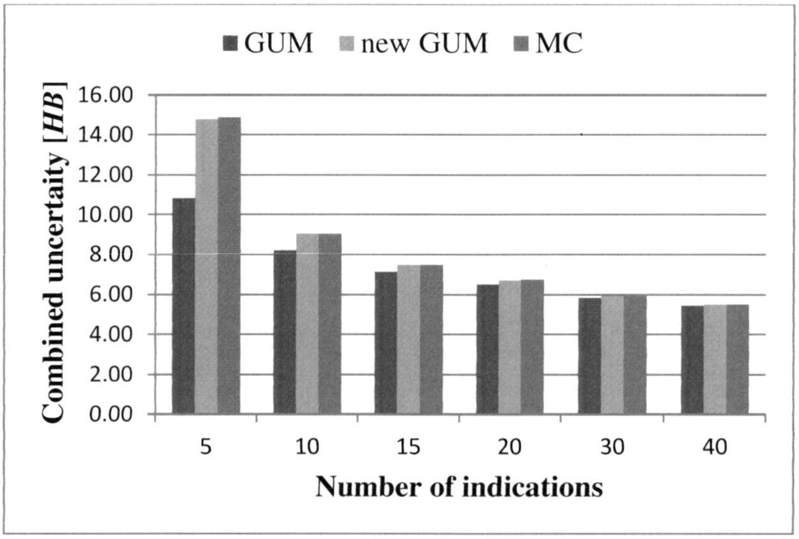

Figure 1. Results for the GUM, new draft GUM and Monte Carlo approaches for the combined uncertainty of Brinell hardness tests.

As can be noted, for all sample sizes the combined uncertainty obtained for the new draft GUM approach was very similar to the one obtained with the Monte Carlo simulation. On the other hand, for small sample sizes, the classic GUM approach result in underestimated combined uncertainties in comparison with the other two methods. This indicates that the new draft GUM proposal is really a better option then the classic GUM, particularly for small sample sizes, considering that the Monte Carlo simulation is the most realistic evaluation as it is free from many approximations due to the law of propagation of uncertainties.

Table 2 shows the comparison for the expanded uncertainties. In this case, confidence intervals were constructed as follows: (a) as per the classic GUM approach, using the effective degrees of freedom (*t*-distribution); (b) using the corrected equation for the repeated indications (equation 2) and multiplying the combined uncertainty by $k = 1.96$ (normal distribution); (c) using the new GUM proposal of $k = 2.98$ (unimodal and symmetric distribution); (d) using $k = 4.47$ (for a completely unknown distribution); and (e) using Monte Carlo simulation. In this last case, for the sake of comparison the expanded uncertainty was calculated as the semi-interval covered by the distribution endpoints for $p = 0.95$, considering a symmetric distribution.

Table 2. Expanded uncertainty results as a function of sample size for a Brinell hardness example. All values are in HB units.

n	Classic GUM (a)	New GUM $k = 1.96$ (b)	New GUM $k = 2.98$ (c)	New GUM $k = 4.47$ (d)	MC (e)
5	27.85	28.91	43.98	65.96	28.98
10	17.39	17.70	26.93	40.40	17.93
15	14.49	14.69	22.34	33.51	14.75
20	13.03	13.17	20.04	30.06	13.21
30	11.53	11.62	17.67	26.51	11.64
40	10.74	10.81	16.44	24.65	10.86

Table 2 shows that there is a good agreement between the GUM corrected for the sample size (b) and the Monte Carlo simulation (e), for all sample sizes. The values for the classic GUM (a) are also very close, but for small sample sizes there is a relatively small underestimation. This indicates that the use of the corrected equation (equation 2) for the *t*-distribution is a good choice for this model. Coverage intervals for the two proposals in the new GUM (c and d) overestimated the expanded uncertainty in comparison with the Monte Carlo result.

5. Conclusions

A revision of the GUM is in progress by the JCGM. The draft of the new version presents concepts that are more consistent with the GUM supplements and includes several new proposals.

Study of an example of Brinell hardness measurement in this work showed that the use Student's *t*-distribution for repeated indication of values in a type A uncertainty is a better option as it compensates small sample sizes in a better way than the use of the Welch–Satterthwaite formula to calculate effective degrees of freedom.

Coverage intervals proposed by the GUM draft are very conservative, raising the expanded uncertainty in least 50% in comparison with the Monte Carlo propagation.

References

[1] BIPM, IEC, IFCC, ILAC, ISO, IUPAC, IUPAP e OIML, Evaluation of Measurement Data – Guide to the Expression of Uncertainty in Measurement (GUM 1995 with minor corrections). Joint Committee for Guides in Metrology, JCGM 100:2008

[2] BIPM, IEC, IFCC, ILAC, ISO, IUPAC, IUPAP and OIML, Evaluation of Measurement Data – Supplement 1 to the Guide to the Expression of Uncertainty in Measurement – Propagation of distributions using a Monte Carlo method, Joint Committee for Guides in Metrology, JCGM 101:2008

[3] BIPM, IEC, IFCC, ILAC, ISO, IUPAC, IUPAP and OIML, Evaluation of Measurement Data – Supplement 2 to the Guide to the Expression of Uncertainty in Measurement – Models with any Number of Output Quantities, Joint Committee for Guides in Metrology, JCGM 102:2011

[4] BIPM, IEC, IFCC, ILAC, ISO, IUPAC, IUPAP and OIML, Evaluation of Measurement Data – An introduction to the Guide to the Expression of Uncertainty in Measurement and related documents, Joint Committee for Guides in Metrology, JCGM 104:2009

[5] BIPM, IEC, IFCC, ILAC, ISO, IUPAC, IUPAP and OIML, Evaluation of Measurement Data – The Role of Measurement Uncertainty in Conformity Assessment, Joint Committee for Guides in Metrology, JCGM 106:2012
[6] Bich W, Cox M and Michotte C, 2016, *Metrologia* **53** S149-S159

IOP Conf. Series: Journal of Physics: Conf. Series **975** (2018) 012023 doi:10.1088/1742-6596/975/1/012023

The state of the art of the impact of sampling uncertainty on measurement uncertainty

V J Leite[1] and E C Oliveira[2]
[1]Pontifical Catholic University of Rio de Janeiro - PUC-Rio, Posgraduate Programme in Metrology, Marquês de São Vicente Street, 225 - Gávea, Rio de Janeiro - RJ, Brazil; [2]Technology Management, Petrobras Transporte S.A., Rio de Janeiro – RJ, Brazil.

E-mail: vi.leit@gmail.com

Abstract: The measurement uncertainty is a parameter that marks the reliability and can be divided into two large groups: sampling and analytical variations. Analytical uncertainty is a controlled process, performed in the laboratory. The same does not occur with the sampling uncertainty, which, because it faces several obstacles and there is no clarity on how to perform the procedures, has been neglected, although it is admittedly indispensable to the measurement process. This paper aims at describing the state of the art of sampling uncertainty and at assessing its relevance to measurement uncertainty.

1. Introduction

Lately, estimating the measurement uncertainty with higher reliability has been a goal, in order to guarantee a measurement process with quality. The measurement process starts with the selection of the sample, so the measurement uncertainty is constituted by the association of the uncertainty from both the sampling and the analytical processes.

The whole process involving sampling and its uncertainty has been neglected by analytical chemists and metrologists due to the lack of knowledge regarding the impact on the quality of the result [1-2]. Thus, everything that involves the control and supervision of this step, and which allows assessing the associated uncertainty, is rarely considered, even with ISO/IEC 17025 stating that sampling is an element of paramount importance for total measurement uncertainty [4].

The difficulties found in the sampling process usually mean that only the analytical uncertainty process is computed for the measurement uncertainty. This approach can offset the value of the uncertainty informed from its real value, since it underestimates the measurement uncertainty by disregarding the sampling uncertainty. This in turn leads to decisions based on quality control and assurance that not always are the most reliable. It is imperative that there are methodologies which are capable of estimating the measurement uncertainty taking into account each part of the measurement process and all associated uncertainties [5].

Thus, this article aims at presenting the state of the art of sampling uncertainty with regard to measurement uncertainty, examples that proves the need to consider both sampling uncertainty and analytical uncertainty in the process of measurement uncertainty and pointing out the possible gaps to be filled.

2. Measurement uncertainty

Measurement uncertainty is a "non-negative parameter that characterizes the dispersion of values assigned to a measurand, based on the information used." For this, it is necessary to know the sampling target with its location in space and time [5,6].

The simplest way to categorize the various sources that can influence the uncertainty is to separate them into two groups: sampling uncertainty and analytical uncertainty.

Although they have the same purpose with well-defined roles, the estimates of each uncertainty group are independent. Thus, the sampling manager should be able to reliably define the sampling uncertainty and the analyst process manager must be able to define the analytical uncertainty, composing the total measurement uncertainty [5].

The standard measurement uncertainty, u, can be estimated by [8]

$$u = S_{med} = \sqrt{S_{sample}^2 + S_{anal}^2} \quad (1)$$

where S_{med} is the estimate of the measurement uncertainty; S^2_{sample} is the estimate of the variance of the sampling and S^2_{anal} is the estimate of the analytical variance.

The standard JCGM 100:2008 [6] cites two approaches, empirical (top-down) and modeling (bottom-up), to quantify the sources of uncertainty. Oliveira and Aguiar [9] present three more, namely Barwick and Ellison, study of variability and fuzzy arithmetic, among others.

The measurement uncertainty value cannot be expressed in a more reliable way if the characteristic part of the sampling uncertainty is missing, in other words, considering only the analytical uncertainty, regardless of the procedure used by the laboratory. However, since the sampling uncertainty is not recognized, it is performed carelessly by inexperienced professionals, although this uncertainty may significantly interfere with the final result [7].

For the analytical chemist with poor knowledge about quality management systems, the uncertainty of interest is the one coming only from the analytical process; however, for the end user and the customer, the uncertainty of interest is the one compounded by all the uncertainties in the measurement process [1,7].

3. State of the art

The demand for studies involving sampling uncertainty is not yet expressive, since conducting research in the area is time and cost consuming. The literature reports the importance of the uncertainty evaluation regarding the sampling process, however, not showing how significant it may be.

The works carried out in the sampling sector are recent and did not delve into this issue. It has been convenient to consider the sampling target only as a representative sample of the population, and don't describe or express the representativeness of the sample and how relevant it is. Thompson goes so far as to use the expression "awakening the sleeping dogs of sampling" [10] to describe how difficult it is to consider this process since there are no clear guidelines and professionals interested in developing research.

From 2001 to 2004, the National Environmental Protection Agency of Italy (ANPA) funded a project to assess practically the uncertainty of sampling [4].

In 2005, the sampling workshop was initiated at the request of the Nordic Innovation Center aimed at highlighting and emphasizing the importance of adequate sampling for environmental or chemical purposes, promoting actions to improve sample reliability, among other actions [7].

In 2007, a seminar about sampling uncertainty in Denmark was held to disseminate a manual on sampling and uncertainty assessment. A set of techniques and tools available for the calculation and control of uncertainty related to sampling was presented [5].

To further justify this relevant subject, two examples have been drawn from the Eurachem and are discussed in [8], both from the food sector.

In the first example, the ion nitrate measurement process in greenhouse grown lettuce is discussed. In this study, a "bay" of up to 20,000 heads of lettuce was used to estimate the ion nitrate concentrations. Considering each "bay" as a sampling target, 10 heads of lettuce were removed from each "bay" for sample. With an empirical approach (top-down) and duplicate method, the calculations were made from robust ANOVA. So, this study presented 22.6% as contributions from the sampling and 6.3% from the analytical process, in a combined measurement uncertainty of 28.9%, being the sampling a dominant factor, responsible for 78.2% of the measurement variance.

In the second example, evaluation of Vitamin A concentration in baby porridge, with fruits and ground cereals, it has been estimated the measurement uncertainty and the contributions of sampling and analysis. For the sampling target, 10 different batches of porridge were determined, each batch being 1,092 kg and, as a sample, a batch of approximately 400 g powder. Two samples were taken per batch and two portions of approximately 40-50 g were sampled from each sample. With an empirical approach (top-down) and duplicate method, by means of one way ANOVA, and with 95% confidence level, the expanded uncertainty related to the sampling is of 10%, while for the analytical step is of 17% [8].

It was observed that the first example presented a sampling uncertainty more representative than the analytical uncertainty and the second one showed an analytical uncertainty more representative than a sampling uncertainty, being the latter still relevant. Thus, it is not possible to predict which uncertainty step will have the greater representativeness, so it is essential the calculation of measurement uncertainty considering both steps.

4. Breaking paradigms

In some areas, sampling uncertainty is more representative than analytical uncertainty, in special when it involves raw materials (natural or industrial) that are being tested [10].

In the scientific community, it is still common practice to ignore sampling since there is no clear description of how to measure its uncertainty, only the measurement uncertainty from the post-entry process of the material in the laboratory [4].

There are procedures called "best practices" or "suitable for the purpose", which are intended to describe and facilitate the process; however, these are trivial, not having the depth required in practice for the attainment of uncertainty.

There is no detailed description of how to perform a sampling procedure, that is technically complicated when not instructed, and whose validation is much less developed than that of the analysis. That is, there are few regulations that require sampling to be accredited, both in Brazil and abroad and the assumptions for quality assurance when required, are derisory [5].

Developing surveys or adding reliability to the entire measurement process from planning, sampling to completion of the analytical process requires investment. However, by reducing uncertainty, with the possibility of controlling the quality of the process, it could generate much greater benefits [1,10]. Thompson shows industrial cases where investment in the sampling process generated a substantial net gain due to the importance given to sampling errors. Therefore, knowing and defining well each contribution of uncertainty leads to quality control and its efficiency can be guaranteed.

5. Conclusion

It is necessary to develop research and projects that aim at estimating the uncertainties, focusing to the process of the sampling uncertainty with regard to measurement uncertainty.

When one knows all the phases that compose the uncertainty, one can control and evaluate where, in what particularity of the process it is more profitable to invest for profit-seeking. Management process becomes more tangible.

However, there is a need for investment from interested organizations involved in the process, breaking paradigms for metrologists and analytical chemists, in order to leave their comfort zone and deepen in this area.

As future work, other real examples related to effluent, food, oil and ore industries are being studied to establish patterns of the impact of sampling uncertainty on measurement uncertainty for each matrix and specific analytes.

6. References

[1] Thompson M 1998 Uncertainty of sampling in chemical analysis, **3**, Accred Qual Assur, pp 117-21.

[2] Guigues N, Desenfant M, Lalere B, Vaslin-Reimann S, Eyl D, Mansuit P and Hance E 2016 Estimating sampling and analysis uncertainties to assess the fitness for purpose of a water quality monitoring network, **21**, Accred Qual Assur, pp 101-12.

[3] ISO/IEC 17025:2005 *General requirements for the competence of testing and calibration laboratories.*

[4] Zorzi P, Belli M, Barbizzi S, Menegon S and Deluisa A 2002 A practical approach to assessment of sampling uncertainty, **7**, Accred Qual Assur, pp 182-8.

[5] Grøn C, Hansen J B, Magnusson B, Nordbotten A, Krysell M, Andersen K J and Lund U 2007 Uncertainty from sampling: workshop to launch a Nordtest handbook on sampling uncertainty estimation and control, **12**, Accred Qual Assur, pp 377-81.

[6] JCGM 100:2008 *Guide To The Expression Of Uncertainty In Measurement.* ISO, Geneva, (1993) (Reprinted 1995: Reissued as ISO Guide 98-3).

[7] Schreiber M P, Komppa V, Wahlström M and Laine-Ylijoki J 2006 Chemical and environmental sampling: quality through accreditation, certification and industrial standards, **10**, Accred Qual Assur, pp 510-4.

[8] Ramsey M H and Ellison S L R 2007 Measurement uncertainty arising from sampling: A guide to methods and approaches. Eurachem/CITAC Guide, Eurachem.

[9] de Oliveira E C and Aguiar P F 2009 Comparação de diferentes abordagens para avaliação da incerteza na cromatografia gasosa do gás natural, **32**, Quim. Nova, pp 1655-60.

[10] Ramsey M H and Thompson M 2007 Uncertainty from sampling, in the context of fitness for purpose, **12**, Accred Qual Assur, pp 503-13.

IOP Conf. Series: Journal of Physics: Conf. Series **975** (2018) 012024 doi:10.1088/1742-6596/975/1/012024

Top down arsenic uncertainty measurement in water and sediments from Guarapiranga dam (Brazil)

M G Faustino, C N Lange, L R Monteiro, H A Furusawa, J R Marques, T B Stellato, S M V Soares, T B S C da Silva, D B da Silva, M E B Cotrim and M A F Pires

Instituto de Pesquisas Energéticas e Nucleares, IPEN-CNEN/SP, São Paulo, Brazil

E-mail: mainarag@usp.br

Abstract. Total arsenic measurements assessment regarding legal threshold demands more than average and standard deviation approach. In this way, analytical measurement uncertainty evaluation was conducted in order to comply with legal requirements and to allow the balance of arsenic in both water and sediment compartments. A top-down approach for measurement uncertainties was applied to evaluate arsenic concentrations in water and sediments from Guarapiranga dam (São Paulo, Brazil). Laboratory quality control and arsenic interlaboratory tests data were used in this approach to estimate the uncertainties associated with the methodology.

1. Introduction

Despite the focus of interest, understanding some concepts can be worth and add value to the work. When analyte concentration is reported with its appropriate analytical method uncertainty, much more information is provided than a simple replicate standard deviation. It does not matter now, but analytical laboratories realized the benefits or were forced to apply the concepts of analytical measurement uncertainty in reporting analyte concentration. After several years of harmonization documents have been published, some measurement uncertainties approaches pop up such as bottom-up, top-down, based on validation studies, empiric and *ad-hoc* methods, the very popular control charts, manufacturers specifications, and others complex theoretical models. Some with general application, others targeting specific situation, and others as a time saving approach [1, 2]. In the TR 537 Ed. 3.1 handbook by Nordtest, Magnusson *et al.* [3] used internal and external quality control data for ammonia assessment in water and estimated the analytical data uncertainties based on a top-down approach. That procedure also checked if the method performance is under control with no significant change or bias. In this study, the top-down uncertainty estimation approach was considered in analyzing analytical data from arsenic concentration in water and sediment from Guarapiranga dam, São Paulo, Brazil. Arsenic values measured at Guarapiranga dam associated to its expanded uncertainty were compared with Brazilian legal requirements to water and to bottom sediment.

2. Methods/Experimental

2.1. Arsenic measurement and quality control

Sample preparation and analytical measurements were carried out at Laboratório de Análises Química e Ambiental, Centro de Quimica e Meio Ambiente in IPEN-CNEN, São Paulo, Brazil. Since 2010, this laboratory has been working under ABNT NBR ISO/IEC 17025 standard principles. Total arsenic was measured by GF-AAS (AAnalyst/HGA 400, Perkin-Elmer, USA) at 193.7 nm as presented at [4] and [5].

2.2. Top down uncertainty estimation
Based on the described in the handbook by Magnusson *et al.* (2011) [6], top-down approach was used to estimate the expanded uncertainty in measuring arsenic in water and sediment. Two components were considered:

2.2.1. Repeatability within laboratory. From 2011 to 2013, arsenic measurements were conducted in NIST 1643e Standard Reference Material – Trace Elements in Water with certified value of 60.45±0.72 μg L^{-1} (U, k = 2) for arsenic together with routine water samples analyses. Control chart with obtained arsenic results from this RM was part of laboratory quality system performance, and was used in the uncertainty estimation of repeatability standard uncertainty, u_{Rw}. In routine analyses, information from repeatability is attributed to short-term variations on uncertainty.

2.2.2. Method and laboratory bias. Uncertainty second component used in this approach was the laboratory bias or bias standard uncertainty, u_{bias}, obtained from 2010 to 2013 performance in interlaboratory tests (n = 10), with a high number of participant laboratories (m = 50). Z-score value is the most common tool to assess the laboratory technical competence no matter the target value is a conventional quantity value or an agreement to a consensus value between the reported values by the participant laboratories. All percentage bias results (not z-score values) for arsenic concentration were considered in this component, where RMS_{Bias} and u_{Cref} are calculated as in (1) and (2).

$$RMS_{Bias} = \left(\sum bias^2 /n\right)^{1/2} \qquad (1)$$

$$u_{Cref} = \frac{S_r}{m^{1/2}} \qquad (2)$$

2.3. Water and bottom sediment sample treatment
Water samples were acidified to pH < 2 and directly measured. Bottom sediment samples were dried for 5 days at 40-60 °C and 0.5 g of each sample was twice digested accordingly US-EPA 3051 method [7] lead to 50 mL final volume. Sediment samples were treated as described by [5].

3. Results and discussion
Uncertainty component, u_{Rw}, associated to repeatability within laboratory was obtained from the control chart of arsenic values in NIST 1643e (n = 60, p = 0.05) measured before each sample batch with 15 % control limits. The estimated u_{Rw} for that standard along three years of routine conditions measurements is 4.5 μg L^{-1} (7.5 %). Once the arsenic certified value is 60.45 μg L^{-1} with an expanded uncertainty of 0.72 μg L^{-1} (k = 2), the obtained u_{Rw} was considered in the uncertainty calculation. Measurement of total arsenic in water was evaluated by laboratory regular participation in proficiency tests (PT) organized by Rede Metrologica do Rio Grande do Sul [8]. Laboratory Z-score performance from PT, in two (X and Y) rounds, from 2010 to 2013, are presented in figure 1. Consensus value, robust standard deviation and laboratory bias (in μg L^{-1} and %) are presented in table 1.

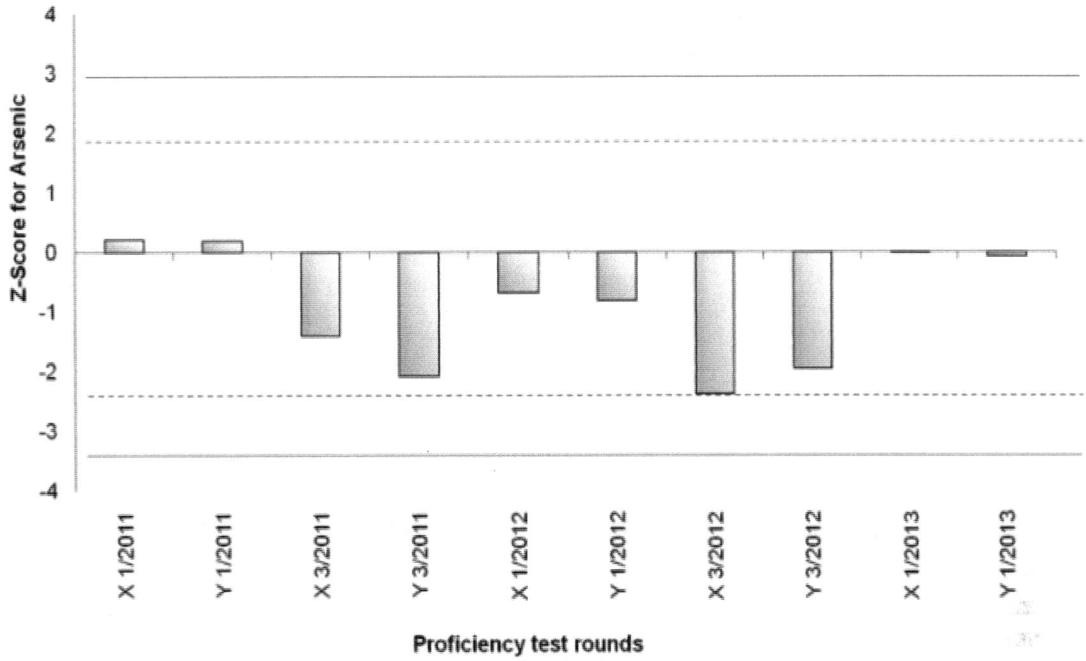

Figure 1: Z-Score for arsenic measurements by GF-AAS in PT rounds from 2011 to 2013.

Table 1. Arsenic interlaboratory results.

Total As	Value ($\mu g\ L^{-1}$)	S_R ($\mu g\ L^{-1}$)	Bias ($\mu g\ L^{-1}$)	Bias (%)	Z-Score
X1/2011	40	8	7	18	0.2
Y1/2011	60	6	0	0	0.2
X3/2011	20	20	-10	-50	-1.4
Y3/2011	45	25	-30	-67	-2.1
X1/2012	30	15	-5	-17	-0.7
Y1/2012	50	30	-10	-20	-0.8
X3/2012	20	7	-17	-85	-2.4
Y3/2012	50	11	-22	-44	-1.9
X1/2013	40	20	0	0	0.1
Y1/2013	80	20	0	0	0.1
$\bar{X}$	43	16	9	26	0.9

From table 1 data, laboratory root mean square bias (RMS_{Bias}) was estimated as 5.9 %. Reference value or nominal value uncertainty u_{Cref} was estimated as PT robust standard deviation (m = 50, p = 0.05) as the second u_{bias} component. u_{Cref} was estimated as 3.7 % and this value was considered also adequate to the measurement purpose. It is important to mention that u_{Cref} component is very dependable of PT provider, how the provider establishes consensual average and robust standard deviation. If these components are evaluated as too large, the laboratory solely action must be the change of PT program to one with smaller S_R. These components combined uncertainty (u_c) corresponded to 7.9 %. To complete uncertainty estimation see table 2.

Table 2. Data from uncertainty estimation from As in water and bottom sediment by GF-AAS.

Action	As in water and bottom sediment by GF-AAS
Specify measurand	Total As measurement by GF-AAS with a ±7.5 % from LQM to 100 µg L^{-1} concentration range
Quantify R_w component: A - Control Chart B - Steps not covered by Control Chart	A: Control limit set as ±4.5 µg L^{-1} or ±7.5 % (95 % confidence limit) B: The control sample includes all analytical steps.
Quantify Bias component	Interlaboratory proficiency tests results from 2011 to 2013 are presented at table 1. The root mean square (RMS) of bias was 5.9 %. The uncertainty of nominal values is $u_{Cref} = 3.7$ %
Convert components to Standard uncertainty $u(x)$	$u_{Rw} = \pm 7.5 \%/2 = 3.75 \%$ * $u_{bias} = \text{Raiz}(RMS_{bias}^2 + u_{Cref}^2) = \text{Raiz}(5.9^2+3.7^2) = 6.96 \%$
Calculate combined standard uncertainty, u_c	$u_c = \text{Raiz}(u_{Rw}^2 + u_{bias}^2) = \text{Raiz}(3.75^2+6.96^2) = 7.9 \%$
Calculate Expanded uncertainty, $U = 2 \times u_c$	$U = 2 \times 7.9 \% = 15.8 \% \sim 16 \%$

*from control chart

3.1. *Total As content in water samples*

In this study, total As measurement expanded uncertainty in water samples considering top-down approach in the concentration range of 20 to 80 µg L^{-1} was estimated as 16 % (U, $k = 2$), which is two to three times higher than expanded uncertainty found with bottom-up approach [2]. Water Framework Directive, European requirements and directives regarding drinking water, groundwater, freshwater, and coastal seawater, establishes that the maximum standard expanded uncertainty with 10 % precision and 10 % trueness (As requirement conditions at 10 µg L^{-1} concentration range) should be 15 %. Therefore, according to this requirement, the top-down uncertainty estimation approach would not be suitable. However, Magnusson et al. [6] present data that shows As requirements are frequently not met in several water sample PT.

3.2. *Total As content in bottom sediment*

In 14 dried bottom sediments samples, collected at Guarapiranga dam, presented a Total As concentration from 2.1 to 8.5 mg kg^{-1} (figure 2).

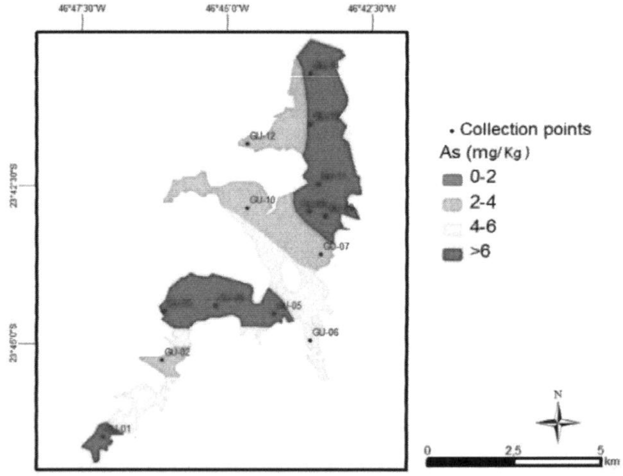

Figure 2. Total As in 14 sediment samples of Guarapiranga dam.

With an associated expanded uncertainty of 0.2 mg kg^{-1} (95 % confidence interval; 14 degrees of freedom and k = 2). For sediment samples, reproducibility within-laboratory standard deviation was considered a good uncertainty estimation. This uncertainty value was considered adequate to bottom sediment samples when compared with: Canadian Sediment Quality guidelines to fresh water sediments are ISQG-interim sediment quality guideline is 5.9 mg kg^{-1} and PEL – probable effect level is 17 mg kg^{-1}. The maximum As uncertainty associated with Guarapiranga 33 core sediment samples digestion and dilution step (u_{dil}) was estimated as *U 12 %, (k* = 2) for the working range of 0.5 to 20 mg kg^{-1} [5].

3.3. Legal compliance

Considering that 10 μg L^{-1} of total As in water is the current legal limit [9], the concentration range (20 to 80 μg L^{-1}) must be improved to reach lower values in the range of 10 μg L^{-1}. In addition, expanded uncertainty estimated by top-down approach in the present scenario did not fit the purposed use of Total As in water measurement.

Considering As Interim Sediment Quality Guideline (ISQG) value of 5.9 mg kg^{-1} in sediments [10] the current method expanded uncertainty of 0.2 mg L^{-1} (U, k = 2) is appropriated to the intended use of the Total As measurement. Any larger dispersion must be related to the sample inhomogeneity. The largest uncertainty contribution to u_c was by far RMS_{Bias}. Therefore, laboratory performance improvement in further PT´s must be a goal to reduce total As measurement uncertainty.

4. Conclusion

Considering that water and sediment samples collection represent a good estimate of the entire Guarapiranga dam scenario for arsenic, top-down procedure to estimate uncertainty on total As measurement was much simpler than bottom-up approach. With top-down procedure, the measurement was proved stable and under controlled conditions. Estimated uncertainty to measure total As in bottom sediment was adequate to the concentration range found at Guarapiranga dam and to meet Brazilian legal requirements and Canadian sediment quality guidelines. In order to comply with current legal requirements, uncertainty estimation on As measurement in water by top-down approach was not adequate, and the lower concentration range must be improved, once concentrations < 20 μg L^{-1} were not reached. However, this could be associated with top down uncertainty estimation procedure, which usually cover many more uncertainty components.

Acknowledgments

Authors wishing to acknowledge assistance from Mr. Stephan and also financial support from FAPESP Project n° 2009/53898-9 and CNPq process n° 134537/2013-0.

References

[1] Ellison L R and Willians A 2012 Eurachem/CITAC guide: *Quantifying Uncertainty in Analytical Measurement* Available from https://www.eurachem.org/index.php/publications/guides/quam (United Kingdow: Eurachem) 3° ed.

[2] Avila A, Araujo T, Couto P and Borges R M 2005 *Experimental design applied to the optimization of pyrolysis and atomization temperatures for As measurement in water samples by GFAAS* Metrologia Available from http://iopscience.iop.org/0026-1394/42/5/005 (United Kingdow: Institute of Physics Publishing) Vol **42** p 368-375

[3] Magnusson B, Naykki T, Hovind H and Krysell M 2012 *Handbook for calculation of measurement uncertainty in environmental laboratories - Report NT TR 537* Available from http://www.nordtest.info/index.php/technical-reports/item/handbook-for-calculation-of-measurement-uncertainty-in-environmental-laboratories-nt-tr-537-edition-3.html (Espoo, Finland: Nordest) 3.1 ed.

IOP Conf. Series: Journal of Physics: Conf. Series **975** (2018) 012024 doi:10.1088/1742-6596/975/1/012024

[4] Faustino M G, Silva D, Monteiro L, Dantas E, Pires M and Cotrim M 2013 *Regulatory quality control in the metal and semi metal environmental monitoring program at IPEN/CNEN-SP* International Nuclear Atlantic Congress ABEN (Recife PE Brazil)

[5] Simoes G, Jesus T, Ortiz N, Dantas E, Monteiro L, Bicudo C and Pires M 2012 *Trends of metal pollution evaluated by Guarapiranga sediment testemonial* SHEWC - Safety, Health and Environment World (Sao Paulo Brazil) pp 44-8

[6] Magnusson B and Koch M 2011 *3.07– Measurement Quality in Water Analysis* Treatise on water Science Elsevier Vol **3** pp 153-169

[7] US EPA 2007 *Method 3051: Microwave assisted acid digestion of sediments, sludges, soils and oils* Test Methods For Evaluating Solid Waste (Washington DC: US Environmental Protection Agency)

[8] Rede Metrologica RS *Ensaios de Proficiência* February 14 2013 Available from http://www.redemetrologica.com.br/2010/# (Porto Alegre Brazil)

[9] US EPA 2001 *National primary drinking water regulations: arsenic and clarifications to compliance and new source contaminants monitoring Final Rule* Code of Federal Regulations (Washington DC: US Environmental Protection Agency) pp 141-142

[10] Canadian Council of Ministers of The Environmental 1999 *Canadian sediment quality guidelines for the protection of quality life: Arsenic* Available from http://ceqg-rcqe.ccme.ca/download/en/230/ (Winnipeg Canada)

IOP Conf. Series: Journal of Physics: Conf. Series **975** (2018) 012025 doi:10.1088/1742-6596/975/1/012025

Ultrasonic properties of a four years old tissue-mimicking material

R M Souza, R M Monteiro, R P B Costa-Felix and A V Alvarenga

Laboratory of Ultrasound (Labus), Directory of Scientific and Industrial Metrology (Dimci), National Institute of Metrology, Quality and Technology (Inmetro), Brazil, RJ.

E-mails: raquel.ms.eng@gmail.com; rebecasouzam@gmail.com; rpfelix@inmetro.gov.br; avalvarenga@inmetro.gov.br

Abstract: This paper presents the metrological study of long-term stability of the ultrasonic properties of a tissue-mimicking material (TMM) manufactured in 2013. The thickness, speed of sound and attenuation coefficient of TMM were estimated by the pulse-echo technique. The Student's paired t-test was applied and the averages of the speed of sound and the attenuation coefficient in the four years-old TMM did not differ statistically from the reference values presented in the standard.

1. Introduction

Tissue-mimicking materials (TMM) have been manufactured for characterization, performance testing and calibration of ultrasonic diagnostic and physiotherapy equipment. According to the standard IEC 60601-2-37 [1], the TMM intended for these applications shall have thermal and ultrasonic properties similar to human soft tissue. The most critical ultrasonic properties of TMM are the speed of sound, acoustic impedance, attenuation coefficient, backscattering coefficient and nonlinearity parameter [2].

According to Culjat et al. [2], the longevity is also important for TMM, or the period of time over which the ultrasonic and mechanical properties are stable and consistent. Longevity can vary widely, depending on the selected materials and preparation technique.

That said, there are some works in literature concerning for long-term stability of mechanical, elastic and ultrasonic properties of TMM [2][5].

Soft tissues are composed of muscles, tendons, ligaments, fascia, fat, fibrous tissue, synovial membranes, nerves and blood vessels [2][6]. There are a variety of soft tissue-mimicking materials, as agarose-based, gelatin-based, zerdine, among others.

In 2013, Petrella et al. [7] fabricated agar TMM to study the influence of subcutaneous fat in surface heating of ultrasonic diagnostic transducers, following the recipe described in [1]. After that, Souza et al. [8] present a more detailed standard operating procedure (SOP) to prepare agar TMM. Since then, some works in the literature [9][10] used the SOP described in [8] for fabrication of agarose phantoms with different objectives.

It is important to mention that the standard [1] describes the way the TMM should be stored and maintained, as well as the ultrasonic and thermal properties of tissues and materials. Also, according to [1], TMM shelf life if it is preserved without air contact should be at least one year.

Considering the foregoing, this paper presents the metrological study of long-term stability of agarose phantom's ultrasonic properties (speed of sound and attenuation coefficient) fabricated in 2013 (see [7]). This study was performed by comparing the ultrasonic properties of a four years-old TMM's [7] with those described in [1].

2. Materials and methods

2.1. Agar phantom

The soft TMM proposed in [1] has appropriated ultrasonic properties similar to that of human soft tissue and are presented in Table 1. The TMM fabricated in 2013 is presented in figure 1.

Table 1. TMM's ultrasonic properties proposed in [1].

Speed of sound [m s^{-1}]	Attenuation Coefficient [dB cm^{-1} MHz^{-1}]
1540	0.5

2.2. Measurement system

The measurement system was composed by: an arbitrary waveform generator model 33250A (Agilent, USA), an oscilloscope model DSO-X 3012A (Agilent, USA), a 5 MHz central frequency transducer model A309S (Olympus-NDT, USA) [11][12], a cylindrical water tank (Ø 300 mm × 110 mm) filled with deionized water, and a stainless steel cylindrical reflector (Ø 63 mm × 10 mm). A positioning system with three linear and two goniometric stages was used for transducer alignment to maximize the ultrasonic reflected pulse.

2.3. Measurement procedure

The pulse-echo technique was used to perform the speed of sound and attenuation coefficient measurements. The signals were acquired using the oscilloscope, and the time of flight measurements were carried out using the oscilloscope's electronic cursors. The transducer (T_X) was positioned pointing directly and normal to the front surface of reflector (stainless steel) to acquire a reference signal. Then, the time of flight in the water (t_w), without the TMM, was measured (see figure 1 for illustrative representation).

Figure 1. Agar TMM fabricated in 2013 (42 mm × 49 mm × 42 mm).

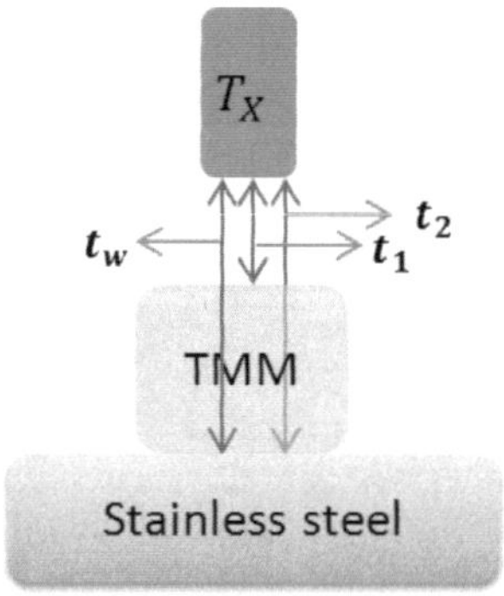

Figure 2. Pulse-echo technique.

Subsequently, for speed of sound calculation, the time of flight on the surface of the TMM (t_1) and the time of flight in reflector with the TMM inserted (t_2) were measured (see figure 2). For the attenuation coefficient calculation, besides all procedure previously described, it was necessary to measure the amplitude of the signals mentioned.

2.4. Measurement uncertainty
The measurement uncertainty was expressed based on the BIPM/JCGM:100 (GUM) [13][14][15]. The sources of uncertainty considered on the measurement of the speed of sound and attenuation coefficient are described in 2.4.1 and 2.4.2.

2.4.1. Speed of sound. The speed of sound (c) is calculated by (1):

$$c = \frac{2\,.\,d}{(t_2 - t_1)} \tag{1}$$

in which c is the speed of sound, d is the thickness of the TMM in meters and is calculated by (2):

$$d = c_w.\left(\frac{t_w - t_1}{2}\right) \tag{2}$$

in which c_w is the speed of sound in the water and is calculated as a function of temperature as described in [16]. The contribution of each source of uncertainty is presented in the Ishikawa's diagram in figure 2.

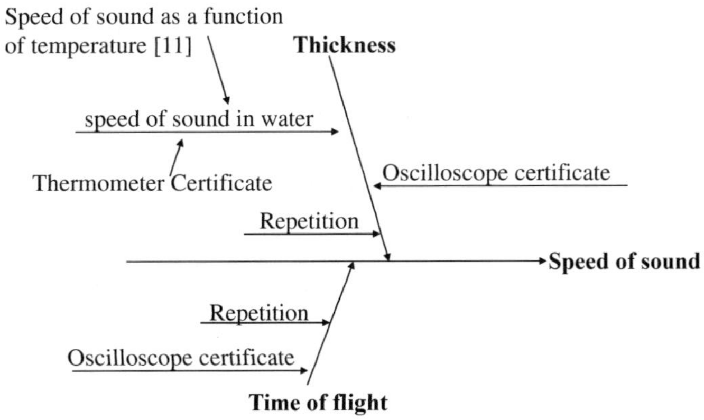

Figure 3. Ishikawa's diagram presenting the sources of uncertainty for speed of sound.

The following sources of uncertainty were considered with respect to the measurement of the group velocity of the TMM: the uncertainty of the temperature measured during the water bath experiments (0.01 ºC), leading to an uncertainty of 0.03 m s^{-1} in the speed of sound; the uncertainty of the determination of the ultrasound propagation velocity in water (0.2 m s^{-1}) [16]; the uncertainty of the oscilloscope time base (2.4×10^{-8} s), given an uncertainty of 0.7 m s^{-1}; the uncertainty of the measurement of phantom's thickness (0.02 mm), contributing with an uncertainty of 0.7 m s^{-1}; and the dispersion of speed of sound measurements led to an uncertainty of 0.091 m s^{-1}. By combination of these figures, the combined uncertainty estimated for speed of sound was 1.0 m s^{-1}.

2.4.2. Attenuation coefficient. In a simple way, attenuation coefficient was determined as the ratio between the amplitudes of the signals reflected over the reflector with and without the TMM. The reflection coefficient of the reflector, as well as the attenuation in the water, was considered in calculation.

The contribution of each source of uncertainty in estimating the attenuation coefficient is presented in the Ishikawa's diagram in figure 3, and the following sources of uncertainty were considered: the uncertainty of the attenuation coefficient in the water (4.4×10^{-4} dB cm^{-1}); the uncertainty of the TMM thickness (0.02 mm), leading to an uncertainty of 1.2×10^{-3} dB cm^{-1}; the reflection coefficient in the stainless steel cylindrical reflector, which accounts for an uncertainty of 5.8×10^{-5} dB cm^{-1}; the uncertainty of the oscilloscope time base (2.4×10^{-8} s), given an uncertainty of 2.2×10^{-4} dB cm^{-1}; the uncertainty concerning the signal amplitude measurement (5.2×10^{-3} V), contributing with an uncertainty of 1.8×10^{-3} dB cm^{-1}. Finally, the dispersion of attenuation coefficient measurements led to a combined uncertainty of 0.028 dB cm^{-1}, which was the most important source of uncertainty.

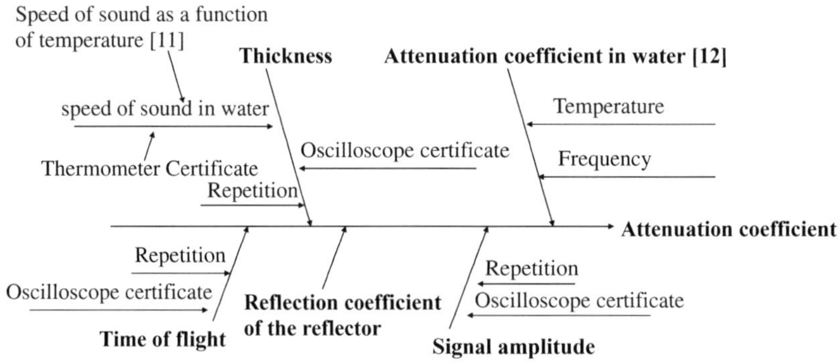

Figure 4. Ishikawa's diagram presenting the sources of uncertainty for attenuation coefficient.

2.4.3. Significance t-test. The t-test was applied to assess the statistical significance between the measurement results of the speed of sound and attenuation coefficient concerning the values presented in [1]. The t-test was applied to assess the statistical significance between the measurement results of the speed of sound and attenuation coefficient concerning the values presented in [1].

$$t_{calc} = \frac{X - \mu_O}{u_c} \tag{3}$$

in which t_{calc} (3) is the test statistic and u_c is the combined uncertainty. Based on a defined significance level ($\alpha = 0.05$) and the number of effective degrees of freedom, a test critical value (t_{crit}). If $t_{calc} < t_{crit}$, H_0 can not be rejected.

3. Results

The phantom's thickness (e) was estimated as 42.817 mm (U = 0.066 mm; p = 0.95; k = 2.2). Table 2 provides the result for speed of sound (c), whilst table 3 presents the result for the attenuation coefficient (a_c). For both, type A (u_A) and type B (u_B) standard uncertainties, coverage factor (k) and expanded uncertainty (U) are also presented. Table 4 shows the result of the t-test.

Table 2. Result for the speed of sound measurements carried out on the four years old TMM.

c [m s^{-1}]	u_A [m s^{-1}]	u_B [m s^{-1}]	k	U [m s^{-1}]
1538.9	0.091	1.017	1.96	2.0

Table 3. Results for the attenuation coefficient measurements carried out on the four years old TMM.

a_c [dB cm^{-1}]	u_A [dB cm^{-1}]	u_B [dB cm^{-1}]	k	U [dB cm^{-1}]
2.573	0.028	0.003	3.18	0.091

Table 4. Significance t-test.

	μ_O	X	t_{calc}	t_{crit}	$t_{calc} < t_{crit}$
	[m s^{-1}]				
c	1540	1538.9	1.12	1.96	Yes
	[dB cm^{-1}]				
a_c	2.5	2.573	2.56	3.18	Yes

4. Discussions and Conclusion

There is no statistical evidence to reject the hypothesis that the velocity is 1540 m s^{-1} ($\alpha = 0.05$). In other words, the mean of measurements of the speed of sound in the four years old TMM does not differ statistically from the reference value in the standard. In turn, the same occur with the attenuation coefficient, where $t_{calc.}$ was lower than $t_{crit.}$ indicating that the mean and the reference value do not differ significantly.

Glycerol is responsible for provide the speed of sound required to the TMM, while the attenuation coefficient is determined by the concentration of 0.3 mm alumina (Al2 O3) .

Taking into account the TMM was maintained without air contact in the solution proposed by [1] (water and glycerol) over the last four years, and based on the results, one can conclude that the TMM's shelf life, if it is adequately preserved, can be larger than two years as described in [1].

References

[1] International Electrotechnical Commission (IEC). IEC 60601-2-37 2015 Ed 2.1 Genève p 208.
[2] Culjat M A O C, Oldenberg D A G, Ewari PRT and Ingh, RASS 2010 *Ultrasound in Med. & Biol.* v **36** (6), pp 861–873.
[3] Manuscript A 2013 NIH Public Access, v **50**(23), pp 5597–5618.
[4] Yasukawa K, Kunisue T and Tsuta K 2007 *IEEE Ultrasonics Symposium*, New York, pp 2501–2502.
[5] Vogt WC, Jia C, Garra B S and Pfefer T J 2014 *Imaging and Sensing*, v **9107** Baltimore. p 1–11.
[6] Abrunhosa V M, Soares C P, Possidonio A C B, Alvarenga A V, Costa-Felix R P B, Costa, M L P S, Mermelstein, C S. 2014 *Ultrasound in Medicine & Biology* **40** pp 504-512.
[7] Petrella LI, Maggi L E, Souza R M, Alvarenga A V and Costa-Félix R P B 2014 *Ultrasonics*, v **54**(6), pp 1476–1479.
[8] Souza RM, Santos T Q, Oliveira D P, Souza R M, Alvarenga AV and Costa-Felix R P B 2016 *Journal of Physicis: Conference Series*, v **733**, paper 12044.
[9] Santos T Q, Alvarenga AV, Oliveira D P and Costa-Felix R P B 2017 *Ultrasound in Med. &*

Biol, v **43**(1), pp 323–331.
[10] Oliveira D P, Santos T Q, Mayworm R C, Souza R M, Alvarenga AV and Costa-Félix, R P B. 2016 Anais do XXV CBEB, pp 231-234.
[11] Silva C E R, Alvarenga A V, Costa-Felix R P B. 2011 *Journal of Physics. Conference Series* pp 279 012025.
[12] Maggi L E, Silva C E R, Alvarenga A V, Costa-Felix R P B. 2011 *Journal of Physics. Conference Series* pp 279 012029.
[13] BIPM, JCGM:100 2008 Paris: BIPM.
[14] Costa-Felix R P B. 2006 Measurement 39 pp 169-175.
[15] Costa-Felix R P B, Machado J C. 2015 Measurement 69 pp 146-154.
[16] Lubbers J and Graaff R 1998 *Ultrasound Med & Biol.*, v **24** (7), pp 1065–1068.
[17] International Electrotechnical Commission (IEC). IEC 62127-2 2017 Ed 1.2. Genève. 388p.
[18] Chatfield C, *Statistics for Technology* 1983 3rd Ed. London: Chapmal & Hall. pp 140-142.
[19] Ferreira E, Vicente J, Silva R C D, Nele M, Nunes E, Costa-Felix R P B. 2014 *Polymer Testing* **33** pp 16-20.

Acknowledgments

Research reported in this paper has been partially supported by the Carlos Chagas Filho Research Support Foundation (Faperj, grant number E-26/201.563/2014), by the National Council for Scientific and Technological Development (CNPq, grant number 310.392/2014-4) and Pronametro/Inmetro.

IOP Conf. Series: Journal of Physics: Conf. Series **975** (2018) 012026 doi:10.1088/1742-6596/975/1/012026

Ultrasound measurement apparatus for liquids characterization

R C Vieira and R P B Costa-Felix

Laboratory of Ultrasound (Labus), Directory of Scientific and Industrial Metrology (Dimci), National Institute of Metrology, Quality and Technology (Inmetro), Av. Nossa Sra. das Graças 50, Duque de Caxias, RJ, Brazil, ZIP 25250-020.

rodrigo_cesar_vieira@yahoo.com; rpfelix@inmetro.gov.br

Abstract: The present paper discloses the validation of an experimental ultrasound apparatus and method for liquids characterization. The research aims to stablish a simple, reliable, accurate and portable way to identify contaminants in hydrocarbon substances, such as adulteration in gasoline. The results depicted so far demonstrated a general uncertainty of speed of sound assessment less than 10 m s^{-1}, and distance accuracy of less than 1%. Those figures are good enough for an in-site device to evaluate possible contamination of fuels or other liquids.

1. Introduction

Despite of the rigid control of Brazilian authorities, gasoline adulteration is much more often than reasonably. This practice results in loss of money for users and for the Government. The Brazilian regulations for gasoline [1] adopted many types of tests to assess its quality, but most of them are expensive, not self-conclusive or demand too much time to get reliable results. Gupta and Sharma **Erro! Fonte de referência não encontrada.** worked on this subject and proposed a new experimental approach using ultrasonic parameters. In the work of Wiedermann et al. [3], the tests used in some fuel quality assurance regulations were tested and the results suggest a lack of confidenciality to be investigated further. It was decided to use ultrasonic measuring facility because it would allow saving time, as ultrasonic parameter determination is straightforward. So, they could differentiate one sample from another, readily. The wave propagation velocity (WPV) and the acoustic attenuation through materials are the so called ultrasonic parameters [4][5]. They are dependent or directly related to variation of temperature, density and viscosity. The ultrasonic wave's energy attenuation, during propagation through the medium is an important parameter to help on material's characterization [6].

Ultrasound has been proved so far to be an interesting and reliable parameter to be used to assess fluid characterization, as well as other technical applications [7][8][9]. Within the research to which this paper belongs, a new apparatus was developed and evaluated regarding the precision and accuracy. The aim of this apparatus is to determine ultrasound parameters such as sound velocity and attenuation. Within this paper, the results regarding sound velocity assessment will be disclosed.

2. Theoretical background

The wave propagation velocity (WPV) is calculated from taking the specific medium's extension value and the time taken by the acoustic wave to pass through it, named time of flight (**t_{of}**) **Erro! Fonte de referência não encontrada.**.

$$c = \Delta_x . (\Delta_t)^{-1} \tag{1}$$

in which, **c** is the wave propagation velocity (WPV), expressed in m.s^{-1}. The Δ_t is the time of flight (t_{of}), expressed in s and Δ_x is the distance in m.

In the present pulse reflection scheme, the ultrasonic wave behavior is specular and normal, i.e., the incident and the reflected waves direction is perpendicular to the incident surface. The following figure illustrates the phenomena.

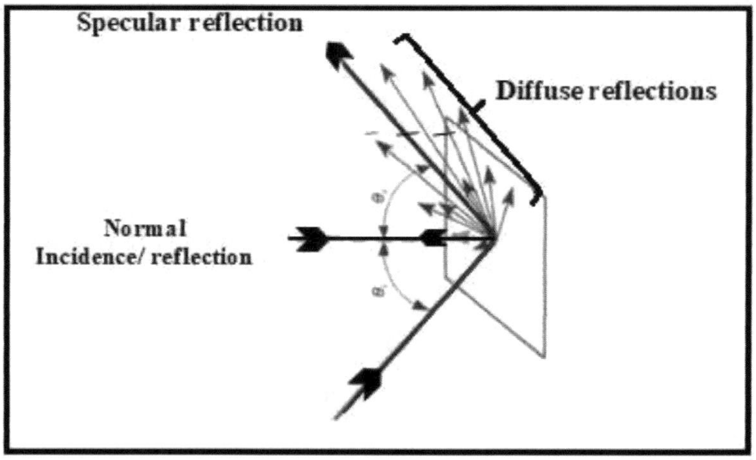

Figure 1. Acoustic wave's Behavior (adapted from [10]).

Water's WPV is a function of temperature, and Lubbers and Graaff's simplified equation resulting in the equation (2) [11].

$$c = 1404.3 + 4.7T - 0.04T^2 \tag{2}$$

in which c is the wave propagation velocity (WPV) expressed in $m.s^{-1}$ and T is the temperature in the medium, in °C.

Water's density can be estimated as function of temperature according to equation (3) [12].

$$\rho = 1000.1 + 0.0134T - 0.0058T^2 + 2e^{-5}T^3 \tag{3}$$

in which ρ is the density in the medium expressed in $kg.m^{-3}$ and T is temperature in the medium, in °C.

3. Methodology

The experimental apparatus consists in a metal tube (diameter equals 31.775 mm) which is filled with the liquid under test (LUT). As reflecting target, a specially developed cylinder made of aluminum was constructed. The cylinder length was defined so that a 20-cycle, 5 MHz, ultrasonic burst could propagate through its extension without interfering, as reflected wave, in the incident wave (about 85 mm long). The ultrasonic signal was delivered by an arbitrary waveform generator model 33500B series (Agilent, USA) to an ultrasonic transducer model A307S- 5 MHz (Olympus-NDT, USA). The reflected signal was digitalized with an oscilloscope model DSO-X 3024A (Agilent, USA), and all calculations were performed in a dedicated spreadsheet.

3.1 Dimensional measurement

The system was designed to allow the ultrasonic propagation inside the cylinder with two different distances. The sound velocity is assessed with better accuracy if two different propagation paths are used with the same emitting and reflecting system. The difference is accounted in the time domain subtracting the two reflected echoes (sequentially, not simultaneously measured).

A mechanical gauge was used to determine the difference of the two propagations paths, as disclosed in figure 2(d). Its length was determined with the aid of a calibrated Vernier caliper model CD8" AX-B (Mitutoyo, Brazil), with expanded uncertainty of 30 micrometers (p = 0.95).

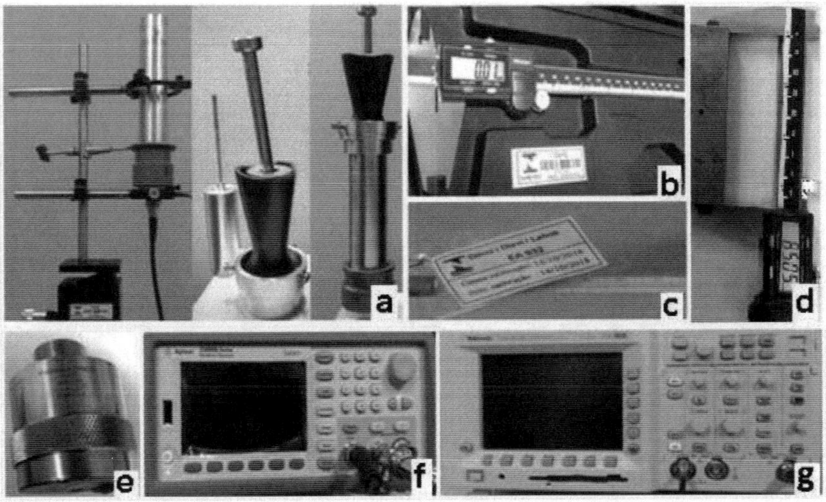

Figure 2. (a)Homemade measurement apparatus. (b) The calibrated Vernier caliper. (c) Analogic thermometer, calibrated. (d)The distance gauge. (e) The 5 MHz Transducer. (f) The Waveform generator. (g) The calibrated Oscilloscope.

3.2 Ultrasonic measurement
For the system and the method validation, water at room temperature was used. The speed of sound was assessed using equation (2). The reflector was positioned in two different distances from the transducer. Increasing the distance, the time of flight will increase accordingly, as far as the speed of sound remains the same. The results were statistically compared. The measurement uncertainty was determined following the GUM guidelines [13][14].

3.3 Fluid density measurement
The system will be used to characterize fluids of dissimilar physical properties, for instance density and viscosity. Within the validation experiment described in this paper, the density was assessed according to equation (3) to evaluate the amount of mass to be filled in the tube. The results were compared with the amount of substance measured with a gravimetric method.

4. RESULTS
The dimensional measurements are presented in table 1.

Table 1. Tube and gauge dimensional measurements.

	Dimensions (mean value) [mm]	**Unc. [mm]**
Tube's diameter	31.775	0.031
Gauge's length	85.010	0.020

Density assessed after temperature measurement is disclosed in table 2.

Table 2. Density value and uncertainty for T = 21.2°C.

Sample's T [ºC]	**Source**	**Result**
21.2	**Equation (3)ρ**	997.97 **[kg.m^{-3}]**
	Equation (2) c	1485.96 **[m.s^{-1}]**
	Sample's mass	
Mean value	81.312**[g]**	
Uncertainty	0.005**[g]**	

Tube					
Diameter (Vernier caliper)		**Time of flight(t_{of}) (through the sample)**		**Distance** c ≡ (2)	**Basis area**
Mean value	**Std. dev.**	**Mean value**	**Std. dev.**	($x = t_{of}$.c. 2^{-1})	a=π.r²
0.0318 [m]	3.064e^{-5} [m]	1.37e^{-4}[s]	1.47e^{-7}[s]	0.1016 [m]	7.93e^{-4} [m²]
Sample's volume (V= area . h)				0.0000806	[m³]
Sample's density (ρ= m.v^{-1})				1008.8	[kg.m^{-3}]
Diff [kg.m^{-3}]	10.83			**Diff [%]**	1.07

The volume was assessed with both a gravimetric method and dimensional method. The length was measured with the Vernier caliper and the ultrasonic time of flight method. All results are disclosed in table 3.

Table 3. Volume measurement.

Measurement by mass					
Temp [°C]	**Equation (2) [11] WPV [m.s^{-1}]**	**Max. error[m.s^{-1}]**		**Equation (3) ρ [kg.m^{-3}]**	**Vol [m³]**
21.2	1485.96	0.18		997.97	0.0815e^{-5}
Measurement by ultrasound					
c (2) [m.s^{-1}]	**Méd.Δ_t [s]**	**Std . dev.**	**Δ_x(1) [m]**	**Tube basis area [m²]**	**Vol[m³]**
1485.96	0.000137	1.473e^{-7}	0.01016	0.000793	0.0000806
Difference [m³]	9e-7			**Difference [%]**	1.1

The dimensional gauge length was assessed with the pulse echo ultrasonic measurement method, and the result is disclosed in table 4 along with its uncertainty. In the table 4, t_1, t_2 and t_3 are the times of arrival for the three first echoes.

Table 4. Gauge's length measurement using ultrasound.

	t_1[µs](d_0)	**t_2 (d_{gauge}) [µs]**	**$\Delta_{t(3-1)}$ [µs]**	**$T_{(water)}$ [°C]**	**WPV(3) [m.s^{-1}]**	**$x = c.\Delta_t.2^{-1}$ [m]**
Repet. 1	36.465	150.927	114.462	22.9	1491.0	0.0853
Repet. 2	36.428	150.989	114.057	22.9	1491. 0	0.0850
Repet. 3	36.439	150.496	114.561	23.0	1491.2	0.0854
Mean values			114.462			0.0853
Std. dev.			0.267			0.000205

Speed of Sound Through the Gauge							
	Repet. 1		**Repet. 2**		**Repet. 3**		
t_1[µs]	29.432		29.434		29.436		
t_2[µs]	58.324		58.334		58.330		
t_3[µs]	86.972		87.036		87.033		
	$\Delta_{t(2-1)}$	$\Delta_{t(3-2)}$	$\Delta_{t(2-1)}$	$\Delta_{t(3-2)}$	$\Delta_{t(2-1)}$	$\Delta_{t(3-2)}$	
	28.89	28.65	28.90	28.70	28.89	28.70	
Mean value Δ_t [µs]	28.77		28.80		28.80		
Mean Δ_t (Repetitions)	(28.79[µs])						28.79e^{-6} [s]
Gauge length							0.08501 [m]
Mean speed of sound (c)							5905.5 [m.s^{-1}]
Uncertainty velocity [m.s^{-1}]	17				**Uncertainty velocity[%]**		0.579

The final results comprise the speed of sound measurement in comparison with the temperature dependent formula (2). Table 5 discloses the results and the comparison of them.

Table 5. Speed of sound measurements.

	Ultrasound measurements Equation (1) Table 4		**Vernier caliper measurements Mean value**	
Gauge's length	0.08525		0.08501	**[m]**
	Equation (2)			
	21.2°C	**22,9°C**		
Water sample´s WPV	1485.96	1490.95		**[m.s^{-1}]**
	Ultrasound measurements(1), gravimetric scale and $\rho = m.v^{-1}$		**Equation (3) at 22°C**	
Water sample´s density	1008.8		997.70	**[kg.m^{-3}]**
Difference [kg.m^{-3}]	**10.83**			
Difference [%]	**1.07**			

5. Discussion

Ultrasound is a very usual tool to identify materials properties, such for fluids (liquid or gases) as for solids. However, specific application demands dedicated assemblies of devices, transducers, accessories and equipment. The objective of the present research is to develop and validate an apparatus able to deliver ultrasound to a liquid medium and, further, assess its physical characteristics.

To evaluate the accuracy of the device, we used pure water as known medium. The dimensional quantities of both the recipient and the mechanical gauge were firstly measured with a Vernier caliper, and in sequence some were estimated using ultrasound waves. The results agreed of 0.24 mm (see Tables 1 and 4). The speed of sound through the gauge was determined with uncertainty of less than, 0.6% (Table 4).

Another way to validate the system is to consider the speed of sound after a well stablished relation between water temperature and speed of sound. The quantities to be assessed are linear and volumetric dimensions, indirectly determined. According to the results disclosed in Tables 2, 3 and 5, one can observe that the ultrasound method was able to achieve about 1,07% for density measurement with the aid of a gravimetric method (Tables 2 and 5) and 1.1% for volume (Table 3).

6. Conclusions

Applying ultrasound for liquids characterization shows confidence and accuracy which is desirable for technical application. Further work will deal with gasoline and other hydrocarbon substances to evaluate the applicability of this approach to identify the presence of contaminants.

References

[1] Brasil, Agência Nacional do Petróleo, Gás natural e Bio Combustíveis (ANP) 2013 *ANP Normative* n°**40**.
[2] Gupta A, Sharma R K 2010 *A new method for estimation of automobile fuel adulteration* Air Pollution Vanda Villanyi (Ed) ISBN: 978-953-307-143-5 InTech.
[3] Wiedermann L S M et al 2005 *Fuel* **84** p 467-473.
[4] Silva C E R, Alvarenga A V and Costa-Felix R P B 2011 *Journal of Physics:Conference Series* **279** pp 012-025.
[5] Maggi L E, Silva C E R, Alvarenga A V and Costa-Felix R P B 2011 *Journal of Physics: Conference Series* **279** pp 012-029.
[6] Costa-Felix R P B and Machado J C. 2015 *Measurement* **69** pp 146-154.
[7] Petrella L I, Maggi L E, Souza R M, Alvarenga A V and Costa-Felix R P B 2014 *Ultrasonics* **54**

pp 1476-1479.
[8] Abrunhosa V M, Soares C P, Possidonio A C B, Alvarenga A V and Costa-Felix R P B, Costa M L P S and Mermelstein C S 2014 *Ultrasound in Medicine & Biology* **40** pp 504-512.
[9] Ferreira E, Vicente J, Silva R C D, Nele M, Nunes E and Costa-Felix R P B. 2014 *Polymer Testing* **33** pp 16-20.
[10] Grimaldi Domenico 2006 *IEEE Transactions on Instrumentation and Measurement* v **55**(1) pp 5-13.
[11] Lubbers J and Graaff R 1998 A simple and accurate formula for the sound velocity in Water *Ultrasound Med. Biol.* v **24** n 7 pp1065-8
[12] Density Accessed on SEPT 13 2017 URL: https://pt.wikipedia.org/wiki/Densidade.
[13] BIPM J C G M 2008 *Evaluation of measurement data – Guide to the expression of uncertainty in measurement GUM* JCGM **100.**
[14] Costa-Felix R P B 2006 *Measurement* **39** pp 169-175.

Acknowledgments
Research reported in this paper has been partially supported by the Carlos Chagas Filho Research Support Foundation (Faperj), grant number E-26/201.563/2014), by the National Council for Scientific and Technological Development (CNPq, grant number 310.392/2014-4) and Pronametro/Inmetro.

IOP Conf. Series: Journal of Physics: Conf. Series **975** (2018) 012027 doi:10.1088/1742-6596/975/1/012027

Measurement uncertainty evaluation of cellular spheroids surface tension in compressing tests using Young-Laplace equation

Anderson Beatrici[1,2], Leandra Santos Baptista[2,3] and José Mauro Granjeiro[2]

[1] National Institute of Metrology, Quality and Technology (Inmetro) - Scientific and Technological Metrology Division (Dimci), RJ, Brazil.
[2] National Institute of Metrology, Quality and Technology (Inmetro) - Division of Metrology Applied to Life Sciences (Dimav), RJ, Brazil.
[3] Multidisciplinary research center in Biology - Numpex-Bio, Federal University of Rio de Janeiro, Campus of Duque de Caxias; Graduate Program in Biotechnology, Brazil.

Abstract. Regenerative Medicine comprises the Biotechnology, Tissue Engineering and Biometrology for stem cell therapy. Starting from stem cells extracted from the patient, autologous implant, these cells are cultured and differentiated into other tissues, for example, articular cartilage. These cells are reorganized into microspheres (cell spheroids). Such tissue units are recombined into functional tissues constructs that can be implanted in the injured region for regeneration. It is necessary the biomechanical characterization of these constructed to determine if their properties are similar to native tissue. In this study was carried out the modeling of the calculation of uncertainty of the surface tension of cellular spheroids with the use of the Young-Laplace equation. We obtained relative uncertainties about 10%.

1. Introduction

Regenerative therapies seek to improve the condition of patients affected by injuries that limit or prevent the normal activities of a healthy individual. One of the most promising regenerative therapies is the cartilage implant in injured joints. For these implants to result in a permanent cure, it is necessary to determine procedures and techniques to provide these implanted tissues similar characteristics to or better than the native tissue, for instance its resistance to compressive forces.

Thus, the determination of the biomechanical properties of these constructs is fundamental for the success of these therapies. Metrological traceability can provide a more precise characterization of the cellular spheroids properties. Comparative research guarantees only qualitative results, but for comparisons with international results of researches we need a quantitative characterization of biomechanical properties of these constructs, it is mandatory a quality control in these results [1, 2].

The determination of the uncertainties associated with the measurements will lead to quantitative results of the biophysical properties analyzed. Therefore, it will be possible to attest that the conditions exhibited by these new 3D constructs are compatible with native tissues and that the techniques of cellular spheroid biofabrication and their subsequent bioprinting [3, 4, 5], this will contribute decisively to provide precise parameters to compare the different biofabrication techniques. In this study, we present a measurement uncertainty evaluation model for the measurement results to cellular spheroids surface tension in micro compression tests. Compressive tests were performed on a parallel plates micro tester where the spheroid is placed and a micro newton scale compression is exerted. The resulting strain is analyzed by the Young-Laplace equation [6, 7, 8] which gives the surface tension resulting from the applied pressure. The uncertainties involved were obtained by the classical model and validated by the Kragten method. The final values of the relative uncertainties were less than 10%, which means an excellent result in biometrology.

2. Compression technique

A cellular spheroid is placed between the parallel plates device and a compression force is exerted on it by a tungsten (W) microbeam (figure 1 and 2). The force measurement was performed with an analytical balance.

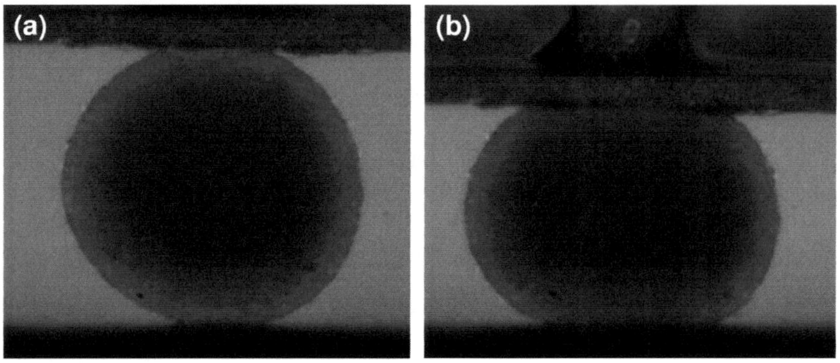

Figure 1. Cellular spheroid in the mechanical tester, uncompressed (a), compressed (b).

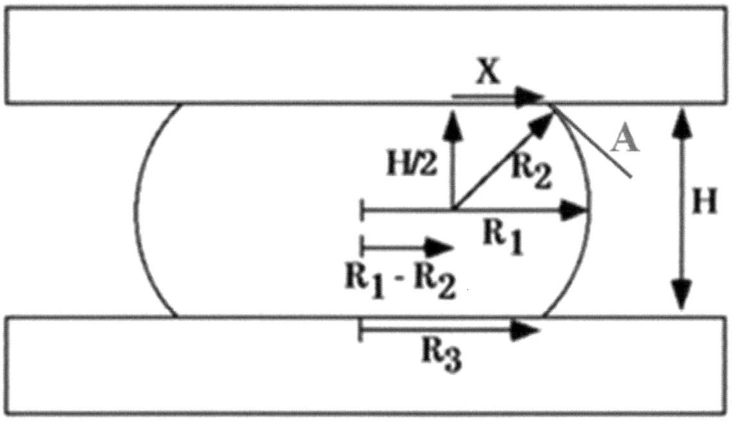

Figure 2. Cellular spheroid compressed between two parallel plates scheme (adapted from [9]).

The surface tension, under these conditions, can be obtained directly from the Young-Laplace equation (1).

$$T = \frac{F}{\pi R_3^2(\frac{1}{R_1} + \frac{1}{R_2}) - 2\pi R_3 sinA} \tag{1}$$

Where,

$T \rightarrow$ Cellular spheroid surface tension.
$F \rightarrow$ Applied force.
$R_1 \rightarrow$ Equatorial curvature radius.
$R_2 \rightarrow$ Border curvature radius.
$R_3 \rightarrow$ Circular contact area radius.
$A \rightarrow$ Angle (rad).

The A angle is the angle between the up spheroid surface and the parallel plate (figure 2). This angle can be rewritten in principal radii terms according to equation (2).

$$\sin(A) = \frac{(R_3 + R_2 - R_1)}{R_2} \tag{2}$$

Substituting (2) into (1) we have,

$$T = \frac{F}{\pi R_3^2(\frac{1}{R_1} + \frac{1}{R_2}) - \frac{2\pi R_3}{R_2}(R_3 + R_2 - R_1)} \quad (3)$$

Thus the equation (3) not depends explicitly of *A* angle. The compressing test is performed by setting the equipment with the chosen parameters and applying a 25% maximum displacement of the initial non-loaded spheroid diameter, recording the maximum force values in 3 up to 5 repetitions per cycle, figure 3. This maximum displacement was chosen because in this limit the spheroid compression do not able to deform permanently the biological samples.

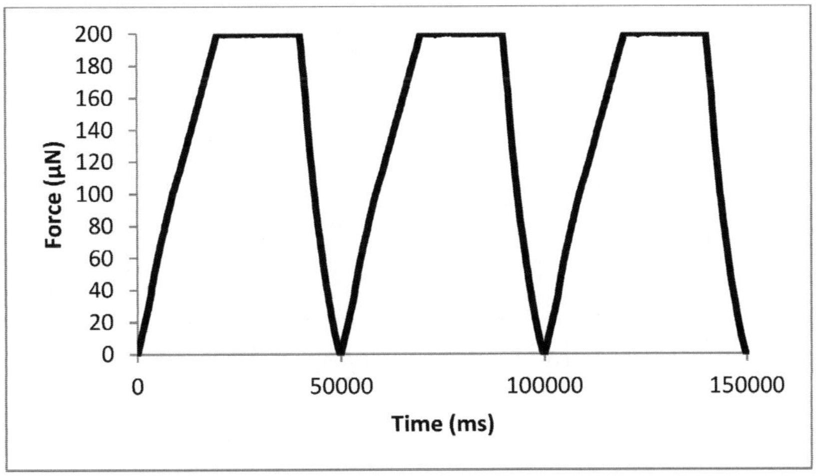

Figure 3. A typical load test in an elastic (non-biological) sample.

3. Uncertainty modeling

The identified uncertainty contributions are shown in the Ishikawa diagram, figure 4. The *A* angle uncertainty contributions was taking account in terms of the three main curvature radius as shown in equation (2). The curvature radius and force uncertainty contributions were obtained directly by the calibrations certificates and measurement results analyses, as well as by mathematical simulations.

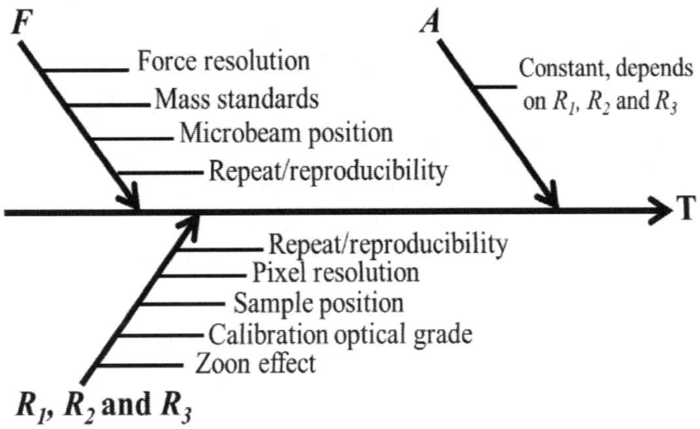

Figure 4. Ishikawa diagram with the entrance uncertainty quantities.

Each of the uncertainty components is obtained from the product of the sensitivity coefficient c_y by the value of the uncertainty u_y of the input magnitude and the combined uncertainty to the surface tension $u_c(T)$ is given by the sum of modulus of uncertainty components (4)

$$u_c(T) = \sqrt{\sum {c_y}^2 \, u_y^2} \tag{4}$$

with $y = F$, R_1, R_2 or R_3.

The coefficients of sensitivity are given by,

$$c_y = \frac{\partial T}{\partial y} \tag{5}$$

Here we will show the equation for the simple form (1), without the *sin A* explicit equation (2), thus,

$$c_F = \frac{1}{\pi R_3^2 (\frac{1}{R_1} + \frac{1}{R_2}) - 2\pi R_3 sinA} \tag{6}$$

$$c_{R_1} = \frac{\pi F \frac{R_3^2}{R_1^2}}{\left[\pi R_3^2 (\frac{1}{R_1} + \frac{1}{R_2}) - 2\pi R_3 sinA\right]^2} \tag{7}$$

$$c_{R_2} = \frac{\pi F \frac{R_3^2}{R_2^2}}{\left[\pi R_3^2 (\frac{1}{R_1} + \frac{1}{R_2}) - 2\pi R_3 sinA\right]^2} \tag{8}$$

$$c_{R_3} = \frac{-2\pi F \left(R_3 (\frac{1}{R_1} + \frac{1}{R_2}) - sinA\right)}{\left[\pi R_3^2 (\frac{1}{R_1} + \frac{1}{R_2}) - 2\pi R_3 sinA\right]^2} \tag{9}$$

$$c_A = \frac{2\pi R_3 F \; cosA}{\pi R_3^2 (\frac{1}{R_1} + \frac{1}{R_2}) - 2\pi R_3 sinA} \tag{10}$$

4. Results

Using the following measurements results (obtained from a cartilage spheroid sample),

$A =$ 35.5° or 0.6196 rad
$T =$ 0.621 N/m
$R_1 =$ 245.9 μm
$R_2 =$ 175.8 μm
$R_3 =$ 172.1 μm
$F =$ 174.1 μN

with the uncertainties (figure 4) and traceability to the kilogram [11, 12] and meter [13, 14] we have the force entrances uncertainties values (table 1).

Table 1. Force uncertainties values.

Entrance uncertainty	Uncertainty value
Force resolution [12]	0.565 μN
Mass standards [11]	0.048 μN
Microbeam position [14]	1.256 μN
Repeatability	0.626 μN
Reproducibility	0.447 μN

and applying in the equation (4) and (5) to the force uncertainty case we have,

$$u(F) = 1.578\ \mu\text{N} \tag{11}$$

Performing the same procedure to the radii R_1, R_2 and R_3 we have (table 2),

Table 2. Radii uncertainties values to R_1 case.

Entrance uncertainty	Uncertainty value
Zoon effect	0.223 μm
Sample position [14]	0.089 μm
Optical grade calibration [13]	0.060 μm
Pixel resolution	0.577 μm
Repeatability	0.670 μm
Reproducibility	0.894 μm

In the same way applying in (4) and (5) we have,

$$u(R_1) = 1.282\ \mu\text{m} \tag{12}$$

$$u(R_2) = 1.138\ \mu\text{m} \tag{13}$$

$$u(R_3) = 1.136\ \mu\text{m} \tag{14}$$

$u(A)$ is not shown here because the uncertainties calculations was performed with the explicit equation for the A angle (2). To obtain the surface tension expanded uncertainty we need to determine the k coverage factor which depends on effective degrees of freedom. In this case effective calculated degrees of freedom is,

$$\nu_{eff} = 34.04 \tag{15}$$

Considering a Gaussian distribution this value correspond to a coverage factor,

$$k = 2.032 \quad (p = 95\%) \tag{16}$$

Putting the results (11) (12) (13) and (14) in (4) we obtain the spheroid surface tension combined uncertainty,

$$u_c(T) = 0.02014\ \text{N/m} \tag{17}$$

The expanded uncertainty is

$$U(T) = k\ u_c(T) \tag{18}$$

Putting (16) and (17) in (18) we have,

$$U(T) = 0.040937 \cong 0.041\ \text{N/m} \tag{19}$$

The result to the spheroid surface tension in this approach is,

$$T = 0.621\ \text{N/m} \pm 0.041\ \text{N/m} \tag{20}$$

or

$$T = (0.621 \pm 0.041)\ \text{N/m} \tag{21}$$

Corresponding to 5.59 % in the relative expanded uncertainty. We verify the modeling and calculations using the Kragten method to uncertainty and it was obtained the same results,

$$U(T)_{Kragten} = 0.040941\ \text{N/m} \tag{22}$$

When we compare the finals uncertainties without approximations to both approaches we have a difference of -2.6×10^{-6} N/m, showing the agreement in our results.

5. Conclusions

We conclude that by considering all input identifying uncertainties and applying the Kragten approach or the classical model, we achieve a relative expanded uncertainty less them 10% in the presented case. These values are smaller than the variability of cellular spheroid properties. Therefore, this uncertainty calculation model can be used to determine the biomechanical properties of cellular spheroids. This modeling was performed for cellular spheroids with less than 2% sphericity. The next steps are to evaluate the sphericity effect on measurement uncertainty because are common samples with sphericity greater than 2% and to apply the same approach for the Young's modulus uncertainties calculation.

6. References

[1] Baker M 2016. How quality control could save your science. *Nature* **529** pp.456-58
[2] Armbruster D and Miller R R 2007. The Joint Committee for Traceability in Laboratory Medicine (JCTLM): A Global Approach to Promote the Standardization of Clinical Laboratory Test Results. *Clin Biochem Rev* **28** pp.105–14
[3] Murphy S V and Atala A 2014. 3D bioprinting of tissues and organs. *Nature Biotechnology* **v.32** pp.773–85
[4] Mironov V *et al* 2011. Organ printing: from bioprinter to organ biofabrication line. *Curr. Opin. Biotechnol.* **22(5)** pp.667-73
[5] Mironov V, Visconti R P, Kasyanov V, Forgacs G, Drake C J, Markwalf R 2009. Organ printing: *tissue spheroids as building blocks. Biomaterials*, **30** pp.2164–74
[6] CellScale Biomaterials Testing (2016). MicroSquisher Micro-scale Tension-Compression Test System. *User Manual*. Version **3.1**
[7] Norote A C, Marga F, Neagu A, Kosztin I, Forgacs G 2008. Experimental evaluation of apparent tissue surface tension based on the exact solution of the Laplace equation. *EPL (Europhysics Letters)* **81** pp.46003p1-p6
[8] Shima S *et al* 1993. Large Deformations of a Rubber Sphere under Diametral Compression: Part 2. *JSME international journal*. **Ser. A**, *Mechanics and material engineering* **v.36**, n. 2, pp.197-205
[9] Butler C M and Foty R A 2011. Measurement of Aggregate Cohesion by Tissue Surface Tensiometry. *J Vis Exp* **50** 2739 pp.01-06
[10] JCGM 2008 Evaluation of measurement data - Guide to the expression of uncertainty in measurement (GUM): *JCGM 100:2008* (Sevres: BIPM)
[11] Standards Weighs Calibration Certificate: DIMCI 1247/2015. Lamas/Inmetro
[12] Balance Calibration Certificate: DIMCI 0800/2017. Lamas/Inmetro
[13] Optical Grade Calibration Certificate: DIMCI 0926/2016. Lamed/Inmetro
[14] Caliper Calibration Certificate: DIMCI 1126/2017. Lamed/Inmetro

Acknowledgments

This work was funded by the National Council for Scientific and Technological Development (CNPq) through a National grant 457541/2013-0. Postgraduate Program in Biotechnology of Inmetro (PPGBiotec). Scientific and Technological Metrology Division (Dimci/Inmetro) and Division of Metrology Applied to Life Sciences (Dimav/Inmetro).

IOP Conf. Series: Journal of Physics: Conf. Series **975** (2018) 012028 doi:10.1088/1742-6596/975/1/012028

Characterization of potassium dichromate solutions for spectrophotometercalibration

F C Conceição[1], E M Silva[2], J F S Gomes[2], P P Borges[3]

[1]Farmanguinhos, Fiocruz, Rio de Janeiro, Brazil
[2]Inmetro, Optical Metrology Division (Diopt), Laboratory of Radiometry and Photometry (Laraf), Duque de Caxias, Rio de Janeiro, Brazil
[3]Inmetro, Chemical and Thermal Metrology Division (Dimqt), Electrochemical Laboratory (Label), Duque de Caxias, Rio de Janeiro, Brazil

E-mail: jfgomes@inmetro.gov.br

Abstract. Spectrophotometric analysis in the ultraviolet (UV) region is used in the determination of several quantitative and qualitative parameters. For ensuring reliability of the analyses performed on the spectrophotometers, verification / calibration of the equipment must be performed periodically using certified reference materials (CRMs). This work presents the characterization stage needed for producing this CRM. The property value characterized was the absorbance for the wavelengths in the UV spectral regions. This CRM will contribute to guarantee the accuracy and linearity of the absorbance scale to the spectrophotometers, through which analytical measurement results will be provided with metrological traceability.

1. Introduction

It is usually recommended testing the spectrophotometers' performance using certified reference materials (CRMs) and it is also important that the technical analyst finds ways to validate if the instrument is operating properly or if it is capable to provide acceptable analytical results [1].

Due to the importance of verifying the spectrophotometers' performance in a regular interval of time, the Electrochemical Laboratory from Chemical and Thermal Metrology Division (Dimqt), working together with the Laboratory of Radiometry and Photometry from Optical Metrology Division (Diopt) developed the CRM of crystalline potassium dichromate ($K_2Cr_2O_7$). From this CRM, some acid solutions with specific concentrations for the wavelengths in the UV spectral regions are prepared. Then, these standard solutions will be used to evaluate the performance of spectrophotometers in the UV regions [2].

Acid solutions from this CRM must have their optical properties properly characterized to be used for this purpose. The characterization stage is very important to the CRM development [3]. In this stage, the value of the property to be certified is obtained and its measurement uncertainty is evaluated. The uncertainty of the characterization (u_{char}) will contribute to the uncertainty of the CRM.

The characterization stage involves the preparation of five different concentrations of dichromate acid solutions from the batch of the candidate CRM of $K_2Cr_2O_7$for measuring the solutions in different equipment, in this case, in two laboratories with different methodologies [3] and by qualified technicians. At the end, the results obtained are compared and an average value from the two laboratories is chosen in order to assure the metrological reliability of the results from the use of

$K_2Cr_2O_7$ as a reference material for the verification of the photometric scale of the spectrophotometer in the UV spectral region.

1.1. Objective

This work aims to present the methodology for the characterization of the candidate CRM of $K_2Cr_2O_7$, in order to fulfil one of the production steps for this material to be recognized as a CRM.

2. Experimental

2.1. Batch of candidate CRM of $K_2Cr_2O_7$

Reference material (RM) developed in a batch contained 64 bottles of the candidate CRM of $K_2Cr_2O_7$ was prepared (Figure 1). Firstly, 1 kg of the salt of $K_2Cr_2O_7$ was mixed and manually homogenized; after, 15 g of the material were weighed in an analytical balance and were filled in each glass bottle, numbered in the order of filling. Finally, 4 bottles were at random selected from the batch for the homogeneity, characterization and stability studies. Before preparing the acid solutions, the purity of the $K_2Cr_2O_7$was determined by coulometric titration (99.921 %) and this value was taken into account for calculating the concentration of the solutions.

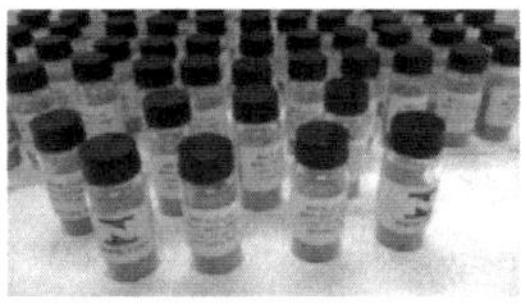

Figure 1. Batch of the candidate CRM of $K_2Cr_2O_7$[4].

2.2. Potassium dichromate acid solutions

Potassium dichromate solutions from the batch of the candidate CRM of $K_2Cr_2O_7$ in 0.001 mol L^{-1} perchloric acid ($HClO_4$) medium were prepared gravimetrically. The concentrations of the $K_2Cr_2O_7$ were 20 mg kg^{-1}, 40 mg kg^{-1}, 60 mg kg^{-1}, 80 mg kg^{-1} and 100 mg kg^{-1}. These concentrations will provide absorption in the specific wavelength range needed to verify the performance of the spectrophotometers in the UV regions. With the measurement results of absorbance for each solution, it can be calculated the apparent absorptivity according to the Lambert-Beer's law (equation 1), where the concentrations are correlated with the absorbance.

$$A_a=a_a \times b \times c \tag{1}$$

Where, A_a is the apparent absorbance; a_a apparent absorptivity; b, length of path and c, the concentration. The term named apparent was introduced together with the parameters absorbance and absorptivity because the effect related to the buoyancy correction for mass and also internal reflections of light in the quartz cuvette were not considered in the measurements.

Thus, acid solutions prepared from different samples of the candidate CRM of $K_2Cr_2O_7$ were measured in two spectrophotometers at Inmetro in different laboratories and the results were compared each other. For guaranteeing the reliability and metrological traceability of the measurement results, all instruments, which had an important effect in the accuracy and reliability of the measurement results of testing, were calibrated before being used.

3. Results and Discussion

3.1. Characterization

Acid solutions from sample 23 and 57 from the batch of the candidate CRM of $K_2Cr_2O_7$ prepared at different concentrations were measured in two different spectrophotometers, with different technicians.

The values obtained in the absorbance measurements are showed in table 1. The final result is the average absorbance (A) measured for each concentration, as well as the expanded uncertainties *U*(A) associated with each measurement, according to the GUM [5].

Table 1. Samples 23 and 57 for comparing of the absorbance results.

	λ (nm)	Sample 23		Sample 57		Error[a]	En[b]
		A	*U*	A	*U*		
20 mgkg^{-1}	235	0.252	0.003	0.262	0.018	0.010	0.56
	257	0.290	0.003	0.303	0.018	0.014	0.75
	313	0.100	0.003	0.105	0.016	0.004	0.28
	345	0.215	0.004	0.225	0.018	0.010	0.55
	350	0.217	0.004	0.227	0.018	0.010	0.57
40 mg kg^{-1}	235	0.511	0.006	0.519	0.010	0.008	0.73
	257	0.593	0.007	0.601	0.012	0.007	0.53
	313	0.200	0.004	0.203	0.008	0.003	0.29
	345	0.440	0.005	0.440	0.010	0.000	0.03
	350	0.444	0.005	0.443	0.010	0.000	0.03
60 mg kg^{-1}	235	0.743	0.009	0.748	0.014	0.006	0.34
	257	0.865	0.012	0.871	0.014	0.006	0.30
	313	0.290	0.003	0.293	0.009	0.002	0.25
	345	0.640	0.007	0.641	0.012	0.002	0.12
	350	0.645	0.008	0.648	0.012	0.002	0.16
80 mg kg^{-1}	235	0.999	0.017	1.011	0.016	0.011	0.49
	257	1.164	0.025	1.176	0.021	0.012	0.38
	313	0.390	0.004	0.394	0.010	0.004	0.37
	345	0.856	0.012	0.862	0.014	0.006	0.32
	350	0.865	0.012	0.872	0.014	0.007	0.36
100 mg kg^{-1}	235	1.270	0.032	1.255	0.016	0.015	0.43
	257	1.480	0.051	1.460	0.021	0.021	0.38
	313	0.493	0.005	0.489	0.010	0.005	0.42
	345	1.085	0.021	1.067	0.014	0.018	0.73
	350	1.097	0.021	1.079	0.014	0.018	0.69

[a] Error = $A_{sample\ 57} - A_{sample\ 23}$; [b] En, normalized error.

The results in table 1 are in accordance with the ASTM E925 [1]. At the table 1 can also be seen the error and the normalized error obtained. The normalized errors (E_n) found were less than 1.0 ($E_n < 1.0$), which attesting the metrological reliability of the results obtained from using of crystalline potassium dichromate as a reference material for photometric scale of spectrophotometers in the UV region.

The environmental conditions during the measurements were T = (25.1 ± 0.8) °C and RH = (58.2 ± 2.6) %. Each laboratory calculated its own uncertainty.

3.2. Apparent absorptivity uncertainty

If it were necessary to use the results in apparent absorptivity, the laboratories need to determine the apparent absorptivity uncertainty that will be composed by absorbance (A), optical path (b) [cm] and concentration (c) [g kg^{-1}] derived from equation (1).

The influence due to the optical path can be calculated from measurements in a set of *n* cuvettes chosen at random from those used in the analyses. Each set cuvette can be measured *n* times and the mean of *n* measurements to be use for calculating the optical path value.

In this case, as the solution was prepared gravimetrically, so the influence due to the concentration considers only the uncertainties of the balance and the purity. The value to be considered as influence due to the concentration will be 0.0001 g kg^{-1}. Table 2 shows the apparent absorptivity uncertainty budget to 20 mg kg^{-1} of $K_2Cr_2O_7$ solution as an example.

Table 2. Apparent absorptivity uncertainty budget to the 20 mg kg–1 solution.

Variable	Unit	Value	Uncertainty Contribution [kg.cm-¹.g-¹]
Absorbance	1	0.0028	2.76×10^{-5}
Light path	cm	0.0021	-1.00×10^{-2}
Concentration	[gkg^{-1}]	0.0001	-5.01×10^{-4}
$u_C(a_a)$	0.010026 kg m-¹ g-¹		
k	2.09		
U (20 mg kg^{-1})	**0.021 kg m-¹ g-¹**		

4. Conclusion

In the characterization stage of the candidate CRM of $K_2Cr_2O_7$, measurements of the solutions were carried out at concentrations of 20 mg kg^{-1}, 40 mg kg^{-1}, 60 mg kg^{-1}, 80 mg kg^{-1} and 100 mg kg^{-1} in two spectrophotometers in different laboratories. The results were compared in order to ensure reliability of the studied material. The analysis of the results shows that the measurements are compatible, taking into account the different measurement methodologies used in each instrument, and also by different technicians.

The characterizations of potassium dichromate solutions considering absorbance as property value will be declared in the certificate.

This stage proves the technical feasibility of Inmetro to produce and to supply to the national and international market a CRM of potassium dichromate for verifying the linearity of commercial spectrophotometers in the ultraviolet region, thus guaranteeing the traceability and reliability of the results of the measurements made by laboratories using the spectrophotometric technique in the country.

On the basis of the results obtained, Inmetro is able to produce the CRM of potassium dichromate, similar in quality to that of international producers, sparing the plaintiffs from the bureaucratic drawbacks of imports. In addition, being the only one available in the domestic market, it is reasonable to expect that the demand for the product will gradually increase.

References

[1] ASTM E925 – 092014 *Standard Practice for Monitoring the Calibration of Ultraviolet Visible Spectrophotometers whose Spectral Bandwidth does not Exceed 2 nm.*

[2] NBS Special Publication 260-54 1977*SRM Certification and use of acidic potassium dichromate solutions as an ultraviolet absorbance standard - SRM 935.*

[3] ISO 17034 2016 *General requirements for the competence of reference material producers.*

[4] CONCEIÇÃO F C 2005 Monografia do Curso Técnico em Metrologia do Inmetro SEEDUC-RJ CECO

[5] GUM 2008 Guia para a expressão de incerteza de medição - Avaliação de dados de medição (Rio de Janeiro: Inmetro)

IOP Conf. Series: Journal of Physics: Conf. Series **975** (2018) 012029 doi:10.1088/1742-6596/975/1/012029

Development of a LED based standard for luminous flux

André Sardinha[1], Ivo Ázara[1], Miguel Torres[1], Thiago Menegotto[1], Hans Peter Grieneisen[1], Giovanna Borghi[1], Iakyra Couceiro[1], Alexandre Zim[2] and Filipe Muller[2]

[1] Divisão de Metrologia Óptica, Instituto Nacional de Metrologia, Qualidade e Tecnologia, Duque de Caxias, RJ, Brazil

[2] Elo Sistemas Eletrônicos, Porto Alegre, RS, Brazil

E-mail: laraf@inmetro.gov.br

Abstract: Incandescent lamps, simple artifacts with radiation spectrum very similar to a black-body emitter, are traditional standards in photometry. Nowadays LEDs are broadly used in lighting, with great variety of spectra, and it is convenient to use standards for photometry with spectral distribution similar to that of the measured artifact. Research and development of such standards occur in several National Metrology Institutes. In Brazil, Inmetro is working on a practical solution for providing a LED based standard to be used for luminous flux measurements in the field of general lighting. This paper shows the measurements made for the developing of a prototype, that in sequence will be characterized in photometric quantities.

1. Introduction

For a long time, laboratory standards for photometry have been incandescent lamps calibrated at the correlated color temperature (CCT) of 2 856 K, which corresponds to the so called illuminant A[1,2]. The spectral power distribution of light emitted by incandescent lamps is very similar to that of a blackbody radiator, which was mathematically described by Planck. Following Planck's law, the measurement of the spectrum of an incandescent lamp allows for the calculation of its CCT and, conversely, the knowledge of the CCT allows the determination of its spectrum.

Since the spectral response of instruments do not match perfectly to the V(λ) function, the spectrum of the standard to be used should be similar to the spectrum of the lamp to be measured, or of the existing luminous source where a calibrated instrument will be used. Thus, the development of LED based standards is crucial.

Spectral power distributions of LED sources used in general lighting (phosphor-based white LED) have great similarity, with a rather sharp peak emission in the blue region accompanied by a broad peak in the yellow, what permits the use of a small number of different spectra to cover the entire range of non-special LED light sources [3]. This boosted the application of LED to general lighting.

LED standards generally require control of the junction temperature because of its strong influence on the luminous emission [4]. Alternatively, this can be achieved by waiting the stabilization of temperature and emission, within the bounds specified by the LED manufacturer. The first solution has greater complexity and costs, and the second leads to long waiting periods (sometimes, in excess of one hour), to obtain the stabilization, which implies in waste of time and reduced number of utilizations of the standard between calibrations, and depends upon the ambient temperature.

IOP Conf. Series: Journal of Physics: Conf. Series **975** (2018) 012029 doi:10.1088/1742-6596/975/1/012029

At Inmetro, we are working on an alternative solution without controlling the temperature. Instead, we monitor it during a fixed amount of time in which the LED is supplied with constant electric current. Using this approach avoids the need for an active electronic temperature control, simplifying the project and reducing costs.

2. Development steps of the prototype

Aging tests were conducted with several LED chips of different technologies. Due to easier integration in the project, the COB (chip on board) type LED was selected. COB's performance dependence on temperature was specified by the manufacturer. The temperature of a COB is easily monitored, in a point designed for this in the chip support, named Tc, and shown in figure 1.

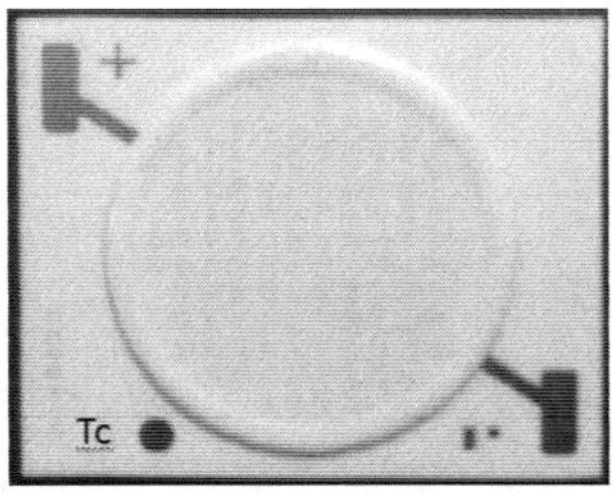

Figure 1. Selected model of COB LED.

The selected COB is supplied with an electrical current so that the luminous flux is around 1 000 lm, which is the typical luminous flux found in commercial lamps tested in integrating spheres. The electrical current is significantly smaller than the nominal value of the COB. Additionally, a housing was designed to fit the COB, composed by: a copper baseplate, a diffusing glass enclosure, thermocouple wires soldered to Tc, and electric wires.

Three units of the selected COB mounted in the housing were aged and monitored during more than 6 000 h, to check its stability. The luminous flux was measured each 100 h or 200 h. These measurements were made using an active temperature control and the results are in figure 2.

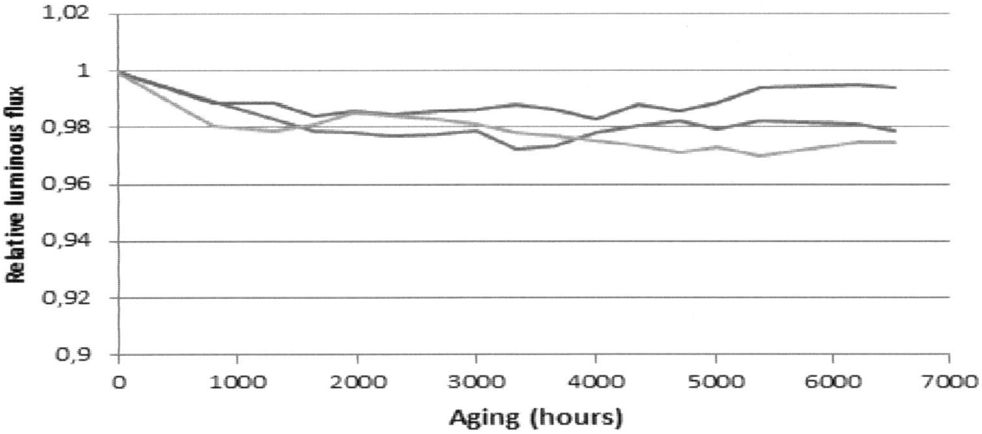

Figure 2. Relative luminous flux of 3 COBs used in the tests.

To select the heat sink, the temperature at Tc was monitored as a function of the operation time with several models of heat sinks. The curve for the selected heat sink is shown in figure 3.

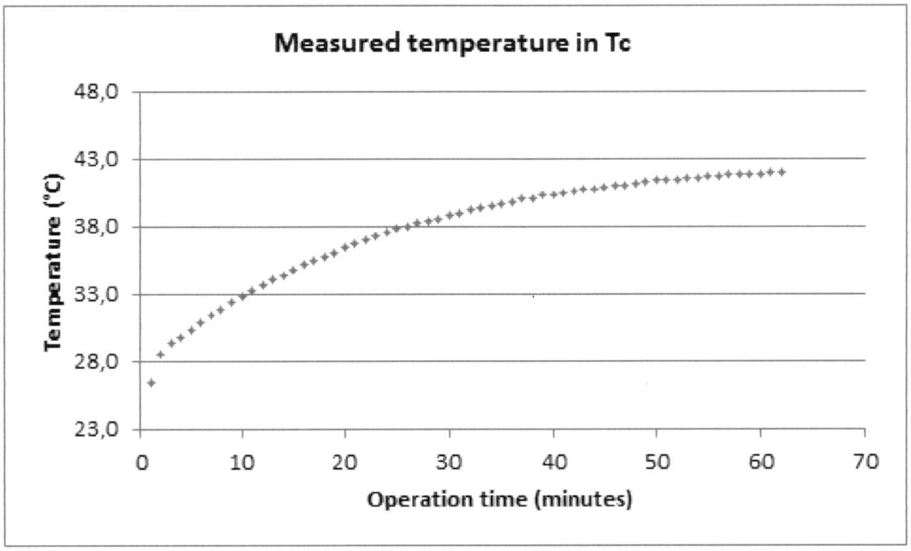

Figure 3: Tc temperature versus operating time.

Figure 4 shows an image and a schematic drawing of the prototype.

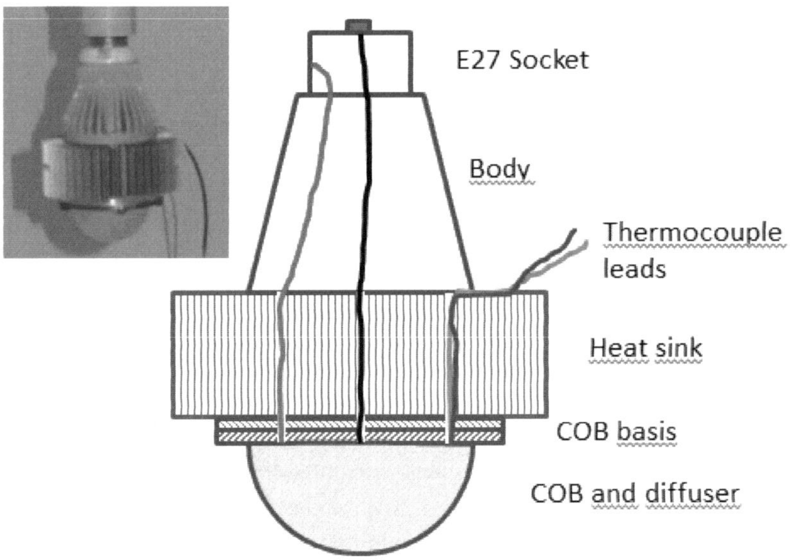

Figure 4 – Image and schematic drawing of the prototype.

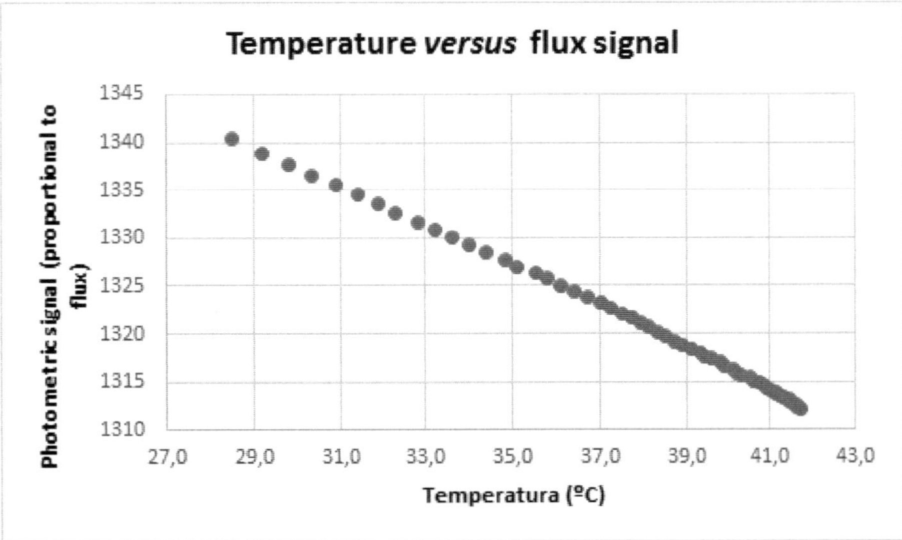

Figure 5 – Luminous flux vs Tc temperature during an interval of 60 minutes (Current 0.202 A).

The luminous flux was measured as a function of the temperature at Tc at the selected current of 0.202 A (figure 5). This allowed the determination of the best temperature interval to perform the measurement.

The measurement of the temperature at Tc is performed using a thermocouple. The electric wiring is shown in figure 6.

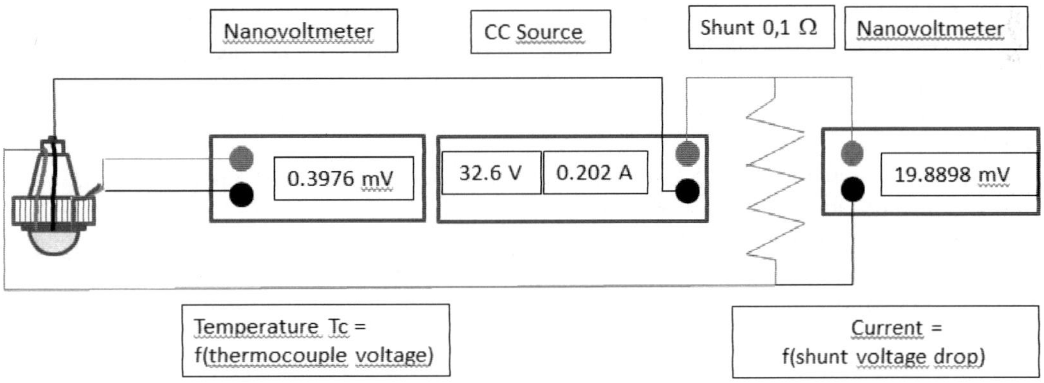

Figure 6 – Instruments and connections to the DC supply, and measurements of the current and Tc temperature

3. Analysis of results

Luminous flux standards are not routinely used. At Inmetro, we estimated that 50 hours of use, which is a typical interval in between calibrations, allows at least 150 measurements, each taking until 20 minutes. After the 6 000 h aging, the maximum observed drift in luminous flux was 0.04 % during 50 h, that is suitable for luminous flux standards.

The evolution of the temperature at Tc (figure 3) shows that it could be necessary more than 60 minutes for full stabilization. However, in figure 5 one can note that the luminous flux varies uniformly with Tc temperature. This allows the selection of a Tc temperature, that should trigger the beginning of the measurements, and a time interval, that describes the duration of the measurement.

A numerical data analysis, based on R^2 parameter of straight line segments fitted to the data, showed that, for this prototype, when the Tc temperature reaches 34 °C, the measurement could be started, with readings being taken in a time interval of 5 minutes. During this period, luminous flux variation is approximately linear and the temperature at Tc changes less than 1 °C.

This process, i.e., supplying electric current to the lamp and waiting for the temperature reaching the suitable value, takes from 8 to 10 minutes due to the dependence on the initial temperature of the prototype.

Once the specified Tc temperature is reached, a series of readings of luminous flux signal are made during a 5 minutes period (figure 7).

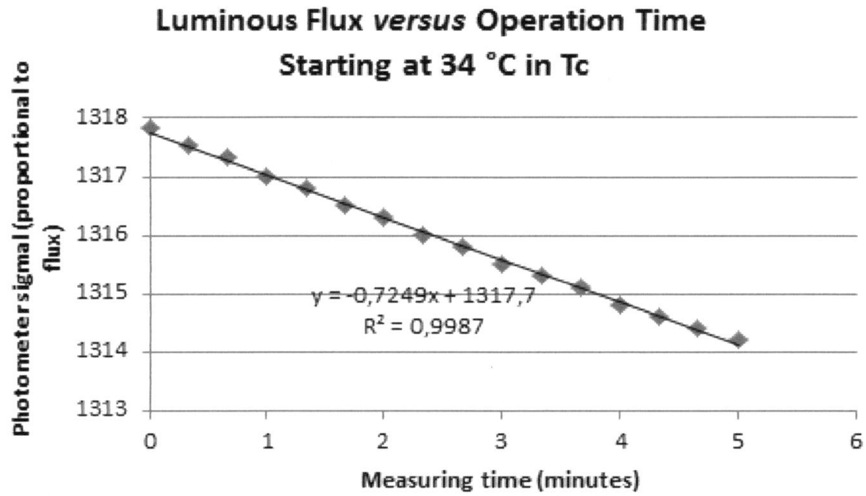

Figure 7 – 16 readings of luminous flux made during a 5 minutes period.

The value of the luminous flux is obtained by the average of the readings. Though it is possible to analyze the dispersion considering the interpolated line, this was not necessary because the standard deviation of the readings in the 5 minutes interval was only 0.1 %, which is negligible to the measurement uncertainty, at least in the current phase.

The change in the prototype's colorimetric properties was also evaluated as a function of the temperature at Tc. This parameter can be considered constant, within the instrument uncertainty, which is 17 K for CCT, for the range of Tc temperature in the measurements.

Uncertainty estimation was based in the Brazilian translation of Guide to the Expression of Uncertainty in Measurements – GUM 2008 [5]. The uncertainty components are presented in Table 1. These components are related to the measurement of the prototype using an integrating sphere with a standard incandescent lamp.

The estimated expanded uncertainty U of the measurement was 1.7 %. The more important contributions to the uncertainty were the working standard lamp (U_{ws} = 1.1 %) and the V(λ) filter ($U_{V(\lambda)}$ = 1.0 %).

Table 1. Uncertainty budget for the measurement of luminous flux of the prototype.

Entries	Symbol	Contribution [%]
Standard lamp uncertainty	ΦN	33.82
Standard lamp repeatability	YN	0.08
Instrument resolution w/standard lamp	δYN-res	0.09
Test lamp repeatability	YT	5.28
Instrument resolution w/test lamp	δYT-res	0.14
Auxiliary lamp repeatability w/standard	YHN	3.04
Non-stability auxiliar lamp w/standard	δYHN-ins	5.31
Auxiliary lamp repeatability w/test lamp	YHT	2.83
Non-stability auxiliary lamp w/test lamp	δYHT-ins	5.35
V(λ) filter uncertainty	δYf1	35.50
Reproducibility	δΦT-rpr	8.56

4. Conclusion

This paper presents an alternative to mount, characterize and use of LED based standard for luminous flux, by the determination of a Tc temperature suitable for the measurement, instead of waiting for the stabilization or using active temperature control. The calibration of the standards will be made using an integrating sphere and a reference luminous flux standard.

Acknowledgements

We would like to thank Inmetro for providing scholarships through the program Pronametro; Elo Sistemas Eletrônicos for providing material and expertise to the project; Renato N. Teixeira, Robson Chagas and Klaus N. Quelhas for the technical support in the temperature measurements.

References

[1] *CIE 18.2 The Basis of Physical Photometry* 1983 (Paris, Commission Internationale de l'Éclairage)
[2] Ohno Y 1997 *Handbook of Applied Photometry* C. DeCusatis (Poughkeepsie, AIP Press) 56-98
[3] Sperling A, Penda S, Taddeo M, Revtova E, Renoux D, Blattner P, Poikonen T and Jordan W 2016 *CIE x041:2016 Tutorial and Expert Symposium on LED Measurements* (Braunschweig, PTB)
[4] Schubert E. 2006 *Light-emitting diodes*. (Cambridge, Cambridge University Press)
[5] *Avaliação de dados de medição: Guia para a expressão de incerteza de medição 2012– GUM 2008*. (Rio de Janeiro, Inmetro)

Experimental study of phase correction determination in steel by stacking method experiment

E Beer, L Álvarez, K Bastida, G Giarmana and L Di Lillo

INTI, Instituto Nacional de Tecnología Industrial, Av. Gral. Paz 5445 CP 1650 San Martin, Buenos Aires, Argentina

E-mail: ethelb@inti.gob.ar

Abstract. The phase correction determination by stacking method experimentation is carried out by comparing the interferometric length of a set of gauge blocks wrung to each other, with the individual interferometric lengths of each gauge block. In this work the repeatability of the phase correction in steel and the uncertainties associated with this correction by using the stacking method are analyzed. Some variables considered in the analysis are the number of gauge blocks that form the stack, the nominal length of the set and the difference of the interferometric lengths obtained for both sides of each gauge block.

1. Introduction

Metre materialization is achieved by using interferometric techniques by means of gauge blocks of different nominal length. The length of the gauge blocks is traceable to standard laser frequencies. These are recommended by the Bureau International des Poids et Mesures, BIPM for the realization of the metre.

In agreement with the definition of the length of a gauge block, and in order to measure its central length, it is necessary to wring the block to an auxiliary platen which has the same surface characteristics and quality. This is not always possible to accomplish in practice. To compensate for differences in reflection properties, a phase correction is added to the measurement result of interferometric calibration of gauge blocks. Reference literature [1,2] has proven that this correction is influenced by material, surface finish and optical properties of gauge blocks and platen. The aim of this work is to study the phase correction obtained by means of the stack method which depends on a variety of parameters.

Our purpose is to establish adequate criteria in the determination of the correction. This has high impact on length measurement quality related to industrial processes, which require tight values of mechanical tolerance, like automotive industry and nuclear technology.

2. Phase correction by stack method

Because of optical properties and superficial quality of materials, the localization of the plane where incident laser radiation is reflected, does not match the real mechanical surface. Such is the origin in the need for the phase correction. The surface quality is also a critical factor in wringing film quality between gauge block and platen.

There are different methods to determine the phase correction from reflectivity or superficial roughness measurements [3,4]. In this work, the stack method is used. The phase correction is experimentally determined for a set of gauge blocks. This method consists in selecting n short gauge blocks from the set, which have been previously measured interferometrically, and the length result for both faces of each block are in good agreement. The blocks are wrung to each other to form a larger one, the stack, (Figure 1).

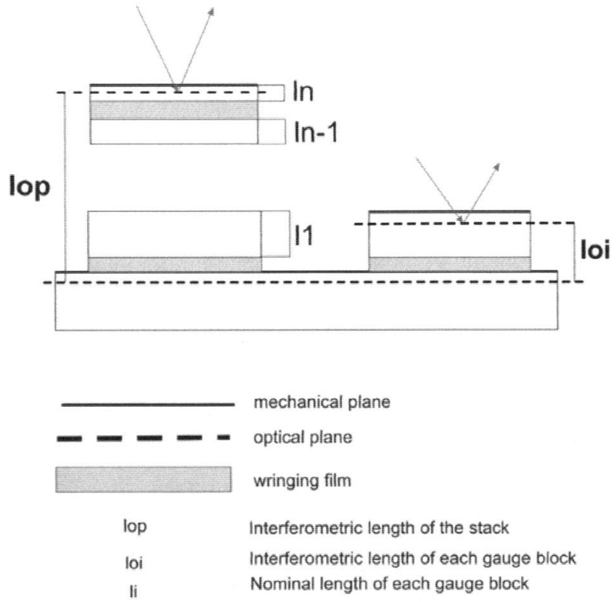

Figure 1. Stack method: localization of mechanical and optical planes.

The length of the pack, $l_{o,p}$,is measured interferometrically, and is compared with the sum of the individual lengths of the blocks $l_{o,i}$ .The difference yields the phase correction ,φ, equation 1

$$\varphi = \frac{1}{n-1}\left[l_{o,p} - \sum_{i=1}^{n} l_{o,i}\right] \qquad (1)$$

If the nominal lengths of the gauge blocks that form the stack are short (< 30 mm), the phase correction combined measurement uncertainty is [4]:

$$u_c^2(\varphi) \cong \frac{(n+1)}{(n-1)^2}\left[u_c^2(F_i) + u^2(l_w) + u^2(l_A) + u^2(l_G)\right] \qquad (2)$$

The contributions to this uncertainty, equation (2) are due to the fringe fraction F_i , wringing film l_w, wavefront errors l_A and form defects l_G.

3. Experimental method

The phase correction is determined for steel gauge blocks, taking into account the following influence parameters: number of blocks in the stack, total nominal length, difference between interferometric calibration lengths for each of the faces of each block in the stack, Table 1

Interferometric calibration is made in a Twyman Green-TESA system, which uses two stabilized lasers in 633 nm and 543 nm wavelength. Interference pattern are analyzed using the fringe fraction method. Table 2 shows the contributions to the phase correction measurement uncertainty.

Table 1. Influence parameters taken into account in the determination of φ in steel.

Stack nominal length	N° of stacks	Manufacture	N° of blocks in the stack	Platen
< 5 mm	2	#1	4	106
5 mm < L ≤ 11 mm	5	#1 (3) #2 (2)	3, 4, 5 3, 4	106 84 and 83
≅ 15 mm	3	#1 (3)	2, 3	83 and 105
> 20 mm	6	#1 (4) #2 (1) #3 (1)	3, 4, 5 3 3	106 83 105

Table 2. Contributions to the measurement uncertainty of φ.

Component	u_i (nm)
Fringe fraction, $u(F_i)$	2,2
Wringing film, $u(l_w)$	5,2
Wavefront errors, $u(l_A)$	2,5
Form defects, $u(l_G)$	1,1

4. Results and discussion

Figures 2 to 5 show the results of the phase correction determination as a function of the parameters previously described: number of gauge blocks in the stack, the stack nominal length, the mean length differences between faces of each gauge block in the stack and platen used in the measurements.

It is verified that the uncertainty in φ is reduced by increasing the number of gauge blocks in the stack. Although, in practice, it is better to keep the number of blocks in less than four. This is because of the difficulty in the wringing process in the stack construction. A minor dispersion for the phase correction is observed for stack nominal lengths between 10 mm and 25 mm. The packs wrung on platen 105 and 106 show major dispersion than the rest. In relation to the length difference between faces of individual blocks, it is advisable to take differences between 2 nm and 15 nm.

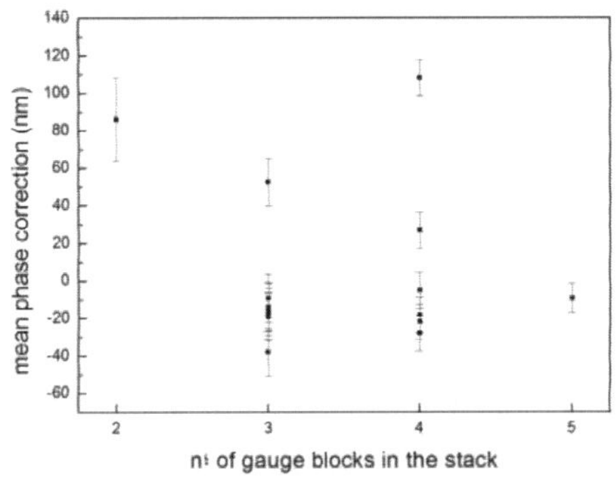

Figure 2. Phase shift as a function of the number of gauge blocks in the stack.

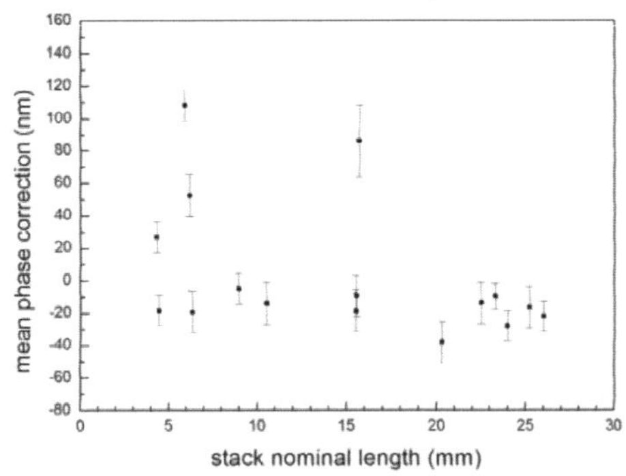

Figure 3. Phase correction as a function of nominal length.

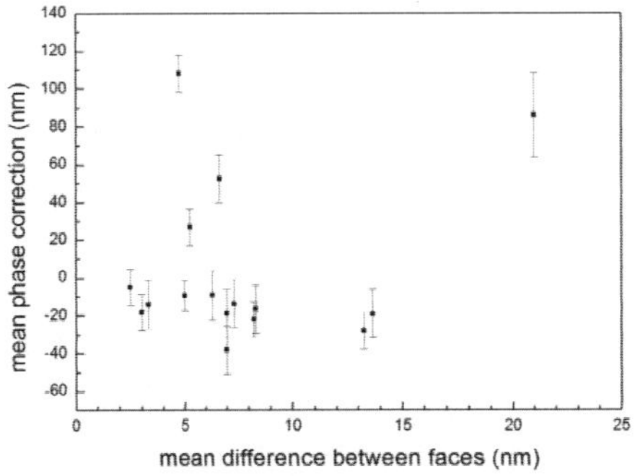

Figure 4. Phase correction as a function of length difference between faces.

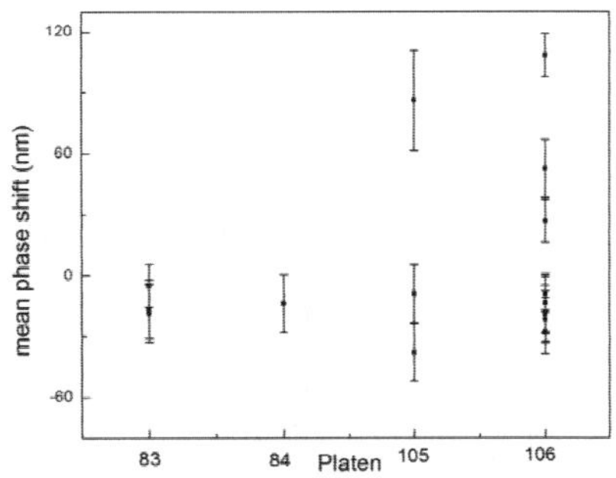

Figure 5. Phase correction as a function of number of platen.

5. Conclusions

The phase shift correction has been experimentally determined using the stack method experiment with different configurations to analyze influence parameters. This has allowed to establish adequate criteria in the determination of the phase shift correction in the interferometric calibration of gauge blocks. This is, in fact, a significative improvement in the materialization of the national length standard at INTI.

Furthermore, it is considered convenient to make a statistical analysis of stacks wrung to different platens. A similar analysis is necessary for the different manufactures; in order to count with an empirical reference that can be used in individual gauge block calibrations or when a particular set of gauge block does not include adequate blocks to build a stack.

6. References

[1] Ramotwski Z and Salbut L 2012 *Meas. Sci. Technol.* **23**

[2] Thwaite E 1978 *Metrologia* **14** pp 53-62

[3] Leach R, Jackson K and Hart A 1997 NPL Report MOT 11

[4] Decker J and Pekelsky J 1997 *Metrologia* **34** pp 479-493

IOP Conf. Series: Journal of Physics: Conf. Series **975** (2018) 012031 doi:10.1088/1742-6596/975/1/012031

Trilateral interlaboratory with SSL (WLEDi) luminaire

E C Burini Junior [1], E R Santos [2,3] and L O Assaf [4]

[1]IEE - Instituto de Energia e Ambiente da Universidade de São Paulo, 05508-010, São Paulo-SP, Brazil;
[2]EPUSP - Departamento de Engenharia Metalúrgica e de Materiais da Universidade de São Paulo, 05508-900, São Paulo-SP, Brazil;
[3]FATEC-SP - Departamento de Sistemas Eletrônicos da Faculdade de Tecnologia de São Paulo, 01124-060, São Paulo-SP, Brazil;
[4]DLLyV/UNT - Departamento de Luminotecnia, Luz y Visión/Universidad Nacional de Tucumán, Tucumán,Argentina.

E-mail: elvo@iee.usp.br

Abstract. The IEE/USP laboratory and two others, all belonging to RBLE (Brazilian Network of Test Laboratories) participated in a trilateral comparison performed from measurement independently of participants interaction. The results from electric and photometric measurements carried out on samples of Solid State Lighting - SSL, Inorganic White Light Emitting Diode (WLEDi) luminaires by three accredited laboratories were considered in order to point out mutual deviations and to verify the confidence in a bilateral comparison. The first analysis revealed a maximum deviation of 4.2 % between the luminous intensity attributed by one laboratory and the arithmetic mean value from three laboratories. The largest standard uncertainty value of 1.9 % was estimated for Total Harmonic Distortion of electric current THDi and the lowest value, 0.4 %, to the luminous flux. The extreme deviation for one parameter results was 7.2 % at maximum luminous intensity and the lowest was 1.7 % for luminous flux.

1. Introduction

According to relevant documents [1, 2] the comparison of results from measurements carried out in different laboratories is a valid instrument of metrology and can be used for quality assessment. Although statistic is an important tool for assessment of laboratory practices, the present paper does not embrace sampling protocols, it has focus on measurement results obtained by three laboratories belonging to the RBLE (Brazilian Network of Test Laboratories accredited by Inmetro), looking for critical points which will require treatments. Assertive decisions are facilitated when the results are properly presented.

In the bidding process conducted by USP (University of São Paulo) to provide Inorganic White Light Emitting Diode (WLEDi) technology for outdoor lighting on their campi, at CUASO campus there was an initial stage in which report containing electric and photometric data on each type of luminaire offered were submitted and a second stage, in which samples were received and tested at the IEE/USP (Institute of Energy and Environment, USP) laboratory.

The results obtained from IPT (Instituto de Pesquisas Tecnológicas) report, considered as results from test of type for the WLEDi 50 W luminaire [3] in the USP bidding process were compared with parameter assessed at the IEE/USP laboratory from four luminaires removed from the product batch supplied (before field installation). From four test reports results [4], where there is indication on the use of goniometer during the measurement process, the assessed data were processed (arithmetic average and the standard uncertainty was estimate). The results from those measurements were compared with data contained from a type test report supplied by the USP Campus Prefecture - PUSP-C [4]. The results from a bilateral experiment is been considered at the first part of this article, which results are presented at section 3.1. The trilateral methodology used is considered at section 2.

2. Methodology (one luminaire and three Labs)

The second part of this article was based on measurements of the same luminaire performed in three laboratories identified as Lab I, Lab II and Lab III to obtain electric power; electric current; total harmonic distortion of electric current (THDi); total harmonic distortion of voltage (THDv); power factor (PF); luminous flux (LF); luminous efficiency (LE) and maximum luminous intensity (MLI). Figure 1 shows the WLEDi 120 W luminaire used at three laboratories for all tests.

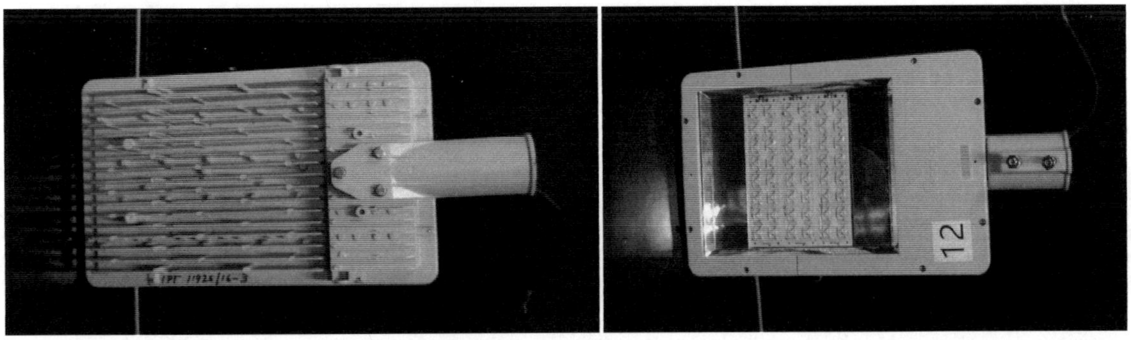

Figure 1. The WLEDi 120 W luminaire supplied on electric power line of 220 V.

At each Lab the luminous flux was determinated with the aid of the goniophotometer procedure. The arithmetic average for each parameter was determined and the uncertainty estimated. In the case of THDv and THDi, due to the results of Lab. II, the average value determination was calculated with only two elements. The WLEDi 120 W luminaire used was energized at the rated electric voltage of 220 V, a stabilization period should be applied as a precondition before the measurement starts according to the particular procedure of each laboratory.

Note: The IESNA LM-79 [5] was used as the reference document at the IEE/USP Lab. for the light output and electric power stabilization period.

Both electric and photometric data were sampled during each luminaire test.

Relative standard deviation from laboratories results was calculated. For the uncertainty estimation or the dispersion around the assigned value (average) the standard uncertainty was used [6]. The results are presented in the section 3.2 and 3.3 by seven tables with assigned value associated, absolute uncertainty and relative deviation from each Lab in relation to the arithmetic average obtained.

3. Results and Discussion

3.1. Parameter from type and samples tests (luminaires and two Labs)

Table 1 (at the IEE column) presents arithmetic average (four samples) and average standard deviation for seven parameters quantities considered of the WLEDi 50 W luminaries from USP CUASO campus installation [4], and at the IPT column presents data from the report assessed [3] and each uncertainty declared is for luminaire code AP-SL-50W, all luminaires are from Alper (a brazilian brand). The relative deviation, the column at the right, table 1, was calculated considering the difference between the results from both laboratories in relation to the IPT result.

On the results, using at least four parameters (luminous flux, electric current, luminous efficiency, THDi), it does not indicate conformity consistency between the type and the samples extracted from the batch.

IOP Conf. Series: Journal of Physics: Conf. Series **975** (2018) 012031 doi:10.1088/1742-6596/975/1/012031

Table 1. Absolute value for seven parameters of the WLEDi 50 W luminaire, the estimated uncertainty and the relative deviation from one Lab result.

Luminaire Parameter	IEE [a]	IPT [b]	Relative deviation (%)
Electric Current (mA)	250 ± 2	230 ± 20	8.8
Electric Power (W)	53.6 ± 0.4	50.45 ± 0.05	6.2
Power Factor	0.9725 ± 0.0005	0.99 ± 0.01	-1.8
THDi (%)	7.34 ± 0.29	7.99 ± 0.05	-8.1
Luminous Flux (klm)	4.97 ± 0.04	4.30 ± 0.15	15.6
Luminous Efficiency (lm/W)	92.75 ± 1.30	85.3	8.7
Correlated Color Temperature - T_{cp} (K)	4236 ± 57	3822 ± 50	10.8

[a] arithmetic average and expanded standard deviation were estimated from result of four luminaries sampled [4]; and

[b] Data from test report, luminaire model AP-SL-50W (luminaire submitted to test, whose, document/report was presented at bidding phase) [3].

3.2. Electric parameter (one luminaire and three Labs)

3.2.1. Electric Current. Table 2 presents the arithmetic average value, estimated uncertainty of the arithmetic average to the electric current for WLEDi 120 W luminaire, the relative standard deviation was calculated for each laboratory in relation of the arithmetic average value.

Table 2. Relative deviation from the average for each laboratory measurement of the electric current, values in %.

Laboratory Identification	Electric Current **(578.5 ± 8.0) mA;** (± 1.4 %)
Lab. I	- 3.1
Lab. II	0.43
Lab. III	2.7

The range of maximum variation between the three labs for the electric current is estimated at 5.8 %.

3.2.2. Electric Power. Table 3 presents the arithmetic average value and uncertainty for the electric power of WLEDi 120 W luminaire, relative deviation was calculated for each laboratory in relation of the arithmetic average value.

Table 3. Relative deviation from the average for each laboratory measurement result of electric power, values in %.

Laboratory Identification	Electric Power **(123.0 ± 1.5) W;** (± 1.2 %)
Lab. I	- 2.9
Lab. II	1.6
Lab. III	1.3

The maximum variation range between three laboratories for the electric power is estimated at 4.5 %.

3.2.3. Total harmonic distortion of the electric current (THDi) and total harmonic distortion of the electric voltage sources (THDv). Table 4 presents the arithmetic average value calculated to estimate the THDi of WLEDi 120 W luminaire (and THDv of each luminaire-source combination) and each relative deviation calculated in relation to the arithmetic average value.

Table 4. Relative deviation from the average for each laboratory measurement result of THDi and THDv, values in %.

Laboratory Identification	THDi **(9.25 ± 0.18)** (± 1.9 %)	THDv **(0.25 ± 0.03)** (± 11 %)
Lab. I	2.7	- 16
Lab. II [c]	---	---
Lab. III	- 2.7	16

[c] Laboratory II do not supplied THDv measurement result and for this reason the THDi (value 1.6 %) was not incorporated at table 4.

The maximum range variation between the three laboratories for the measurement of the THDv is estimated at 32 % and 5.4 % (for THDi).

The algebraic sum of the relative deviation of each parameter is null, this is a strong indication of consistency in the procedure used.

3.2.4. Power Factor (PF). Table 5 presents the arithmetic average value calculated to estimate the PF of WLEDi 120 W luminaire and each relative deviation in relation to the arithmetic average value.

Table 5. Relative deviation from the average for each laboratory measurement result of the PF, values in %.

Laboratory Identification	Power Factor **(0.966 ± 0.007)** (± 0.7 %)
Lab. I	0.2
Lab. II	1.4
Lab. III	- 1.7

The maximum range variation between the three laboratories for the measurement of the power factor is estimated at 3.1 %.

3.3. Photometric parameter (one luminaire and three Labs)

3.3.1. Luminous Flux (LF). Table 6 presents the arithmetic average value, the uncertainty of the average luminous flux of WLEDi 120 W luminaire, the relative deviation calculated for each laboratory determination in relation to the average value.

Table 6. Relative deviation from the average for each laboratory determination of the luminous flux, values in %.

Laboratory Identification	Luminous Flux **(13.15 ± 0.05) klm**; (± 0.4 %)
Lab. I	- 0.9
Lab. II	0.07
Lab. III	0.8

The maximum range variation between the three laboratories for the determination of the luminous flux is estimated at 1.7 %.

3.3.2. Luminous Efficiency (LE) [7]. Table 7 presents the arithmetic average value, the uncertainty of the average luminous efficiency of WLEDi 120 W luminaire and each relative deviation in relation to the average value.

Table 7. Relative deviation from the average for each laboratory determination of luminous efficiency [7], values in %.

Laboratory Identification	Luminous Efficiency [7] **(107 ± 1) lm/W;** (± 0.9 %)
Lab. I	2.2
Lab. II	- 1.7
Lab. III	- 0.5

The maximum range variation between the three laboratories for the determination of the luminous efficiency is estimated at 3.9 %.

3.3.3. Maximum Luminous Intensity (MLI). Table 8 presents the arithmetic average value, the uncertainty for the average value to the maximum luminous intensity (I_{MAX}) of WLEDi 120 W luminaire and each relative deviation in relation to the average value.

Table 8. Relative deviation from the average for each laboratory measurement result of the maximum luminous intensity (I_{MAX}), values in %.

Laboratory Identification	MLI (I_{MAX}) **(8.98 ± 0.16) kcd;** (± 1.8 %)
Lab. I	3.0
Lab. II	1.2
Lab. III	- 4.2

The range of maximum variation between the three laboratories for the (I_{MAX}) is estimated at 7.2 % (the wider range obtained).

Due to the coexistence and possible use of two different systems (European, CIE and North America, IESNA) for photometry of public luminaire, the vertical and horizontal angles were not here considered.

4. Conclusions

According to Table 1, and for at least four parameters (luminous flux, electric current, luminous efficiency, THDi), the difference showed between the type and samples test results of WLEDi 50 W luminaires, it does not support conformity and suggest to be very different the lighting products.

The result from laboratory measurement of three accredited laboratories (RBLE), eight parameters of one SSL luminaire (WLEDi, 120 W) were considered in this work. For the analysis of results the relative deviations of each Lab and in relation to the arithmetic average value determined were used. The smaller amplitude of variation in relation to the average value occurred for luminous flux and was 1.7 %; followed by PF 3.1 %; luminous efficiency 3.9 %; electric power 4.5 %; THDi 5.4 %; electric current 5.8 %; maximum luminous intensity 7.2 % and for THDv 32 %.

As each Lab used different electric power supply for the measurements further investigations are suggested to THDv consideration.

IOP Conf. Series: Journal of Physics: Conf. Series **975** (2018) 012031 doi:10.1088/1742-6596/975/1/012031

The luminous flux determination resulted in the lowest relative standard uncertainty (0.4 %) and THDi the largest (1.9 %).

If THDv parameter is not taken in consideration, the largest amplitude of the relative deviations for all the magnitudes from one laboratory was 6.9 % (Lab. III); the lowest 3.3 % (Lab. II).

The estimated values for standard uncertainty (non-expanded) for the results from WLEDi 120 W luminaire and three laboratories considered are less than 2 %, a level that can be considered acceptable without the THDv been considered.

The algebraic sum of the relative deviation of each parameter from all Labs results was found close to zero, this is an indication of consistency in the procedure used.

The large deviation observed on THDv suggest a better knowledge development and control over the type of electric source been used.

The wide differences shown for the maximum luminous intensity, electric current and for others electric parameters suggest further investigations.

Acknowledgments
The authors are grateful to the financial support from the Brazilian agency CAPES (Coordenação de Aperfeiçoamento de Pessoal de Nível Superior), two country project nº 048/2013; to the Eng. Enea from the PUSP-C for the luminaires samples (50 W) and test reports [4] and to Eng. Mauro from the Demape Company who supplied reports from two laboratories and one luminaire (120 W).

References
[1] ASSOCIAÇÃO BRASILEIRA DE NORMAS TÉCNICAS - ABNT ISO/IEC 17025:2005. *Requisitos Gerais para Competência de Laboratórios de Ensaio e Calibração.*
[2] Draft prEN ISO/IEC 17025:2016 (ISO/IEC DIS 17025:2016). *General requirements for the competence of testing and calibration laboratories.*
[3] Test Report from IPT n. 1 043 934 – 203 April, 10, 2013 luminaire submitted to test (the document was presented at bidding), Alper brand, model AP-SL-50W, 6 pp.
[4] Test Reports from IEE/USP, codes RE (77.162; 77.165; 77.168; and 77.171), identification of samples: ID (1223; 1196; 1238; 1244), reports received from the PREFEITURA DO CAMPUS USP DA CAPITAL – PUSP-C, http://www.puspc.usp.br/.
[5] ILLUMINATING ENGINEERING SOCIETY - IES LM-79-08 (2008) *Approved Method for the Electrical and Photometric Measurements of Solid-State Lighting*, Illuminating Engineering Society, chapter 5.0 Stabilization on SSL product, pp 3. At https://www.nist.gov/sites/default/files/documents/2017/04/05/iesna_lm_79-iesna_lm_79_1.pdf.
[6] ASSOCIAÇÃO BRASILEIRA DE NORMAS TÉCNICAS - ABNT ISO/IEC GUIA 98-3. 2014 *Incerteza de medição, Parte 3: Guia para expressão de incerteza de medição* (GUM: 1995), 154 pp.
[7] ASSOCIAÇÃO BRASILEIRA DE NORMAS TÉCNICAS. (TB-23) NBR 5461 1991 – *Iluminação, terminologia*, 68pp, item 3.1.55 (Eficiência luminosa de uma fonte), pp13.

IOP Conf. Series: Journal of Physics: Conf. Series **975** (2018) 012032 doi:10.1088/1742-6596/975/1/012032

Measurements and uncertainty budget for chiral liquids optical rotation at 633 nm

L P Souza, M B Guedes, K C Cunha and A P D Alvarenga

National Institute of Metrology, Quality and Technology, 25250-020, Brazil

adalvarenga@inmetro.gov.br

Abstract. Here is presented the assembling of an extension of the polarimeter for optical rotation measurements in chiral liquids at 633 nm and 1 cm optical path. It was tested using a standard sucrose solution, and by comparison considering the assessed uncertainties on both the calculated expected and the experimental values. The normalized error criterion showed a very good agreement, and some discrepancies are possibly due to the cuvette material variation.

1. Introduction

Optical polarization of light was discovered in 1808 by Étienne-Louis Malus and has evolved into one main field of study [1]. One practical application is the study of the optical rotation of substances and its use in metrology. Quartz control plates (QCP) having their optical rotation well determined are standards used to calibrate polarimeters [2]. The Optical Metrology Division of Inmetro assembled a reference polarimeter for QCP calibration service [3, 4]. In order to extend the capabilities of the instrument for the measurements of chiral liquids, a device was designed and built in order to support and thermalize optical cuvettes. Optical rotation measurements in a solution of sucrose reference material were realized and the results were compared to the calculated expected values [2] considering the uncertainties on both measured and calculated values. Here we report on the first results from this system for the measurement of optical rotation at 633 nm of small amounts of chiral liquids [5].

2. Experimental set-up

2.1. The thermalization device

The optical rotation of a substance depends on its temperature and on the wavelength of the incident radiation, optical path and concentration. Temperature control is achieved by a circulating thermal bath through the polarimeter chambers, monitored by inserted thermistors. A developed device, based on a tested prototype [6], was assembled at the polarimeter, as shown in Figure 1: (a) the topped inserted cuvette (white lid visible), the inserted thermistor for the cuvette temperature control, and the air temperature measurement, near the device; (b) filled cuvette.

2.2. The preparation of the standard solution

In order to verify the accuracy of measurement in the modified polarimeter, a standard sucrose solution was prepared at the Electrochemistry Laboratory of Inmetro following ICUMSA [2]: 26.0160 g of pure sucrose dissolved in pure water to a final volume of 100.000 cm^3 under normal conditions.

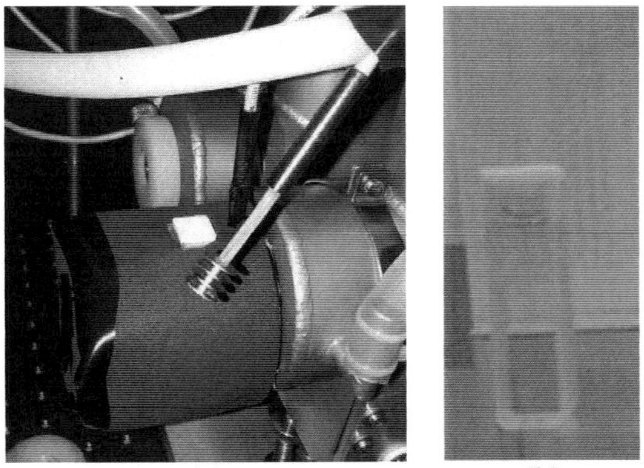

(a) (b)

Figure 1. Cuvette thermalization device

The analytical balance was calibrated; Type I water was used to dissolve the commercial sucrose reference material of 99.5 % purity. The sucrose was slowly poured in a beaker and weighed, water added to the final desired weigh and a magnetic agitator further homogenized the solution. Four clean and identified cuvettes of 1 cm optical path and 3.5 mL volume (as in Figure 1b) were filled and immediately taken to the Laboratory of Optical Applications (Laopt) for the polarimetric measurements. Four months later, the standard solution was prepared again and the measurements repeated.

2.3. *The optical rotation measurements*

The polarimeter has been described elsewhere [3, 4, 5]. It has two thermalized chambers: one contains a standard QCP for the polarimeter verification, and the other (Figure 1a) has the device with the cuvette. The whole instrument is automated with LabVIEW programs that also acquire, register and analyze data giving the mean values of the optical rotation and temperature. Each register consists of six repetitions of the measurement of the angle of rotation of the polarization plane of the radiation from a calibrated He-Ne laser of (632.990978 ± 0.000027) nm, from the QCP and sucrose solution; the temperature is controlled by the thermal circulating bath set at 20 °C, and is sensed by a calibrated thermistor inserted near the cuvette and registered during the acquisition. The four cuvettes were measured in the same day, delay time between them only due to optical alignment and temperature stabilization.

3. Experimental set-up

The measured optical rotation values were compared to the calculated expected values by the equations given in the ICUMSA standard [2] where the input parameters are the measured mass fraction and the temperature, and thus uncertainties were evaluated for the standard sucrose solution expected value. The polarimeter had already been characterized [4] and a new uncertainty budget was obtained for the modified setup [5]. The uncertainty analysis calculation was realized in an Excel worksheet.

3.1. *Uncertainty evaluation of the measured mass fraction*

The mass fraction, w, was calculated from the mass values measured during the preparation, corrected by the balance calibration certificate and purity of the sucrose reference material. In (1) m is the corrected sucrose weighed mass, M is the corrected sucrose plus water weighed masses, and p is purity.

$$w = \frac{m \times p}{M \times 100} \tag{1}$$

The uncertainty evaluation:

$$u_w^2 = \frac{p^2}{(100\times M)^2} \times (u_m)^2 + \frac{m^2}{(100\times M)^2} \times (u_p)^2 + \frac{m^2 \times p^2}{(100\times M)^2} \times (u_M)^2 \tag{2}$$

where u_m, u_M and u_p are the contributions due to *m, M* and *p*.

3.2. Uncertainty evaluation of the calculated optical rotation value

The ICUMSA equation in (3) establishes the value of the optical rotation, α, of a standard sucrose solution under the radiation of 546.2271 nm, for an optical path of 20 cm, as a function of the mass fraction (valid from 0 to 0.6), and at the temperature interval 18 °C to 30 °C:

$$\alpha_{546} = a_{01} \times w + a_{02} \times w^2 + a_{03} \times w^{23} + a_{04} \times w^4 + a_{05} \times w^5 + (a_{11} \times w + a_{12} \times w^2 + a_{13} \times w^3)(t - 20.005) \tag{3}$$

where w is the mass fraction, t is the measured temperature, and the a_{ij} are given constants. For notation simplicity the wavelengths values are depicted as 546 nm and 633 nm.

In order to compare with the measurements, the calculated value needs to be converted to the wavelength used at the polarimeter using the ICUMSA equation, where $\lambda = 0.633$ μm, and *a, b, c* and *d* are given constants:

$$\alpha_\lambda = \frac{\alpha_{0.546}}{(a + b\lambda^2 + c\lambda^4 + d\lambda^6)} \tag{4}$$

The temperature and mass fraction dependences of the optical rotation are independent in the validity range of the equation [2]. For the uncertainty evaluation in the calculated optical rotation, it was applied the Kragten method, considering variations of the magnitude of the uncertainties in w and in t. For a fixed t, the difference in calculated (3) for w and $w + u_w$ resulted in the contribution to the optical rotation uncertainty due to the uncertainty in the mass fraction, $u_{\alpha w}^2$. Considering a fixed w, (3) was calculated for t and $t + u_t$, resulting the contribution to the optical rotation uncertainty from the uncertainty in the measured temperature $u_{\alpha t}^2$. The resulting uncertainty in the calculated optical rotation in 546 nm is:

$$u_{\alpha 546}^2 = u_{\alpha w}^2 + u_{\alpha t}^2 \tag{5}$$

The uncertainty in the calculated optical rotation in 633 nm was obtained by applying (4).

3.3. Uncertainty evaluation of the measured optical rotation value

The influence factors were studied and the main components for the uncertainty evaluation come from the traceability of the standards, type B, normal, the optical alignment, type A, normal, acquisition and analysis methods considered under the repeatability, type A, normal. As a numerical example, the components c_i already in units for optical rotation measurements are: c_l from the laser wavelength calibration certificate, $(2.97 \times 10^{-6})°$; c_{tr} from the QCP calibration, 0.0005°; c_{temp} for the temperature sensor calibration, $(3.15 \times 10^{-5})°$; c_{axis} from the optical axis of the polarimeter, 0.0004°; c_{rep} from the repeatability, 0.0002°. Following the GUM [7], it was calculated the effective degrees of freedom, the standard combined uncertainty $U_c = 0.0005°$, and the coverage factor, $k = 2.00$, resulting the expanded uncertainty $U = 0.001°$.

4. Results and Discussion

In order to do the comparison with the calculated values the measured values and uncertainties were multiplied by 20, as the optical path in the cuvette is 1 cm and the equation (3) is for a 20 cm optical path. The mass fraction values and uncertainties of the first preparation was $w_1 = 0.25882 \pm 0.00026$ and for the second was $w_2 = 0.25879 \pm 0.00026$. Figure 2 shows the calculated (squares) and measured (circles) values of the optical rotation of standard sucrose solutions prepared on two distinct occasions on July, 27, 2016 (2a) and on November, 11, 2016 (2b) for the four identified cuvettes. The uncertainties were plotted as error bars in the graphs.

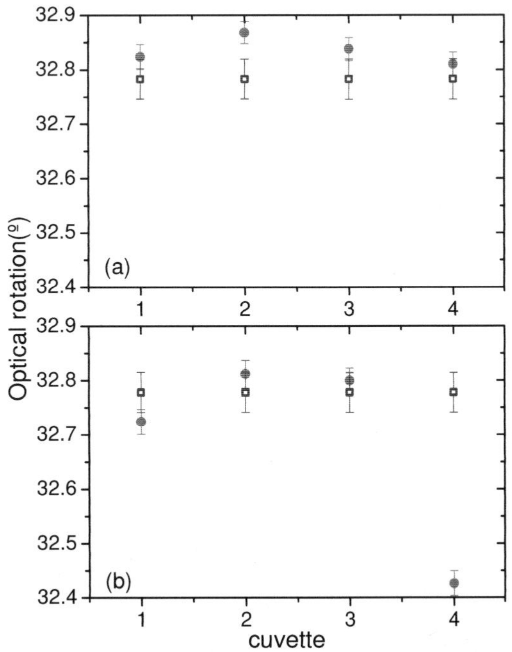

Figure 2. Calculated (squares) and measured (circles) optical rotation values of sucrose solution: (a) on July, 27, 2016 and (b) on November, 11, 2016.

The calculated values are practically constant in each day and very repeatable, their averages and standard deviations are (32.7829 ± 0.0002)° and (32.7781 ± 0.0002)°. Those values although very precise still do not present the necessary reliability for the comparison, what was achieved with the uncertainty assessment which values are presented on table 1. The measured values are more spread, and in the second day one was considered outlier. The cuvette quality appears to be the driving factor of the discrepancies and a systematic study is a future improvement.

The acceptance criterion for the comparison was the normalized error $E_n < 1$ taken between the calculated and measured averaged values and respective uncertainties, presented in Table 1. The obtained E_n values were very small even considering the outlier, 0.0013 for the first day and 0 or 0.0019 for the second, attesting the good performance of the modified polarimeter for optical rotation measurements on chiral liquids under 633 nm radiation.

Table 1. Average optical rotation values and uncertainties of a standard sucrose solution under radiation of 633 nm, at 20 °C and 20 cm optical path.

Date	Calculated, U		Measured, U		E_n
July, 27	32.78°	0.04°	32.84°	0.02°	0.0013
November, 11	32.78°	0.04°	32.78°	0.02°	0
outlier			32.69°	0.02°	0.0019

5. Conclusions

Here is presented the assembling and first measurements of an expansion of the reference polarimeter of Laopt for the measurements of the optical rotation of chiral liquids under 633 nm and 1 cm optical path. The system was tested with a standard solution of a commercial sucrose RM for which the calculated expected values of the optical rotation were compared to the measured ones taking into account the uncertainties. The agreement was good and pointed to improvements in the system, and to the use of a purest reference material. A new result brought by this work was the assessment of the uncertainty on the calculated expected value.

Acknowledgements

A P D Alvarenga thanks CNPq, and L P Souza thanks Professional Master's Course in Metrology and Quality of Inmetro.

References

[1] Azzam R M A 2011 *Thin Solid Films* **519** 2584
[2] *ICUMSA Method SPS*-1 2009 Polarimetry and the International Sugar Scale-Official
[3] Alvarenga A P D, Pereira N C E, Tarelho L V G, França R S and Belaidi H 2010 *July OIML Bulletin* 5 www.oiml.org/bulletin
Schulz M, Fricke A, Stock K, Alvarenga A D and Belaid H 2006 High-accuracy polarimetric calibration of quartz control plates *Proc. XVIII IMEKO World Congress* Rio de Janeiro Brazil
[4] Gomes B S 2011 *Realização de testes para a caracterização metrológica do polarímetro de referência para calibração de placas de controle de quartzo* final course dissertation at CTM Inmetro SEE- RJ CECO Duque de Caxias Brazil
Alvarenga A D, Pereira N C E, Gomes B S and Grieneisen H P H 2011 Evaluation of measurement uncertainties for polarimetric calibration of quartz control plates *Proc. 6th Brazilian Congress on Metrology* Natal Brazil
Ferreira G L, Guedes M B, Pereira N C E, Souza L P and Alvarenga A P D 2015 Medições da dependência com a temperatura da rotação óptica em placas de controle de quartzo *Proc. 8th Brazilian Congress on Metrology* Bento Gonçalves Brazil
[5] Souza L P 2017 *Montagem de um sistema para a medição da rotação óptica em líquidos quirais* Master degree dissertation PGMQ Inmetro Duque de Caxias Brazil
[6] Souza L P and Alvarenga A P D 2015 Montagem de um sistema para medição da rotação óptica em líquidos *Proc. 8th Brazilian Congress on Metrology* Bento Gonçalves Brazil
[7] *GUM 2008* 2012 INMETRO/CICMA/SEPIN ISBN 978-85-86920-13-4

New method for characterization of retroreflective materials

O S Junior[1],E S Silva[1,2], K N Barros[1,3],J G Vitro[1]

[1]Institute for Technological Research of Sao Paulo; Av. Prof. Almeida Prado 532 Cid. Universitária - Butantã. São Paulo - SP, 05508-901,Brazil.
[2] University of São Paulo, Av. Prof. Almeida Prado, nº1280 - Butantã. São Paulo, 05508-070. Brazil.
[3]Federal University of ABC. Av. dos Estados, 5001 - Bangú, Santo André - SP, 09210-580,Brazil.

E-mail: osanchez@ipt.br; emerson@ipt.br; kelvis@ipt.br; jgvitro@ipt.br.

Abstract: The present article aims to propose a new method of analyzing the properties of retroreflective materials using a goniophotometer. The aim is to establish a higher resolution test method with a wide range of viewing angles, taking into account a three-dimensional analysis of the retroreflection of the tested material. The validation was performed by collecting data from specimens collected from materials used in safety clothing and road signs. The approach showed that the results obtained by the proposed method are comparable to the results obtained by the normative protocols, representing an evolution for the metrology of these materials.

Keywords: Assay, Retroreflexion, Goniophotometry, Comparison.

1. Introduction

The determination of properties of the retroreflective materials has been benefiting from the evolution of the positional photometric technologies, with lower uncertainty and automation for positioning, which reduces times and costs. Although this fact is relevant in the various methods applied, practices that require a considerable amount of time to perform tasks and insufficient amount of data are still perceptible. The evolution of metrology has brought new possibilities to minimize uncertainty, which allows a continuous work improvement. Thus, the method presented here aims to broaden the range of data collected, allowing a broader analysis of retroreflective material and, at the same time, increase the accuracy. In this way, a more secure definition of the conformity of these products is possible. Another objective is increasing productivity with the method, which makes it possible to reduce the number of professionals required to perform the test. This generates cost reduction, time reduction and greater dynamism in laboratory operations.

1.1 Definition of Retroreflection

As defined by [1], retroreflection is the superficial phenomenon of reflection, in which the reflected rays are returned preferentially in directions close to the opposite direction of the incident rays and this property is maintained over wide variations in the direction of the incident rays. Thus, the retrorefletance is the proportion of the radiating flux (or luminous flux) emerging relative to the incident flow, within the restriction angle of the geometric conditions of incidence and reflection. [1], as can be verified in figure 1.

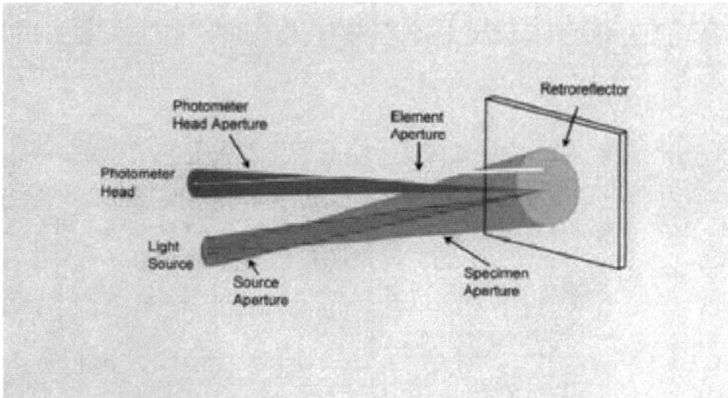

Figure 1. Illustration of aperture angles for Retroreflection [1].

Retrorefletor is a material with a surface or device that exhibits retroreflection [1]. Retroreflective material is a material that has a thin continuous layer of small retroreflective elements on or very near to its surface (e.g., retroreflective sheeting, beaded paint, highway sign surfaces, or pavement striping).

1.2 Definition of Goniophotometry

The goniophotometers are equipments used for measurements of the optical characteristics of luminous bodies. The name derives from traditional goniometers that are single-purpose instruments to angular measurements. In addition to the photometric component contained in the name, it is added to the significance of angular measurements of the optical characteristics of luminous bodies.

The goniometers allows to determine primarily the luminous flux (given in lumens), and the light distribution graph of the emitted light. The item to be evaluated can be a luminaire, a lamp or other light emitting devices [2]. With this equipment, all the optical characteristics of the products can be evaluated accordance to the angular measurements of the light emitting spectrum of the luminous bodies [3]. There are varied types of goniophotometers constructions, so it is necessary to use classifications. Among the goniophotometers, there are those of type C. These are ideal for measurements where the object to be measured is stationary [2]. See a typical arrangement in Figure 2.

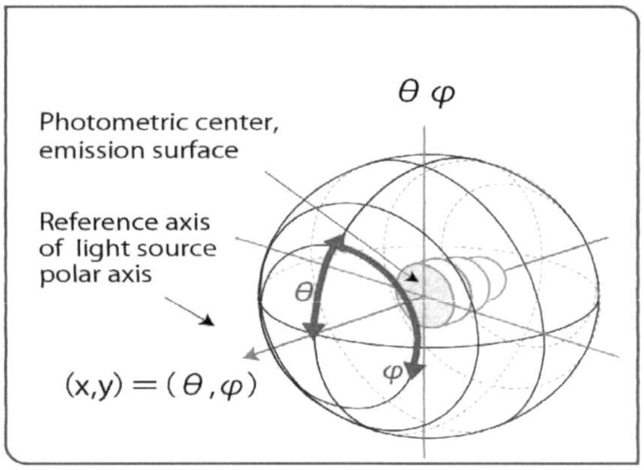

Figure 2. Graphical representation of the Goniophotometers type C, in the center is positioned the luminous body to be measured, and on the axes in red are represented the axis of the sensor's offset. Credit: OTSUKA ELECTRONICS Co.,LTD Retrieved from: <https://www.otsukael.jp/upload/files/GP%281%29.jpg> Accessed on: ago 2017.

For measurements of intensity distributions in a goniometer type C, the object to be measured is positioned in the geometric center of the goniometer. In one of the arms of the goniometer, a CCD camera is positioned and this accompanies the movement on the imaginary spherical surface by which the movement of the sensor occurs. the radius of this sphere is fixed, varying only the angle of the camera [4].

2. Test Method

In this method of retroreflectance proposed, the equipment was used according to the traditional method described in the standard [5-7] which can be summarized as: Goniometer; Photometer; Light source; Surface of the test body and dark photometric room. The validity of the measurements was guaranteed by the verification of the minimum instrumentation requirements laid down by the above mentioned standards, namely:

- Goniometer: Minimum resolution of 1° for angular step;
- Photometer Receiver: 1931 CIE standard Photopic Observer, insensitive to light polarization, linearity of photometric and stable scale;
- From the light Source: correlated color temperature (CCT) of 2856 K, non-polarizing light and proportional distance of the test body;
- From the proof body surface: variation of Iluminance below 61% and uniformity above 65%;
- The proof body: dimensions proportionate to the distance of the light source;
- The environment: A level of background light below 5% to the incident.

Some adjustments were needed to meet the requirements. With regard to the goniometer, photometer receiver, body of evidence and body of proof, all requirements were met in function to the infrastructure already existing in the laboratory. In the light source item, a LED flashlight with emission spectrum in the 5000 K range was used with application of filters to reach a CCT of 2856 K ± 20k determined by [6]. This adaptation can be observed in figure 3.
After adjusting the color temperature in the light projector, the goniometer was used to characterize the light source as its emission curve. For this, the projector of light was positioned in the geometric center of the goniometer and the measurements were carried out to generate the table of distribution of the emission intensity of the projector for each given angle.

Figure 3. Measuring assembly diagram of the CCT of the light source. In the right corner the filters used for adjusting the CCT of the light source (flashlight).

Finally, the support assemblies were carried out for the positioning of the body of evidence and the source of light, seeking to always maintain proportionality to the arrangement of the ASTM E810 procedure, as shown in figure 4.

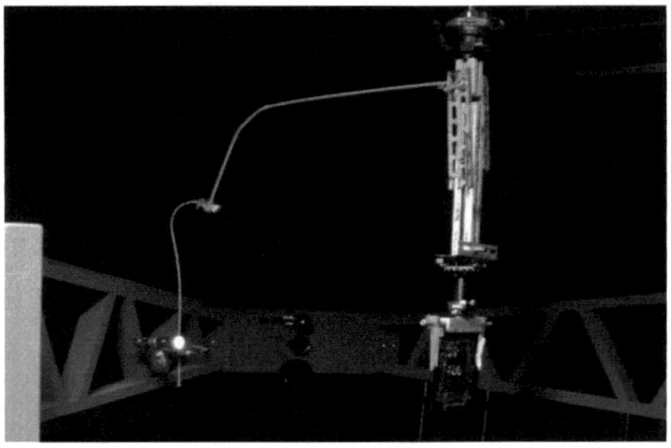

Figure 4. Mounting diagram for measurements of the retroreflective characteristics. On the right side we have the surface for placement of the body of evidence and, on the left, the light source.

In this case it was positioned a plate with retroreflector (square body of dimensions of 44 mm x 44 mm) in the geometrical center of the Goniophotometry and perpendicularly to this axis, the projector of light was positioned to a distance of 70 cm. The measurements were carried out seeking to determine the distribution of light intensities reflected by the retroreflective material, maintaining the position of the axis of the incident light. The norms demand the determination of the retrorefletance coefficient of these subjects in candelas(cd) divides by lux times meter squared (cd/lux. m^2). The distribution of intensities determined by the goniophotometer allows to calculate the illuminance (in Lux) to the observer in the sensor direction. The source intensity (on cd) is determined by a photometer positioned on the sample plane and the sample reflection area (in m^2) is measured directly.

3. Results

In figures 5a-b a three-dimensional representation of the data obtained may be verified.

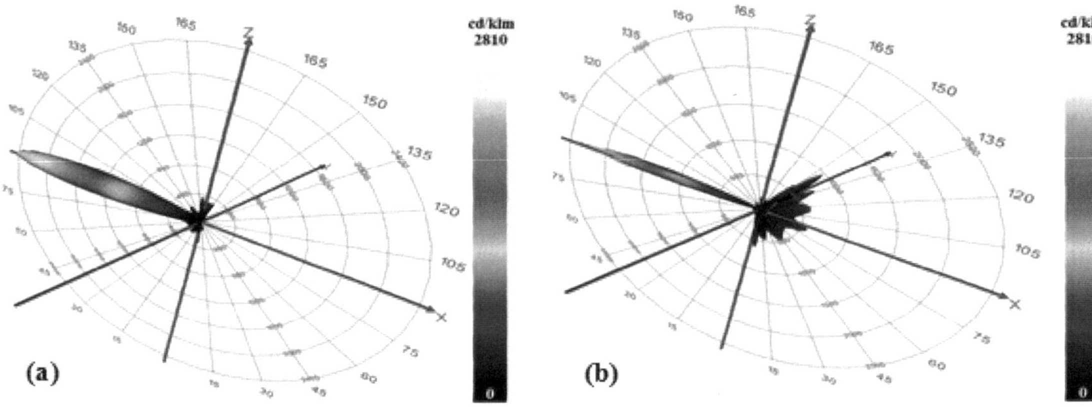

Figure 5. Distribution of luminous intensity obtained from (a) projector and (b) Retroreflective material.

In the Figure 5a-b, "leakage" of residual light is observed for the posterior portion of the specimen, which can be resolved placing the test bodie on a larger mounting bracket.

Measured in the goniophotometer, the data was analyzed in order to understand and define the optical characteristics of the retroreflective material. It was also sought, as in the norms, to determine the coefficient of retrorefletance in the various angles that the distribution of intensities allows. This implies an analysis capability with a higher resolution and higher breadth than determined by the standards, because it is not restricted to just a few angles but it allows to evaluate the coefficient of retrorefletance for all angles of the frontal hemisphere of the sample.

It is estimated that the data has an uncertainty of the order of 3%, compatible with the data obtained with the conventional method (Goniophotometry in a photometric tunnel). However, simulating a vibration in the bracket, it was observed that this uncertainty could reach 7%. Therefore, a critical portion of this method is the rigidity of the support that maintains the projector at distance and fixed position relative to the tested body. This aspect should receive more attention in the experimental assembly.

4. Conclusions

The method proved to be practical, allowed less time spent on both mounting and data acquisition and allowed a better angular resolution for measurements. The norms and documents that specify requirements for retrorefletive materials follow parameters limited by a lesser amount of angles. In this case, the angles and the number of information were shown more comprehensive for analyses. It can additionally generate three-dimensional graphs, which provides very rich and precise information for the characterisation of the retoreflective materials, which provides better conditions for verification of the requirements laid down in the DENATRAN resolutions.

Acknowkedgments

This work was prepared in collaboration of Electrical and Optical Equipment Laboratory (LEO) of the Institute for Technological Research Sao Paulo (IPT-SP).

References

[1] CIE 54.2 2001 *Retroreflection: Definition and measurement* (Paris: Commission Internationale de L'éclairag) p 1-60.

[2] IES LM-79 2008 *Electrical and Photometric Measurements of Solid-State Lighting Products* (New York: lluminating Engineering Society of North America) p 1-23.

[3] CIE 121 1996, *The Photometry and Goniophotometry of Luminaires*(Paris:Commission Internationale de L'éclairag) p-1-51.

[4] TECHNOTEAM 2015 *RIGO 801 – 300, specification*, (Ilmenau: TechnoTeam.de) p 1-23.

[5] ASTM E810-03 2013 *Standard test method for coefficient of retroreflection of retroreflective sheeting utilizing the coplanar geometry*(West Conshohocken: American Society for Testing and Materials) p 1-8.

[6] ASTM E809-08 2013 *Standard Practice for Measuring Photometric Characteristics of Retroreflectors*(West Conshohocken: American Society for Testing and Materials) p 1-11.

[7] ASTM E808-01 2016 *Standard practice for describing retroreflection,*(West Conshohocken: American Society for Testing and Materials) p 1-10.

IOP Conf. Series: Journal of Physics: Conf. Series **975** (2018) 012034 doi:10.1088/1742-6596/975/1/012034

Software for imaging phase-shift interference microscope

I Malinovski, R S França, I B Couceiro

Instituto Nacional de Metrologia, Qualidade e Tecnologia – INMETRO, Av. N.S. das Graças, 50, Xerém, RJ, 25250-020, Brazil

E-mail: rsfranca@inmetro.gov.br

Abstract. In recent years absolute interference microscope was created at National Metrology Institute of Brazil (INMETRO). The instrument by principle of operation is imaging phase-shifting interferometer (PSI) equipped with two stabilized lasers of different colour as traceable reference wavelength sources. We report here some progress in development of the software for this instrument. The status of undergoing internal validation and verification of the software is also reported. In contrast with standard PSI method, different methodology of phase evaluation is applied. Therefore, instrument specific procedures for software validation and verification are adapted and discussed.

1. Introduction

The automated measurement systems with computerized processing have virtually revolutionized physical metrology in recent years. Adequate software (SW) often provides significant improvement in usability and accuracy of the instrument. However this process immediately raises important issues about validation and verification of the software as an integral part of metrology systems. We present here our ongoing work on development and validation/verification of the software for automated interference microscope (IM) that was recently upgraded with some new hardware (HW) [1]

2. General system description

Our IM is based on phase-shifting (PSI) technique combined with interference pattern digital imaging. In standard PSI methods [2] 3-5 fixed step interferograms are used for phase extraction. In our algorithm we use multiple phase-stepped interferograms (image stack) to calculate the height from waveforms that carry all necessary phase information. Each pixel of the interferometric image stack produces one waveform corresponding to one surface point (Figure 1). The disadvantage of this approach is multiple interferograms processing (significant amount of data). The advantage of this method is the possibility of the phase step self-calibration. Indeed, if the whole waveform of fringe signal is available, the phase step information is simply deducted from such waveform. As more advantages or this method we should mention: less sensitivity to mechanical errors in step-shifting and higher resolution in phase determination. Specifically for this interferometric microscope we have developed dedicated custom software package that includes data acquisition, data processing and visualization. In figure 1 we demonstrate details of the phase-shifting algorithm.

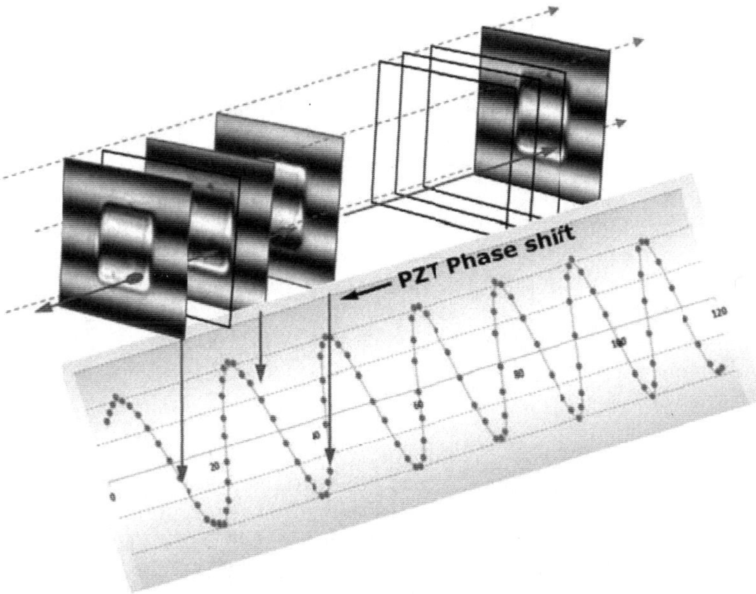

Figure 1. Image stack of interferograms (selected frames) and phase waveform signal for one pixel digitized. Each frame produces one point data on sinusoidal waveform as shown with arrows. The whole waveform results in a constant phase for one surface point

One complete measurement with full phase-shifting scan (measurement cycle) typically consist about 100 points that is much more than Nyquist criterion requirement. We repeat cycles with final result averaging if noise reduction is desirable. It is clear from figure 1 that one measurement cycle results in one 4D data image stack. The image stack after processing should yield into 3D surface map of the object. The software works both with individual pixels or group of neighboring pixels (binned pixels) as option. We use pixels within XY range of interest (ROI) of CCD that is manually selected by markers on 2D screen.

3. Physics of measurement and model

For necessary specification of the physical processes we should start with background physics of the measurement involved. From general physical principles the pseudo-continuous phase-shift results in approximately sinusoidal wave form at each pixel. Therefore, the simple sinusoidal waveform can be used as the first approximation model. Next step is to correct this model to possible phase-shifter nonlinearity that results in second order correction term of the argument of sinusoidal function. So we use the following model function:

$$F(x) = Off + A \sin(\omega x + \Omega_m x^2 + \Phi), \qquad (1)$$

where x is the value proportional to phase-shifter actuator voltage, Off is the offset of the sinusoidal fringe, A is the amplitude, ω is the frequency, Ω_m is the frequency modulation coefficient, Φ is the phase that carries height information. This model function is used to find best possible fit between measured sinusoidal waveform and model with all 5 parameters as variables. Least square criteria of the fit quality are used. We fit individually all pixels in ROI.

4. Software specifications and structure

From management point of view validation of the software starts with correct specifications. In our case specifying software is easy because physical part of the measurement process is clear. The general requirement is that the software must provide automated measurement with adequate calculation for obtaining meaningful result in comprehensible SI unit quantified form. The software suitable to operate automated PSI imaging instrument requires minimum 3 main functionalities as most important: (i) automation and data acquisition, (ii) data processing (phase extraction) (iii) Error control/ reporting. We will concentrate on those in next sections.

From coding point of view the implementation of above scenario it is more or less straight forward programming task. However, making the software foolproof and suitable for so called "stress tests" is a much more demanding process. We have analyzed most critical parts of the software from verification point of view. Following practice of PTB and NPL [3-4] we have defined the list of most important testable requirements with suitable test criteria for each item. We will follow the list in following sections and sub-sections.

5. Automation and data acquisition

The automation of the SW requires synchronized and robust input-output data flow control. This requirement was validated by both static analysis and white-box testing. As a first step of static program analysis the data communication code was checked in order to provide predictable code execution with any data input. This includes trapping values outside the expected range, interception of sudden jumps in values as well as flat (not changing) data input detection. The most conservative action to correct situation was adapted as a full reset of the measurement cycle in case of problem detection.

The data readout was also verified by dynamic white-box testing in which we executed the SW code together with background monitoring of the whole communication input-output calls between the SW and HW. As an auxiliary step we could trace the exact code execution with inserted check points and detailed log files. This feature was specifically implemented to ensure confidence in white-box testing. Those tests were performed in both normal operational mode and simulated problematic situations (such as low light level on CCD, saturated illumination, malfunction of phase-shifter unit, etc.)

6. Data processing / computation

The data received from the hardware is subject of further processing. *Correct computation* is required for each step of the processing. This requirement is subdivided in several topics due to complexity of the process.

6.1. Algorithm numerical stability

The data processing is based on numerical fit as the most critical part of the calculation methodology. Therefore, the fitting algorithm is the most important part of the processing to be verified. We should start with the curve fit criterion that is in our case the most obvious minimum least square difference between measured and fit points. Any fitting algorithm works with certain trade-off between calculation speed and numerical stability of the output. Our fastest two-pass iteration algorithm described in [1] unfortunately was found to be insufficiently stable for some test cases (specifically cases of 2 or more local minima in multi-parameter space). Thus, the fitting procedure has been modified to avoid instability at expense of processing speed decrease. We have added one more preliminary phase of the search in which presence of several minima is verified. A resulting approximate solution is fed to same two-pass iterations as before. Final SW performance is about 0.5 ms per fit on 3.5 GHz processor.

IOP Conf. Series: Journal of Physics: Conf. Series **975** (2018) 012034 doi:10.1088/1742-6596/975/1/012034

6.2. Data filtering
There are multiple sources of noise during the measurement. Fortunately since the signal of interest lies on quite narrow frequency range we can achieve significant signal-to-noise improvement by data filtering. We apply forward-reverse Fast Fourier Transform (FFT) with intermediate Gaussian shaped filter. This filter is supposed to produce no significant phase distortions. Correctness our FFT implementation module was separately verified by black-box test comparison with published C++ code [5]. Numerical stability was tested with our test cases. The whole data filtering procedure has been verified in conjunction with fitting as described in the next section.

6.3. Dynamic black-box testing
Compliance of calculated and expected results was verified with dynamic program black-box test using test case data. Since ready test cases for our type of data are not available we have produced our own test cases by computer simulations of data sets with known function and parameters. Each test case consists of input values, expected output values, and the predicted program behavior with corresponding warnings/reporting via messages. Some of the test cases were designed to estimate type A and type B errors. Typical signal distortions were simulated in our test cases to evaluate possible deviation of the results (figure 2). We have validated our test cases with commercial off-the- shelf and open source software packages such as TableCurve 2D (systatsoftware.com) and SimFit (simfit.org.uk).

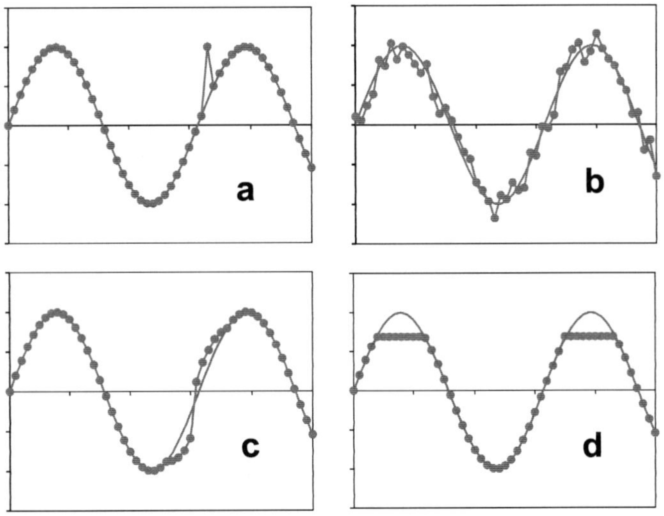

Figure 2. Numerically simulated test cases for black-box testing examples: a) "Hot pixel" peak defect, b) random (white) noise, c) phase jump (jitter), d) CCD sensor saturation. Dots are the simulated signal points and line is the model function (fit).

6.4. Back-to-back testing
The filtering and fitting part of the software was compared to corresponding parts of our previous SW package developed for the gauge block interferometer [6]. This instrument with its automation SW has participated at an international comparison and it can be considered as "end-to-end" tested unit [7]. No significant differences were detected during this testing.

IOP Conf. Series: Journal of Physics: Conf. Series **975** (2018) 012034 doi:10.1088/1742-6596/975/1/012034

7. Conclusions

The software package suitable for imaging phase-shift IM is under development. The SW is currently under validation and verification process. Most important observations are as follows: validated physical background is used for adequate model construction, implementation of the model is verified with both static and dynamic testing, back–to-back tests successfully passed, and insufficient numerical stability was detected and removed using more conservative minima search. We found that validation and verification performed in parallel with SW code improvements is very useful tool for SW development in its final stage.

References

[1] Malinovski I, França R S, Bessa M S, Silva C R and Couceiro I B 2016 *Surf. Topogr.: Metrol. Prop.* **4** 024006
[2] Malacara D 2007 *Optical Shop Testing* y (Hoboken: John Wiley & Sons, Inc.) p 862
[3] Wichmann B, Parkin G and Barker R 2007 *NPL Report* DEM-ES 014
[4] Greif N, Schrepf H, Richter D 2006 Software validation in metrology: A case study for a GUM-supporting software Measurement **39** (9) 849
[5] Press W H, Flannery B P, Teukolsky S A, and Vetterling W T 2007 *Numerical Recipes. The Art of Scientific Computing* (Cambridge: Cambridge University Press) p 1256
[6] Titov A, Malinovsky I and Massone C 2000 *Metrologia*, **37** 121
[7] Decker J E, et al. 2007 Final Report on SIM.L-K1 (SIM.4.2) Regional Comparison. Stage One: Calibration of gauge blocks by optical interferometry *Metrologia* **44** 04001

IOP Conf. Series: Journal of Physics: Conf. Series **975** (2018) 012035 doi:10.1088/1742-6596/975/1/012035

Determination of uranium and thorium using gamma spectrometry: a pilot study

D M M Olivares*, E S Koch, M V M Guevara and F G Velasco

Research Center in Sciences and Technologies of Radiation (CPqCTR), State University of Santa Cruz (UESC), 45662-900, Ilhéus-BA, Brazil.

*Email-address: diango.mo87@gmail.com

Abstract. This paper presents the results of a pilot experiment aimed at standardizing procedures for the CPqCTR/UESC Gamma Spectrometry Laboratory (LEG) for the quantification of natural radioactive elements in solid environmental samples. The concentrations of ^{238}U, ^{232}Th and ^{40}K in two sediment matrix belonging to the Caetité region were determined, by using the absolute method with uncertainties about 5%. The results were obtained using gamma spectrometry with a high-resolution p-type HPGe detector. As a closure, the absorbed dose, radium equivalent activity and the annual effective dose were calculated.

1. Introduction

Environmental assessments are one of the permanent activities, in which universities and their research centers are involved, and they can impact on a positive way in a given region. Gamma-ray spectrometry is one of the most widely used techniques for environmental radioactivity assessment [1, 2]. It allows to perform a precise quantitative analysis of the radionuclide concentration (natural or anthropogenic) in different types of environmental samples, without the application of laborious sample preparation [3].

However, the quantification of radionuclide in extended sources, by using the absolute method, requires the knowledge of the efficiency detection. The efficiency curve can be experimentally obtained by using a "poly-energetic standard source" with geometry and a matrix composition similar to the studied samples. When these "standard sources" are not available, the efficiency curve for volumetric sources can be obtained from point-like sources of calibration using the efficiency transfer method [4]. Two sediment samples belonging to Caetité region, where it is concentrated nowadays the uranium ore activity in Brazil were analyzed. The efficiency curve determined in [5] was used for quantifying the amount of uranium, thorium and potassium in both samples.

This paper presents the results of a pilot experiment aimed at standardizing procedures from the Research Center in Sciences and Technologies of Radiations (CPqCTR/UESC) LEG for the quantification of natural radioactive elements (^{238}U, ^{232}Th and ^{40}K) in solid environmental samples.

Based on the fact, that mining is catalogued by the United Nations Scientific Committee on the Effects of Atomic Radiation (UNSCEAR) as a potential source of exposure to natural radioactive elements [6] different radiological magnitudes were also calculated.

2. Materials and methods

2.1. Sample collection and preparation

The sediment samples (SS1 and SS2) were collected from two specific sites of the influence region of uranium mining in Caetité, Bahia.

The samples were stocked and sealed in plastic bags in order to avoid any external contamination. After that, the samples were sent to our Environmental Samples Preparation Laboratory at the CPqCTR and prepared according to the procedure established in [7]. The granulometric reduction of the collected samples was done using a sieve of 170 mesh (90 μm). After the homogenization procedure, samples were dried in the oven at a temperature of 110 °C. At intervals of 2 hours, the mass of the samples was verified, until achieving a stable difference (lesser than 5 mg) in the dry weight, in temperature and humidity conditions about 20 °C and 40%, respectively. The dried samples were hermetically sealed in two plastic containers (vessels) with a cylindrical geometry (6.2 cm^2 x 2.7 cm), previously identified. Finally, both samples were stored during 30 days in a desiccator in order to reach the secular equilibrium of radon with other elements of the natural radioactive series.

2.2. Experimental Setup

The experimental measurements were carried out using a coaxial p-type HPGe detector (model **GC2518**), manufactured by Canberra. The geometrical dimensions and spectrometric characteristics of the detector were verified by [5] and compared with those provided by the manufacturer. This detector has a nominal relative efficiency about 25% compared to 7.62 cm x 7.62 cm NaI (Tl) detector. The measured resolution (FWHM) and the Peak-to-Compton ratio (P/C) for the ^{60}Co gamma-ray energy of 1.33 MeV were 1.72 keV and 58:1, respectively. These characteristics were in disagreement with those reported by the manufacturer for the same energy (1.62 keV and 61.6:1, respectively). The detector is installed in a 10-cm-thick lead shield, model 747 [8]. The HPGe is connected to a pre-amplifier (model **2002C**) and a Digital Spectrum Analyzer (DSA-200), with a shaping time of 4 μs. The course and fine gain were fixed in 40x and 1.0502x, respectively. A high voltage of 4000 V was applied to the detector.

The energy calibration function, which establishes the relation between gamma-ray energies and the electronic channels, was obtained in the energy range from 59-1408 keV. Adjustment of the experimental data was achieved by an excellent linear fit (r^2=0.999), whose parameters are: m=0.2302 keV $channels^{-1}$ and b= -0.6077 keV.

The full-energy peak efficiency calibration was performed in the same energy range with point-like standard sources of: ^{22}Na, ^{60}Co, ^{133}Ba, ^{137}Cs, ^{152}Eu, ^{155}Eu and ^{241}Am. The spectrum acquisition time was chosen depending on the activity of the source to keep the statistical uncertainty in the peak area below 1%. Background radiation was measured for 72 h, and its contribution was subtracted from the corresponding standard source spectra.

3. Results

The samples were measured at 5 cm from the top of the detector for 72 h. Spectra were acquired using the software Genie 2000 [9]. Gamma-ray energies of 351 keV (^{214}Pb), 1120 and 1764 keV (^{214}Bi) were used for ^{238}U calculations. The energies of 583 keV (^{208}Tl) and 968 keV (^{228}Ac) were used for ^{232}Th calculations, and for ^{40}K quantification was used its own gamma-ray energy (1460 keV). The measured spectrum of the sample SS2, is shown in figure 1. Due to the analytic lines, used in the quantification procedure (highlighted in red), were above 350 keV, the spectrum is shown for energies greater than 300 keV.

The specific activity of ^{238}U, ^{232}Th and ^{40}K, corresponding to both sediment samples from Caetité region, were obtained using the Absolute Method (Eq. 1):

$$A = \frac{CPS_s - CPS_g}{\varepsilon \cdot I_\gamma \cdot m} (Bq.kg^{-1}) \quad (1)$$

IOP Conf. Series: Journal of Physics: Conf. Series **975** (2018) 012035 doi:10.1088/1742-6596/975/1/012035

where CPS_s and CPS_g represent the count rates in the full-energy peak and the background count rates, ε is the efficiency corresponding to energy E of interest, I_γ represents the photon yield or photon-emission probabilities to energy E, and m is the dried mass of the sample in kg.

Energy and photon yield values were taken from the database of the International Atomic Energy Agency [10].Background spectra were used for the calculation of the detection limits (*DL*) and the Minimum Detectable Activity (*MDA*). In this work was used the Currie modified criterion for the calculation of detection limit and therefore to MDA [11].

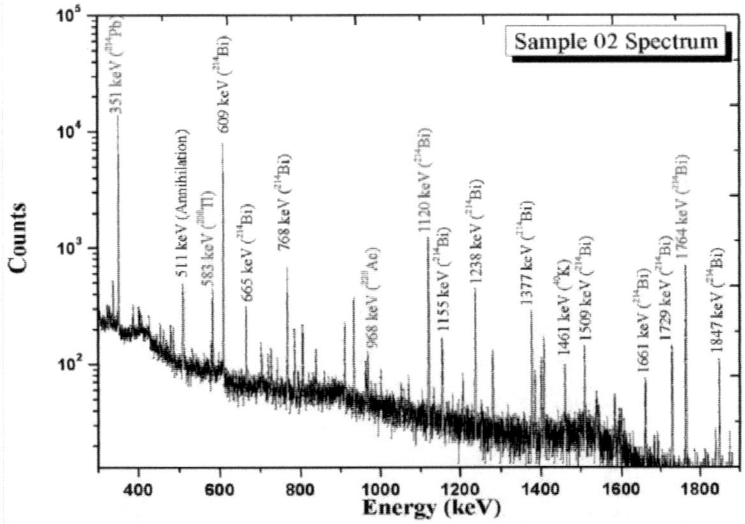

Figure 1: Measured spectrum of the sample SS2 acquired using GENIE 2000.

The efficiency curve was obtained after adjusting the nominal detector parameters using a Monte Carlo N-Particle eXtended (MCNPX) transport code [12], and using the efficiency transfer method. Correction factors such as solid angle for the volumetric sample and self-absorption correction factors ($F_{self\text{-}abs}$) were also taken into account to calculations. If the reader is interested, additional details can be found in [5]. Figure 2 shows the efficiency curve in the energetic range of 53-1764 keV for the studied volumetric sample. Red points represent the efficiency values for the analytical gamma lines of interest.

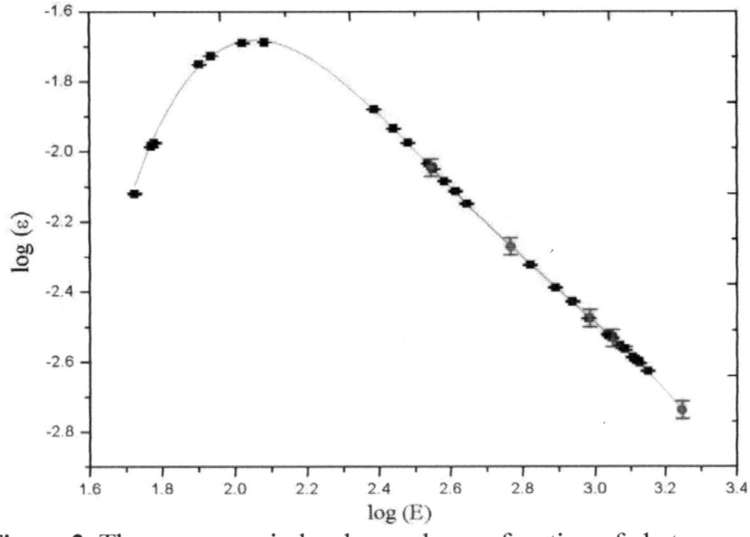

Figure 2 :The ε_{vol} curve in log-log scale as a function of photon energies (E). Solid line represents polynomial logarithmic fit (r^2=0.998).

The ^{238}U, ^{232}Th and ^{40}K concentration values (table 1) were estimated from the specific activity Eq. (1), with a mean relative error of 5%. Due to the spectrum acquisition time was chosen to keep the statistical uncertainty in the peak area below 1% (Type A uncertainty), and the photon yield of this lines are lower than 0.3 % (Type B uncertainty); the concentration values' uncertainties were given mainly by the uncertainties in the estimative of the efficiency. For all the efficiency values, the uncertainty was computed as the quadratic sum of the individual uncertainties of the factors used in the efficiency calculation. These factors and their relative errors, are listed in table 2.

Table 1: Main results for ^{238}U, ^{232}Th and ^{40}K concentrations for both sediment samples.

Energy	**SS1**		**SS2**	
(keV)	A (kBq kg^{-1})	C (mg kg^{-1})	A (kBq kg^{-1})	C (mg kg^{-1})
		^{238}U		
351	2.8 ± 0.2	22 ± 1	8.1 ± 0.4	64 ± 3
1120	2.7 ± 0.2	21 ± 1	7.7 ± 0.4	61 ± 3
1764	3.1 ± 0.2	24 ± 1	8.5 ± 0.5	67 ± 4
		^{232}Th		
583	0.20 ± 0.01	68 ± 4	0.35 ± 0.02	117 ± 7
968	0.22 ± 0.02	74 ± 5	0.33 ± 0.02	111 ± 8
		^{40}K		
1460	0.68 ± 0.04	-	0.61 ± 0.03	-

Table 2: Uncertainty sources contributing to the total uncertainty of the efficiency (ε).

Uncertainty source	**Type**	**Relative error**
Count rate	A	< 1 %
Radioactive decay	A	< 1 %
Photon yield	B	< 0.7 %
Geometric factor	B	~3 %
Self-Absorption factor	B	~2-3 %
Error interpolation	A	~2 %

As a closure for this work, radiological protection magnitudes such as: the absorbed dose rate (D), the radium equivalent activity (Ra_{eq}) and the annual effective dose equivalent (H) for 1 m from the surface for adults were calculated according to [13].

The absorbed dose, due to gamma radiation resulting from exposure to the primordial radionuclide for both samples were: 1.4 ± 0.1 μGy h^{-1} and 3.7 ± 0.1 μGy h^{-1} for SS1 and SS2 respectively. The Ra_{eq} values for the two samples were: 3204 ± 127 Bq kg^{-1} (for SS1) and 8624 ± 260 Bq kg^{-1} (for SS2) and the annual effective dose values were: 1.7 ± 0.1 mSv y^{-1} for SS1 and 4.5 ± 0.1 mSv y^{-1} for SS2.

The annual effective dose values do not exceed the accepted as safe limit for occupationally exposed persons (20 mSv y^{-1}) [13]. They are even included in the radiation dose limits to mine workers at different uranium operations around the word (< 4.8 mSv) [14].

4. Conclusions

The pilot experiment carried out in the CPqCTR/UESC LEG established a methodology to perform quantitative procedure of radionuclides by using the Absolute Method. It successful application

IOP Conf. Series: Journal of Physics: Conf. Series **975** (2018) 012035 doi:10.1088/1742-6596/975/1/012035

allowed to determine the concentration of ^{238}U, ^{232}Th and ^{40}K in two sediments samples belonging to the Caetité region and the assessment of the outdoor dose, which do not exceed the regulated levels. This work itself do not constitute an environmental study, because there are not enough sampling points for characterizing the region.

Acknowledgments
The authors acknowledge Funding Authority for Studies and Projects (FINEP 338/2009) for the financial support to acquire the spectrometer, and also Fundação de Amparo à Pesquisa do Estado da Bahia (FAPESB) and Chemical Engineering Academic Centre of the State University of Santa Cruz for the financial support to participate in the Metrologia 2017 event.

References
[1] Hannan M, Wahid K and Nguyen N 2015 Assessment of natural and artificial radionuclides in Mission (Texas) surface soils *J. Radioanal. Nucl. Chem.* **305** 573-82
[2] Santos J J A, Amaral R S, Menezes R S C, Álvarez J R E, Santos J M N, Fernández Z H, Bezerra J D, Silva A A, Damascena K F R and Maciel N J A 2017 Influence of terrestrial radionuclides on environmental gamma exposure in a uranium deposit in Paraíba, Brazil *Ecotoxicol. Environ. Saf.* **141** 154-9
[3] Agbalagba E O, Avwiri G O and Chadumoreh Y E 2012 Sprectroscopy measurement of natural radioactivity and assessment of radiation hazard indices in soil samples from oil fields environment of Delta State, Nigeria *J. Environ. Radioact.* **109** 64-70
[4] Vargas M J, Díaz N C and Sánchez D P 2003 Efficiency transfer in the calibration coaxial p-type HPGe detector using the Monte Carlo method *Appl. Radiat. Isot.* **58** 707-12
[5] Montalván-Olivares D M, Manso-Guevara M V and Velasco F G 2017 Determination of the HPGe detector efficiency in measurements of radioactivity in extended environmental samples *Appl. Radiat. Isot.* **130** 34–42
[6] United Nations Scientific Committee on the Effects of Atomic Radiation (UNSCEAR) 2000 *Sources and Effects of Ionizing Radiation* (New York: United Nations)
[7] International Atomic Energy Agency (IAEA) 2003 *IAEA - TECDOC -1360 Collection and preparation of bottom sediment samples for analysis of radionuclides and trace elements* (Vienna: International Atomic Energy Agency)
[8] Canberra 2013 *Model 747 and 747E Lead Shield* (Meriden: Canberra Industries, Inc)
[9] Canberra 2006 *Genie™ 2000: Spectroscopy Software* (Meriden: Canberra Industries, Inc)
[10] International Atomic Energy Agency (IAEA) 2007 *Update of X- Ray and Gamma-Ray Decay Data Standards for Detector Calibration and Other Applications* (Vienna: International Atomic Energy Agency
[11] Knoll G F 2010 *Radiation Detection and Measurements* (United States of America: John Wiley & Sons, Inc)
[12] Pelowitz D B (ed) 2008 *MCNPX™ User's Manual Version 2.6.0* (Los Alamos: Los Alamos National Laboratory)
[13] United Nations Scientific Committee on the Effects of Atomic Radiation (UNSCEAR) 2010 *Annex B: Sources and Effects of Ionizing Radiation- UNSCEAR 2008 - Report to General Assembly With Scientific Annexes* vol 1:Sources (New York: United Nations)
[14] Nielsen S P, Ross P and Andersson K G 2015 Predicted Radiation Exposure from Mining at Kvanefjeld: Introduction to Radiation, Review of Baseline Information, and Predicted Radiation Exposures from Kvanefjeld Mining, Mineral Processing and Refining (Roskilde: DTU-Nutech)

IOP Conf. Series: Journal of Physics: Conf. Series **975** (2018) 012036 doi:10.1088/1742-6596/975/1/012036

Evaluation of a LiF:Mg,Ti thermoluminescent ring dosimeter according to the IEC 62387:2012 Standards

Edyelle L. B. Oliveira[1], Vinícius S. M. de Barros[1], Viviane K. Asfora[1], Helen J. Khoury[1]

[1] Nuclear Energy Department, Federal University of Pernambuco, Av. Prof. Luiz Freire, 1000, Recife-PE, CEP 50.740-545

vsmdbarros@gmail.com

Abstract. This work shows results of type testing of a ring radiation dosimeter system under IEC 62387:2012. The personal dosimeter investigated in this work consists of a commercial one element plastic ring which contains an LiF:Mg,Ti thermoluminescent pellet. By applying requirements for statistical fluctuations and linearity, a minimum measurable dose in Hp(0.07) was established. Energy and angular dependence aided in determining energy correction factors and fading requirements were used to select the most appropriate preheat scheme. Type testing of passive radiation monitors was performed in the Radiation Metrology Laboratory (LMRI-DEN/UFPE) of the Federal University of Pernambuco and is a major step in Brazil for the independent evaluation of these dosimeters, currently not available in the country.

1. Introduction

Due to an increasing demand for extremity radiation dosimetry for medicine, the Radiation Protection Laboratory (LPR-DEN/UFPE) of the Federal University of Pernambuco has been investigating an appropriate system for ring dosimetry. There are few publications on testing of ring dosimeters according to newer IEC 62387:2012 [1] standards. Reports [2,3] usually are based on older ISO 6280 [4] ISO/DIS 12794-1 [5] standards. The scope of this paper is to present the results down to the limits of acceptability of tests performed on a commercial ring dosimeter (RADOS) with LiF:Mg,Ti crystals according to the IEC 62387:2012 [1] standards for Hp(0.07) (personal dose equivalent for 0.07 mm depth). Testing was independently performed in the Radiation Metrology Laboratory (LMRI-DEN/UFPE) of the Federal University of Pernambuco.

2. Dosimetry system design

The personal dosimeter investigated in this work consists of a commercial one element plastic ring dosimeter (RADOS) which contains an LiF:Mg,Ti with 4.55 mm in diameter (MTS, Poland) detector and 0.93 mm in thickness. The front cover of the ring has a measured thickness of 0.38 mm. Figure 1 shows the front and back covers of the ring and the TL element placement on the back cover. The back cover has a wall thickness surrounding the edge of the crystal with 0.98 mm in thickness.

Measurements were performed in a Harshaw-Bicron 3500 TL reader using 165 °C / 16 s pre-heat. Indicated values were corrected for individual element sensitivity variations (Element Correction

Factors) and Reader Calibration Factors. The ring dosimeters were previously calibrated in Hp(0.07) at ISO-Narrow Series N-150 quality using ICRU rod phantom. Air kerma was determined using a standard ion chamber (PTW LS-01), traceable to the primary standards laboratory Physikalisch-Technische Bundesanstalt (PTB).

3. Non-linearity and statistical fluctuations

IEC 62387:2012 [1] requires that statistical fluctuations of the indicated value to be performed together with the testing of non-linearity. A total of w = 10 dose values with n = 7 dosimeter per group were tested. Irradiations were performed on a rod phantom using Hp(0.07) values of 0.2 mSv to 2.0 Sv. Figure 2 shows the results and the required limits (solid lines). The maximum coefficient of variation (CV) of 24% was found for the dose value of 0.2 mSv which is just outside of the 23% limit. All other CV values are within the c_1/c_2 corrected limiting values of (16 – H/0.1 mSv) % for Hp(0.07) values within 0.1 and 1.1 mSv and 5% for values higher than 1.1 mSv [1].

The results for linearity (Figure 2) show that values above 1.0 mSv were considered acceptable. The two lowest values of 0.2 and 0.4 mSv were outside the limits due to proximity to background, which consists of nearly 50% of the raw reading at these low dose values. By using the IEC linearity criteria, instead of the commonly used three standard deviations above background, the minimum measurable dose (MMD) was set to 1.0 mSv. Since dose limits for skin, hands and feet are 500 mSv/yr, and Investigation Limits in Brazil are set to 20 mSv/month according to national regulations [6], the value of 1.0 mSv for the MMD was considered acceptable.

Figure 1. Ring dosimeter front and back covers.

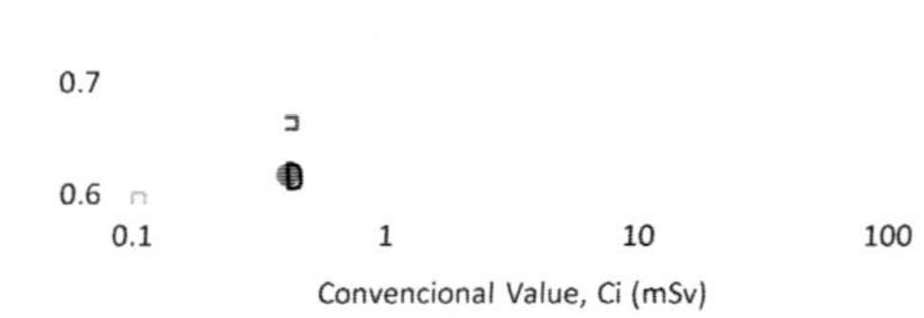

Figure 2. Linearity and statistical fluctuations results. Dashed and solid line limits are the required limits without and with the uncertainties for irradiation, respectively.

IOP Conf. Series: Journal of Physics: Conf. Series **975** (2018) 012036 doi:10.1088/1742-6596/975/1/012036

4. Energy and angular response

LiF:Mg,Ti has an energy dependence that must be corrected using an appropriate algorithm in order to assess the Hp(0.07) dose values over the rated energy range within acceptable tolerances. An algorithm for a single element dosimeter consists of a multiplying factor.

Testing was performed using n = 6 dosimeters per energy value in the range between 20 keV (ISO N30) and S-Cs-137 quality. Figure 3 shows the results of the energy dependence of the ring dosimeters and the IEC 62387:2012 acceptance criteria (red solid lines). The upper curve refers to the values using only element correction coefficients (ECC) and the reader calibration factors. An energy correction factor f_E of 0.90 (lower curve in Figure 3) is a sufficient correction so the values are within the IEC acceptable limits, over the energy range from 20 keV to 662 keV.

Angular dependence of the ring dosimeters was assessed at 0°, 15°, 30°, 45° and 60° and for the three lowest effective energies (20, 24 and 33 keV). The results are within the IEC requirements [1], as shown in Figure 4.

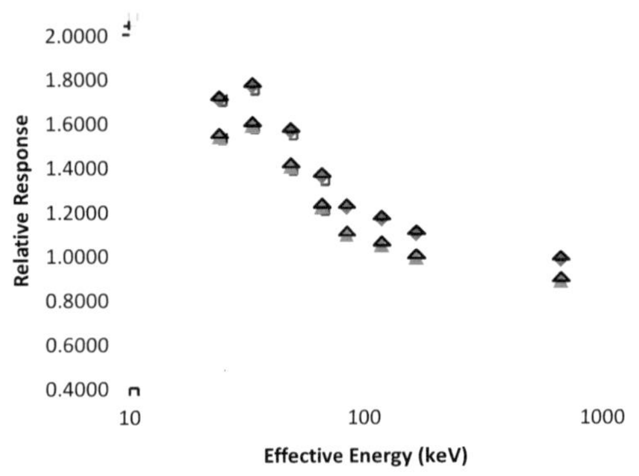

Figure 3. Energy dependence of the ring dosimeters with an energy dependence factor applied (lower curve) and without (upper curve).

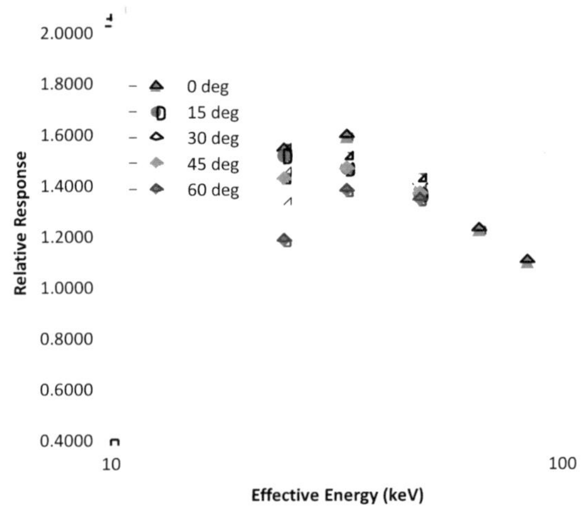

Figure 4. Angular and energy dependence of the ring dosimeters.

5. Over response to side irradiation

The assessment of the overresponse due to side irradiation prevents the acceptance of a detector with high atomic number and without sufficient shielding [1]. Dosimeters were irradiated free in air using an H*(10) of 3 mSv in air from 60° to 120° (α_{max} = 60°) for Cs-137 and for the three lowest energies (20, 24 and 33 keV). Side irradiation and the over response is required to be within 2.0 times of the response at 0° for Hp(0.07).

Since LiF:Mg,Ti has a relatively low energy dependence, an over response was not found but instead a decrease in the response as the angle is varied from 60 to 120°. For Cs-137 gamma rays a decrease of -8% was found, followed by a 12% decrease for 33 keV (N60), 29% for 24 keV (N40) and up to 46% for 20 keV (N30). These results meet the IEC requirements. The decrease in response observed can explained by the fact that the back/side cover of the dosimeter is thicker than the front cover (see Figure 1) and shields more effectively the lower energy photons.

6. Dose build-up, fading, self-irradiation and natural irradiation

The relative response and deviation due to dose build-up and fading was tested using eight groups of dosimeters. Groups 1 to 3 (n=6 each) were irradiated at an Hp(0.07) of 7 mSv and Group 4 (n=25) with 1 mSv. Groups 5 to 7 (n=6 each) and 8 (n=25) were reserved for background measurements. Groups 1 to 5 were read out at 1 h after irradiation and Groups 2 and 6 after one week. Groups 3, 4, 7 and 8 were read after one month. For these measurements a preheat of 160 °C/16s was used. Results showed that the response is better than 2% and therefore within the 0.91 to 1.11 tolerance limit [1].

In addition to these tests, it is known that due to shallow trap instability in LiF:Mg,Ti at room temperature that an appropriate preheat is necessary prior to the dose measurement. In order to evaluate the appropriateness of preheat schemes with respect to IEC requirements, three different preheats were tested at 1 h and at 7 days after irradiation: (a) 160 °C/16 s; (b) 150 °C/16 s; and (c) 160 °C/10 s.

The results showed a small 0.7 % decrease in response for the 160 °C/16s preheat scheme after 7 days; followed by a 4.7% decrease for 160 °C/10s; and a 7.7% decrease for 150 °C/16s. Therefore, the 160 °C/16s preheat was considered more appropriate for routine use.

7. Conclusions

The assessment of the dosimeter ring system was considered appropriate, according to IEC 62387:2012 for $H(0.07)_{low} \geq 1.0$ mSv, which is considered appropriate for extremity dose evaluations. IEC Standards also aided in the refinement of the dose algorithm by determining energy correction factors within the stated range and preaheat setups.

The implementation of type testing of passive radiation monitors performed in the Radiation Metrology Laboratory (LMRI) of the Federal University of Pernambuco is a major step in Brazil for the independent evaluation of these dosimeters, currently not available in the country.

8. References

[1] International Electrotechnical Commission 2012 Passive integrating dosimetry systems for personal and environmental monitoring of photon and beta radiation. IEC 62387.

[2] Brasik N, Stadtmann H., Kindl P. 2007 *Rad. Prot. Dosim.* 125 p. 331.

[3] Squair P. L. 2007 *Int. J. Low Radiation* 4 p. 88.

[4] International Organization for Standardization 1996 Reference beta radiations for calibrating dosimeters and dose-rate meters and for determining their response as a function of beta-radiation energy - International Standard ISO 6980.

[5] International Organization for Standardization (ISO) 1997 Individual Thermoluminescence Dosimeters for Extremities and Eyes, ISO/DIS 12794-1.

[6] CNEN Regulatory Statement of 3.01 2011 *POSIÇÃO REGULATÓRIA* 3.01/004.

Sensitometric curve of radiographic films by X-ray fluorescence

E A S Gonçalves[1], D F Oliveira[2,3], M J Anjos[2], L F Oliveira[2], R T Lopes[3]

[1] LISComp - Federal Institute of Rio de Janeiro, Paracambi, Brazil
[2] Physics Institute - State University of Rio de Janeiro, Rio de Janeiro, Brazil
[3] COPPE - Federal University of Rio de Janeiro, Rio de Janeiro, Brazil

E-mail: elicardo.goncalves@ifrj.edu.br

Radiographic film exposure is traditionally measured by the transmittance of a beam of light through the film. There are many mathematical and computational models to characterize the curve behaviour and its properties, but almost none of them considers the limitations caused by the equipment used. As long as exposure in film increases, light intensity measured after the film decreases in a way that from a certain exposure, light couldn´t be distinguished from any kind of noise. This work aims to propose x-ray fluorescence as a solution for better measure high exposed films and show how it could be modelled mathematically.

1. Introduction

Radiographic film is a well known and still very used way to measure ionizing radiation exposure both in research and in individual dosimetry.[1,2] The main propriety is the metallic silver concentration adhered to film after development due reaction with photons [3-6].

As the metallic silver is opaque, its concentration increase turn the film also opaque. A way to measure this is by light transmission across the film and how it is attenuated. The main models consider radiation exposure proportional to silver concentration which is proportional to light attenuation. So, read the optical density is indirectly measure the exposure before development [5,6].

For high values of optical density, the light that crosses the film is to low to be read, and it makes a upper limit for equipment, even if the silver concentration still growing. This fact brings a question about if another way to read silver concentration could be better for high exposures [7,8].

If the new method to measure silver had no limit, the H&D curve could have a range bigger than made by light measurement [7,8].

This work aims to show a little correction in mathematical model for radiographic film due the light limitation and also suggest the x-ray fluorescence (XRF) as a way to read films avoiding this limitations.

2. Mathematical Model and instrument correction

2.1. The traditional models

Most of the models to the sensitometric curve are similar and take two considerations to make a mathematical model:

• The probability to a photon to interact to a halide silver grain in film (turning it into a black metallic silver grain) is proportional to the number of grains and the number of photons (which is proportional to the total exposure)[9,10];

• To increase the optical density, a unsensitized grain must interact with a photon, and once sensitized, no matter if a new photon hits it again. It creates a saturation behavior, once number of the unsensitized grain decreases due exposure[9,10]

This behavior could be represented by the equation:

$$dN=-N\alpha QX \quad (1)$$

Where N is the number of unsensitized grains per area, α is the cross section of a single grain to x-ray photons, Q is a constant and X is the exposure. Solving this, we can find [9,10]:

$$N_a=N_0\left(1-\exp^{-\alpha QX}\right) \quad (2)$$

N_a and N_0 are blackened and total number of grains per area, recpectively. As optical density is defined by reading values I_0 of light intensity entering and I crossing the film [9,10]:

$$OD=\log\left(\frac{I_0}{I}\right) \quad (3)$$

If we treat light photon interaction with matter like x-ray photons,

$$I=I_0\exp^{n\sigma s} \quad (4)$$

$$OD=n\sigma s\log(e) \quad (5)$$

Where n is the grain concentration per volume, σ is the cross section for light and s is the emulsion thickness.

As said before, OD is proportional to N_a [7,8]:

$$OD=N_0\sigma\log(e)\left(1-e^{\alpha QX}\right) \quad (6)$$

This equation gives the OD in function of exposure X. The behavior can be seen in figures (1) and (2).

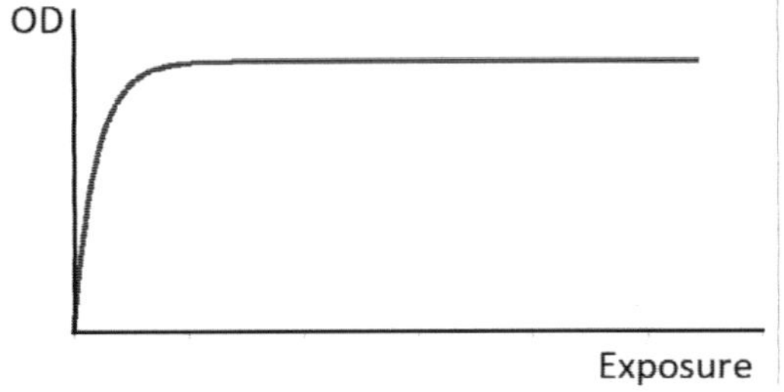

Figure 1. OD x Exposure by equation (6). It starts increasing fastly but saturates soon.

Figure 2. OD x log(Exposure) by equation (6). It starts increasing slowly, became fastly but saturates soon. Between the slow part and the saturated part there is a linear.

2.2. Instrument correction

In equation (3), I is the light read in OD measurement after crossing the film. If we add a noise, the equation will be:

$$OD = \log\left(\frac{I_0}{I+\varepsilon}\right) \qquad (7)$$

Doing same as did before the new final equation is:

$$OD = \log\left(\frac{I_0}{I_0 \exp\left(-\sigma N_0\left(1-e^{-\alpha QX}\right)\right)+\varepsilon}\right) \qquad (8)$$

For low values of X, ε is insignificant. For high' values, it gives a limitation in OD. This is shown in figure (3).

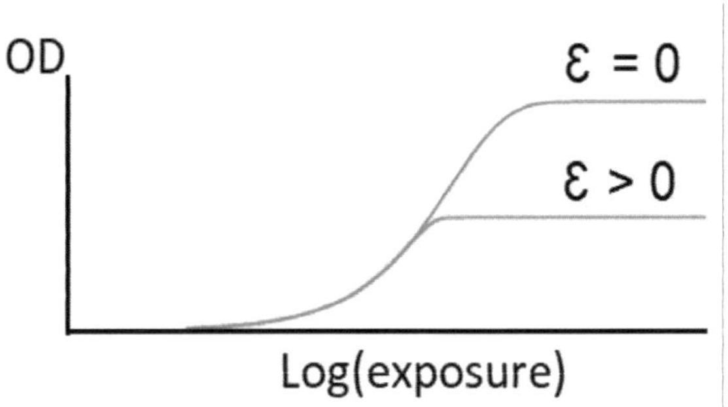

Figure 3. OD x log(Exposure) by equation (8). When $\varepsilon > 0$, the saturation is in low values, hiding a significant range from characteristic curve.

3. Materials and Methods

3.1. Exposing the films

Odontological KodaK E type films were exposed in a 40kV 80μA x-ray been, from an Amptek mini-X tube, in different exposures controlled by time. A non exposed and a film exposed to light for a day were also made for create a “zero” and “infinite” exposures. The films were developed using a portable Kodak developing chamber and chemicals, using times and temperatures suggested by the manufacturer. The optical density by light was obtained using a X-rite densitometer, model 341.

3.2. XRF measurement
After read optical densities, each film was also read in a Amptek XRF equipment, 40 kV and 80 μA. The area of each Kα peak was taken to build a XRF sensitometric curve.

3.3. Light and X-ray measurement fitting
The next step is to fit Light and X-ray fluorescence reads, at least the low values. in the same graph , finding the proportion between them.

3.4. Finding parameters to the equation corrected.
Fitting curves and, theirs limits and proportion, its possible to find all the parameters to build the film equation corrected by instrument limits. After, the theoretical optical density for all the exposures were calculated, to plot together.

4. Results

4.1. Sentitometric curves by X-ray, light and theoretical
The graph showing the three curves is shown below.

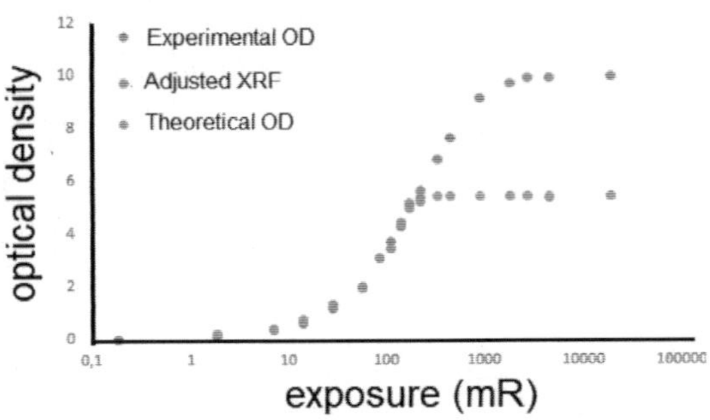

Figure 4. OD x log(Exposure) and XRF measurement. XRF behaves like the total range without limitation.

4.2. The corrected equation
The proportion between read optical density and Kα peak area (in number of counts) is β = 0,0012. Using this value, o better plot equation is given below:

$$OD = \log\left(\frac{4495}{4495^{\left(-22,9\left(1-e^{-0,0043X}\right)\right)}+0,0176}\right) \quad (9)$$

5. Discussion

The mathematical model was created with a small correction of pre-existing models, giving rise to a slightly larger expression because there were no simplifications previously used. Within a small statistical variation this model seems to be at least a good approximation for the relationship between the H & D curve and the limitation for optical density reading. The model still suggests ways to overcome the problem in some situations, requiring minimum and maximum values. For the light readers used in the curve construction, and consequently in the scanning of radiographic films. The model was successful in its experimental validation. The data used in this work has a pretty good fit. It was thus possible to construct the equation of the H&D curve for the film used in this work and predict how far it would go without limitation. The use of the X-ray fluorescence peak in place of the optical density was very successful avoiding the existence of a maximum value of the concentration of the plot. A characteristic curve made through XRF are due to the adjustment of the geometry and correction, showing purely proportional silver concentration, and convenient for use, besides proving the fact that the limiting density of the H&D curve is in the measure by optical density.

References

[1] Ting N A, Broadbent, J M and Duncan W J 2013 *N. Z. Dent J* **109** 107–14
[2] Low D 2015 J. *Phys.:Conf. Ser* **573** p. 1–7
[3]Knoll G F 2010 *Radiation Detection and Measurement*, (New York: John Wiley & Sons) p 864
[4]Attix F H. 2004 *Introduction to Radiological Physics and Radiation Dosimetry*, (Weinheim: Wiley Verlag GmbH & Co. KgaA) p 628
[5] Moslehi A, Hamidi S, Raisali G, Gheshlaghi F. 2010 *Radiat Meas*, **45**, 98–102.
[6] Webb S. 1988 The Physics of Medical Imaging, Taylor & Francis Group, New York,.
[7] Goncalves E A S, Oliveira D F, Anjos M J, Assis J T, Oliveira L F, Lopes R T. 2017 *X-Ray spectrometry*, **46**(5), 361-6.
[8] Goncalves E A S, Oliveira D F, Anjos M J, Assis J T, Oliveira L F, Lopes R T. 2017 *Digitizing Radiographic Films Through X-Ray Fluorescence Mapping, (*Dusseldorf: Lambert AP) p 64
[9] Dixon RL and Ekstrand KE. 1976 *Med. Phys* **3** 340–45
[10] ZHU XR, Yoo S, Jursinic PA, Grimm DF, Lopez F, Rownd WW and Gillin MT. 2003 *Med. Phys* **30** 912–19,

Stereotactic radiotherapy for choroidal melanoma: analysis of eye movement during treatment

Souza F M L[1,a], Gonçalves O D[2], Batista D V S[3] and Cardoso S C[2]

[1]Escola Politécnica, Universidade Federal do Rio de Janeiro, 21941-909 Rio de Janeiro, Brazil
[2]Instituto de Física, Universidade Federal do Rio de Janeiro, 21941-909 Rio de Janeiro, Brazil
[3]Instituto de Radioproteção e Dosimetria, Comissão Nacional de Energia Nuclear, 22780-160 Rio de Janeiro, Brazil

[a]Corresponding author, e-mail: felipemlucas@poli.ufrj.br

Abstract. This study aims to analyse the eye's movement during radiotherapy treatment for choroidal melanoma, as well as the methodology used in the repositioning of the patient between treatments sessions. For this purpose, the procedures used by the hospital staff were analysed on site and videos were recorded during the treatments. The methodology for the fixation of the eye is correct in its objective. However, the repositioning needs improvements in its reproducibility. It is recommended the study to fix the eye by the healthy eye and the feasibility study for the development of a software that assists in patient repositioning.

1. Introduction

Malignant choroidal melanoma is a tumor arising in the layer of blood vessels, choroidal, positioned under the retina. It is a rare occurrence, affecting five to every million inhabitants, but rapidly evolving to metastasis [1, 2].

The procedure of removal of the tumor by enucleating it does not offer advantage regarding the occurrence of metastasis, when compared to treatments that preserve the eye [3].
One of the treatments for choroidal melanoma is the application of small photon bundles directed to the volume to be treated, the prescribed dose is applied in multiple sessions [4]. The most common side effects in this type of treatment are retinopathy, optic neuropathy, development of cataract, neovascular glaucoma and problems in the corneal epithelium [5].

1.1 Treatment success

Since the patient needs to be repositioned between sessions on different days according to the planning reference, a lot of care should be taken to ensure positional reproducibility. For the repositioning of the table, the treatment software reads, by infrared, six spheres positioned in the array, above the patient's chest. In addition, before each field, the table can also be adjusted by comparing two x-rays and a reference image performed by tomography.

In the case of intraocular radiotherapy, a non-invasive device is necessary to fix the patient's eye. Therefore, a LED lamp has been added to the original configuration, which one the patient is instructed to maintain eye contact during treatment.

The success of the treatment depends directly, among other biomedical factors, on the precision of the application and, above all, on the accuracy of the repositioning of the patient between sessions. Thus, as radiotherapy has evolved to increase beam precision, this work consists of studying the feasibility of reducing uncertainties associated with patient repositioning and fixation of patient's eye, identifying critical points of imprecision.

2. Methods and materials

Description of methods and materials used for radiotherapy treatment, patient's eye non-invasive fixation and patient's gaze motion analysis during treatment.

2.1. Radiotherapy equipments

The treatment was performed on a Novalis radiotherapy equipment of the Brainlab brand. The patient's skull was immobilized using a stereotactic mask made of thermoplastic material (a), positioned under a structure called array (b), both manufactured by Brainlab.

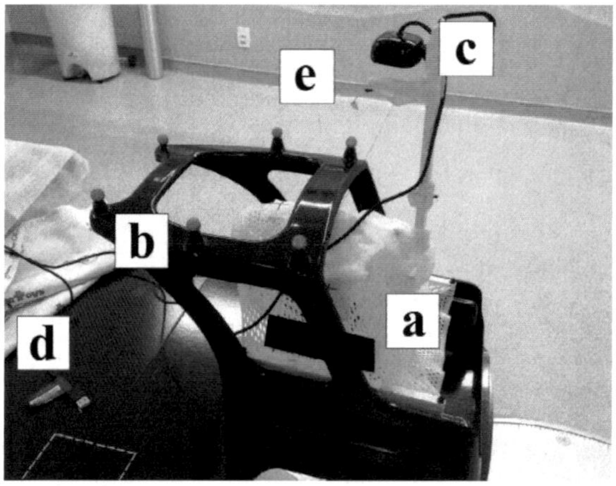

Figure 1. Setup of the equipments.

2.2. Devices for analysis and fixation of the patient's gaze

In order to analyse and control the iris displacement during the radiotherapy treatment, it was added to the standard configuration, a webcam (c), which one with a ten meters HDMI cable (d), took the image of the patient's eye in real time to a laptop located on the outside of the treatment room. The figure 1 shows the set-up.

In addition, as previously mentioned, the patient was instructed to look at the LED light (e), located on the same vertical axis of the isocenter and positioned above the webcam. For this procedure, the patient used the diseased eye.

2.3. Motion analysis methodology

A hospital professional was responsible for looking at the image in real time during treatment and visually comparing it with a reference photograph taken on the day of the CT scan used in treatment planning.

In case of movement of the eye, the professional intuitively should guide the operator to turn off the equipment, thus stopping the irradiation.

At the same time, the image of the patient's eye was recorded and later analysed by its displacement. The videos were transferred to the software Tracker [6], a motion analysis software, and separated as their lighting quality. Some videos could not be used for analysis due to the fact that the gantry passes between the light source and the camera, plus it is not a place with constant lighting.

In the analysis software, the video was digitally treated to increase the contrast between the iris and the conjunctiva. Consequently, it was possible to follow the movement of the center of the iris in an automated way, returning the values of the position in the axes x (horizontal displacement) and y (vertical displacement).

3. Results

Table 1 shows the sample analysed for this study.

Table 1. Study samples.

	Total
Videos	5
Fields	16
Frames	1791
Time (s)	215.7

3.1. Displacement of the iris center relative to the orthophoria point

After analysing the positions of the center of the iris in the software and calculating its displacements relative to the orthophoria point, the values found were grouped in table 2.

Table 2. Displacements statistics values.

	X-axis (mm)	Y-axis (mm)
Mean	-0.018	-0.020
Standard deviation	0.368	0.364
Standard uncertainty	0.009	0.009
Variance	0.136	0.133

In order to analyse the absolute displacements, these values were also grouped as shown in table 3.

Table 3. Absolute displacements statistics values.

	X-axis (mm)	Y-axis (mm)
Mean	0.228	0.227
Standard deviation	0.290	0.286
Standard uncertainty	0.007	0.007
Variance	0.084	0.082

So that the effects of accumulated displacements could be analysed, the cumulative displacement distribution and displacement histogram relative to the orthophoria point were also plotted as shown in figure 2(a) and figure 2(b).

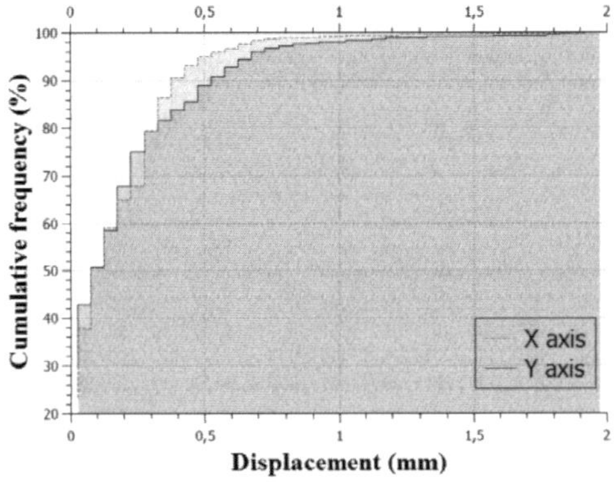

Figure 2(a). Cumulative distribution of displacements relative to the orthophoria point.

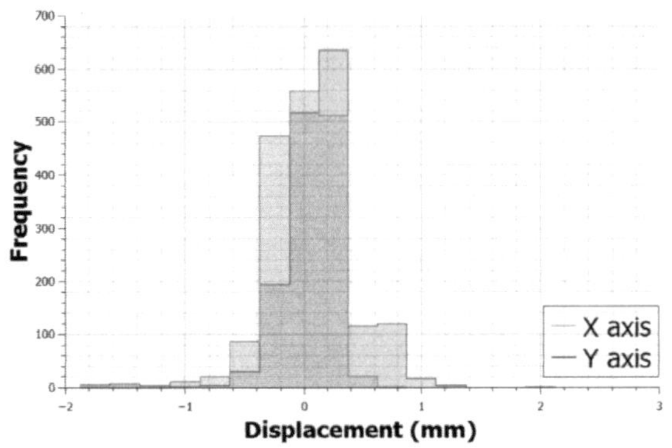

Figure 2(b). Displacement histogram relative to the orthophoria point.

3.2. Displacement of the iris center relative to the first day of treatment – repositioning accuracy
The displacements of the iris center were analysed by taking the first day of treatment as reference; this means that this statistic measures the capacity of repositioning of the patient by the accuracy of the equipment used. Table 4 shows the statistic values for the displacements of the iris center relative to the first day of treatment and table 5 shows the absolute values for the same analytical reference.

Table 4. Displacements statistics values.

	X-axis (mm)	Y-axis (mm)
Mean	-0.43	1.90
Standard deviation	0.62	1.85
Standard uncertainty	0.02	0.05
Variance	0.38	3.42

Table 5. Absolute displacements statistics values.

	X-axis (mm)	Y-axis (mm)
Mean	0.54	2.06
Standard deviation	0.53	1.66
Standard uncertainty	0.01	0.04
Variance	0.28	2.77

Figure 3(a) and figure 3(b) show the cumulative displacement distribution and the displacement histogram relative to the first day of treatment, so that the effects of accumulated repositioning errors could be analysed.

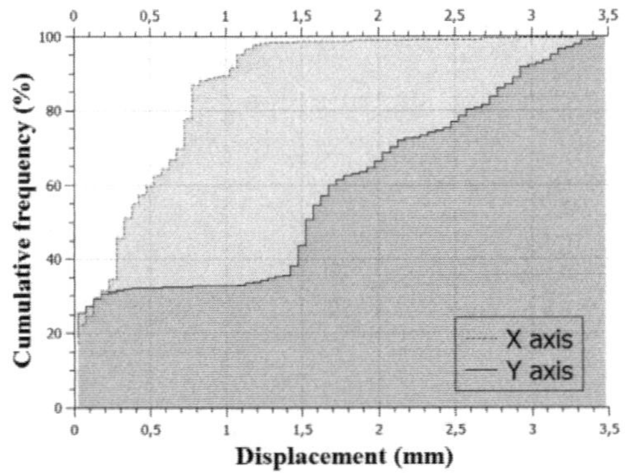

Figure 3(a). Cumulative distribution of displacements relative to the first day of treatment.

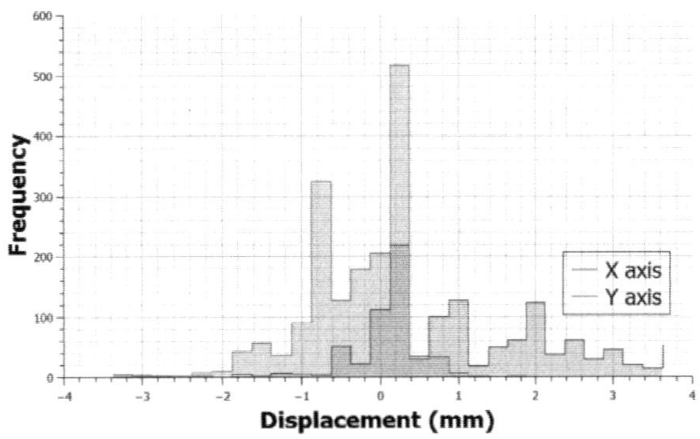

Figure 3(b). Displacement histogram relative to the first day of treatment.

4. Discussion

As could be observed during the study, the LED is enough for noninvasive fixation of the patient's eye during treatment. Besides that, it is certain that during the treatment the visual acuity of the patient suffers worsening as a side effect, consequently his ability to locate the LED lamp decreases as the therapy progresses.

Concerning the current beam disconnection system in case of unplanned movement of the patient's eye, it is necessary to understand that there are multiple sources of possible systematic errors in the case the equipment needs to be stopped. Because it is a non-automated method and depends on the reaction time of two people, first to visual stimulus, and then the reaction time to audio stimulus, which have values of 331 ms and 284 ms [7], respectively. Therefore, it is not the best methodology for the procedure, which allows further studies in order to determine the best method of this procedure.

In the analysis of movement that was performed after treatment, it was evident the poor quality of the videos (resolution, lighting and frame per second rate). However, it is worth mentioning that the result obtained in the study is compatible with analysis carried out by other groups [8].

Uncertainties of type B were not used in the results for the following reasons: the manufacturer of the product informs that the accuracy of the positioning in radiotherapy is submillimetric; besides the poor quality of the acquired images, the uncertainty due to positioning reading (pixel resolution), is too small compared to uncertainties of type A.

By interpreting the results, it is clear that the greatest problem of the system lies in the repositioning accuracy of the patient between different sessions, in other words, it lacks reproducibility of the patient's reference position that is used as basis for treatment planning. Such problem can lead to a dose depot outside the planning field.

5. Conclusion

The LED lamp is essential for fixing the patient's eye, so an alternative to the current system is necessary due to the loss of visual acuity of the eye affected by the disease. It is recommended to study the feasibility of positioning the LED above the healthy eye, and obstructing the unhealthy eye, contrary to the current method. In addition to that it is also recommended the image acquisition during the treatment by a high definition camera with the possibility of capturing in the range of the infrared light spectrum.

IOP Conf. Series: Journal of Physics: Conf. Series **975** (2018) 012038 doi:10.1088/1742-6596/975/1/012038

Besides that, the procedure for repositioning the patient between sessions can be studied so that its accuracy can be improved, reducing the dose outside the planning field for radiotherapy treatment. Simultaneously, a software can be developed for performing real-time analysis of the movement of the patient's eye which, in the case of previously established and established displacement detection, can disconnect the beam without further damage to the patient being treated.

7. References

[1] Chauvel P, Sauerwein W, Bornfeld N, et al. Clinical and technical requirements for proton treatment planning of ocular diseases. In: Wiegel T, Bornfeld N, Foerster MH, et al. editors. Radiotherapy of ocular disease. Basel: Karger; 1997. p. 133–142.

[2] Dieckmann K, Bogner J, Georg D, et al. A LINAC-based stereotactic irradiation technique of uveal melanoma. Radiother Oncol 2001;61:49–56.

[3] Seddon JM, Gragoudas ES, Albert DM, et al. Comparison of survival rates for patients with uveal melanoma after treatment with proton beam irradiation or enucleation. Am J Ophthalmol 1985;99:282–290.

[4] Shields CL, Shields JA, Gunduz K, et al. Radiation therapy for uveal malignant melanoma. Ophthalmic Surg Lasers 1998;29: 397–409.

[5] Dunavoelgyi R, Dieckmann K, Gleiss A, et al. Radiogenic side effects after hypofractionated stereotactic photon radiotherapy of choroidal melanoma in 212 patients treated between 1997 and 2007. Int J Radiation Oncology Biol Phys 2012;83:121-128.

[6] Loo W, Tze L. Video Analysis and Modeling Performance Task to Promote Becoming Like Scientists in Classrooms. American Journal of Educational Research. 2015; 3(2):197-207.

[7] Shelton J, Kumar G. Comparison between Auditory and Visual Simple Reaction Times. Neuroscience & Medicine 2010;1:30-32.

[8] Joachim B, Bernhard P, Dietmar G, et al. A noninvasive eye fixation and computer-aided eye monitoring system for linear accelerator-based stereotactic radiotherapy of uveal melanoma. Int J Radiation Oncology Biol Phys 2003;56:1128-1136.

IOP Conf. Series: Journal of Physics: Conf. Series **975** (2018) 012039 doi:10.1088/1742-6596/975/1/012039

Reliability of an x-ray system for calibrating and testing personal radiation dosimeters

M C Guimarães[1], C R E Silva[1], P H G Rosado[2], P G Cunha[2] and T A Da Silva[1]

[1] Centro de Desenvolvimento da Tecnologia Nuclear, CDTN, Belo Horizonte, Brazil
[2] Instituto de Radioproteção e Dosimetria, IRD, Rio de Janeiro, Brazil

E-mail: margaretecristinag@gmail.com, silvata@cdtn.br

Abstract. Metrology laboratories are expected to maintain standardized radiation beams and traceable standard dosimeters to provide reliable calibrations or testing of detectors. Results of the characterization of an x-ray system for performing calibration and testing of radiation dosimeters used for individual monitoring are shown in this work.

1. Introduction

X and gamma reference radiations were internationally established for calibrating and determining the energy dependence of dosimeters that are used for radiation protection purpose [1].

Dosimetry of the reference radiations in terms of air kerma in air requires reliable and traceable standard dosimeters. Reliable dosimetry is the basis of a system to derive the operational quantity for individual monitoring, the personal dose equivalent in 10 mm depth, Hp(10). Calibration and performance testing conditions of personal dosimeters are well established to assure their metrological reliability [2].

Several passive solid-state or direct-reading electronic type dosimeters have been used for individual monitoring in Brazil in compliance with radiation protection legal requirements. All dosimeters are required to be calibrated and have their response to different radiation field conditions well- known [3].

The aim of this work was to verify the reliability of the x-ray calibration system of the Dosimeter Calibration Laboratory, Centro de Desenvolvimento da Tecnologia Nuclear, LCD/CDTN, for performing calibration and testing of radiation dosimeters used for individual monitoring in terms of personal dose equivalent, Hp(10).

2. Methods and results

2.1. Repeatability and reproducibility of the 2575 NE standard dosimeter

The stability of a 600cc 2575 NE standard ionization chamber was studied based on its response from a $^{90}Sr/^{90}Y$ radiation source-chamber fixed geometry.

Results from 2013 to 2016 showed that its repeatability (the variation coefficient of many cycles of ten measurements) varied from 0.04% to 0.21%. In the same time interval, its reproducibility represented by its relative response varied within ±1.0%. Both results are acceptable values for a standard chamber.

2.2. Traceability of the 2575 NE standard dosimeter

The traceability of the 2575 NE standard ionization chamber was verified by calibrating it against a similar chamber that is the national standard of the Brazilian Laboratory of Ionizing Radiation Metrology, Instituto de Radioproteção e Dosimetria, LNMRI/IRD.

Air kerma rates in air were measured by both chambers by substitution method in N60, N80, N100 and W60, W80 e W110, narrow and wide ISO spectrum series, respectively (Fig. 1)

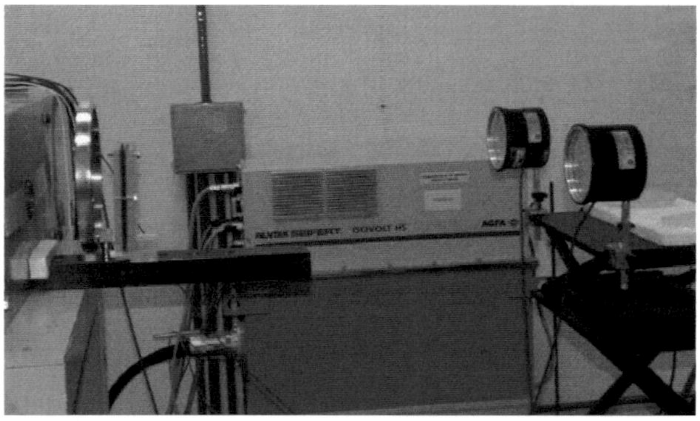

Figure 1. LCD/CDTN x-ray calibration set-up for calibrating the 2575 NE chamber against the LNMRI/IRD national standard.

Calibration coefficients of the LCD/CDTN 2575 NE ionization chamber (Table 1) showed that its maximum energy dependence was about 2% in the energy range; this is a typical value for a standard chamber.

Table 1. Calibration coefficients of the LCD/CDTN 2575 NE ionization chamber.

Reference radiation	Calibration coefficient (10^4 Gy.C^{-1})	Expanded uncertainty, k=2 (%)
N60	4.30	2.28
N80	4.27	2.28
N100	4.22	2.28
W60	4.34	2.25
W80	4.31	2.25
W110	4.25	2.25

2.3 Reliability of the LCD/CDTN dosimeter calibration procedure in terms of air kerma

The reliability of the LCD/CDTN calibration procedure in terms of air kerma in air was verified through a cross-comparison with the LNMRI/IRD by calibrating a TK30 model EXRADIN travelling ionization chamber in both laboratories (Fig.2).

Results were given in terms of the calibration coefficients with their respective expanded uncertainties for k equal to 2 (Table 2).

Analysis was done in terms of the Zeta score that establishes that results are consistent for $|z| \leq 2{,}0$, acceptable but investigation is needed for $2{,}0 < |z| < 3{,}0$ and unacceptable for $|z| \geq 3{,}0$ [4].

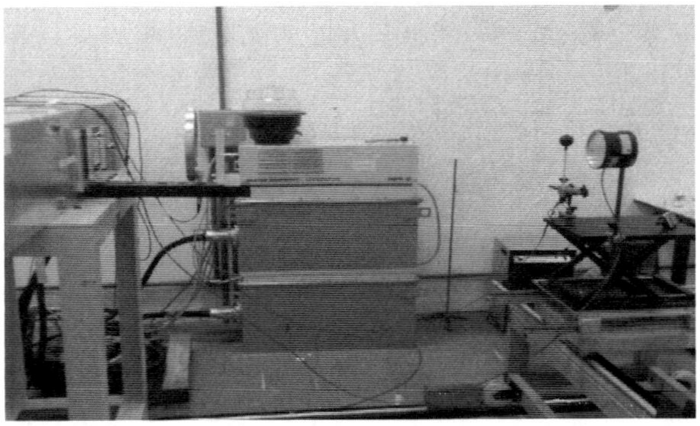

Figure 2. X-ray set-up for calibrating the TK30 travelling chamber against the CDTN standard chamber.

Table 2. Cross-comparison between the LCD/CDTN and LNMRI/IRD calibration procedures of a TK30 ionization chamber in terms of air kerma in air.

Reference radiation	Calibration coefficient (10^6 Gy.C^{-1})		Zeta score
	LCD/CDTN	LNMRI/IRD	[4]
N60	1.023 (2.84%)*	1.022 (2.28%)*	0.05
N80	1.035 (2.76%)	1.039 (2.28%)	0.20
N100	1.036 (2.92%)	1.028 (2.28%)	0.40
W110	1.037 (2.68%)	1.040 (2.25%)	0.15

* *Expanded uncertainty, k=2.*

In comparison to the values obtained in the LNMRI/IRD, the calibration coefficients of the TK30 chamber determined in the LCD/CDTN showed a small increase in the uncertainty values due to its calibration procedure. Considering that Zeta score [4] were lower than 2.0, the results confirmed the metrological coherence between both laboratories and, consequently, the reliability of the LCD/CDTN calibration procedure in terms of air kerma.

2.4 Reliability of the LCD/CDTN to irradiate dosimeters in terms of Hp(10)

The metrological ability of the LCD/CDTN to irradiate dosimeters in terms of Hp(10) was compared with two others Brazilian metrology laboratories. A set of five thermoluminescent (TL) dosimeters was irradiated by the three laboratories on the ISO standard slab phantom in N60, N80, N100 and W110 radiations with a requested Hp(10) value. Figure 3 shows the LCD/CDTN x-ray set-up where TL dosimeters were irradiated.

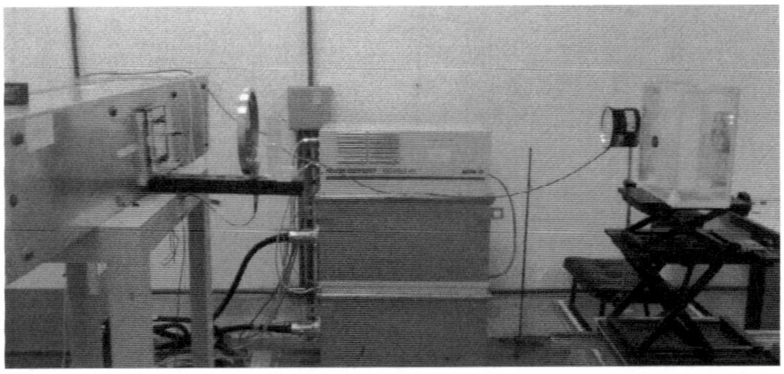

Figure 3. LCD/CDTN x-ray set-up used to irradiate TL dosimeters on the ISO slab phantom.

Dosimeters were provided and evaluated by an independent dosimetry laboratory in terms of Hp(10). Table 3 shows the Hp(10) evaluated values and the analysis in terms of the Zeta score.

Table 3. Comparison between the LCD/CDTN and two other metrology laboratories on the ability to irradiate dosimeters in terms of Hp(10).

Reference radiation	Personal dose equivalent, Hp(10) (mSv)			Zeta score [4]		
	CDTN	Lab. 2	Lab. 3	CDTN to Lab. 2	CDTN to Lab.3	Lab.3 to Lab.2
N60	2.77 ± 0.14*	2.94 ± 0.16*	2.62 ± 0.12*	1.6	1.1	2.2
N80	2.39 ± 0.14	2.33 ± 0.16	2.33 ± 0.15	0.6	0.4	0.0
N100	2.11 ± 0.13	2.33 ± 0.11	2.04 ± 0.09	2.6	0.6	2.7
W110	2.18 ± 0.13	2.30 ± 0.16	2.02 ± 0.10	1.2	1.3	2.2

* *Expanded uncertainty, k=2.*

Results show that the metrological coherence exists between the LCD/CDTN and Lab.3, since Zeta score is lower than 2,0. Related to Lab.2, result for N100 is acceptable but it should be investigated. Investigation should also be done to identify the cause of the differences between Lab.2 and Lab.3.

3. Conclusion

Since the metrological tests showed acceptable results, one can state that the reliability of the LCD/CDTN x-ray system to calibrate and test personal dosimeters in terms of Hp(10) was proved.

References

[1] International Organization for Standardization ISO 1996 X and gamma reference radiations for calibrating dosemeters and doserate meters and for determining their response as a function of photon energy - Part 1 -Radiation characteristics and production methods *ISO International Standard 4037* Geneva.

[2] International Organization for Standardization ISO 2012 2015 Reference radiation fields for radiation protection — Definitions and fundamental concepts *ISO International Standard 29661:2012 +A1:2015* Geneva.

[3] Guimarães M C, Silva C R E, Oliveira P M C and Da Silva T A 2016 On the response of electronic personal dosimeters on constant potential and pulsed x-ray beams *J. Phys. Conf. Ser. 733 012094 pp 1-5 doi:10.1088/1742 -6596/733/1/012094.*

[4] British Standard, BS 2010. Conformity assessment – General requirements for proficiency testing *ISO/IEC International Stanfard 17043* Geneva.

Acknowledgments - M C Guimarães is thankful to CNEN for her doctorate fellowship and to IPEMED for supporting her participation in the CBMRI 2018 congress. We thank C Maurício (IRD) for providing and evaluating the TL dosimeters. This work was supported by METRORADI/SIBRATEC Network, FAPEMIG (Universal 2015) and CNPQ (PQ Fellowship). It is part of the Project INCT Radiation Metrology in Medicine.

IOP Conf. Series: Journal of Physics: Conf. Series **975** (2018) 012040 doi:10.1088/1742-6596/975/1/012040

^{65}Zn and ^{133}Ba standardizing by photon-photon coincidence counting

Jamir S. Loureiro[1], Paulo A. L da Cruz[1], Akira Iwahara[1], José U. Delgado[1], Ricardo T. Lopes[2]

[1]Laboratório Nacional de Metrologia das Radiações Ionizantes/Instituto de Radioproteção e Dosimetria/Comissão Nacional de Energia Nuclear (LNMRI/IRD/CNEN), Av. Salvador Allende n° 9, Barra da Tijuca, Rio de Janeiro, RJ, Brasil, CEP22783-127;

[2]Programa de Engenharia Nuclear/Coordenação de Programas de Pós-Graduação em Engenharia/Universidade Federal do Rio de Janeiro (PEN/COPPE/UFRJ), Rua Moniz de Aragão, 360, Bloco 1, Ilha do Fundão – Cidade Universitária, Rio de Janeiro, RJ, Brasil, CEP 21941-594.

palcruz@ird.gov.br

Abstract. The LNMRI/Brazil has deployed a system using X-gamma coincidence technique for the standardizing radionuclide, which present simple and complex decay scheme with X-rays of energy below 100 keV. The work was carried on radionuclide metrology laboratory using a sodium iodide detector, for gamma photons, in combination with a high purity germanium detector for X-rays. Samples of ^{65}Zn and ^{133}Ba were standardized and the results for both radionuclides showed good precision and accuracy when compared with reference values. The standardization differences were 0.72 % for ^{65}Zn and 0.48 % for ^{133}Ba samples.

1. INTRODUCTION

The importance of the development and implementation of an absolute measurement technique for the radionuclide standardizing is highlighted. It is, in short, the state of the art, due to the complexity and specificity of the radionuclide decay-schemes.

The existence of several techniques enriches and makes the metrological chain robust, in order to provide results with traceability and reliability related to the determination of atomic and nuclear data and the activity quantity.

Techniques for absolute radionuclide standardizing are still few and rare, because their recognition and adoption by the international metrology system requires de confirmation of experimentation and theoretical formulation for years, finally enshrined through key-comparisons promoted by Bureau International des Poids et Mesures (BIPM).

The coincidence $4\pi\beta$(PC)-γ(NaI) counting based on gas flow proportional counter have been developed [1,2,3,4], rising to $4\pi\beta$(LS)-γ(NaI) based on liquid scintillation [5,6]. The anticoincidence $4\pi\beta$(PC)-γ(NaI) also based on gas flow proportional counter have been developed [7,8,9] rising to $4\pi\beta$(LS)-γ(NaI) based on liquid scintillation [10,11], in the actually.

With respect to the beta-emitting radionuclides only from the 80's the theoretical and experimental formulation was developed using the liquid scintillation technique by CNET (CIEMAT/NIST Efficiency Tracing) [12,13,[14] and TDCR (Triple-to Double Coincidence Ratio) [15,16] methods.

The development of solid Germanium and Lithium detectors has given new possibilities to gamma-spectrometry technique such as in radionuclide absolute standardizing by the sum-peak [17,18] and photon-photon [19,20] coincidence counting methods.

This work presents the implementation of absolute radionuclide measurement by photon-photon coincidence counting in LNMRI/Brazil, applied to the standardization of ^{65}Zn and ^{133}Ba, examples of simple and complex decays, respectively. In the specific case, the uses of X-gamma emissions do not involve corrections due to angular correlation and simplifies the experimentation and theoretical formulation.

The ^{65}Zn is used in nuclear medicine and serves to calibrate detectors, as a sodium iodide. The ^{133}Ba is a radionuclide with half–life relatively long and therefore used to calibrate germanium detectors in low energy region, serving as a source for consistency tests in radionuclide calibrator.

2. EXPERIMENTAL PROCEDURE

The photon-photon coincidence system was implemented in the LNMRI to standardize radioactive solution with nuclear transitions accompanied by X-rays w emissions below 100 keV. The reduction of HPGe detector efficiency was performed moving away the radioactive sample from detector or introducing absorbers between the sample and the detector. In the figures 1 and 2 are displayed a drawing and a block diagram symbolizing the electronics modules used in the system.

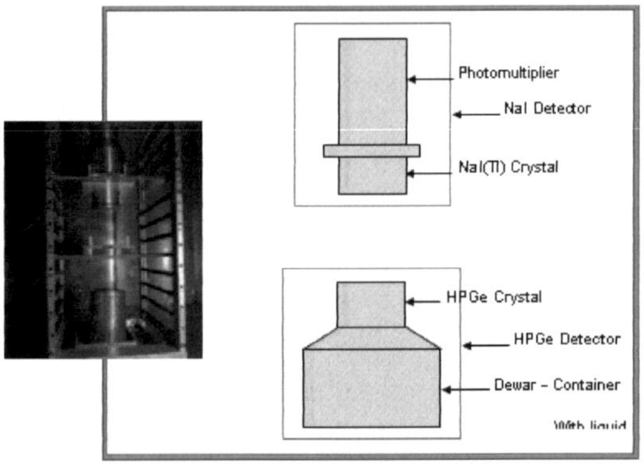

Figure 1 X-γ coincidence counting NaI(Tl) and HPGe detectors.

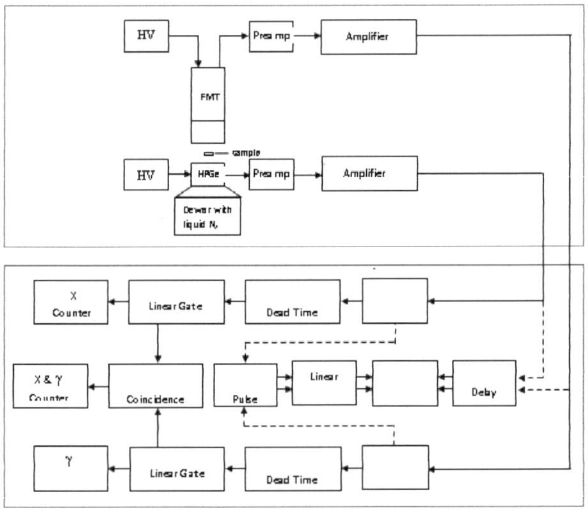

Figure 2 Block diagram of X-γ coincidence system.

The ^{65}Zn disintegrates by electron capture to the 1115 keV^{65}Cu excited level and beta plus emission to the ground state level of ^{65}Cu, as shown in figure 3.

The ^{133}Ba disintegrates by electron capture, mainly to two ^{133}Cs excited levels of 437 keV (85.4 %) and of 383 keV (14.5 %) with three minor branches to the 160 keV, 81 keV excited levels and the ground state, as shown in figure 4.

The activities were determined by extrapolation of apparent activities curves in function of the relative inefficiency factor [3], reducing HPGe detector efficiency. To decrease X-rays near 8 keV of intensity, was increased the distance between of ^{65}Zn samples and HPGE detector by steps of 0.5 cm until the maximum of 5 cm. This movement was done keeping unchanged the distance between the sample and NaI detector, while at the same time the distance from both and HPGe detector was being increased. To decrease X-rays near 30 keV of intensity, were inserted between of ^{133}Ba samples and detector until 9 aluminum discs, each one with 0.5 cm thick.

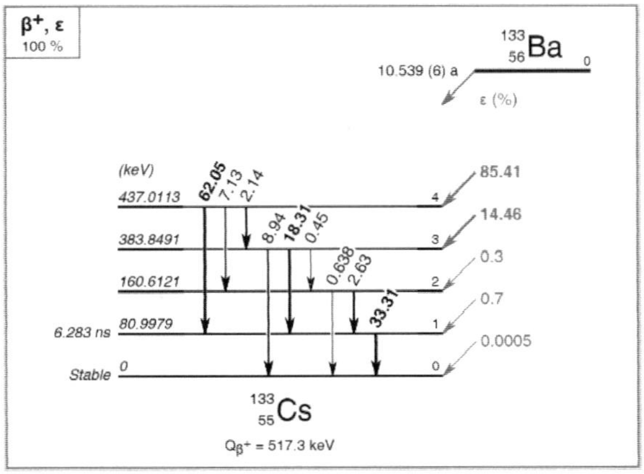

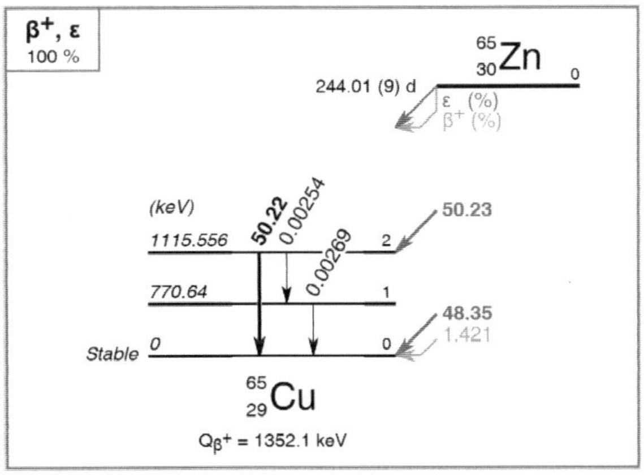

Figure 3 Simplified ^{65}Zn decay scheme with the selected highlighted events for counting

Figure 4 Simplified ^{133}Ba decay scheme with the selected highlighted events for counting.

The activities were determined by extrapolation of apparent activities curves in function of the relative inefficiency factor [3], reducing HPGe detector efficiency. To decrease X-rays near 8 keV of intensity, was increased the distance between of ^{65}Zn samples and HPGE detector by steps of 0.5 cm until the maximum of 5 cm. This movement was done keeping unchanged the distance between the sample and NaI detector, while at the same time the distance from both and HPGe detector was being increased. To decrease X-rays near 30 keV of intensity, were inserted between of ^{133}Ba samples and detector until 9 aluminum discs, each one with 0.5 cm thick.

Three ^{65}Zn samples were prepared by deposition of standard radioactive solution on a plastic tape fixed at the bottom of an acrylic ring, covered by another similar plastic tape. A total of four ^{133}Ba samples were prepared on a thin plastic film of polyvinyl chloride (VYNS) and measured by photon-photon system. The validation of this system was carried out with the measuring of the same ^{133}Ba samples by $4\pi\beta$–γ(NaI) coincidence counting system.

The measurements were done adjusting each Single Channel Analyzer – SCA with pulse height suitable to detect the selected events. For ^{65}Zn samples were considered gamma photons of 1115 keV (SCA windows from 7.40 V to 10 V) and X-rays from 8 keV to 9 keV (SCA window from 0.15 V to 0.30 V). For ^{133}Ba samples were considered gamma photons of 356 keV (SCA windows from 2.60 V to 3.32 V) and X-rays from 30 keV to 36 keV (SCA window from 0.60 V to 0.98 V).

3. RESULTS

The apparent activities were calculated for each detection efficiency variation by software deployed in the LNMRI/IRD, which uses the coincidence method [21, 22, 23].

As the selected events for counting are mono-energetic gamma photons and X-rays produced by electron capture, originated in the same decay, and considering too that there's no overlapping in the energy range of the two kinds of selected events, each channel of the instrumentation certainly will register only their respective events and no other one. In these conditions, the N_β of the BAERG equation (1) for ideal detectors can be substituted by N_X to determine all apparent activities with good accuracy without the need to introduce any correction factor in the value found after the experimental curve extrapolation:

$$N_0 = \frac{N_X N_Y}{N_C} \tag{1}$$

where, N_0 is apparent activity; N_X is -ray count rate of HPGe detector; and $N\gamma$ is gamma count rate of NaI(Tl) detector; and N_C is the X-gamma count rate coincidence. For each experimental curve, the sample activities were determined through linear regression and extrapolation to zero value of the relative inefficiency factor.

The figure 5 displays the apparent activity curve in function of relative inefficiency factor for the ^{65}Zn sources. The mean activity concentration found in these samples was 646.02 kBq/g ± 0.56 %, with the uncertainty determination showed in Table 1.

The figure 6 displays the apparent activity curve in function of relative inefficiency factor for the ^{133}Ba sources.The mean activity concentration found in these samples was 504.23 kBq/g ± 0.39 %, with the uncertainty determination showed in Table 2.

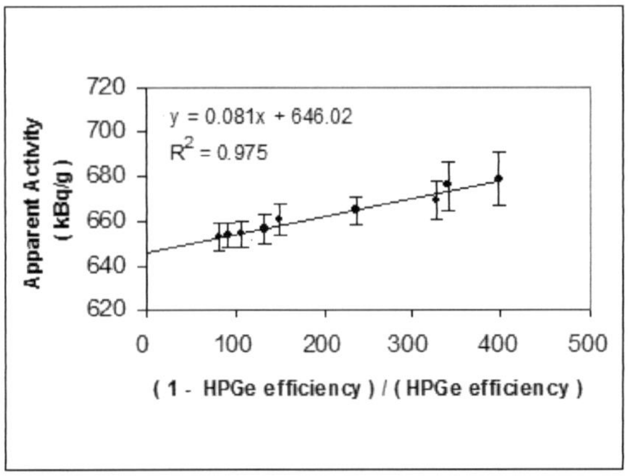

Figure 5 Apparent activity produced by relative inefficiency factor for the ^{65}Zn samples.

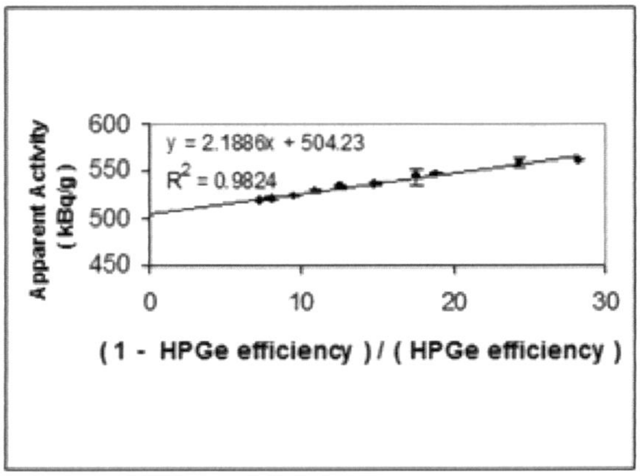

Figure 6 Apparent activity produced by relative inefficiency factor for the ^{133}Ba samples.

Table 1 Uncertainty budget (k = 1) of the ^{65}Zn solution standardization.

Components	Type	Uncertainty (%)
Statistic counts	A	0.45
Half-life	B	0.04
Background	A	0.03
Mass	B	0.05
Extrapolation	A	0.033
Combined (k = 1)		0.56

Table 2 Uncertainty budget (k = 1) of the ^{133}Ba solution standardization.

Components	Type	Uncertainty (%)
Statistic counts	A	0.38
Half-life	B	0.04
Background	A	0.02
Mass	B	0.05
Extrapolation	A	0.004
Combined (k = 1)		0.39

The Tables 3 and 4 display the comparisons between standardization result for ^{65}Zn and ^{133}Ba samples using the X-gamma coincidence system and their registered reference values for the same solutions.

Table 3 Comparison ^{65}Zn standardization (k = 1)

Measurement system	Activity (kBq/g)	Uncertainty (%)	Δ (%)
Reference (Ionization chamber)	641,43	0.80	-
Photon-photon coincidence	646.02	0.56	0.72

Table 4 Comparison ^{133}Ba standardization (k = 1)

Measurement system	Activity (kBq/g)	Uncertainty (%)	Δ (%)
Reference (Ionization chamber)	501,84	0.25	-
Photon-photon coincidence	504.23	0.39	0,48

4. CONCLUSION

The deployment work of this new X-gamma coincidence system was well succeeded, as shown in table 4 by the comparisons of the ^{133}Ba samples, which produced difference 0.48 % between the results obtained by X-gamma and 4πβ(PC)-γ(NaI) coincidence counting, showing that the new system has introduced an important improvement in the LNMRI routines for standardization of samples with radionuclides of complex decay scheme. By the other side, the difference of 0.72 % obtained with the ^{65}Zn samples shows that is the possibility to standardize radionuclide samples with X-rays in order of 8 keV.

REFERENCES

[1] Barnothy J and Forro M 1951 Coincidence Methods of Measuring Disintegration Rates of Radioactive Sources The Review of Scientific Instruments **22** 415.

[2] Campion PJ 1959 The standardization of radioisotopes by beta-gamma coincidence method using high efficiency detectors, Int. J. Applied Radiation and Isotopes **4** 232.

[3] Baerg A P 1973 The Efficiency Extrapolation Method in coincidence counting, Nuclear Instruments and Methods **II2** 143.

[4] Iwahara A 1979 Técnica de traçador para medida específica de ^{63}Ni, utilizando um sistema de coincidência 4πβ–γ Tese de Mestrado *Instituto Militar de Engenharia* Brasil.

[5] Christmas P, Judge S M, Ryves TB and Smith D 1983 The decay scheme of ^{64}Cu Nuclear Instruments and Methods **215** 397.

[6] Simpson B R S and Meier BR 1994 Activity measurement of ^{204}T1 by direct liquid scintillation methods Nuclear Instruments and Methods in Physics Research A **369** 340.

[7] Bryant J 1962 Anticoincidence counting method for standardizing radioactive materials. Applied Radiation and Isotopes **13** 273.

[8] Bryant J 1996 Advantage of anticoincidence counting for standardization of radionuclides emitting delayed gamma *IEEE Symp. Standardization of Radionuclides* SM **79**21.

[9] Baerg AP, MUNZENMAYER K and BOWES GC 1976 Live-Timed Anticoincidence Counting with Dead Time Circuitry Metrologia **12** 77.

[10] Bobin CH, Bouchard J, Hamon C, Iroulart MG and Plagnards J 2007 Standardization of ^{67}Ga using a 4π(LS)β–γ Anti-Coincidence System Applied Radiation and Isotopes **65** 757.

[11] da Silva CJ 2008 Implementação de um Sistema de anti-coincidência 4πNaI(Tl)-CL com cronometragem em tempo vivo e tempo morto extendível Tese de doutorado *Universidade Federal do Rio de Janeiro* Brasil.

[12] Coursey BM, Mann WB, Malonda G, Torãno EG, Los Arcos JM, Gibson JAB and Reher E 1986 Standardization of C-14 by 4πβ Liquid Scintillation Efficiency Tracing with Hydrogen-3Applied Radiation and Isotopes **37** 403.

[13] Malonda AG 1995 Modelos de Parâmetro Libre em Centelleo Liquido *Editorial CIEMAT* Madrid.

[14] Loureiro JS 2000 Calibração de radionuclídeos que decaem por emissão beta ou por captura eletrônica pela técnica de cintilação líquida Tese de Mestrado *Instituto Militar de Engenharia* Brasil.

[15] Cassette P and Broda R 1992 Experimental evaluation of TDCR models for the 3 PM liquid-scintillation counter Nuclear Instruments and Methods A **312** 95.
[16] da Cruz PAL, da Silva CJ, Iwahara A, Loureiro JS, Oliveira AE, Tauhata L and Lopes RT 2016 TDCR and CIEMAT/NIST Liquid Scintillation Methods applied to the Radionuclide Metrology *IOP Journal of Physics: Conf. Series 733/012099* doi:10.1088/1742-6596/733/1/012099.
[17] Brinkman GA, Aten Jr. AHW 1963 Absolute standardization with a NaI(Tl) crystal III. Int. J. Applied Radiation and Isotopes **4** 503.
[18] da Silva, RL, Almeida MCM, Delgado JU, Poledna R, Veras AS, Rangel J, Trindade OL 2016 Metrological activity determination of ^{133}Ba by sum-peak absolute method 8th Brazilian Congress on Metrology *IOP Journal of Physics: Conferences Series 733/012095* doi:10.1088/1742-6596/733/1/012095.
[19] Iwahara A, Marechal MH, da Silva CJ and Poledna R 1990 Determination of the activity concentration of a ^{125}I solutions by X-(X,γ) coincidence counting and efficiency extrapolation curve Nuclear Instruments and Methods A **286** 370.
[20] Schrader H 2006 Photon-photon coincidences for activity determination: I-125 and other radionuclides Applied Radiation and Isotopes **64** 1179.
[21] Baerg AP 1966 Measurement of Radioactive Disintegration Rate by Coincidence Method Metrologia **2** 166 23.
[22] Kawada Y 1972 Extended Applications and Improvement of the 4βπ–γ coincidence Method in Standardization of Radionuclides *ETL report 730.*
[23] Grigoresku L 1973 Accuracy of coincidence measurements Nuclear Instruments and Methods **112** 151.

IOP Conf. Series: Journal of Physics: Conf. Series **975** (2018) 012041 doi:10.1088/1742-6596/975/1/012041

Assessment of Ethylene Vinyl-Acetato Copolymer (EVA) Samples Bombarded by Gamma Radiation via Linearity Analyses

L N de Oliveira[1,3], E O do Nascimento[1], F Schimidt[1], P L Antonio[3] and L V E Caldas[3]

[1]Instituto Federal de Educação, Ciência e Tecnologia de Goiás-IFG, Campus Goiânia, Goiânia-GO, Brazil
[2]Polícia Militar de Goiás, Goiânia-GO, Brazil
[3]Instituto de Pesquisas Energéticas e Nucleares-IPEN/CNEN, São Paulo-SP, Brazil

E-mail: lucas@ifg.edu.br, eriberto.on@gmail.com, schimidt99@gmail.com, patrilan@ipen.br and lcaldas@ipen.br

Abstract. Materials with the potential to become dosimeters are of interest in radiation physics. In this research, the materials were analyzed and compared in relation to their linearity ranges. Samples of ethylene vinyl-acetate copolymer (EVA) were irradiated with doses from 10 Gy to 10 kGy using a ^{60}Co Gamma-Cell system 220 and evaluated with the FTIR technique. The linearity analyses were applied through two methodologies, searching for linear regions in their response. The results show that both applied analyses indicate linear regions in defined dose interval. The radiation detectors EVA can be useful for radiation dosimetry in intermediate and high doses.

1. Introduction

EVA means Ethylene Vinyl Acetate. This product is in the form of a synthetic foam, and it is produced from its thermoplastic copolymer. It is a very flexible material, suitable for various purposes [1]. When this copolymer has more than 40% vinyl acetate in its composition it becomes very flexible, and it is called EVA rubber. This copolymer may be vulcanized and the rubber can increase the number of its applications in the industry. It can be used in the adhesive industry both associated with epoxy producing commercial hot-melt adhesives. It also may be used for traffic paints and marine paint applications [2]. The ethylene–vinyl acetate copolymers with ammonium polyphosphate are also used as enhanced flame-retardants, and their performance are related to the formation of a thermally stable phosphocarbonaceous structure in the char, and the blend even shows a slight improvement in mechanical behavior. Poly vinyl alcohol is widely used in connection with spray-drying of emulsions, in particular ethylene–vinyl acetate copolymer emulsions. The PVA is added both as a protective colloid during the polymerization and as a redisposing aid prior to the actual spray drying. Therefore, the objective of this research is to investigate EVA for its linearity after bombardment with gamma rays.

2. Materials and Methods

The ethylene vinyl acetate (EVA) samples used in this work were composed by small cubic pieces, with dimensions of 1 x 1 x 0.5 mm^3 and of three colors: black, green and white. The EVAs were irradiated with absorbed doses between 10 Gy and 10 kGy using a ^{60}Co Gamma Cell-220 system (dose rate of 1.089 kGy/h); then each sample was evaluated, and it presented an absorbance spectrum acquired on a Fourier Transform Infrared (FTIR) Spectrometer (Frontier/Perkin Elmer). The spectra were acquired with the crude samples directly on the PIKE GladiATR accessory without any preparation. All spectra were obtained in triplicate for each sample, in the absorbance mode, in the region between 4000 and 450 cm^{-1}, with 16 acquisitions, 4 cm^{-1} resolutions and 1cm^{-1} spectral ranges.

3. Results

Figure 1 presents the spectra from black, green and white EVA samples. In these results it is possible to infer the characteristics of each sample along the associated wavenumbers, as well as the regions and peaks in which the irradiation was significant. The total area under the curves from 450 to 4000 cm^{-1} is presented in Figure 2. Linear and nonlinear regions from methods can be seen using the FTIR technique. A linear relationship between integrated area and the absorbed dose of the samples can be observed in Figure 3. The results obtained were R^2=0.30, R^2=0.46 and R^2=0.22 respectively for black, green and white EVA samples. In Figure 4, the dose-response curves of the ethylene vinyl acetate, using a ^{60}Co gamma radiation source in black color for a wavenumber of 2940 cm^{-1}; green color for a wavenumber of 430 cm^{-1}; and white color for a wavenumber of 2233 cm^{-1} are shown. The results obtained were R^2=0.51, R^2=0.89 and R^2=0.30 respectively for black, green and white EVA samples. In this way the green EVA shared a greater linearity than the other analyzed materials. In Figure 5 the cumulative probability as a function of linearity is shown. Specifically in the region of the zoom shown in the figure it is possible to infer that the material that presents greater linearity is the green one, in relation to the white and black ones. For green samples, the linearity begins around 0.5 and remains in the region cumulative probability larger than the other materials.

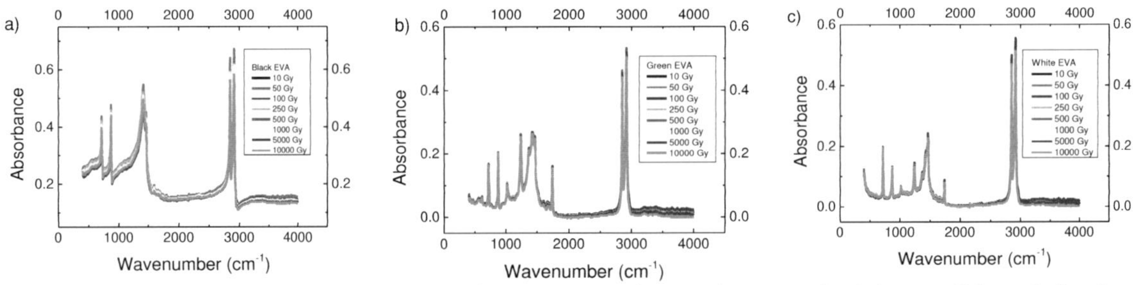

Figure 1. Spectra from black, green and white EVA samples. In these results it is possible to infer the characteristics of each sample along the associated wavenumbers, as well as the regions and peaks in which the irradiation was significant.

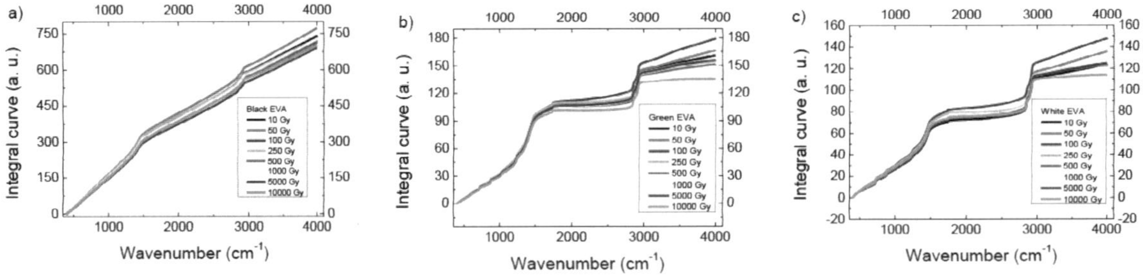

Figure 2. Integral area *versus* wavenumber, using the FTIR technique and irradiation from 10 Gy up to 10 kGy of absorbed doses. (a) Black; (b) Green; and (c) White samples.

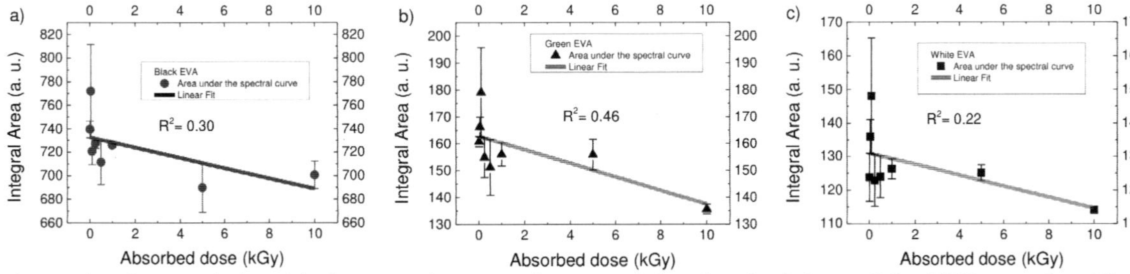

Figure 3. Linear relationship between integrated area and the absorbed dose of the EVA samples. The results obtained were R^2= 0.30, R^2 = 0.46 and R^2= 0.22 respectively for black, green and white EVA samples using the Integrated Area Under the Curve Method.

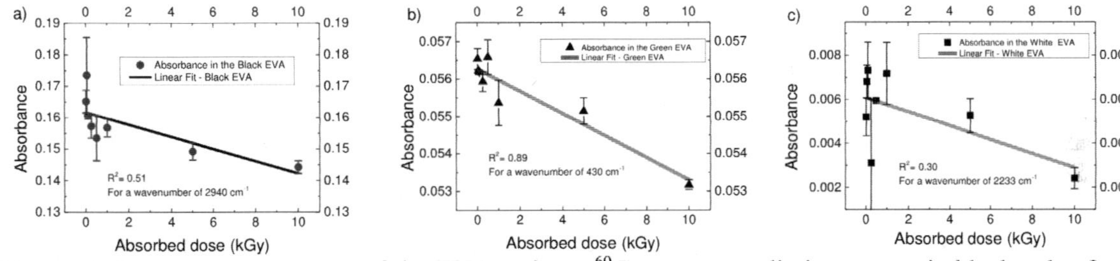

Figure 4. Dose-response curves of the EVA, using a ^{60}Co gamma radiation source in black color for a wavenumber of 2940 cm^{-1}; green color for a wavenumber of 430 cm^{-1}; and white color for a wavenumber of 2233 cm^{-1}. The linearity results obtained were R^2=0.51, R^2=0.89 and R^2=0.30 respectively for black, green and white EVA samples, using the Wavenumber Method.

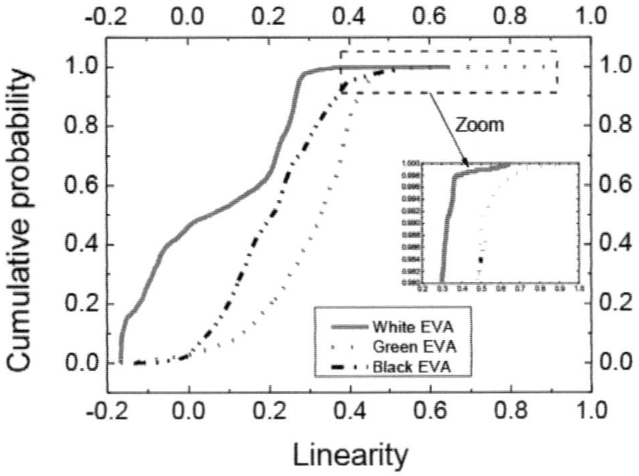

Figure 5. Cumulative probability *versus* linearity, for white, green and black EVA samples for wavenumber shown 400 up to 4000 cm^{-1}.

IOP Conf. Series: Journal of Physics: Conf. Series **975** (2018) 012041 doi:10.1088/1742-6596/975/1/012041

4. Conclusions

The FTIR measurements may be useful in dosimetry using the spectra of irradiated EVAs. The dose-response curves showed a non-linear relationship for all wavenumbers investigated. The highest linearity was in order respectively for green, white and black EVAs. The radiation detectors present potential use for radiation dosimetry in intermediate and high doses.

Acknowledgements

This work was supported in part by the Brazilian agencies: CNPq, under Grants No. 165466/2015-4, 151013/2014-4, and 301335/2016-8, FAPESP and MCTI.

References

[1] Herman F M 2005 *Encyclopedia of Polymer Science and Technology* (John Wiley & Sons, Inc.)
[2] Whelan T 1994 *Polymer Technology Dictionary* (Springer Science)
[3] Boguski J, Przybytniak G, Łyczko K 2014 *Radiat. Phys. Chem.* **100** 49-53

IOP Conf. Series: Journal of Physics: Conf. Series **975** (2018) 012042 doi:10.1088/1742-6596/975/1/012042

Comparison for Air Kerma from Radiation Protection Gamma-ray Beams with Brazilian Network - 2016/2017

TS Cabral [1], CNM da Silva[1], MPA Potiens[2], CMA Soares [3], RR Silveira[4], H Khoury[5], V Saito[5], E Fernandes [6], WF Cardoso[7], HPS de Oliveira[7], MA Pires[8], AS de Amorim[9], M Balthar[9].

[1] Instituto de Radioproteção e Dosimetria, Rio de Janeiro, Brazil;
[2] Instituto de Pesquisas Energéticas e Nucleares, São Paulo, Brazil;
[3] Centro de Desenvolvimento de Tecnologia Nuclear, Minas Gerais, Brazil;
[4] Centro Regional de Ciências Nucleares do Nordeste, Pernambuco, Brazil;
[5] Departamento de Energia Nuclear da UFPE, Pernambuco, Brazil;
[6] Laboratório de Ciências Radiológicas da UERJ, Rio de Janeiro, Brazil;
[7] Eletronuclear – Eletrobrás Termonuclear S.A., Rio de Janeiro, Brazil;
[8] MRA Comércio de Instrumentos Eletrônicos Ltda, São Paulo, Brazil.;
[9]Centro Tecnológico do Exercito, Rio de Janeiro, Brazil.

E-mail: tschirn@ird.gov.br

Abstract. The results of the comparison involving 9 laboratories in Brazil are reported. The measured quantity was the air kerma in ^{137}Cs and ^{60}Co, at the level of radioprotection. The comparison was conducted by the National Laboratory Metrology of Ionizing Radiation (LNMRI/IRD) from October 2016 to March 2017. The largest deviation between the calibration coefficients was 0.8% for ^{137}Cs and 0.7% for ^{60}Co. This proficiency exercise proved the technical capacity of the Brazilian calibration network in radiation monitors and the results were used by some in the implementation of the standard ISO/IEC 17025.

1. Introduction

One of the activities inherent in calibration laboratories is the periodic participation in Comparison programs where, among others, it is possible to assess the ability of the qualified technician to carry out a calibration service, in accordance with instructions laid down in procedures documented in appropriate environmental conditions and under control, getting a result consistent with the measurement accreditation as stated by the labs Participation in exercises of this nature, proficiency testing, is one of the requirements of norm ABNT ISO/IEC 17025 BR.

The use of calibrated instruments for the purposes of radiological protection is a requirement that ensures the safe use of the various areas in which we work with ionizing radiation. These calibrations should be made based on measurement standards and following written procedures. The National Laboratory Metrology of Ionizing Radiation (LNMRI/IRD/CNEN) organized and conducted this

IOP Conf. Series: Journal of Physics: Conf. Series **975** (2018) 012042 doi:10.1088/1742-6596/975/1/012042

comparison exercise from October 2016 to March 2017. The protocol was structured according to the ISO 17043.

The quantity compared is the calibration coefficient in air kerma in ^{137}Cs and ^{60} Co beams. Nine laboratories participated in the comparison exercise. These exercises are important for increasing confidence in the ability of the results of measurements carried out by the Brazilian network of radiation monitors calibration laboratories.

1.1. Participating Laboratories.

- Laboratório de Calibração de Monitores de Radiação – LCMR/LNMRI/IRD
- Instituto de Pesquisas Energéticas e Nucleares - IPEN
- Centro de Desenvolvimento de Tecnologia Nuclear- CDTN
- Centro Tecnológico do Exercito - CTEx
- Departamento de Energia Nuclear da UFPE - DEN/UFPe
- Laboratório de Ciências Radiológicas da UERJ - LCR/UERJ
- Eletronuclear – Eletrobrás Termonuclear S.A.
- MRA Comércio de Instrumentos Eletrônicos Ltda.
- Centro Regional de Ciências Nucleares do Nordeste – CRCN

The Calibration Laboratory of Radiation Monitors LNMRI/IRD determined the calibration factor reference. It been the mean of the calibrations performed during the exercise.

2. Objective

The purpose of the comparison exercise was:

a) Calculate the coefficient of calibration in air kerma and compares them;

b) Determine the performance of the calibration of the qualities S-Cs and S-Co in terms of air kerma of participating laboratories;

c) Identify problems and propose corrective action.

2.1. Instrument Submitted For Comparison - The Transfer Chamber

The ionization chamber PTW LS01 was used as the transfer chamber is spherical graphite walls. Just as the chamber was sent, the laboratories were instructed to make their own measurements as their usual procedures.

The main features of the transfer chamber are:
Manufacturer: PTW
Model: TN32002
Type: Ionization chamber
Series Number: 490
Sensitive volume: 1,000 cm^3
Outside diameter: 140 mm
Electrode diameter: 50 mm
Wall thickness: 3 mm
Potential of high voltage electrode: - 400 V

IOP Conf. Series: Journal of Physics: Conf. Series **975** (2018) 012042 doi:10.1088/1742-6596/975/1/012042

3. Determination of the calibration factor

Comparison for air kerma was performed indirectly by comparing the calibration coefficients in air kerma the transfer chamber, N_k. The calibration coefficient is determined using the expression:

$$N_k = \frac{\dot{K}_a}{\left(\frac{M}{t}\right).\Phi(T, P).f^e.3600} \quad [Gy/C] \qquad (1)$$

Where:

N_k is the factor of chamber calibration in terms of air kerma (Gy /C).

$\dot{K}_a$ is the rate of air kerma measured with a standard chamber

$\overline{\left(\frac{M}{t}\right)}$ is the average reading of the chamber you want to calibrate divided by time;

$\phi(T,P)$ is the correction factor for temperature and pressure;

f^e is the calibration factor of the electrometer.

The LNMRI used the standard chamber to determinate the reference value of the calibration coefficient, $N_{k,LNMRI}$ against which the results of the measurements of the calibration factor by participants, $N_{k,part}$, were compared. The $N_{k,LNMRI}$ in air kerma was determined with a standard chamber LS01 model TM32002 serial number 104 calibrated in June 2013 at the Physikalisch-Technische Bundesanstalt (PTB).

The calibration coefficient for the transfer chamber was provided in terms of air kerma, in units of Gy /C in referenced standard conditions of temperature, pressure and relative humidity at T = 293.15 K, P = 101.325 kPa h = 65%. Participants did not apply any correction for ion recombination and the relative humidity should be between 30% and 75% during calibration, followed the normal routine of the laboratory with respect to the operator and the equipment used.

4. The standard ionization chambers used

Participated in the comparison exercise seven ionization chambers, as shown in Table 1 All chambers are reference standards calibrated to the energies of ^{60}Co and ^{137}Cs.

Table 1. The standard ionization chambers used by the participating laboratories and by LNMRI.

Laboratories	Standard ionization chambers
LNMRI	PTW TN32002#104
LCR/UERJ	PTW N23361#2923
LCR/UERJ	PTW TN32002#489
ETN	PTW N32002#86
MRA	PTW 32002#309
IPEN	PTW W32002#25
DEN/UFPE	NE 2575#526
DEN/UFPE	PTW TW32002#550
CRCN	NE 2575#518
CDTN	PTW TN32002#382
CTEx	PTW TW32002#527

5. Transfer chamber calibration

The transfer chamber circled among the LNMRI and participants through the planning done by LNMRI together with laboratories. The LNMRI calibrated the transfer chamber to verify the performance, before sending it to the next participating laboratory. Control measures were obtained by LNMRI thus proving the proper functioning of the transfer chamber.

The following terms of reference were recommended:

1 Distance source - Chamber: 200 cm.

2 Operating Voltage: - 400 V.

No deviations from these conditions were reported.

Leakage currents of measurement systems, composed of electrometer and chamber were measured in accordance with the procedures of each laboratory.

The percentage deviation for the calibration coefficient in terms of air kerma, and LNMRI between the participating laboratories should be equal to or less than 2.0%.

6. Results

6.1. Physical quantity to be measured

The quantity chosen to be measured for comparison was the air kerma (K_{ar}), since all the chambers were calibrated in this quantity. The value of air kerma, K_{AR}, was determined for the irradiation conditions with the transfer chamber in each participating laboratory.

The Calibration coefficients of the transfer chamber, N_k obtained by participating laboratories and by the LNMRI are in Table 2 along with their respective uncertainties expanded standard.

Table 2. The calibration coefficient and uncertainties determined by the participating laboratories.

Laboratories	**Calibration Factor** $N_k \pm U$ **(%)**	**Calibration Factor** $N_k \pm U$ **(%)**	**Quantity/Unity**
	Cs-137	Co-60	
LNMRI	2,486E+04 ± 2,7%	2,436E+04 ± 2,6%	K_{ar} /Gy.C^{-1}
LCR/UERJ	2,488E+04 ± 4,7%		K_{ar} / Gy.C^{-1}
LCR/UERJ	2,460E+04 ± 4,7%		K_{ar} / Gy.C^{-1}
ETN	2,516E+04 ± 5,1%		K_{ar} / Gy.C^{-1}
MRA	2,504E+04 ± 2,7%	2,458E+04 ± 2,7%	K_{ar} / Gy.C^{-1}
IPEN	2,523E+04 ± 3,7%	2,426E+04 ± 4,2%	K_{ar} / Gy.C^{-1}
DEN/UFPE	2,468E+04 ± 3,4%	2,409E+04 ± 4,6%	K_{ar} / Gy.C^{-1}
DEN/UFPE	2,481E+04 ± 3,4%	2,438E+04 ± 4,6%	K_{ar} / Gy.C^{-1}
CRCN	2,470E+04 ± 2,0%		K_{ar} / Gy.C^{-1}
CDTN	2,470E+04 ± 5,5%		K_{ar} / Gy.C^{-1}
CTEx	2,463E+04 ± 3,0%	2,415E+04 ± 2,8%	K_{ar} / Gy.C^{-1}

OBS: LNMRI – reference

In Figure 1, we can see the values determined for the calibration coefficient of the LNMRI and by the participating laboratories. Error bars are the values of the standard expanded uncertainties submitted by the laboratories.

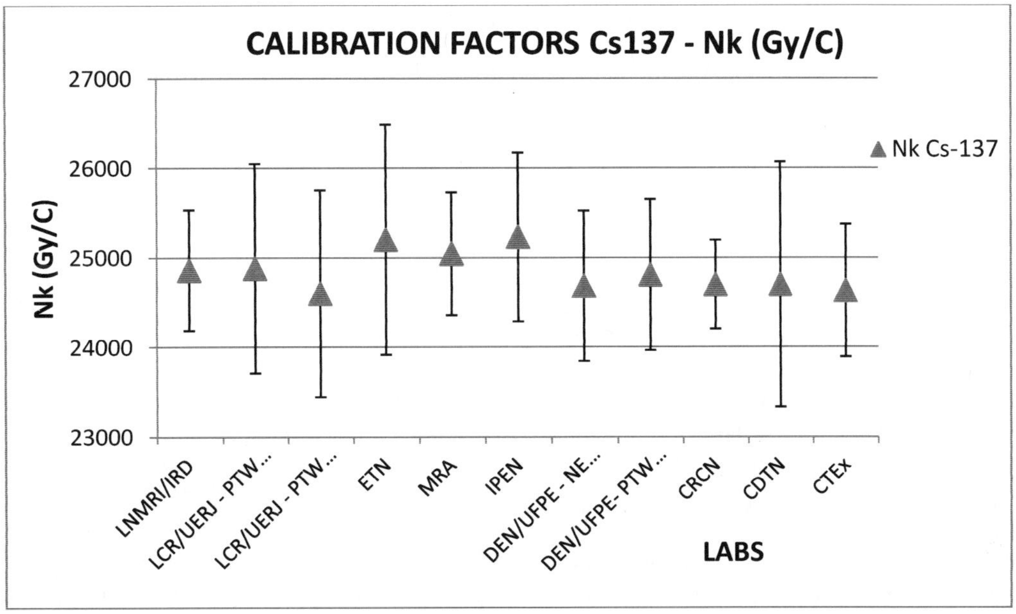

Figure 1. Calibration Factors determined for the transfer chamber by LNMRI and participating laboratories for quality S-Cs

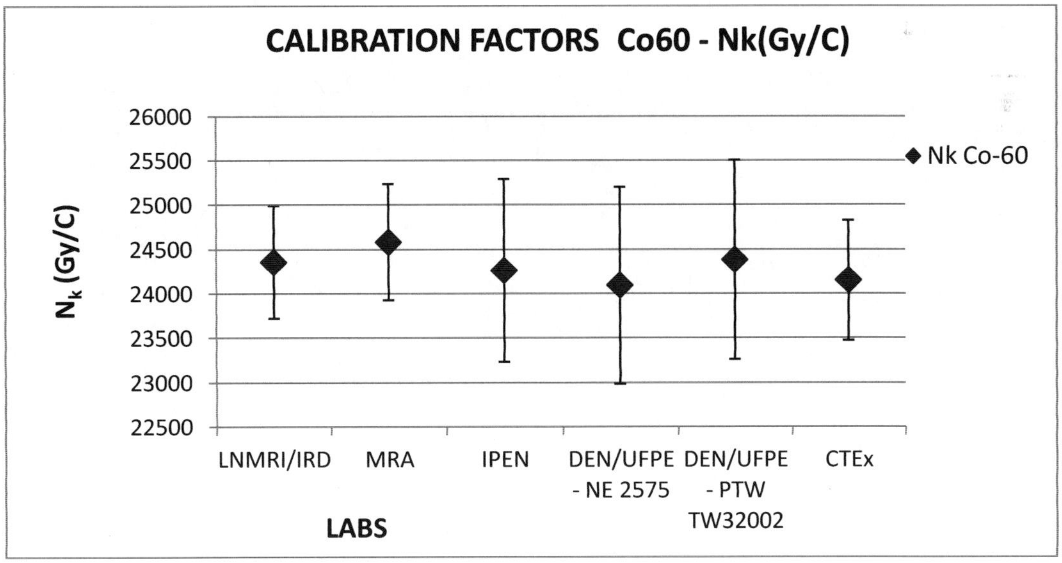

Figure 2. Calibration Factors determined for the transfer chamber by LNMRI and participating laboratories for quality S-Co.

6.2. Normalized response - $N_{k,part}/N_{k,\ LNMRI}$.

All values were normalized to the value of the calibration coefficient in air kerma obtained by LNMRI; the values can be seen in Table 3 and Figure 3 and 4.

Table 3. Resposta normalizada - $N_{k,part}/N_{k,LNMRI}$.

Laboratories	Normalized response $R=N_{k,PART}/N_{k,LNMRI}$	
	Cs-137	Co-60
LNMRI	1,0	1,0
LCR/UERJ - PTW TN32002	0,990	
LCR/UERJ – PTW N23361	1,001	
ETN	1,012	
MRA	1,007	1,009
IPEN	1,015	0,996
DEN/UFPE – PTW TW32002	0,998	1,001
DEN/UFPE – NE 2575	0,993	0,989
CRCN	0,994	
CDTN	0,994	
CTEx	0,991	0,992

The standard deviation between the calibration coefficients is 0.8% for S-Cs quality and 0,7% for S-Co quality. The results are considered to be acceptable for the purpose of this comparison if the expanded uncertainty of the ratio $N_{k,part}/N_{k,LNMRI}$ covers the reference value and $0.98 \leq R \leq 1.02$.

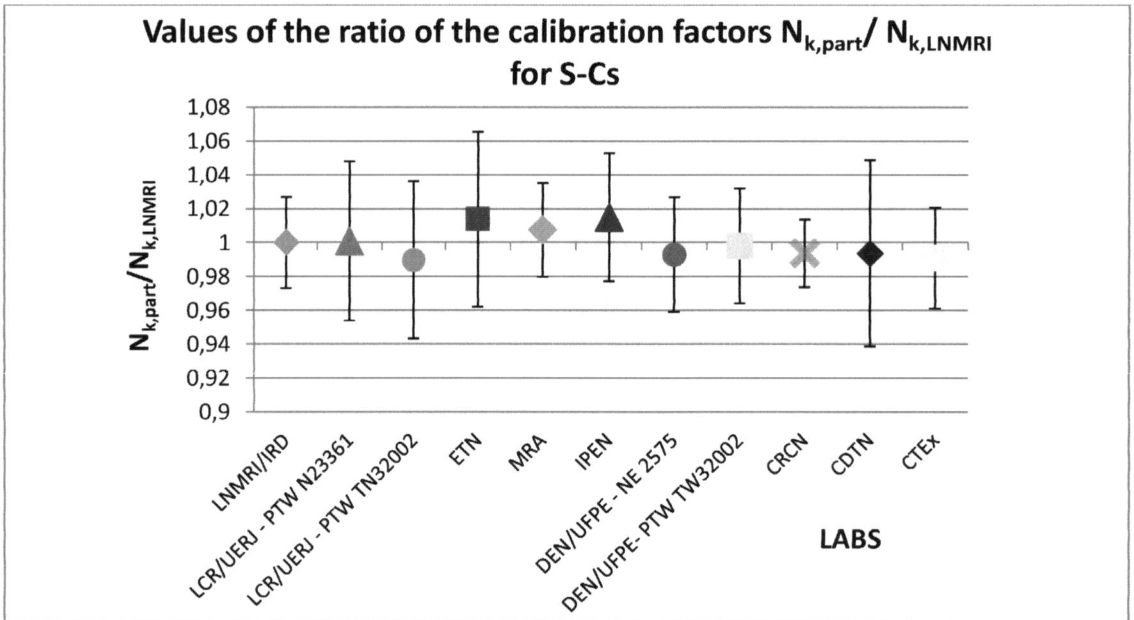

Figure 3. Values of the ratio of the calibration factors $N_{k,part}/ N_{k,LNMRI}$ for S-Cs.

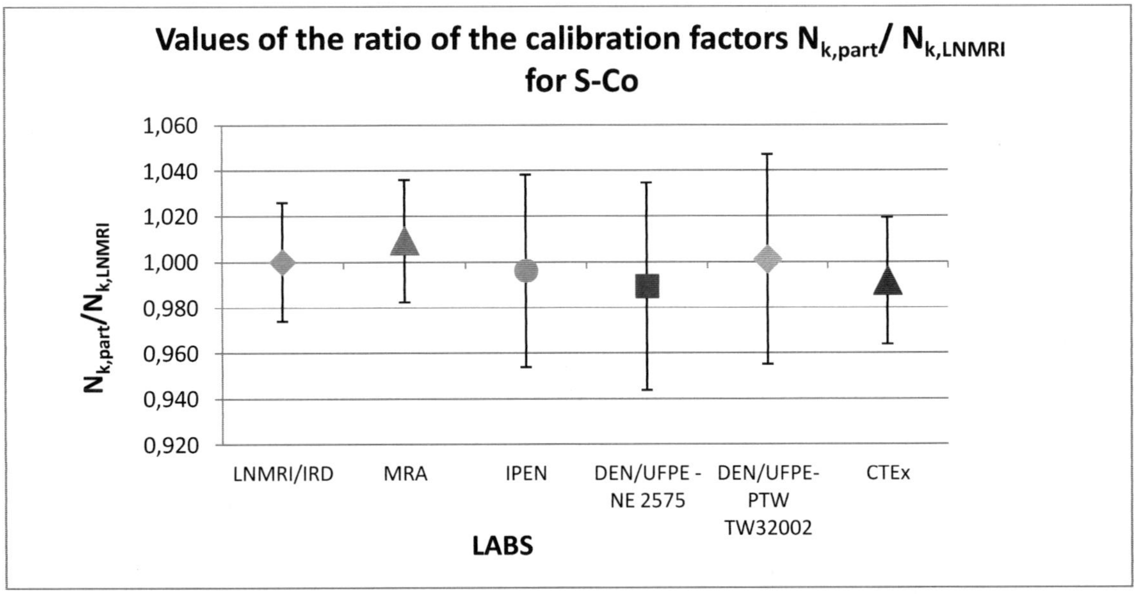

Figure 4. Values of the ratio of the calibration factors $N_{k,part}/ N_{k,LNMRI}$ for S-Co.

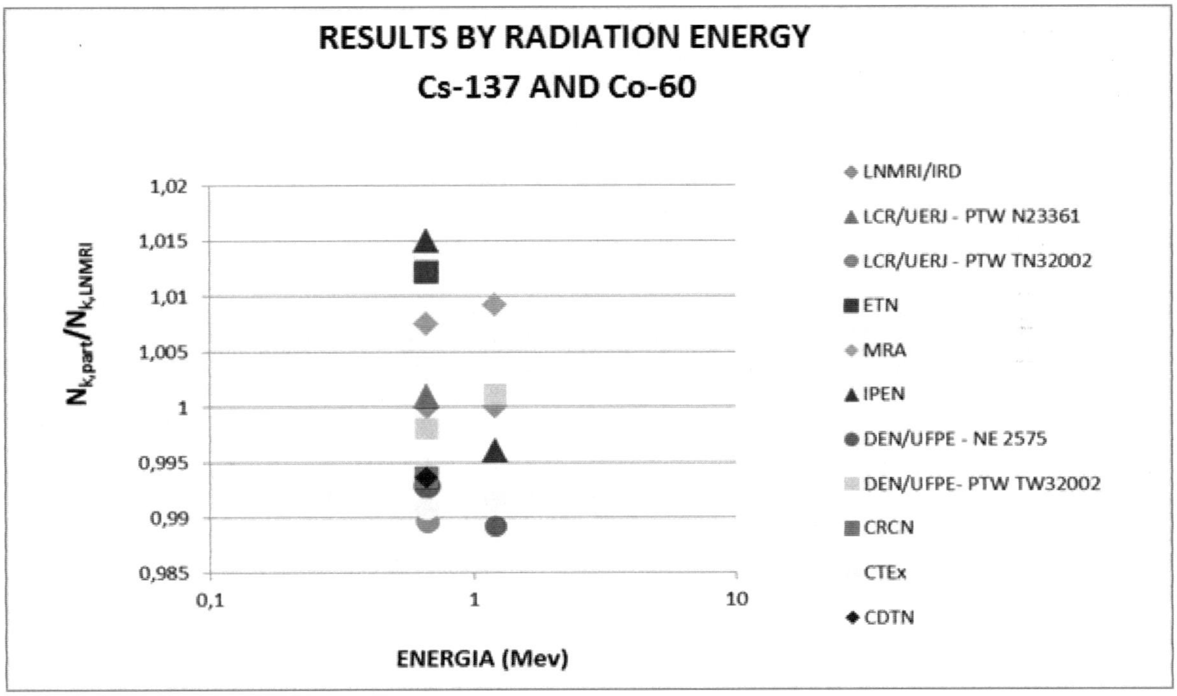

Figure 5. Results by Radiation Energy

6.3. Uncertainties

Uncertainties measures were calculated according to the ISO/IEC Guide 98-3 - "Guide to the expression of uncertainty in measurement". The total uncertainty was obtained by a combination of type A and B uncertainties regarding measurement multiplied by the coverage factor $k = 2$, which corresponds to a confidence level of 95.45%. The combined uncertainty was calculated by:

$$u_c = (u_A^2 + u_B^2)^{1/2} \tag{2}$$

Where u_A and u_B are type A and type B standard uncertainties.

The components of uncertainty that contributed to the combined standard uncertainty of the calibration of the transfer chamber are raised in dosimetry with standard chamber as positioning, irradiation distance, uncertainty of the calibration standard (certified standard chamber), stability, uniformity field, time factors thermometer, barometer and electrometer, energy dependence, leakage current measurement, repeatability, resolutions depending on the equipment and set up some more, and should take most of the components into account again when calibrating the chamber transfer.

7. Discussion and analysis of results

The results were evaluated by the percentage difference D% using the recommended methodology in ISO/IEC 17043-1, as most reference chambers, if not all, used by the laboratories were calibrated by LNMRI.

The percentage difference is calculated by the equation:

$$D_{\%} = \frac{N_{k,participante} - N_{k,LNMRI}}{N_{k,LNMRI}}.100 \tag{3}$$

Where:

$N_{k,LNMRI}$ = Calibration factor N_k obtained by LNMRI

$N_{k,Participant}$ = Calibration factor N_k obtained by the participating laboratory

The percentage difference for the calibration coefficient in terms of air kerma, and LNMRI between the participating laboratories should be equal to or less than 2.0%. None of the results exceeded the tolerance level.

Table 4. Results of the comparison exercise - Percentage Difference between the reported coefficients of participating laboratories with $N_{k,LNMRI}$.

Labs	D(%) – S-Cs	D(%) – S-Co
LCR/UERJ -TN32002	0,1	
LCR/UERJ –TN23361	-1,0	
ETN	1,2	
MRA	0,7	0,9
IPEN	1,5	-0,4
DEN/UFPE –TW32002	-0,2	0,1
DEN/UFPE – NE 2575	-0,7	-1,1
CRCN	-0,6	
CDTN	-0,6	
CTEx	-0,9	-0,8

The results were within the acceptance limit of ±2.0%, established by LNMRI in the protocol. The acceptance limit was calculated taking into account the existence of the dependence between the

IOP Conf. Series: Journal of Physics: Conf. Series **975** (2018) 012042 doi:10.1088/1742-6596/975/1/012042

calibrations of the laboratory standards performed by LNMRI. The highest percentage difference was 1.5%, being below the limit of acceptance established by the LNMRI; the index of this exercise was smaller than the previous comparison, even with the inclusion of a laboratory that started to work recently, in the second half of 2016.

Considering the points discussed, it is concluded that the Comparison Exercise presented an excellent result and this shows the competence of the Brazilian laboratories in performing the calibration services.

Only one laboratory was informed that it needs to review its calculation of the uncertainties and their components.

The Comparison Exercise tested the actual measurement capacity of the laboratories performing calibration services of radiation and irradiation monitors in Brazil, since the measurements are performed in the premises and with the laboratory procedures, which allows a greater reliability in the measurements and services Performed by the Brazilian calibration network.

This work counted on the partial financial support of FINEP/SIBRATEC/METRORADI

References

[1] NORMA ISO/IEC 17043:2010 - Conformity assessment -- General requirements for proficiency testing.
[2] VOCABULÁRIO INTERNACIONAL DE METROLOGIA/INMETRO: Vocabulários Internacional de termos Fundamentais e Gerais de Metrologia - 2012.
[3] NORMA ISO-4037-2: 1997 - X and Gamma Reference Radiation for Calibrating dosemeters and doserate meters and for determining their response as a function of photon energy – Part 2: Dosimetry for Radiation Protection over the energy ranges 8 keV t0 1.3 MeV and 4 MeV to 9 MeV.
[4] NORMA ISO-4037-3:1999 - X and Gamma Reference Radiation for Calibrating dosemeters and doserate meters and for determining their response as a function of photon energy – Part 3: Calibration of area and personal dosemeters and the measurement of their response as a function of energy and angle of incidence.
[5] Radioactive half-lives - Table of recommend values: http://www.nucleide.org/NucData.htm

Construction of the TH-GEM detector components for metrology of low energy ionizing radiation

N F Silva[1]; T F Silva [2]; M C Castro[1]; H Natal da Luz[2]; L V E Caldas[1]

[1] Instituto de Pesquisas Energéticas e Nucleares (IPEN/CNEN-SP)
Av. Professor Lineu Prestes, 2242 - 05508-000, São Paulo, SP.

[2] Instituto de Física da Universidade de São Paulo (IFUSP)
R. do Matão, 1371 - 05508-000, São Paulo, SP.

E-mail: nsilva@ipen.br

Abstract: The Gas Electron Multiplier (GEM) detector was originally proposed as a position sensitive detector to determine trajectories of particles prevenient from high-energy collisions. In order to study the potential of TH-GEM type detectors in dosimetric applications for low energy X-rays, specifically for the mammography standard qualities, it was proposed to construct a prototype with characteristics suitable for such use. In this work the general, structural and material parameters applicable to the necessary conditions were defined, establishing the process of construction of the components of a prototype.

Keywords: TH-GEM, dosimetry, Gaseous detector, low energy.

1. Introduction

The Gas Electron Multiplier (GEM) detector was first proposed by Sauli as a position-sensitive detector, aiming the determination of trajectories of particles provenient from high-energy collisions. Subsequently, these detectors found applications in several areas related to the production of x-ray images [1].

The TH-GEM detector (Thick GEM) is a variation of the GEM detector [2]. Its structure resembles that of an extended scale GEM detector. The advantages of the TH-GEM detector are many, outstanding for its high sensitivity and low cost.

In order to study the suitability of TH-GEM detectors for low energy X-rays dosimetry, specifically for the standardized mammography qualities, it was proposed the construction of a portable prototype of a TH-GEM detector with characteristics that fulfill the dosimetric constraints. This work establishes the process of construction of some components of this prototype, describing their dimensions and fundamental characteristics. The information will be used in the future as a technical guide for the construction of equivalent dosimeters.

1.1 Operation of the TH-GEM detector

The basic constituent of a TH-GEM detector is the amplification plate, which is a thin non-conductor plate (thickness of approximately 0.5 mm) coated on both sides by metallic layer. This plate presents holes of approximately 0.3 mm diameter in a hexagonal pattern, and is immersed in a rarefied gas. A high potential difference applied between the coatings creates an intense electric field configuration within the holes. Each individual hole acts as a proportional counter, multiplying the electrons produced in the ionizations in the gas induced by an incident radiation in an avalanche process, and generates a measurable charge which is collected by an electrode [3]. Figure 1 shows a cross-section of the TH-GEM type detector, where $\vec{E}_d$ is the electric field in the region where the incident radiation interacts with the gas producing an electron, and $\boldsymbol{V_d}$ is the voltage for the collection of this electron (drift region). The subscript *o* represents the region of a orifice of the polymer plate, being $\vec{\boldsymbol{E}}_{\boldsymbol{o}}$ and $\boldsymbol{V_o}$ the electric field in the region where the avalanche of ionizations occurs and the voltage necessary. Lastly, $\boldsymbol{V_t}$ is the voltage needed to guide the electrons in the region of the electric field $\vec{\boldsymbol{E}}_{\boldsymbol{t}}$ to the collector plate.

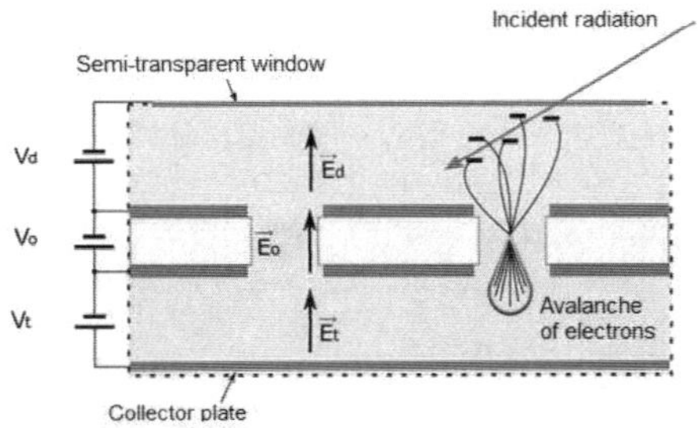

Figure 1 – Schematic illustration of a TH-GEM type detector. Adaptation of [1].

More than one plate can be used in series to further amplify the cascade effect.

In the case of a GEM detector, the plate is perforated in a micrometric pattern with its dimensions being 5 to 20 times smaller than in the case of TH-GEM detectors. Like the TH-GEM detectors, in GEM detectors an electric potential is applied between both sides of the coating and the plates are also emerged in a rarefied gas. As the dimensions are proportionally smaller, the required difference of potential between both sides of the amplification plate are also smaller. A great advantage of the GEM and TH-GEM detectors is their easy adaptation in a adaptable format [5].

2. Materials and methods

The construction of the components was done at the mechanical workshop of the Laboratory of Material Analysis with Ion Beams of the University of São Paulo (LAMFI-USP), which is participating in this project in the form of institutional collaboration (Figure 2).

Figure 2 – Construction of the detector base in the machining workshop of the Laboratory of Material Analysis with Ion Beams of the University of São Paulo.

For the detector base the polyoxymethylene material was used, density 1.41 g/cm^3, because it is a rigid and resistant material, and offers a good electrical insulation and good machining characteristics. In order to simplify the construction of the detector, the squared format for the base, with sides of 11.00 cm, was determined.

For the entrance window, 6 μm thick aluminum foil was used, density 2.6989 g/cm^3.

The amplification and collector plate were made of fiberglass printed circuit boards, density 2.49 g/cm^3, coated by copper, density 8.96 g/cm^3[6]. The copper plates received a chemical treatment that created a thin silver cover on top to prevent for oxidation. The chemical treatment was done on both the amplification plate and the collector.

IOP Conf. Series: Journal of Physics: Conf. Series **975** (2018) 012043 doi:10.1088/1742-6596/975/1/012043

The perforation of the amplification plate was done using a CNC device constructed at the LAMFI-USP laboratory for internal use to fabricate small printed circuits.

The gas inlet and outlet were positioned on opposite sides of the detector to ensure the renew process in the amplification area by a continuous flux of gas. The electrical feedthroughs were provisionally implemented as three 2 mm diameter copper wires, and glue to avoid gas leakage.

3. Results and discussion

It was possible to construct a TH-GEM detector, with sides of 11 cm (Figure 3). The whole system weights approximately 300 g.

(a)

(b)

Figure 3 – Photo of the TH-GEM detector components: a) without the shield; b) with the shield.

The thickness of the 6 mm shield was defined by Monte Carlo simulation with code MCNP5 [7]. The geometry, the materials that compose the detector, the experimental configuration and the characteristics of the radioactive sources were specified. The energy deposited at the sensitive volume was observed, and the thickness of the shield was increased until an efficiency value of 98.5% was achieved. The diameter of the orifices and of the etched rim of the amplifier plate were measured using a USB digital camera with up to 1000x magnification. These dimensions are shown in the caption of Figure 4.

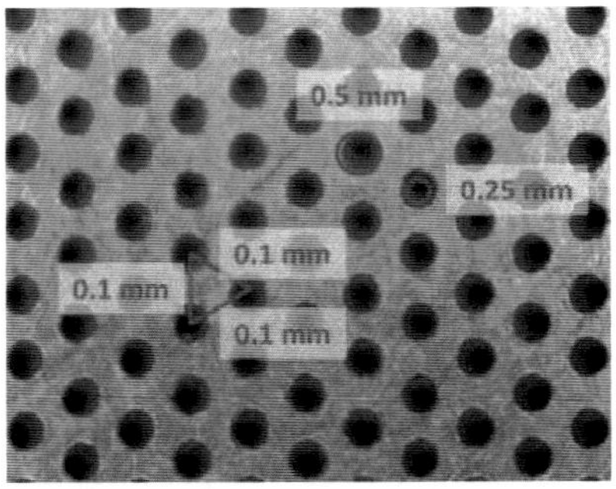

Figure 4 - Amplification plate seen through the digital microscope. The average diameter of the orifices in the fiberglass plate is 0.25 mm, the diameter of the etched rims is 0.5 mm, and the distance between the center of one orifice to the other is 0.1 mm

4. Conclusion

It was possible to construct the components of the TH-GEM portable detector. These processes allowed the elaboration of the geometry of the detector accordingly to previous simulations, which will also guide the choice for the gas that will fill the detector and the values expected in the experimental tests for collected charge.

Acknowledgments

The authors acknowledge the partial financial support from the Brazilian agencies: CNEN, FAPESP, CNPq, CAPES and MCTIC (Project: INCT – Radiation Metrology in Medicine).

References

[1] Cortesi M, Alon R, Chechik R, Breskin A, Vartsky D and Dangendorf V 2007 Investigations of a THGEM-based imaging detector. *J. Instrum.* **9** 9002
[2] Chechik R, Breskin A, Shalem C and Mörmann D 2004 Thick GEM-like hole multipliers: properties and possible applications. *Nucl. Instr. Meth. Phys. Res.* A **535** 303-8
[3] Bouclier R, Dominik W, Hoch M, Labbé J C, Million G, Ropelewski L, Sauli F and Manzin G 1997 New observations with the gas electron multiplier (GEM). *Nucl. Instr. Meth. Phys. Res.* A **396** 50-66
[4] Sauli F 2016 The gas electron multiplier (GEM): Operating principles and applications. *Nucl. Instr. Meth. Phys. Res.* A **805** 2-24
[5] Breskin A, Alon R, Cortesi M, Chechik R, Miyamoto J, Dangendorf V and Dos Santos J M F 2009 A concise review on THGEM detectors. *Nucl. Instr. Meth. Phys. Res.* A **598** 107-11
[6] Willians III R G, Gresh C J, Pagh R T 2006 Compendium of Material Composition Data for Radiation Transport Modeling, *Report PNNL-158870,* Pacific Northwest National Laboratory
[7] LOS ALAMOS LABORATORY, 2008 MCNP – A General Monte Carlo N-particle Transport Code **Version 5** LA-UR-03-1987

Correlation and uncertainties evaluation in backscattering of entrance surface air kerma measurements

GJ Teixeira[1,2,4], CHS Sousa[1], JGP Peixoto[1,3]

[1]Instituto de Radioproteção e Dosimetria (IRD/CNEN);
[2]Curso Superior de Tecnologia em Radiologia (CSTRAD/UNESA);
[3]Laboratório de Ciências Radiológicas (LCR/UERJ); [4]Faculdade de Ciências Médicas (FCM/UERJ)

E-mail: gt@ird.gov.br

Abstract: The air kerma measurement is important to verify the applied doses in radiodiagnostic. The literature determines some methods to measure the entrance surface air kerma or entrance surface dose but some of these methods may increase the measurement with the backscattering. Was done setups os measurements to do correlations between them. The expanded uncertainty exceeded 5% for measurements with backscattering, reaching 8.36%, while in situations where the backscattering was avoided, the uncertainty was 3.43%.

1. Introduction

Entrance Surface Air Kerma measurements (ESAK) are relevant for the verification of the applied doses in radiodiagnostic. These measurements are performed at the corresponding point to the flat surface of a specified object. The object may be any region of a patient or a simulator and the results associated with a backscatter factor [1].

Thermoluminescent dosimeters (TLD) are the most recommended for these measurements because they can be positioned in the patients themselves and the readings do not require corrections [2]. Despite the recommendations, there is the problem of the dynamics of the use of the TLD, being complex because professionals in this field do not always have at their disposal a dosimetry laboratory, which makes this method very expensive. The resolution of this problem can be the use of the ionizing chamber (IC) for the measurements.

The IC should be calibrated by accredited laboratories, making them traceable. This calibration is performed 1 meter from source [2; 3] in a collimated beam and compared to a standard reference chamber. It turns out that the available protocols for entrance surface dose and entrance surface air kerma measurements differ from the calibration method.

TRS 457 [2] recommends that ESAK measurements, when performed by ionization chambers, are obtained from the Incident Air Kerma (IAK) measurements and then corrected by the inverse square law. The IAK, according to the TRS, is obtained by positioning the IC between the simulator and the tube, significantly reducing the distance between the source and the detector (less than 1 meter), so the distance is different from the instrument calibration.

An ESD measurement methodology published by the National Agency for Sanitary Surveillance - Brazil (ANVISA) in the Equipment Safety and Performance Manual [4], recommends that the ionization chamber be placed on a support at a distance corresponding to the plane of the simulator or patient, or on the table. When correct adjustment of the distance between the Xray tube and the detector is not feasible, it is recommended that the measurements be corrected by the inverse law of the square of the distance.

The position of the IC in the examination table may involve the production of backscattering from the interaction of the primary beam with the table itself and, depending on the distance and the radiographic technique employed, can determine if the radiology service is in a "compliance" situation or "Non-compliance" situation with radiation protection recommendations and standards. This effect was not described in any reference literature used in this study.

The measurement uncertainty is a parameter associated with the result of a measurement, which characterizes the dispersion of the values that can be reasonably attributed to the measurand [5]. The sources of uncertainty of a radiological measurement, due to the characteristics of the dosimetric systems and characteristics of X-ray equipment, can be diverse, such as pressure, temperature, energy dependency among others.

The objectives of this work involve the correlation between possible techniques for entrance surface air kerma measuring and the verification of the uncertainties involved in the process, covering the inverse square law for these measurements and relations between tubedetector-table distance of radiodiagnostic practices.

2. Materials and methods

2.1. Equipment and radiographic techniques

A radiodiagnostic equipment of 500 mA and voltage between 30 and 150 kV was used, with a high frequency generator. The lateral lumbar spine exam with radiographic technique used by the staff of the radiology service participating in the research was defined as the reference procedure. The time-to-current ratio was 100 mAs, being 200 mA and 0.5 s, with 80 kVp of voltage. The size of the field varied depending on the positioning of the IC due to the change of its sensitive area by the distance.

The radiographic technique was based on the reference man, with focus-film distance of 1 meter [6] and focus-object distance of 70 cm. Measurements were performed with a Radcal Dosimetric System, model 9015 with 6 cc cylindrical chamber, calibrated in terms of Air Kerma for a beam quality RQR 6 [2].

2.2. Setups

Measurements were made for 5 different experimental setups, varying the camera-source, table-source and table-chamber distances, with five readings for each setup and the average and standard deviations were calculated.

The setup to obtain the reference reading was performed with the X-ray tube in its maximum vertical displacement (Figure 1), establishing as 100 cm the tube-detector distance (detector at 30 cm from the table) avoiding backscatter (Setup 1 - S1).

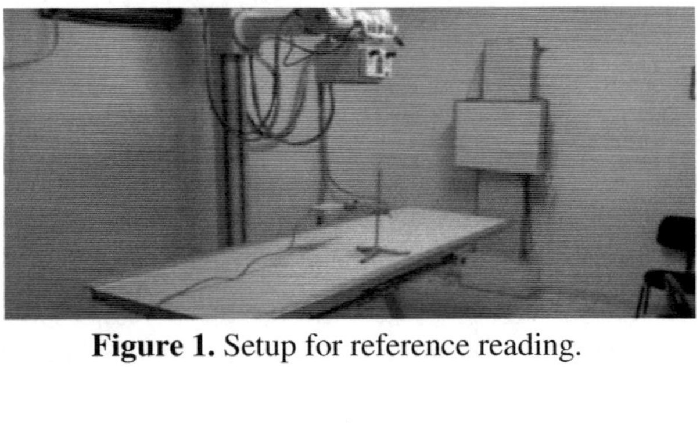

Figure 1. Setup for reference reading.

The chamber was also positioned at 20 cm (Setup 2 - S2) and 10 cm (setup 3 - S3) of the table, keeping the tube within 1 meter of the detector to verify the influence of the backscattered radiation at those distances, that is, with the tube (Focal point) at 120 cm and 110 cm from the table. Another setup was mounted, so that the effective focal point of the chamber was positioned closer to the table (3 cm) (figure 2) with the tube at 100 cm from the chamber (setup 4 - S4).

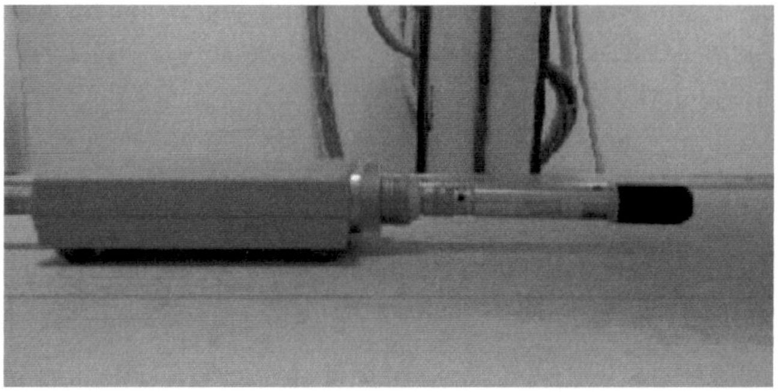

Figure 2. Arrangement with the ionization chamber near 3 cm from the backscattering source.

Setup 5 (S5) consisted of positioning the chamber 30 cm from the table with the tube 70 cm from the chamber (figure 3), according to recommendations [2; 4]. Finally, the measurements obtained with S1 were corrected by the inverse square law to 70cm that we call here by T1.

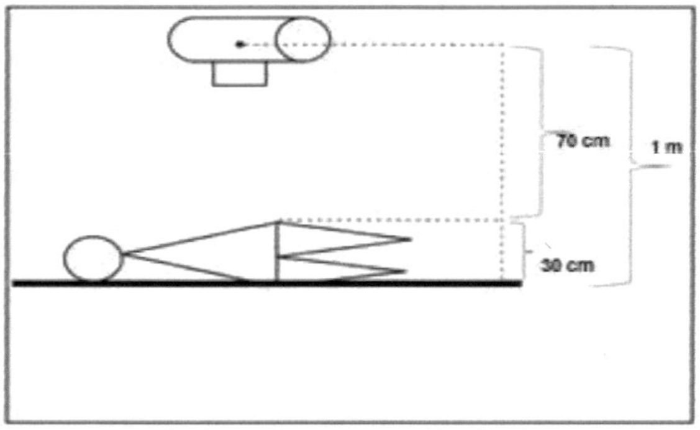

Figure 3. Setup 5.

2.3. Backscattering factor

With the measurements of setup 4 an increase in dose was observed in relation to setups 1, 2 and 3, being defined by:

$$BS = [L_d - L_p] \quad (1)$$

where L_d is the average reading between S1, S2 and S3 and L_p is the average reading with the detector 3 cm from table top (S4) and the backscattering factor (BSF) defined as the ratio between the same average readings (L_d/L_p).

2.4. Correlations

Three correlations were made, one being between the results of setups 1 and 4, another between the results of setup 4 and the same results with the application of BSF (S4B). The third correlation was performed between S5 and T1. The average and standard deviations [7] of the series of measurements for the correlation were determined by applying the Pearson momentum, according to equation

$$r_{xy} = \frac{\left(\left(\sum_{i=1}^{n} x_i y_i\right) - \left(\frac{\left(\sum_{i=1}^{n} x_i\right)\left(\sum_{i=1}^{n} y_i\right)}{n}\right)\right)}{\sqrt{\left(\sum_{i=1}^{n}\left(x_i - \bar{x}\right)^2 \left(\sum_{i=1}^{n} y_i - \bar{y}\right)^2\right)}} \quad (2)$$

where *xi* corresponds to the independent measurements related to each setup and *yi* corresponds to the independent measurements related to the other series of measurements that were correlated with *xi* [7].

Once the correlation between the input quantities was established, the combined variance of the input quantities *xi* and *yi* was determined, being these type A, according to equation 3,

$$S(\bar{x}, \bar{y}) = \left(\frac{1}{n(n-1)}\right)\sum_{i=1}^{n}(x_i - \bar{x})(y_i - \bar{y}) \quad (3)$$

2.5. Uncertainties evaluation

To evaluate the uncertainties, the recommendations of the Guide to the Expression of Uncertainty of Measurement [7] were used, considering as uncertainty type A, the coefficient of variation of the series of measurements, through equation

$$\sigma(\%) = \frac{\sigma}{\bar{x}} * 100 \quad (4)$$

where σ é o standard deviation and x is measurements average. The uncertainties type B are defined as the other contributions that were not obtained by statistical methods, that is, intrinsic data of the measurement system. This work considered only the energy dependence of the system, the resolution of the electrometer, the resolution of the distance measurement and the calibration factor of the ionizing chamber. The combined standard uncertainty (u_c), the expanded uncertainty (U) and the confidence interval (k) were calculated.

3. Results

Table 1 shows the results of the correlation of the measurements between the setups and their variances. It can be observed that all the setups presented a positive correlation, the correlation between S4 and their measurements with BSF (S4B) being an absolute correlation. Figure 4 shows the application of backscatter factor in S4 measurements.

Table 1. Correlation and variance between the setups

Setup Combinations	Correlation (rxy)	Variance (X,Y)
S1 and S4	0,1491	0,1425
S4 and S4B	1,0000	0,0953
S5 and T1	0,9385	0,0506

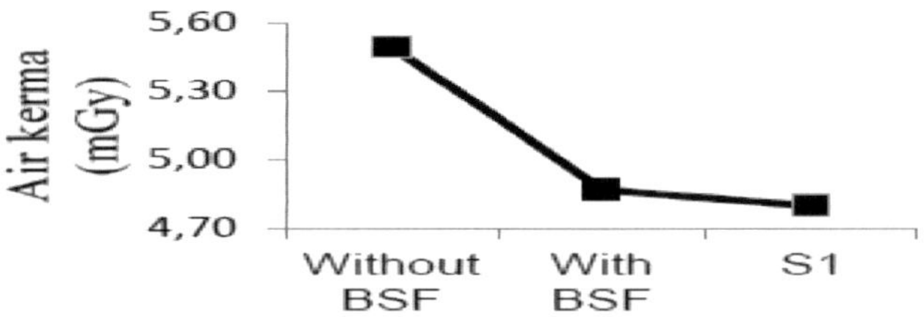

Figure 4. Comparison of the values taken 3 cm from the table top with and without the BSF and the reference reading obtained with the chamber at 30 cm from the table top.

Table 2 shows the combined standard uncertainty (uc), the expanded uncertainty (U), and the confidence level (k) for each measurement setup, and table 3 shows the correction of setup 1 by the inverse square law for 70 cm (T1), making evident the dose difference between the methods.

Table 2. Combined standard uncertainty (u_c), the expanded uncertainty (U) and the confidence interval (k) for each setup or method.

Setup	u_c	U	k
S1	1,7117	3,4335	2
S4	3,4444	8,3658	2,43
S4B	3,4443	8,3654	2,43
S5	1,7996	3,6065	2
T1	1,8065	3,6216	2

Table 3. Comparison between the dose obtained at 100 cm corrected by the law inverse square law to 70 cm and the dose obtained at 70 cm by simulating the distance source-patient, the ratio and difference.

Air Kerma (mGy)	
Average of reference reading taken at 100 cm	4.797
Average reference reading corrected to 70 cm	9.790
Average measurement at 70 cm	10.238
Difference	0.448

4. Discussion

The correlation between S5 and T1 shows that there is a strongly positive relationship between the methods, that is, the change in the measurement setup or the inverse law of the square of the distance are congruent. Observing Table 3, it can be concluded that the dose difference between the measurements is high for radiodiagnostic. The non-positioning of the chamber at 100 cm implies a lack of traceability of the measurements, since the readings of the other distances could not be simply corrected by the law of the inverse of the square, this effect was also observed by Teixeira [8].

The backscattering factor (BSF) is presented and discussed in several articles [8] but always taking into account only the radiation scattered by the tissue or phantom equivalent to tissue, characterized as water or water phantom [2]. Clinical practice shows that the simulator is not used, leaving only the table as a scatter object. BSF is defined by

$$BSF_X^{(W)} = \frac{X^{(W)}\left[\mu_{tr}/\rho\right]_{W,air}^{W}}{X^{(free)}\left[\mu_{tr}/\rho\right]_{w,air}^{(free)}} \quad \textbf{(5)}$$

Where $X^{(w)}$ is the Exposure on the water phantom surface, $X^{(free)}$ is the Exposure at the same point of space in the absence of the simulator and $\mu tr/\rho_{w,air}$ is the proportion of masstransfer energy coefficients for water and air in the presence of disperser and free space.

This factor is important because of the increase in the actual measurement, but in this study only the backscattering produced by the interaction of ionizing radiation with the examination table was evaluated. It has been found that, due to the maximum displacement of the vertical tube of x-rays, it is necessary in some cases to bring the chamber closer to the table, which generates a significant backscattering.

The technique used in this work to determine the BSF was acceptable, since the correlation between S4 and S4B is absolute, since they are directly proportional. Figure 4 shows that the application of the BSF factor was efficient, but the expanded related uncertainty for these 2 setups exceeds 5% acceptable for radiodiagnostic measurements. The correlation between S1 and S4 shows a weak positive value, leading to the conclusion that the backscattering has a significant contribution to the dose and the variance in the measurements shows that the backscatter of the table is not homogeneous, resulting in a high coefficient of variation, 2.98 % .

5. Conclusions

The calibration of the detectors is extremely important for any measurement of dosimetry and protection against radiation. The ionizing chambers, for their readings to be considered reliable, need to be calibrated in a standard reference laboratory for results to be traceable.

Current protocols recommend different ways of performing measurements of entrance surface air kerma and entrance surface dose, which makes the traceability of the absolute quantity air kerma confusing.

The adjustment of the values found with the detector 100 cm and corrected by the inverse square law differs significantly from the results of the measurements with the section of the detector in the plane of the patient or surface of the simulator.

The results obtained with the detector at the patient's position for metrological purposes are not acceptable because ESAK or ESD reference values, according to standards and recommendations, are not subject to fluctuations but limits. It is recommended to do so, that measurements are taken at the same distance from the calibration to trace.

The approach of the ionization chamber to the examination table also generates a significant backscatter, which can be avoided by positioning the instrument away from the table, but is sometimes not feasible due to the maximum vertical displacement of the X-ray tube.

The ESAK or ESD verification with detector ionizing chambers is feasible and reliably replaces the TLD, eliminating even the uncertainty surrounding the thermoluminescent method. The feasibility and accuracy of the IC depend on the appropriate use. A setup of the same conditions as the standard reference laboratory is recommended. If it is not possible to remove the ionization chamber from the examination table, it is recommended to use an adequate backscatter factor.

References

[1] IAEA TECDOC 1423 2004 Optimization of the radiological protection of patients undergoing radiography, fluoroscopy and computed tomography.

[2] IAEA TRS 457 2007 Dosimetry in diagnostic radiology: An international code of practice ISSN 0074-1914; no 457 *Technical Reports Series no 457* (Vienna: IAEA).

[3] Allisy-Roberts P.J., Burns D.T., Kessler, Spasic-Jokic V. 2002 Comparison of the standards of air kerma of the SZMDM Yugoslavia and the BIPM for 60Co γ rays, Rapport BIPM-02/01.

[4] BRASIL ANVISA 2005 Radiodiagnóstico Médico: Desempenho de Equipamentos e Segurança.

[5] Peixoto J.G.P and Almeida C. E. The radiation metrology network related to the field of mammography: implementation and uncertainty analysis of the calibration system. *Meas. Sci. Technol.* 12. UK 15861593. 2001.

[6] Bontrager, Kenneth L. 2010 Manual prático de técnicas e posicionamento radiográfico. São Paulo: Editora Elsevier, 7ª Ed.

[7] Sousa, C.H.S., Teixeira G. J., Peixoto J. G. P., Estimated of the associated incertainties with the linearity test of activemeters. *Journal of Physics*. Conference series 575. 2015.

[8] Teixeira, G. J. Coeficientes de transferência de energia em massa: determinação e análise. Dissertação de Mestrado. IRD. Rio de Janeiro. 2013.

IOP Conf. Series: Journal of Physics: Conf. Series **975** (2018) 012045 doi:10.1088/1742-6596/975/1/012045

Design and development of a geo-referenced database to radionuclides in food

L M E Nascimento[1], A C M Ferreira [1] and S A Gonzalez [1]

[1]Instituto de Radioproteção e Dosimetria/Comissão Nacional de Energia Nuclear – IRD/CNEN-Brazil

E-mail: anacris@ird.gov.br

Abstract. The primary purpose of the range of activities concerning the info management of the environmental assessment is to provide to scientific community an improved access to environmental data, as well as to support the decision making loop, in case of contamination events due either to accidental or intentional causes. In recent years, geotechnologies became a key reference in environmental research and monitoring, since they deliver an efficient data retrieval and subsequent processing about natural resources. This study aimed at the development of a georeferenced database (SIGLARA – **SI**stema **G**eoreferenciado **L**atino **A**mericano de **R**adionuclídeos em **A**limentos), designed to radioactivity in food data storage, available in three languages (Spanish, Portuguese and English), employing free software[1].

1. Introduction

The consumption of contaminated food is considered as the ultimate factor to the intake of radionuclides by the human body. As a consequence, studies are being carried out worldwide, aiming the assessment of both toxic elements concentrations in food and maximum permissible amounts to human consumption. In terms of health effects, some features must be taken into account, such as identification and intake levels of the absorbed radionuclide(s). In a follow up study, levels of toxic elements in human diets must be monitored and the intake of the related food items continuously evaluated, in a way that a specific database can be provided to further comparisons to reference limits established by public health organizations.

Through the *Codex Alimentarius* Commission, the World Health Organization (WHO) sets out limit week values for tolerable ingestion of toxic or potentially toxic elements, based upon the maximum lifetime ingestion exempt of significant health risks, being all known concerned facts taken into account.

The Food and Agriculture Organization (FAO/WHO) is responsible for advising member nations on a broad range of parameters affecting relevant topics as agriculture, food production, processing, storing and trading, including the compliance with regulations referred to food quality and safety [2].

The United Nations Scientific Committee on the Effects of Atomic Radiation (UNSCEAR) comprises the scientific body of WHO, in charge of producing reports approaching overall radiation levels and effects related to exposure to ionizing radiation.

The International Atomic Energy Agency (IAEA) is the WHO branch responsible for setting and disseminating protection and safety standards for ionizing radiation, as well as ensuring the peaceful use of nuclear activities.

IAEA puts a strong emphasis on environmental and public protection and many countries established an independent public supervisory authority responsible for monitoring ionizing radiation deployment, preventing radiological and nuclear accidents, as well as evaluating the impact on society.

In Brazil, such duties are delegated to the National Nuclear Energy Commission (CNEN) [3].

Since the Chernobyl accident in 1986, a mandatory radiological certificate is required in export documentation of food by nearly all importing countries. As a consequence, a comprehensive assessment of radioactive background levels in food items produced worldwide can be useful to such business segment, as well as to scientific society.

Though a growing interest in the dynamics of extensive access and exchange of reliable data and environmental information is arising, simultaneous concerns regarding the strategies to accomplish such tasks – deployment costs and difficulties in the access of new technologies included – are evident too.

The primary objective of the management of an information database related to the environmental assessment is to provide an upgraded data and relevant information access to world community, both as a tool in the decision making process, in case of contamination arising from either accidental or purposeful reasons and to fulfill the documentation requirements regarding food exporting in the international market.

2. Materials and methods

2.1. System Scheduling

Methods were split up in six stages, as follows:

- Database setting and validation;
- System development tools settings;
- Modelling and database creation;
- Use case diagram;
- System development;
- Logo and login page design.

2.2. Tools

In order to accomplish this project, reliable software is requested to ensure the safe achievement, storage, management and display of data.

It must be emphasized, though, that all tools used throughout the different stages of this study meet the free distribution paradigm, which gives a particular feature to the system.

The development and use of free software based systems have been strongly encouraged by government agencies, aiming the cost reduction and the settlement of a program installed base, with the adoption of measures to avoid the illegal use of computational programs.

SIGs are defined as multitasking computational tools, that is, they can be used by diverse users communities. However, most of SIG´s adopt third party software interface infrastructures and such approach can frequently lead to interaction problems with users.

Some practical expectations are supposed to be met by the deployment of this system. The interface is simple, intuitive and interactive, being assumed that users do not necessarily have a previous practical experience or theoretical background regarding software and tools in database development, modelling and setting.

IOP Conf. Series: Journal of Physics: Conf. Series **975** (2018) 012045 doi:10.1088/1742-6596/975/1/012045

The most relevant tools used in the development of the SIGLARA system are mentioned as follows, being free distribution attended, whenever feasible:

LINUX operating system, responsible for the computer performance, connecting hardware (printer, monitor, mouse, keyboard) to software (application programs in general).

APÁCHE server for the interaction between client and server, by means of protocols like the *Hyper Text Transfer Protocol* (HTTP), which standardizes data input and output, regardless of the platform. APACHE server was chosen due to some positive aspects. It is a free web server, provides versions to several operational systems, as well as a convenient installation. Furthermore, APACHE offers a practical file setting, which allows further settings without a following mandatory server reboot.

The implementation of a database requires the choice of a database management system (DBMS) with Structured Query Language (SQL) as interface. The most popular open code database worldwide, MsSQL meets such prerequisite, besides all further key requirements for a DBMS, like strength, safety and reliability, providing a high performance affordable supply of reliable database applications, based upon web.

An applicative development platform that generates PHP (*Hypertext Preprocessor*) codes and is compatible with all web navigators, SCRIPTCASE was deployed as a development tool in the interface implementation process. PHP is a wide range programming language, remarkably efficient to web developments, in so far as it allows developers to write pages supposed to be dynamically generated later.

Google Earth is a free program developed and issued by Google, that enables the user to "flight over" a terrestrial globe 3D tridimensional model, with high-resolution graphics and designed from satellite images from several sources, aircraft pictures and SIG 3D.

In that way, it can be used basically as a satellite images and bi-dimensional maps generator, as well as a simulator of several landscapes on Earth. Hence, Google Earth allows the identification of places, buildings, cities, monuments, geographical features, etc. Google Earth is similar to another Google product, Google Maps, though more complex. Both are designed on the same foundations: high resolution pictures from all feasible and available places, taken by satellites and private airplanes, hired by Google.

2.3. Database Design

In this study, an approach based upon data representation in the relational model was adopted. The entity-relation model has the perception of the real world as a cornerstone, conceiving such real world as a set of primary objects named entities and their interactions.

This methodology was formerly proposed by Dr. Peter Pin Shan Chen in 1990 [4], as a tool to facilitate the modelling of a database project, in order to enable the data knowledge of an information system.

The system is laid out in tables, which join together under specific relation rules. Such rules consist of correlating a feature of a table with a tab strip from another table, by means of the inclusion of table key field.

2.4. System features

Table 1 shows a further explanation about the system features.

Table 1 System Features

Name	Descripcion
Register Sample	Inclusion, alteration, exclusion, consultation and printing the analyzed data of samples.
Register Sample Measurement	Inclusion, alteration, exclusion, consultation and analyzed data of radionuclides .
Register FAO Group	Inclusion, alteration, exclusion, consultation and printing the analyzed data of FAO Group .
Register Food Group	Inclusion, alteration, exclusion, consultation and printing the Food Groups.
Register Food	Inclusion, alteration, exclusion, consultation and printing the data of Registers Food .
Register Radionuclidos	Inclusion, alteration, exclusion, consultation and printing the data of register radionuclides.
Register Apresentation	Inclusion, alteration, exclusion, consultation and printing the data of sample apresentation.
Register FAO Zone	Inclusion, alteration, exclusion, consultation and printing the data of FAO Zone (to fish sample).
Register Categories	Inclusion, alteration, exclusion, consultation and printing the data of sample categories.
Register Laboratories	Inclusion, alteration, exclusion, consultation and printing the data of laboratories that analyze the sample.
Register Units of Measurement	Inclusion, alteration, exclusion, consultation and printing the units of measurement.
Register Countries/States	Inclusion, alteration, exclusion, consultation and printing the Counstries and States Registers in the System.
Register Groups and User of System	Inclusion, alteration, exclusion, and consultation of the System users and association with the appropriate profile.
Confer and Generate Sample Map	On line consultation of data and generate maps.

2.5. System tables

The database comprises a set of tables, as can be seen below in Table 2.

Table 2 Tables of System

Name of Table	Descripcion
Sample	Register sample.
Measurement sample	Concentrations of the radionuclides in the sample.
FAO Group	FAO Food Group.
Kind of Food	Food Group.
Food	Food analysed.
Radionuclido	Registerd Radionuclides.
Apresentation	Presentation of the sample: acidified, frozen, bottled, refrigerated, packed or has no.
FAO Zone	Mapping of FAO's fishing zones.
Categories	Sample categories: natural or industrialized
Laboratories	List of laboratories accredited for sample analysis.
Units of Measurement	Units of measurements of radionuclides analyzed, such as: Bq/m^3, Bq/L, Bq/sample, Bq/kg, Bq/kg dry, Bq/kg wet and Bq/kg gray.
Methodologies of analyses	Sample analysis method.
Countrie	Countries (initially from Latin America).
State	States (initially from Latin America).
User Group	User profile of the system, such as: Country Administrator, Technical Manager (Informatics), Technical User and User (for consultation only).
Solicitations	Request new users.
User	System users.
LogSystem	It is the registry of the main movements of the users in the system. It serves to record all the history of navigation by the system made by the users.

IOP Conf. Series: Journal of Physics: Conf. Series **975** (2018) 012045 doi:10.1088/1742-6596/975/1/012045

3. Conclusions

The objective of the present study was the attainment of a georeferenced database through a Geographic Information System (SIG), in order to support incorporation and storage of radionuclides levels data in food issues.

On the basis of the dataset obtained in the development of this project, the key features of the system can be quoted below.

- Creation of a georeferenced database;
- Development of an information system aiming the storage of data related to the occurrence of radionuclides in food;
- Standardization and storage of analytical data from samples;
- Setting of food and radionuclides tables;
- Convenience of data manipulation within the system scope, such as insertion, modification and deletion;
- Deployment of consultation mechanisms and reports access facilities aiming the provision of information to system users;
- User convenience of database info download and backup in the PDF or XLS formats (electronic worksheet).

The SIGLARA system can be defined as an application program designed and developed to embody the resources that enable the remote access to data.

Employing free technologies with open source codes, like Google Earth, MYSQL (SGBD) database management system and Web Services, SIGLARA delivers a practical and efficient communication both in data exchange and system integration online.

It can be considered as objectives of SIGLARA system, too:
(i) to provide updated information regarding radionuclides findings in food.
(ii) to reference samples from the geographic standpoint and
(iii) to integrate the provision of information both to public and scientific communities.

The SIGLARA system is a SIG with a friendly and simple interface that simplifies the access of all users, meeting the specific requirements to the integration of the information provided by the system.

Both the design and development of the system were enabled by the adoption of public free software. The SIGLARA georeferenced database was tested in terms of feasibility, by the input and treatment of analyzed real data. It was concluded, then, that the system is capable of storing, retrieving and simultaneously processing the samples register data.

Designed to connect the users community right to the database, the structured graphic interfaces showed to be efficient, facilitating the system operation.

Along with the integration of designed procedures to data management of registered samples, the application of SIGLARA to the environmental management of radionuclides levels in food yields results already.

Such results are expected to stimulate and promote the extensive use of SIGLARA abroad, including international organisms, public and scientific communities in general, as an useful tool to a quick and safe decision making loop.

In the near future, SIGLARA can be expanded to include dose calculations from food ingestion.

System Adress: http://siglara.gov.br

Figs 1 and 2 show some SIGLARA viewings.

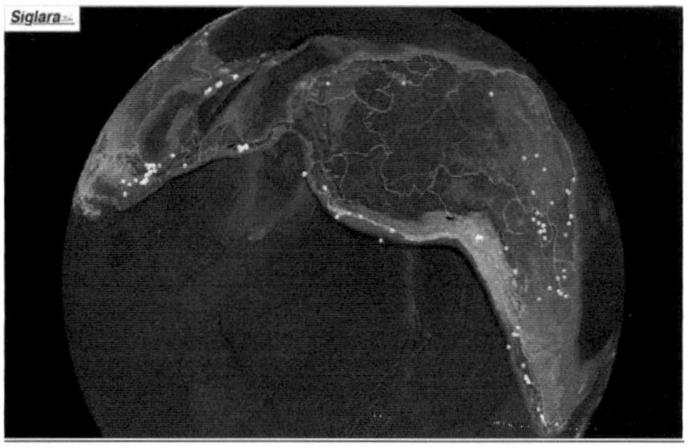

Figure 1. Example map sample in the SIGLARA System

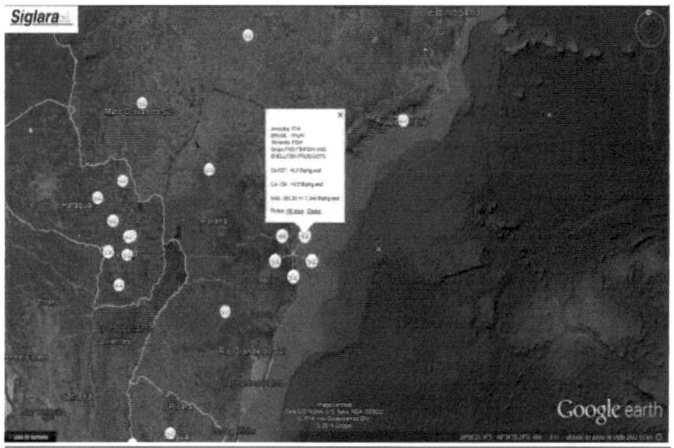

Figure 2. Example of label sample in the SIGLARA System

4. References

[1] Nascimento, LME 2014 Desenvolvimento de um banco de dados georeferenciado de radionuclídeos em alimentos típicos na américa latina – SIGLARA-Instituto de Radioproteção e Dosimetria-Dissertação de Mestrado.
[2] Randell, A 2004 *Radionuclide contamination of foods: FAO recommended limits, Food, Nutrition and Agriculture* - 2/3 - Nutrient Requirements. Roma, pp. 271-284. Disponível em: <www.fao.org/docrep/u5900t/u5900t08.htm>. Acesso em: mar. 2014
[3] Silva, TAA 2007 Simulação de Acidentes Radiológicos Através de Software Baseado em Agentes-78 f- M.Sc Engenharia de Computação- Universidade Estadual do Rio Janeiro- Rio de Janeiro
[4] Chen, P 1990 Modelagem de dados: a abordagem entidade-relacionamento para projeto lógico-São Paulo, Editora Makron Books

IOP Conf. Series: Journal of Physics: Conf. Series **975** (2018) 012046 doi:10.1088/1742-6596/975/1/012046

Development and evaluation of a technique for *in vivo* monitoring of ^{60}Co in human liver

GH Gomes, MC Silva, JQ Mello, ALA Dantas and BM Dantas
Instituto de Radioproteção e Dosimetria- IRD-CNEN

E-mail: bmdantas@ird.gov.br

Abstract: ^{60}Co is an artificial radioactive metal produced by activation of iron with neutrons. It decays by beta particles and gamma radiation and represents a risk of internal exposure of workers involved in the maintenance of nuclear power reactors. Intakes can be quantified through *in vivo* monitoring. This work describes the development of a technique for the quantification of ^{60}Co in human liver. The sensitivity of the method is evaluated based on the minimum detectable effective doses. The results allow to state that the technique is suitable either for monitoring of occupational exposures or evaluation of accidental intakes.

1. Introduction

^{60}Co is a synthetic radioactive isotope with a half-life of 5.27 years produced in nuclear reactors. It decays to the stable isotope ^{60}Ni, emitting β particle. The activated nickel nucleus emits two γ rays with energies of 1.17 and 1.33 MeV [1]. ^{60}Co is a result of multiple stages of neutron activation of iron components of the reactor structure, posing a risk of occupational internal exposure to workers involved in the handling and maintenance of parts and equipment potentially contaminated in nuclear power units. Measurable quantities are also produced as a by-product of typical nuclear power plant operation and may be detected externally when leaks occur. Internal monitoring procedures aimed to identify and quantify intakes by inhalation and ingestion may be accomplished through in-vivo measurements.

The IRD whole-body counter has been recently calibrated to perform lung monitoring using a NaI(Tl)8x4 scintillation detector [2] installed in a shielded room designed originally for low activity measurement [3].

This work describes the standardization and evaluation of the sensitivity of a technique for in-vivo measurement of ^{60}Co in human liver.

2. Materials and Methods

The calibration of the detection system is performed using a LLNL thorax phantom [4] containing a certified ^{60}Co liver phantom supplied by the University of Cincinnati. The procedure allows obtaining a calibration factor which correlates the net count rate in specific regions of interest (ROI) and the activity present in the phantom. This parameter is necessary to calculate the activity content in the liver of an individual potentially exposed and measured *in vivo* using the whole body counter detection system.

2.1. Determination of the calibration factor

The LLNL thorax with a liver phantom (Figure 1a) containing an activity of (34501 +/- 345) Bq of ^{60}Co was placed on the monitoring chair located inside the shielded room of the IRD whole-body counter and the Pb-collimated NaI(Tl)3x3 detector was positioned at a distance of 2 cm between the surface of the phantom and the front face of the crystal (Fig. 1 b). A series of five counts of 300 seconds were performed at the standard geometry and the spectra saved for subsequent calculations.

Figure 1c presents a γ-spectrum of ^{60}Co showing the two characteristic peaks of 1173 and 1332 keV. Three regions-of-interest (ROI) were defined including a range of energy from [1086-1264] for the first peak, [1264-1438] for the second and a third region of [1086-1438].

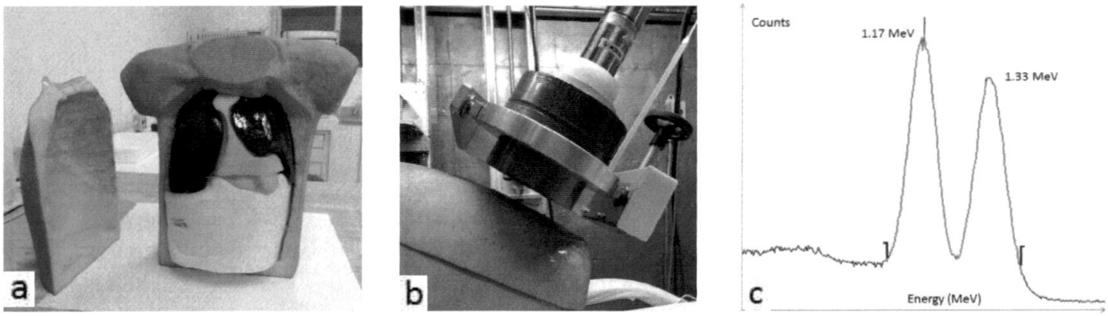

Figure 1. (a) The LLNL thorax containing a liver phantom; (b) Setup for the calibration of the NaI(Tl)3x3 scintillation detector installed in shielded room of the IRD-WBC; (c) Gama spectrum of 60Co obtained with the NaI(Tl)3x3 scintillation detector showing the two photopeaks of 1713 and 1333 keV used for the calculation of the calibration factor.

After the counts have been completed, the calculation of the calibration factors (CF) in each ROI was performed. The CF expresses the ratio between the average net count rate of the five counts, in cps, and the activity of the phantom, in Bq, as follows:

$$CF_{cps/Bq} = cps/A \quad (1)$$

Where: cps is the net count rate (averaged cps of the phantom and A (Bq) = ^{60}Co activity content of the liver phantom.

2.2. *Evaluation of sensitivity*

In order to be considered useful for internal dosimetry purposes, any technique should, at least, be able to detect an activity that would result in a committed effective dose below 1 mSv, taking into consideration the most likely internal exposure scenario [5]. Thus, the evaluation of the sensitivity of the method was based on the calculation of the MDA (minimum detectable activity) and the corresponding MDI (minimum detectable intake) and the MDED (minimum detectable effective dose)

Such calculations were based on the biokinetic and dosimetric models of ^{60}Co in human body applied to a realistic internal exposure scenario. According to the model ICRP 78 [6], ^{60}Co is distributed throughout the whole body and a significant fraction is retained in the liver.

The MDA of the method is calculated as follows [7]:

$$MDA = \frac{4.65\sqrt{N}}{CF \times T} + \frac{3}{CF \times T} \quad (2)$$

Where: N = Total counts of the background in the ROI; CF = Calibration Factor (cps/Bq) and T = count time.

The MDI is a function of the MDA and depends on the exposure scenario and time elapsed between intake and the *in vivo* measurement. It is calculated as follows:

$$MDI = MDA / m(t)_{inh} \quad (3)$$

Where: MDA = Minimum Detectable Activity (Bq) and m(t) = Retention fraction in the compartment of interest, in Bq/Bq.

The last parameter to be calculated is the MDED, which is a function of the MDI and the dose coefficient, e(g), associated to the intake scenario adopted in the simulation. It is calculated as follows:

$$MDED_{Bq} = MDI_{inh} \text{ x } e(g)_{inh} \quad (4)$$

Where: MDI = Minimum Detectable Intake (Bq) and $e(g)_{inh}$ = Dose coefficients (mSv/Bq).

The values of "m(t)", in Bq/Bq, where generated for a period of 1 to 14 days after a single intake of ^{60}Co via ingestion using the software AIDE [7]. The dose coefficient, e(g), in mSv/Bq, correspond to the committed effective dose resulting from an intake of 1 Bq.

3. Results and Discussion

Table 1 presents the calibration parameters (Regions of Interest, Calibration Factors and respective MDAs).

Table 1. Results of the calibration of the NaI(Tl)3x3 detector for *in vivo* measurement of ^{60}Co in the liver, comparing three regions of interest

	Calibration Parameters		
ROI (Channels)	347-418	418-488	347-488
ROI (keV)	1086 - 1264	1264 - 1438	1086 - 1438
CF (Cps/Bq)	0.00142	0.00131	0.00291
MDA (Bq)	114	126	79

ROI #3 (347-488) was elected since it has shown to be the most sensitive based on its corresponding MDA. Therefore, the following calculation where all based on this ROI. The next step was to calculate the MDI and the MDED in the period of 1 to 14 days after the intake,

Table 2 presents the values of MDI and MDED derived from the selected MDA, considering a scenario of a single intake of ^{60}Co via ingestion.

Table 2. Minimum Detectable Intakes and Minimum Detectable Effective Doses over a period of 1 to 15 days after the intake, assuming a scenario of a single intake of ^{60}Co by ingestion. (Dose coefficient = 3.41×10^{-6} mSv/Bq)

T (d)	m(t) (Bq/Bq)	MDI (Bq)	MDED (mSv)
1	3.21×10^{-3}	2.48×10^{-4}	0.084
2	4.20×10^{-3}	1.89×10^{-4}	0.065
3	4.23×10^{-3}	1.88×10^{-4}	0.064
4	4.04×10^{-3}	1.97×10^{-4}	0.067
5	3.81×10^{-3}	2.09×10^{-4}	0.071
10	2.94×10^{-3}	2.70×10^{-4}	0.092
14	2.50×10^{-3}	3.18×10^{-4}	0.108

4. Conclusions

Based on the sensitivity of the method, expressed by the MDED, it is concluded that the technique is able to detect activity levels of ^{60}Co in the liver derived from intakes via ingestion that would result in committed effective doses in the order of microsieverts. It should be note that up to 14 days after the intake, the MDED is still about 10 times below 1 mSv, the suggested Registry Level for occupational exposure, according to the IAEA [5]

The results obtained in this work allow concluding that the proposed technique is sensitive enough for the control of intakes of ^{60}Co via ingestion. The technique is suitable for a routine monitoring program aimed to control this type of exposure in nuclear power stations. The methodology is valid either in routine operations or in special tasks like the ones performed during maintenance of the reactor. It can also be useful for the evaluation of accidental intakes.

5. References

[1] US EPA 2017. https://www.epa.gov/radiation/radionuclide-basics-cobalt-60

[2] Mello JQ, Lucena EA, Dantas ALA and Dantas BM 2016. Development and evaluation of a technique for in vivo monitoring of ^{60}Co in human lungs. *J.of Physics*. Conference Series (Print). , **v.733**, p.012089

[3] Oliveira CAN, Lourenço MC, Dantas BM, Lucena EA and Laurer GR 1989. The IRD/CNEN whole body counter: Background and calibration results. *Radiat. Prot. Dosm.* **v.29** (3) 203.

[4] Spitz HB, Glover S, Liu N, Smith B, Hickman D, Kruchten D and Anderson L 1994. Measurement of the attenuation coefficient for Livermore thoracic phantom lungs fabricated using contemporary materials. *Health Phys*., v.67, n.1, p.39-46.

[5] International Atomic Energy Agency (IAEA) 1999. Assessment of Occupational Exposure Due to Intakes of Radionuclides - Safety Standards Series, no. RS-G-1.2

[6] International Commission on Radiological Protection (ICRP) 1998. Publication 78: Individual Monitoring for Internal Exposure of Workers,.

[7] International Standard Organization. Radiation Protection (ISO) 2001. Performance Criteria for Radiobioassy. Part 1: General principles.

[8] Bertelli L, Melo DR; Lipsztein JL and Cruz-Suarez R 2008. AIDE: Internal Dosimetry Software. *Radiation Protection Dosimetry* **v. 130**(3), p. 358-367.

Acknowledgment

The authors would like to thank the support of FINEP /MCTI (Project 01.10.0650.01)

IOP Conf. Series: Journal of Physics: Conf. Series **975** (2018) 012047 doi:10.1088/1742-6596/975/1/012047

Development of test specimens to obtain the transmission factors to attenuate photons of 0.511 MeV

J J S Costa[1], D D Cardoso[1], S Gavazza [1], C L Oliveira[1], R K Morales[1], A S Amorim[2], M C V Balthar[2], L S R Oliveira[2]

[1]Seção de Engenharia Nuclear, Instituto Militar de Engenharia, Praça General Tibúrcio, N° 80, Praia Vermelha, Urca, Rio de Janeiro, RJ, 22290-270, Brasil
[2]Centro Tecnológico do Exército, Avenida das Américas, N° 28.705, Guaratiba, Rio de Janeiro, RJ, 23020-470, Brasil

Email: jacojulio@gmail.com

Abstract: For designing a shielding, it is necessary, mainly, to determine or have access to the following parameters: transmission factors of the material used and type of radiation to be shielded. Cylindrical test specimens with different thicknesses were developed for experimentally obtaining the material transmission factor for shielding calculation. The cylindrical test specimens were made considering the geometric characteristics of the detector, the ease of production and the energy of 0.511 MeV from the ^{18}F-FDG decay. A type of concrete widely used in Brazil was used in the preparation of the cylindrical test specimens.

1. INTRODUCTION

The breakthrough of nuclear medicine in the last 20 years has caused a singular increase in examinations and treatments. Technologies such as PET and PET/CT were implemented and their use optimized the diagnosis and minimized disease detection time.

This type of technology performs use the radiopharmaceutical 18F-FDG [1,2] that in its decay releases a positron, this positron after a time ends up going through the process of annihilation, thus releasing two photons of 0.511 MeV, a type of highly penetrating radiation.

In order to guarantee the physical protection of the public and worker in these installations, a shielding must be designed for electromagnetic radiation with energy of 0.511 MeV.

In this way the determination of parameters, such as the transmission factor, will contribute to the process of calculating the shielding for this type of facility.

The materials normally used to shield this type of radiation are lead, iron and concrete. For economic and practical reasons, concrete is commonly used in various facilities.

In this work, cylindrical test specimens of different thicknesses were developed to obtain the concrete transmission factor, irradiated with 0.511 MeV electromagnetic radiation [1], and thus, to obtain a fundamental parameter for the calculation of the shield.

Parallel to obtainment this experimental data, it is developed through a computational modeling in the Monte Carlo N-Particle software, a simulation of the experimental context to generate computational data.

Then, a comparison between the computational data and the experimental data was created, also creating another comparison between data generated and data of consecrated literatures, guaranteeing a greater reliability in the parameters for the project of the shield and protection of the general public and workers.

2. MATERIALS AND METHODS

2.1. Trace of concrete
The trace of concrete is defined as the expression of the relative amounts of the concrete components.
The components used in this work are:

- Portland 2 Cement
- Fine, Medium e Coarse Sand
- Crushed Stone number 0
- Water

2.2. Quantities of the material used
From the calculates for the trace of the concrete elaborated according to the reference [3], the quantities expressed in spread sheet 1 was obtained:

Table 1: Quatities of the Materials Used

Materials	Quantities
Portland 2 Cement	2,2 kg
Sand Fine	3,4 kg
Sand Medium	1,75 kg
Sand Coarse	1,75 kg
Crushed Stone N:0	7,6 kg
Water	2 L

The crushed stone type 0 has been sieved to obtain the smallest stones, which will fit into the cylindrical test specimens.

2.3. Cylindrical test specimen
The Cylindrical test specimens were designed considering the ease of production and the geometric characteristics of the experimental apparatus used.

Cylindrical test specimens were made from a PVC tube 10 centimeters in diameter and 2 meter in length. 7 cylindrical test specimens of this tube were cut into the lathe, cuts of 2 to 8 centimeters in length, varying 1 centimeter. Were also cut, three cylindrical test specimens of this tube, each 20 centimeters in length.

2.4. Equipments
The equipment used to make the concrete are:

- Concrete-mixer CSM 145 Litros
- Trowel
- 4" knitted sieve
- 3/8" knitted sieve
- Plastic Sheet

- Sideboard for concrete
- 1 Litre beaker

2.5. procedures

The procedures were performed according to reference [3], as following:

- The concrete mixer was prepared by wetting its interior to receive the materials;
- Crushed stone, cement, fine sand, medium sand, coarse sand and water were introduced;
- The concrete mixer was activated, and waited 30 minutes until the mixture was homogeneous;
- The concrete mixer was turned off, and the concrete mix was poured into the Sideboard. The interior of the concrete mixer was checked to see if there were no traces of the mixture, if it still had, a trowel was used to remove traces of the mixture;
- The homogeneity of the mixture in the sideboard for concrete was observed with the naked eye. if the mixture was not homogeneous, turn the mixture with the trowel until it remained;
- The vibrating machine was prepared by lining the workbench with a plastic sheet;
- The cylindrical test specimens were placed on top of the vibrating machine bed, then the cylindrical test specimens were filled with the concrete;
- The vibrating machine was turned on, so that the concrete was level in each cylindrical test specimens, after leveling the machine was turned off;
- It was waited 24 hours for the concrete to dry;
- After 24 hours, cylindrical test specimens were placed underwater to cure (increase the strength of the concrete);
- After 28 days of cure, according to the literature [3], the cylindrical test specimens are ready for use.

3. RESULTS

The purpose of this work was to demonstrate the care at the moment of designing and manufacturing the cylindrical test specimens, guaranteeing an experimental apparatus according to the context.

And in order to reproduce the ideal conditions for the design of a shield, in which the material to be shielded has its specific characteristics. the results of the cylindrical test specimens according to previously defined criteria, were as shown in figures 1 and 2:

Figure 1: Cylindrical test specimens 2 to 8 centimeters in height and 10 centimeters in diameter.

Figure 2: Cylindrical test specimens 10 centimeters in diameter and 20 centimeters in height.

4. CONCLUSION

The cylindrical test specimens were made, and the experimental data for the experimental transmission factor for the 0.511 MeV energy is now being used. The Monte Carlo simulation is in the final stage of elaboration, spread sheet 2 shows the preliminary results of the simulation, where “Length” is the length of the wall, “Buildup 1” is the buildup in the finite medium, “Buildup 2” is the buildup in the infinite medium, and “Correction Factor” is the factor to correct the buildup, so that compare the computational data with those of consecrated literatures.

Table 2: Results of the simulation.

Length (cm)	Buildup 1	Buildup 2	Correction Factor
2	1.3161	1.4606	0.6862
4	1.7018	2.0015	0.7007
6	2.1762	2.6219	0.7252
8	2.6668	3.3118	0.7210
10	3.2570	4.0938	0.7295
14	4.5961	5.9126	0.7320
18	6.2105	7.7893	0.7674

Tests will also be done to analyze the composition of the concrete, below presents the two tests that will be done:

- With the scanning electron microscope will be done the EDS;
- X-ray Fluorescence.

Relevant points to emphasize:

- This made concrete has different specifications of those proposed in the literature;
- Data will be created in the literature for the energy of 0.511 MeV, because currently there has to be interpolation between the energies of 0.5 MeV and 0.6 MeV.

Acknowledgments
Thanks to the collaborators of the Centro Tecnologico do Exercito (CTEX), for their instructions and cooperation for this work.

Thanks also to the professors and collaborators of the Instituto Militar de Engenharia (IME), for all efforts in carrying out this work.

REFERENCES

[1] Madsen M T, Anderson J A, Halama J R, Kleck J, Simpkin D J, Votaw J R, Wendt III R E, Williams L E and Yester M V 2005 *PET and PET/CT shielding requirements* AAPM Vol **33** N 1 4-15

[2] Hays M T, Watson E E, Thomas S R and Stabin M 2002 *Radiation absorbed dose estimates from ^{18}F-FDG* The Journal of Nuclear Medicine Vol **43** N 2 210-214

[3] Petrucci E G R, Paulon V A 2005 *Concreto de cimento portland* São Paulo Globo 01-307

IOP Conf. Series: Journal of Physics: Conf. Series **975** (2018) 012048 doi:10.1088/1742-6596/975/1/012048

Dosimetry Evolution in Teletherapy: Polimer Gel

J H Hamann, J G P Peixoto

Instituto de Radioproteção e Dosimetria – IRD / CNEN – Rio de Janeiro – RJ – Brazil

E-mail: jhhamann@gmail.com

Abstract: Polymer gels evolution and chemical composition used in dosimetry.

Type	Composition
First gels	Folin's Phenol or Gallic Acid
Polymer Gel	Agarose and N,N´-methylene-bis-acrylamide
BANANA	Bis, acrylamide, nitrous oxide and agarose
BANG-1TM	Bis, acrylamide, nitrogen and gelatin
BANG-2TM	Bis, acrylic acid, sodium hydroxide, nitrogen and gelatin
BANG-3TM	Bis, methacrylate acid, sodium hydroxide, nitrogen and gelatin
MAGIC	Methacrylate acid, ascorbic acid, gelatin and copper sulphate

1. Introduction

Recently developed in radiotherapy techniques, such as the tridimensional planning (3D) treatment, Intensity Modulated Radio Therapy beam (IMRT), radiotherapy (conformational and conventional), Images Guided radiation Therapy (IGTR) and stereotactic radiosurgery have been increase the oncologic treatment complexity by radiation. All these techniques are used to reduce the treatment toxicity generated by means of the dose optimization in the target volume, radiosensitive irradiation minimizing, healthy structures or adjacent to tumor area [1].

However, irradiation reducing area to nearest to the target volume, the treatment accuracy system is increased as well as the dose deposited in the tumor area volume. Therefore reliable programs are necessary to systematically ensure the high quality and confidence in the whole planning process, treatment and dose deposited in the target volume [1].

The current methods are radiotherapy dosimetry standards to dose distribution evaluation used film dosimetry, thermos-luminescent dosimeters (TLDs), ionization chambers or photodiodes.

All of these are two dimensional methods for dose measurements [2]. These dosimeters will not measure the dose distribution in three dimensional space and only at certain points of analysis. Some dosimeters are energy and angular dependent beam radiation positioning for correct measurement [3].

Thus, the use of such detectors is also associated with additional problems: (*i*) large volume relatively, which prevents measurement in high dose gradient regions and (*ii*) may be larger than the treated tumor volume. In certain treatment situations this characteristic limit of spatial dose resolution;

(*iii*) detectors are not equivalent to the human tissues and (*iv*) some treatment types there is the possibility of radiation fields disturbing causing errors in the exposure rate measurement [4].

Therefore, in radiotherapy treatment centers system most frequently used in clinical dosimetry and that is capable of partially tracing 3D dose distributions is a position phantom filled with water and connected to point detector (ionization chambers or semiconductor detectors, for example). Unfortunately this system will be not allow complex dose distributions formed by the overlapping of different radiation beams, also full dose distribution measurement the in space [5–7].

2. Historical review

2.1. First polymer gels dosimetry forms

In 1950s, Day and Stein [8], the three dimensional dosimetry first forms appeared. The dosimeter had been developed observing a coloration alter after ionizing radiation interacting. The gel dosimeter had a chemical formulation, essentially presented a Folin's Phenol (gallic acid) reagent [8 and 9].

In 1954, Alexander et al. [10] discussed the ionizing radiation effects on polymethylmethacrylate. Therefore was first associated idea of gel polymeric systems dosimetry [9].

Andrew et al. [11] performed dose measurements associated with photons and electrons beams using agar gels associated with spectrophotometry [9].

A year later Hoecker and Watkins [12] developed an alternative study for gel dosimetry based on the radio induced monomers and polymers polymerization in aqueous solution [9, 10, 12 and 13].

In 1961, Boni [14] developed a gel dosimeter using only polyacrylamide, that had been used as gamma radiation dosimeter [9 and 14].

2.2. Gel dosimetry

Three dimensional dosimetry presented a great innovation in the 1980s, when Gore et al. [15] associated the Fricke gel with magnetic resonance imaging (MR) for radiotherapy treatments dose distribution evaluation. Therefore this Fricke gel dosimetry form had recording dose values purpose through modified dosimeter that its molecular structure when interactions with X radiation.

The molecular alteration would be proportional to the dose applied, being possible to visualize the gel irradiated volume through the MR. This association has become a promising tool, which has requirements satisfied as ideal dosimetry system. Unlike the other dosimetry methods, MR associated with imaging Fricke gel dosimetry for gel irradiated is completely noninvasive and it is not necessary to remove some of the irradiated material for testing, because a radiation gel detector propriety [9, 13, 15 and 16].

Since then, the use of Fricke gel dosimeter association have been performed. Gum et al. [17] determined that for anthropomorphic phantoms containing inhomogeneous tissues, Fricke gel associated with MR was a powerful tool for the verification of radiotherapy treatment plans. The from phantom developed was anatomical area of the thorax, used of soft tissues simulation (lungs, heart and spinal cord) with Fricke gel with a low density. An average deviation was found less than 5% difference between virtually treatment planning and phantom planning applied, the difference was generated due to low magnetic resonance signal to noise ratio [17].

Oldberg et al. [18] demonstrated that the Fe^{+2} ion diffused in low density Fricke gel dosimeter (simulating lung tissues) was reliable for dose measurements absorbed in thorax simulated phantoms.

The Fricke gel dosimeter conducted by Chu et al. [19] where had a Fe^{+3} ions low diffusion. Thus, the gel was based on the polyvinyl alcohol and new hydrogels additions were developed as three dimensional dosimetry forms in radiotherapy. Due to its transparency, the Fricke hydrogel may have images generated by magnetic resonance imaging and optical tomography [19].

However, the greatest disadvantage in use the Fricke gel dosimetry is the fast diffusion process of ferric ions into dosimetric solution. Through this action occurs loss information process, which means a lower spatial resolution in the dose registered. This loss is greater how much spent time longer between Fricke gel irradiation and sample imaging [4 and 20].

2.3. Polimer gel dosimeter

Kennan et al. [7] describe dosimetric solution irradiation as composed of agarose and N, N'-methylene-bis-acrylamide, observed R1 and R2 ratios relaxation of the material when submitted to MR increased proportionally with the dose deposited.

However Maryanski et al propose the gel characteristics: (*i*) use of monomers derived from acrylic and cross linking agents; (*ii*) gel solution based on agarose; (*iii*) polymerized region dimensional stability [21 – 23].

Therefore a new gel dosimeter formulation was proposed by Maryanski. In chemical formulation the dosimetric solution had two monomers: acrylamide and N-N'-methylene-bis-acrylamide (Bis), which were diluted in aqueous agarose matrix. Thereby monomers would undergo the radio induced polymerization process. This system was called BANANA (due presence of its chemical compounds: N-N'-methylene-bis-acrylamide, Acrylamide, nitrous oxide and agarose).

Polymer gels prepared with monomers such as acrylamide are currently used in biochemistry as field for protein electrophoresis and nucleic acids separation, typically, present: acrylamide monomers and a cross linking agent, known as Bis (N-N'-methylene-bis-acrylamide).

Gels polymerization process used for electrophoresis is generally initiated and chemically controlled using a free radical initiator, such as PSA (ammonium persulfate). However, the polymerization can also be induced by radiation through the production of a free radical generated during water radiolysis [5, 7 and 12].

BANANA gel (composed of N-N'-methylene-bis-acrylamide, acrylamide, nitrous oxide and agarose) was the gel polymer developed first form. In this type of gel polymer, acrylamide and N-N' were added in gelatin of the agarose type. Sequential works utilized gels polyacrylamide (PAG) and BANG^{TM}. The development of the BANG^{TM} gel resulted increasing sensitivity of the gelatinous radiation fields solution. Nitrous oxide or nitrogen is added to the chemical solution expelling oxygen during organic dosimeter preparation period [24].

The BANG 1^{TM} gel uses acrylamide powder form, BANG 2^{TM} gel replaces acrylamide powder by acrylic acid and sodium hydroxide as the buffer solution for the pH of the solution. The response of the BANG 2^{TM} gel when subjected to magnetic resonance imaging is high than the BANG 1^{TM} gel, which means that the relaxation ratio of the spins of the water molecule is higher per dose unit [24 and 25].

BANG 3^{TM} gel type has superior response to the BANG 2^{TM} type gel in magnetic resonance imaging [26]. BANG 3^{TM} gel acrylic acid is replaced by methacrylate acid [27].

After irradiation, the gel may contain polymerized and cross linked regions. This is the spatial dose response origin characteristics of the gels with the polymerization dependent degree on the amount of free radicals generated by the incident radiation and therefore absorbed dose.

Polymerized regions certain water molecules clusters alter their linking state. This can be investigated using the magnetic resonance relaxation time characteristics (in images generated through the longitudinal relaxation time - T1 or through images generated by the transverse relaxation time - T2) [2, 23, 27 and 28]. Computed tomography X rays images (CT) generated can also be used where they are produced through measuring the different attenuation values generated by effective atomic number significant changes of the polymerized region [30].

The sensitivity (i.e., the change in the relaxation time in MR or the value CT effective atomic number) and the polymer gel dosimeters saturation limit critically depend on the conditions during gel preparation. Factors such as: (*i*) exposure to light and oxygen during the preparation period, (*ii*) fraction of monomers and gelatin by total weight final solution and (*iii*) the temperature during the period of acquisition of MR or CT images are some conditions that influencing the sensitivity of the dosimeter [2, 25 e 29].

Normoxics: Due to the free radicals nature produced by the radiation interaction with the gels polymer, these must be a hypoxic material, that is, with an infinitesimal presence of oxygen (1 / 1,000,000 – oxygen / dosimetric solution) in chemical composition. This process is given by oxygen being highly reactive chemical element, which when present in large quantities in hydrogel will inhibit the polymerization process [13].

Through the work produced by Fong et al. [31], the gels polymer dosimetry has undergone a significant development, a new chemical composition produce this new gel dosimeter type as known MAGIC uses (*i*) a chemical agent that realizes the free oxygen capture in aqueous solution and (*ii*) another chemical agent that will form with the oxygen molecule captured in (*i*) chemical bond that will result in an organometallic complex in solution.

In this way, the oxygen does not inhibit the radiation induced polymerization process. Another advantage presented by normoxed gels is that they can be prepared under normal ambient conditions not requiring a glovebox, as it happens with PAG gels for their preparation [30 and 31].

The MAGIC gel formulation consists of methacrylic acid, ascorbic acid, gelatin and copper sulphate. The ascorbic acid binds with free oxygen in the gelatinous matrix, an organometallic complex is then formed with the copper sulphate. The Deene et al. [33] demonstrated that other antioxidants can be used to the capture of the free oxygen present in the dosimetric solution [33].

With new class hydrogels studies introduction by MR have been developed to behavior investigate of these materials when submitted to radiotherapy techniques (e.g. IGRT or VMAT) [9, 33 and 34].

3. Results

Through gel polymer association to simulating object (or phantom) can be ensured quality, confidence and accuracy during the planning, treatment and dose deposited in target volume. The hydrogel presents equivalence to human tissue and greater temporal and dimensional stability in the dose register when compared to Fricke gel dosimetry as observed by MARYANSKI et al. [21, 23 and 25], DEENE et al. [36], BALDOCK et al. [9 and 13].

The gel dosimeter new chemical composition (derived from the nPAG family) and doses absorbed response behavior by the dosimetric solution to the energy ranges used in radiotherapy treatments according ISO 4037 [37].

4. Conclusions

The radiotherapy treatment new modalities bring greater accuracy and safety in the techniques used by teletherapy. However, due to its greater complexity mainly in dose deposition relation in tumoral mass the two dimensional dosimetry is not able to perform this analysis satisfactorily [35].

As such we observed that the dosimetry by polymer gels has also evolved since first applications. Therefore this dosimetry form is a promising tool because its associated with MR or CT. Can perform the isodose mapping curves for volume given according to the techniques used for irradiation demonstrating register capacity for three dimensional dose which can't see at radiographic films and ionization chambers dosimetry.

The gel dosimetry also demonstrates to be efficient dosimeter for registered and complex dose distribution evaluation and can be an efficient tool for validation of treatment plan, dose delivery systems and patient positioning procedures along the linear accelerator.

Another gels dosimeter important feature is the stability after irradiation. The macroscopic diffusion of products generated by X radiation through the gelatinous matrix is negligible and the registered dose distribution is stable for long periods of time (weeks, months and even years), it is not necessary removing gel polymer parts irradiated and polymerized to perform dosimetric tests, because the dosimetric system itself corresponds to detection system [21 and 25].

When simulator object filling with a gel dosimeter is taking: (*i*) tool along with an international standards protocol based ISO 4037 [37] that will evaluate what has been planned through patient treatment software, (*ii*) can be tool for certification of clinics and radiotherapy hospitals in relation to determination radiation oncology treatments and, finally, (*iii*) develop the culture of total volumetric dosimetry with hospitals and radiotherapy clinics, as well as the competent entity and organs.

The final result of the factors (*i*), (*ii*) and (*iii*) presented previously aims better quality control program, where the greatest beneficiary is the patient who will be performing a more reliable and efficient radiotherapy treatment.

References

[1] Schlegel W, Bortfeld T, and Grosu A L, 2006 News technologies in Radiation Oncology, (1° ed.) Germany
[2] Bankamp A and Schad L R, 2003 Comparison of TSE, TGSE and CPGM measurement techniques for MR polymer gel dosimetry. *Magn. Reson. Imaging*, **21**, pp. 929–939
[3] Baras P, Seimenis I, Papagiannis P, Pappas E and Kipouros P, 2002 Polymer gel dosimetry using a three-dimensional MRI acquisition technique. Med. Phys, **29**, pp. 2506–2516
[4] Gambarini G, Brusa D, Carrara M, Castellano G, Mariani M, Tomatis M, Valente M, Vanossi E, 2006 Dose imaging in radiotherapy photon fields with Fricke and normoxic polymer Gels, *J. Phys. Conf. Ser.*, vol. **41**, pp. 466–474
[5] Cruz A, 2003 Desenvolvimento de um dosímetro para análise tridimensional de radiação ionizante utilizando gel polímero, Centro Federal de Educação Tecnológica do Estado do Paraná, Curitiba
[6] Hamann J H, 2005 Análise qualitativa do dosímetro tipo BANG-1 por imagens em ressonância magnética e tomografia computadorizada, Centro Federal de Educação Tecnológica do Estado do Paraná
[7] Hamann J H, 2009 Gel dosímetro tipo BANG-1: aplicação em planejamentos de tratamento por radioterapia e avaliação qualitativa através de imagens obtidas em ressonância magnética, Universidade Tecnológica Federal do Paraná
[8] Day M J and Stein G, 1950 Chemical effects of ionizing radiation in some gels. *Nature*, vol. **166**, pp. 146–147
[9] Baldock C, 2009 Historical overview of the development of gel dosimetry: another personal perspective, *J. Phys. Conf. Ser.*, vol. **164**, pp. 1–9
[10] Dumas A, 1960 of Medicine Editorial Radiation
[11] Andrews H L, Murphy R E and Lebrun E. J, 1957 Gel dosimeter for depth dose measurements, *Ver. Sci. Instrum.*, vol. **28**, pp. 329–332
[12] Hoecker F E and Watkins I W, 1958 Radiation polymerization dosimetry, *Int. J. Appl. Rad. Iso.*, vol. **3**, pp. 31–35
[13] Baldock C, De Deene Y, Doran S, Ibbott G, Jirasek A, Lepage M, McAuley K. B, Oldham M, Schreiner L. J, 2010 Polymer gel dosimeter, *Phys. Med. Biol.*, vol. **55**, pp. 1–63
[14] Boni A L, 1961 Polyacrylamide gamma dosimeter, Radiat. Res., vol. **14**, pp. 374–380
[15] Gore J C, Kang Y S, Schulz R J, 1984 Measurement of radiation dose distribution by nuclear magnetic resonance (NMR) imaging, *Phys. Med. Biol.*, vol. **29**, pp. 1189–1197
[16] Baldock C, HARRIS P J, PIERCY A R, HEALY B, 2001 Experimental determination of the diffusion coefficient in two-dimensions in ferrous sulphate gels using the finite element method, *Australas Phys. Eng. Sci. Med.*, vol. **24**, pp. 19–30
[17] Gum F, Scherer J, Bogner L, Solleder M, Rhein B, Bock M, 2002 Preliminary study on the use of na inhomogenous anthropomorfic Fricke gel phantom and 3D magnetic resonance dosimetry for verification of IMRT tretment plans, *Phys. Med. Biol.*, vol. **47**, pp. 67–77

[18] Oldberg S, Skretting A, Bruland O, Olsen D R, 2000 Dose distribution measurements by MRI of a phantom containing lung tissue equivalent compartments made of ferrous sulphate gel, *Phys. Med. Biol.*, vol. **45**, pp. 2761–2770

[19] Chu K C, Jordan K J, Battista J J, Van Dyk J, Butt B K, 2000 Polyninyl alchohol-Fricke hydrogel and cryogel: two new gel dosimetry systems with low Fe+3 diffusion. *Phys. Med. Biol.*, v. **45**, p. 955-969

[20] Schreiner L J, 2004 Review of Fricke gel dosimeters, *J. Phys. Conf. Ser.*, vol. **3**, pp. 9–21

[21] Maryansky M J, 1994 United States Patent/ number 5.321.357

[22] Maryanski M J, Gore J C, Schulz R J 1992 3-D radiation dosimetry by MRI: solvent próton relaxation enhancement by radiation-controlled polymerisation and cross-linking in gels, *Intl. Soc. Mag. Reson. Med*

[23] Maryanski M J, Gore J C, Schulz R J 1993 Relaxation enhancement in gels polymerized and cross-linked by ionizing irradiation: a new approach to 3D dosimetry by MRI, *Magn. Reson. Imaging*, vol. **11**, pp. 253–258

[24] McJury M, Oldham M, Leach M O, Webb S, 1999 Dynamics of polymerization in polyacrylamide gel (PAG) dosimeters: (I) ageing and long-term stability, *Phys. Med. Biol.*, vol. **44**, pp. 1863–1873

[25] Maryansky M J, Schulz R J, Ibbott G S, Gatenby J C, Xie J, Horton D, Gore J C, et al 1994 Magnetic resonance imaging of radiation dose distributions using a polymer gel dosimeter, *Phys. Med. Biol.*, vol. **39**, pp. 1437–1455

[26] Oldham M, Siewerdsen J H, Shetty A, Jaffray D A, 2001 High resolution gel dosimetry by optical CT and MR scanning, *Med. Phys.*, vol. **28**, pp. 1436–1445

[27] Macdougall N D, Miquel M E, Wilson D J, Keevil S F, Smith M A, 2005 Evaluation of the dosimetric performance of BANG 3 polymer gel, *Phys. Med. Biol.*, vol. **50**, pp. 1717–1726

[28] Meeks S L, Bova F J, Maryansky M J, Kendrick L A, Kanade M K, Buatti J M, Friedman W A, 1999 Image registration of BANG gel dose maps for quantitative dosimetry verification, *Int. J. Radiat. Oncol. Biol. Phys.*, vol. **43**, no. 5, pp. 1135–1141

[29] Appleby A, Christian E A, Leghrouz A, 1987 Imaging of spatial radiation dose distribution in agarose gels using magnetic resonance, *Med. Phys.*, vol. **14**, pp. 382–384

[30] Hilts M, Audet C, Duzinli C, Jirasek A, 2000 Polymer gel dosimetry using x-ray computed tomography: a feasibility study, *Phys. Med. Biol.*, vol. **45**, pp. 2559–2571

[31] Fong P M, Keil D C, Does M D, Gore J C, Polymer gels for magnetic resonance imaging of radiation dose distributions at normal room atmosphere, *Phys. Med. Biol.*, vol. **46**, pp. 3105–3113

[32] Gustasson H, Back S A J, Medin J, Grusell E, El Olsson L, 2004 Linear energy transfer dependence of a normoxic polymer gel dosimeter investigated using proton beam absorbed dose measurements, *Phys. Med. Biol.*, vol. **49**, pp. 3847–3855

[33] De Deene Y, De Wagter C, Van Duyse B, Derycke S S, Mersseman B, De Gersem W, Voet T, Achten E, De Neve W, 2000 Validation of MR-based polymer gel dosimetry as a preclinical three-dimensional verification tool in conformal radiotherapy, *Magn. Reson. Med.*, vol. **43**, pp. 116–125

[34] Venning A J, Hill B, Brindha S, Healy B J, Baldock C, 2005 Investigation of the PAGAT polymer gel dosimeter using magnetic ressonance imaging, *Phys. Med. Biol.*, vol. **50**, pp. 3875–3888

[35] Vergote K, De Deene Y, Duthoy W, De Gersem W, De Neve W, Achten E, 2004 Validation and application of polymer gel dosimetry for the dose verification of na intensity-modulated arc theraphy (IMAT) treatment, *Phys. Med. Biol.*, vol. **49**, pp. 287–305

[36] De Deene Y, Wagter C, Van Duyse B, Derycke S, De Neve W, Achten E, 1998 Three-

IOP Conf. Series: Journal of Physics: Conf. Series **975** (2018) 012048 doi:10.1088/1742-6596/975/1/012048

dimensional dosimetry using polymer gel and magnetic resonance imaging applied to the verification of conformal radiation therapy in head-and-neck cancer, *Radiother. Oncol.*, vol. **48**, pp. 283–291

[37] ISO, 1995 ISO 4037-2 - X and gamma reference radiation for calibrating dosemeters and doserate meters and for determining their response as a function of photon energy -- Part 2: Dosimetry for radiation protection over the energy ranges from 8 keV to 1,3 MeV and 4 MeV

IOP Conf. Series: Journal of Physics: Conf. Series **975** (2018) 012049 doi:10.1088/1742-6596/975/1/012049

Evaluation of the uncertainties in the TLD radiosurgery postal dose system

L T Campos [1,2], S P Leite[2], C E V de Almeida[2], L A G Magalhães[2.]

[1] Universidade do Estado do Rio de Janeiro, Departamento de Física Aplicada e Termodinâmica (DFAT-UERJ), Rio de Janeiro, Brazil.

[2] Universidade do Estado do Rio de Janeiro, Laboratório de Ciências Radiológicas (LCR-UERJ), Rio de Janeiro, Brazil

E-mail: tc_luciana@yahoo.com.br

Abstract: Stereotactic radiosurgery is a single-fraction radiation therapy procedure for treating intracranial lesions using a stereotactic apparatus and multiple narrow beams delivered through noncoplanar isocentric arcs. To guarantee a high quality standard, a comprehensive Quality Assurance programme is extremely important to ensure that the measured dose is consistent with the tolerance considered to improve treatment quality. The Radiological Science Laboratory operates a postal audit programme in SRT and SRS. The purpose of the programme is to verify the target localization accuracy in known geometry and the dosimetric conditions of the TPS. The programme works in such a way those thermoluminescence dosimeters, consisting of LiF chips, are sent to the centre where they are to be irradiated to a certain dose. The TLD are then returned, where they are evaluated and the absorbed dose is obtained from TLDs readings. The aim of the present work is estimate the uncertainties in the process of dose determination, using experimental data.

1. Introduction

Stereotactic radiosurgery (SRS) is a single-fraction radiation therapy procedure for treating intracranial lesions using a stereotactic apparatus and multiple narrow beams delivered through noncoplanar isocentric arcs. The same procedure is called stereotactic radiotherapy (SRT) when used to deliver multiple dose fractions [1]. Both techniques comprise precise delivery of a high dose to a small target volume with the goal of a positive therapeutic gain. SRS and SRT require patient immobilization and repositioning with an invasive frame, a non-invasive frame, or a frameless system to direct precise radiosurgery beam targeting. SRS and SRT require careful error assessment at each step of the treatment process. This includes assessing the integrity of the stereotactic device, the spatial accuracy of the imaging modality, and the accuracy of the radiation beam delivered to the patient. The ultimate accuracy of the treatment is a combination of all these potential sources of error.

The accurate placement of the intended radiation dose is fundamental to these techniques. Small errors in the placement of the radiation dose from individual beams or beamlets can result in an inaccurate estimate of the accumulated dose as well as an inaccurate estimate of the steepness and location of the high dose gradient region that may be designed to protect adjacent critical structures and organs.

To guarantee a high quality standard, a comprehensive Quality Assurance (QA) programme is extremely important to ensure that the measured dose is consistent with the tolerance considered to improve treatment quality.

A QA programme is necessary to guarantee the desired level of accuracy for the dose delivered to each patient [2-4].

A QA programme can be implemented using a head phantom with or without an anthropomorphic form and a dosimeter that can measure the dose delivered during the therapy and its distribution in the volume treated [5-6].

The Radiological Science Laboratory (LCR/UERJ) operates a postal audit programme in SRT and SRS. The purpose of the programme is to verify the target localization accuracy in known geometry and the dosimetric conditions of the TPS. The programme works in such a way those thermoluminescence dosimeters (TLDs), consisting of LiF chips, are sent to the centre where they are to be irradiated to a certain dose. The TLD are then returned, where they are evaluated and the absorbed dose is obtained from TLDs readings. The aim of the present work is estimate the uncertainties in the process of dose determination, using experimental data.

2. Material and Method

2.1 Irradiation procedure

The head phantom was made of PMMA material and constructed with high accuracy. Is consisted of a cylinder with a diameter of 16 cm and a length of 21 cm with several modular components to accommodate spherical imaging markers, radiochromic film and TLDs. Precise instructions for the computed tomography study, dosimetric planning and dose delivery were mailed to the radiotherapy clinic. The institution imaged the phantom, developed a treatment plan and irradiated the phantom according to the plan. The present study used mini-thermoluminescent cylinders and radiochromic films. A mini TLD cylindrical-type TLD 700H (LiF: Mg, Cu, P) with a 3 mm diameter, a 0.38 mm length and a 2.6 g/cm^3 density was also used. The annealing process was performed in a Thermolyne/4700 oven by pre-irradiation annealing at 240°C for 10 min. This pre-annealing was performed to erase all the information due to any previous irradiation from a thermoluminescence material to restore the crystals to their initial conditions before irradiation. The general purpose of this thermal treatment was to restore the trap-recombination centre structure to the former one obtained after the initialization procedure. The post-irradiation procedure at the reader was performed at 165°C for 10 s. This was necessary to erase the low temperature peaks. The readout process was performed with a Harshaw/QS 3500 reader [5].

Measurements were performed to determine the values of the correction factors as a function of the absorbed dose, the beam energy and fading after irradiation.

The TLD readings at the point of interest were corrected by the mean background thermoluminescence response (BG), the individual sensitivity factor (S_i) and the correction for the variation of the dosimeter response as a function of the beam energy (F_{depen}) and for

IOP Conf. Series: Journal of Physics: Conf. Series **975** (2018) 012049 doi:10.1088/1742-6596/975/1/012049

time, and variations of TLDs readings (F_{corr}) according to Equation (1), to obtain the absorbed dose (D).

$$D = \frac{(TL - BG) \cdot 10^2}{S \cdot F_{lin} \cdot F_{depen} \cdot F_{corr}} \tag{1}$$

The TLD response for a given dose depends on the photon energy, due to the dependence of the TL material's mass absorption coefficient ($\mu en/\rho$) on energy [6]. The energy correction factor is defined as the ratio of the response R per unit dose measured at the D=1 Gy in a Co γ ray beam to a response R per unit dose at 1 Gy in the x-ray beam quality $TPR_{20/10}$. The values of the correction factor for different beam qualities are approximated by a linear fit that was made to TL irradiations showed in equation 2.

$$F_{depen} = 0,4736(TPR_{20/10})^2 0,8433(TPR_{20/10})+1,3225 \tag{2}$$

In the dose range of interest (0,5-10 Gy) the dose response exhibits a small supralinearity effect, which requires to be corrected for. This effect arises primarily from competitive mechanisms in the recombination stage [7]. The dose response non-linearity correction factor is the ratio of the TL response per unit dose measured at the dose of 1 Gy to the Tl response per unit dose measured at the dose D. The non-linearity factor was determined by making a linear fit to the experimental data showed in equation 3.

$$F_{lin} = -9,71924 \text{ x } 10^{-10} (D)^3 + 4,74633 \text{ x } 10^{-07} (D)^2 + 2,92386 \text{ x } 10^{-04} (D) + 0,97078 \tag{3}$$

2.2 Expressions of uncertainties

The methodology used in the present work for estimating the uncertainties is based on the recommendations of ISOGUM (ISO Guide to the Expression of Uncertainty in Measurement) [8].

By calling the TL-BG reading simply as L and taking the partial derivates for each variable (L,S, F_{lin}, F_{depen} and F_{corr}, we have the equations 4, 5, 6, 7 and 8:

$$\frac{\partial D}{\partial L} = \frac{10^2}{S.F_{lin}.F_{depen}.F_{corr}} = \frac{D}{L} \tag{4}$$

$$\frac{\partial D}{\partial S} = -\frac{L.10^2}{S^2.F_{lin}.F_{depen}.F_{corr}} = -\frac{D}{S} \tag{5}$$

$$\frac{\partial D}{\partial F_{lin}} = -\frac{L.10^2}{S.F_{lin}{}^2.F_{depen}.F_{corr}} = -\frac{D}{F_{lin}} \tag{6}$$

IOP Conf. Series: Journal of Physics: Conf. Series **975** (2018) 012049 doi:10.1088/1742-6596/975/1/012049

$$\frac{\partial D}{\partial F_{depen}} = -\frac{L.10^2}{S.F_{lin}.F_{depen}^{\;2}.F_{corr}} = -\frac{D}{F_{depen}} \tag{7}$$

$$\frac{\partial D}{\partial F_{corr}} = -\frac{L.10^2}{S.F_{lin}.F_{depen}.F_{corr}^{\;2}} = -\frac{D}{F_{corr}} \tag{8}$$

Assuming that the factors in equation 1 are uncorrelated, the combined relative uncertainty in the dose determined from the TLD readings is the square root of the sum of the squared relative uncertainties.

The uncertainty in the ratio is determined from error propagation according to equation 7:

$$\left(\frac{u_D}{D}\right)^2 = \left(\frac{u_L}{L}\right)^2 + \left(\frac{u_S}{S}\right)^2 + \left(\frac{u_{F_{lin}}}{F_{lin}}\right)^2 + \left(\frac{u_{F_{dpen}}}{F_{depen}}\right)^2 + \left(\frac{u_{F_{corr}}}{F_{corr}}\right)^2 \tag{9}$$

3. Results and Discussion

The calculations of the uncertainties in the LCR TLD postal dose programme were performed as described in the previous section. The uncertainties associated with the dose determination are listed below:

$\left(\frac{u_L}{L}\right)$ = 0.0019. The uncertainty in the mean TLD reading of a single TLD was estimate as the arithmetic mean of the distribution of standard deviations of the mean for a large number of TLDs that have been irradiated with the same dose.

$\left(\frac{u_S}{S}\right)$ = 1%. This was the sensitivity coefficient maximum value estimated.

$\left(\frac{u_{F_{lin}}}{F_{lin}}\right)$ = 0.012. This was the percentage deviation of the linear fit of the curve to the experimental data.

$\left(\frac{u_{F_{dpen}}}{F_{depen}}\right)$ = 0.0005. This is the maximum value to round up.

$\left(\frac{u_{F_{corr}}}{F_{corr}}\right)$ = 0.011. This was the maximum correction to time that was used in this work.

The combined relative uncertainty in the dose determined from the TLD readings is presented by equation 9.

$$\frac{u_D}{D} = \sqrt{(0{,}0019)^2 + (0{,}01)^2 + (0{,}012)^2 + (0{,}0005)^2 + (0{,}011)^2} \tag{10}$$

IOP Conf. Series: Journal of Physics: Conf. Series **975** (2018) 012049 doi:10.1088/1742-6596/975/1/012049

The relative uncertainty in the determined dose was calculated with equation 9 giving a value of 1,93%.

The major components in the uncertainty in the dose determination are the uncertainties in the sensitivity, linearity and time correction.

4. Conclusion

An analysis of the uncertainties in the dose evaluation process in LCR TLD postal dose programme has been performed. The analysis comprises uncertainties in the reading of the dosimeters, uncertainties in the calibration coefficient, and uncertainties in correction factors for energy dependence and dose response non-linearity.

The largest contributors to the uncertainty have been found to be the sensitivity, the energy correction factor and the dose response non-linearity correction factor.

Statistical uncertainty can be reduced by performing repeated measurements.

The statistical uncertainty is a reproducibility index not only of the dosimeters but also of relevant instrumentation (TLD reader, annealing oven, and the ionization chamber used for the calibration) and measurement set-up. Therefore, instruments should be checked for reproducibility and consistency over longer periods of time in order to ensure that their impact on the statistical uncertainty of TLD measurements is of minor importance.

References

[1] Khan, FM. 2010 The Physics of Radiation Therapy. 4th ed. Lippincott Williams & Wilkins. Minneapolis, Minnesota.

[2] Luxton G, Petrovich Z, Jozsef G et al. 1993 Stereotactic radiosurgery: principles and comparison of treatment methods Neurosurgery, 32: 241-58.

[3] Almeida CE, Affonseca M, Calcina SG, et al. 2005 Rastreabilidade das referências metrológicas em dose absorvida na água do Programa de Qualidade em Dosimetria Radiol. Bras. 38 205-208.

[4] Derreumaux, S, Etard C, Huet C, Trompier F et al. 2008 Lessons from recent accidents in radiation therapy in France Radiat. Prot. Dosim. 131: 130-35.

[5] Bilski P, Waligórski MPR, Budzanowski M, Ochab E, Olko P. 2002 Miniature thermoluminescent detectors for dosimetry in radiotherapy. Radiat Prot Dosim. 101: 473-76.

[6] Horowitiz YS 2001 Theory of thermoluminescence gamma dose response: The unified interaction model. Nuclear Instruments and Methods in Physics Research B 184 68-84.

[7] McKeever SWS, Moscovitch M and Townsend PD 1994 Thermoluminescence materials: Properties and use: Nuclear Technology Publishing.

[8] Guide to expression of uncertainty in measurement 2008.

IOP Conf. Series: Journal of Physics: Conf. Series **975** (2018) 012050 doi:10.1088/1742-6596/975/1/012050

Exposure to ^{137}Cs deposited in soil – A Monte Carlo study

Lucas M. da Silveira[1], Marco A. M. Pereira[1], Lucio P. Neves[2,3,4], Ana P. Perini[2,3,4], Walmir Belinato[5], Linda V. E. Caldas[3], William S. Santos[2]

[1]Instituto de Geografia, Universidade Federal de Uberlândia, Monte Carmelo, MG, Brasil
[2]Instituto de Física, Universidade Federal de Uberlândia (INFIS/UFU), Uberlândia, MG, Brazil.
[3]Instituto de Pesquisas Energéticas e Nucleares, Comissão Nacional de Energia Nuclear (IPEN-CNEN/SP), São Paulo, SP, Brazil.
[4]Programa de Pós-Graduação em Engenharia Biomédica, Faculdade de Engenharia Elétrica, Universidade Federal de Uberlândia, MG, Brazil.
[5]Instituto Federal da Bahia (IFBA), Vitória da Conquista, BA, Brazil.

E-mail: william@ufu.br

Abstract. In the event of an environmental contamination with radioactive materials, one of the most dangerous materials is ^{137}Cs. In order to evaluate the radiation doses involved in an environmental contamination of soil, with ^{137}Cs, we carried out a computational dosimetric study. We determined the radiation conversion coefficients (CC) for effective (E) and equivalent (H_T) doses, using a male and a female anthropomorphic phantoms. These phantoms were coupled with the MCNPX (2.7.0) Monte Carlo simulation software, for three different types of soil. The highest $CC[H_T]$ values were for the gonads and skin (male) and bone marrow and skin (female). We found no difference for the different types of soil.

1. Introduction

As a consequence of the radiological accidents of Chernobyl (Ukraine) in 1986 [1], Goiânia (Brazil) in 1987 [2] and, recently, Fukushima (Japan) in 2011 [3], there is a great concern regarding the radiation doses from these accidents. In such accidents, several radioactive materials may have been released in the environment. Among these radioactive materials, ^{137}Cs is one of the most common heavy fission products, and it may spread for several km^2 [4, 5, 6].

It is hard to evaluate the doses from these accidents, mainly because they may spread on a variety of soils and depths. Experimental measurements may also be difficult to be carried out, since ^{137}Cs is a dangerous material. Therefore, one of the most safest and confident way to evaluate these doses is with the use of Monte Carlo simulations.

Some papers previously evaluated the doses from ^{137}Cs in soil, using mathematical phantoms, to represent the exposed individuals [7]. These phantoms, however, present a simplified anatomical structure. Intending a more detailed simulation, we proposed, in this work, the use of a more realistic scenario.

The main objective of this work was to carry out a detailed dosimetric evaluation of exposures from different soils contaminated with ^{137}Cs, using Monte Carlo simulations [8] coupled with adult anthropomorphic phantoms, named FASH3 and MASH3 [9, 10].

2. Materials and Methods

The dosimetric evaluation was carried out using a ^{137}Cs radioactive source, which emits photons with energy of 662 keV. We considered this source as soil contaminant, from spill of radioactive material from a radiological accident [6].

This source emits photons isotropically, which are interacting with the individuals at the surface. The results were obtained using the radiation transport code MCNPX (version 2.7.0) [8], and the absorbed doses were determined over the organs and tissues of the adult virtual anthropomorphic phantoms MASH3 (male, 1.76 m height and 73 kg) and FASH3 (female, 1.63 m height and 60 kg). A detailed description of these phantoms may be found elsewhere [9, 10].

The geometrical arrangement, used in the MCNPX code, is presented in Figure 1. The soil was represented by a cylinder with a depth of 2 cm and 5 m of radius, and the radioactive material was uniformly distributed.

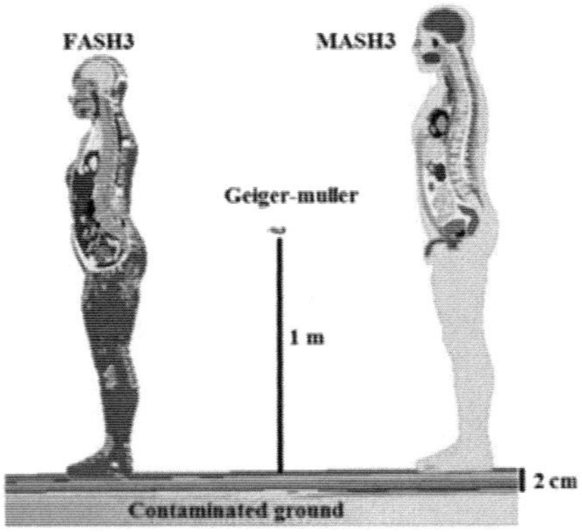

Figure 1. Exposure scenario with the FASH3 and MASH3 phantoms placed on a contaminated soil.

We also investigated the influence of chemical and physical characteristics of the soil, named soils 1, 2 and 3, on the CC values. Their description is presented in Table 1.

The absorbed doses were estimated separately for each source and soil, and calculated for several organs and tissues. The weighting factors w_{T} for each organ and tissue, recommended by ICRP 103 [11] were applied to the equivalent dose (H_{T}) to each organ and tissue. These values were normalized by the air kerma, measured with a Geiger-Müller Pancake Probe (GM), computationally modeled (Figure 2), and positioned 1 m above the ground. A detailed description of the GM may be found at the manufacturer website [12]. The Monte Carlo simulations were carried out with 1E8 particle histories.

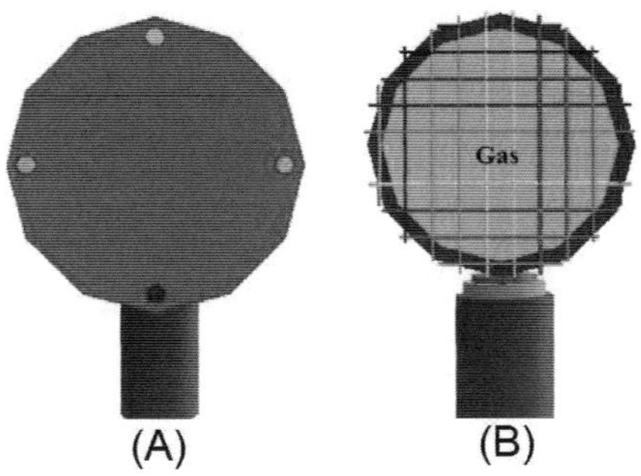

Figure 2. Geometry of the GM counter - external (A) and internal (B) frontal views.

As a result, we obtained the conversion coefficients (CC) for equivalent dose ($CC[H_{\mathrm{T}}]$). With these values, we determined the CC for effective dose ($CC[E]$), as stated in Equation 1.

Table 1. Soil composition used in the Monte Carlo simulations.

Soil	Density (g/cm^3)	Composition (%)	
1	2.00	Si (21.61)	Na (0.6)
		Al (8.2)	K (2.33)
		Fe (4.5)	P (0.02)
		Mn (0.1)	H (3.04)
		C (2.4)	O (51.5)
		Ca (5.1)	
2	1.25	H (16.87)	Al (1.976)
		O (27)	Si (8.963)
3	1.60	H (2.1)	Si (27.1)
		C (16)	K (1.3)
		O (57.7)	Ca (4.1)
		Al (5.0)	Fe (1.1)

$$CC[E] = \sum_T w_\mathrm{T} \left[\frac{CC[H_\mathrm{T}]_\mathrm{Male} + CC[H_\mathrm{T}]_\mathrm{Female}}{2} \right] \qquad (1)$$

were w_T is the weigting factor for each organ and tissue, $CC[H_\mathrm{T}]_\mathrm{Male}$ is the conversion coefficient for equivalent dose for the male phantom, and $CC[H_\mathrm{T}]_\mathrm{Female}$ for the female phantom.

3. Results and Discussion

The $CC[H_\mathrm{T}]$ and $CC[E]$, measured 1 m above the ground, are presented in Tables 2, 3 and 4, respectively. All uncertainties are below 1%.

Table 2. $CC[H_T]$ measured 1 m above the ground (Sv/Gy) for the MASH3 virtual anthropomorphic phantom for the different soils. Remainder tissues are: Adrenals, extrathoracic region, gall bladder, heart, kidneys, lymph nodes, muscles, oral cavity, pancreas, uterus, small intestine and spleen.

Organs	Soil 1	Soil 2	Soil 3
Bone marrow	0.63	0.64	0.62
Colon wall	0.61	0.61	0.60
Lung	0.57	0.58	0.57
Stomach wall	0.55	0.55	0.55
Breast	0.75	0.76	0.76
Reminder tissues	0.05	0.05	0.05
Gonads	0.84	0.86	0.82
Bladder wall	0.66	0.67	0.64
Oesophagus	0.48	0.49	0.48
Liver	0.57	0.57	0.56
Thyroid	0.44	0.45	0.43
Bone surface	0.15	0.15	0.15
Brain	0.55	0.55	0.55
Salivary glands	0.21	0.21	0.21
Skin	0.85	0.85	0.84
Eyes	0.61	0.61	0.60

Table 3. $CC[H_T]$ measured 1 m above the ground (Sv/Gy) for the FASH3 virtual anthropomorphic phantom for different soils. Remainder tissues are: Adrenals, extrathoracic region, gall bladder, heart, kidneys, lymph nodes, muscles, oral cavity, pancreas, uterus, small intestine and spleen.

Organs	Soil 1	Soil 2	Soil 3
Bone marrow	0.86	0.87	0.85
Colon wall	0.60	0.61	0.60
Lung	0.64	0.64	0.64
Stomach wall	0.56	0.56	0.56
Breast	0.77	0.77	0.76
Reminder tissues	0.05	0.05	0.47
Gonads	0.64	0.65	0.63
Bladder wall	0.69	0.71	0.68
Oesophagus	0.57	0.56	0.57
Liver	0.57	0.58	0.57
Thyroid	0.55	0.55	0.55
Bone surface	0.13	0.13	0.12
Brain	0.59	0.59	0.59
Salivary glands	0.23	0.23	0.23
Skin	0.86	0.87	0.85
Eyes	0.67	0.65	0.65

Table 4. $CC[E]$ measured 1 m above the ground (Sv/Gy) for the MASH3 e FASH3 virtual anthropomorphic phantoms.

Soil	$CC[E]$
1	0.57
2	0.57
3	0.56

According to the data listed in Tables 2 and 3, we may note that the $CC[H_T]$ were higher for the gonads and skin, in the male phantom, while the skin and bone marrow presented the highest values for the female phantom. These structures are all located near the ground, were the radioactive sources were inserted.

Considering the different soil constitutions, the $CC[E]$ values presented a small variation, as may be seen at Table 4, which indicates no different shielding effects.

4. Conclusions

In this work we evaluated the doses involved in an environmental contamination of soil, with ^{137}Cs. The results showed that differences in soil composition did not change significantly the CC values. Besides, the highest $CC[H_T]$ values were for the gonads and skin (male phantom) and bone marrow and skin (female phantom). Therefore, the $CC[H_T]$ established in this study will be useful to readily estimate the equivalent dose, during an environmental contamination of soil with ^{137}Cs.

Acknowledgments

This work was also presented at the METROLOGIA 2017 conference. The authors would like to thank Dr. Richard Kramer for kindly providing the virtual anthropomorphic phantoms. This work was partially supported by the Brazilian agencies: Fundação de Amparo à Pesquisa do Estado de Minas Gerais (FAPEMIG Grants Nos. APQ-03049-15 and APQ-02934-15) and CNPq (Grants Nos. 301335/2016-8, 157593/2015-0, 421603/2016-0, 420699/2016-3 and 168947/2017-0).

References

[1] INTERNATIONAL ATOMIC ENERGY AGENCY 1992 The Chernobyl accident: Updating of INSAG-1 – INSAG-7 Tech. rep. IAEA
[2] INTERNATIONAL ATOMIC ENERGY AGENCY 1988 The radiological accident in Goiânia Tech. rep. IAEA
[3] INTERNATIONAL ATOMIC ENERGY AGENCY 2015 The Fukushima Daiichi Accident – Technical volume 1 Description and context of the accident Tech. rep. IAEA
[4] Krstić D, Nikezić D R and Matović M D 2005 *Medicus* **6** 30–3
[5] Krstić D and Nikezić D 2009 *Health Physics* **91** 249–57
[6] Han B, Zhang J, Na Y H, Caracappa P F and Xu X G 2010 *Radiation Protection Dosimetry* **141** 299304
[7] Timms D N, Smith J T, Cross M A, Kudelsky A V, Horton G and Mortlock R 2004 *Journal of Environmental Radioactivity* **72** 323–34
[8] Pelowitz D B 2011 *MCNPX User's Manual, Version 2.7.0* (Report LA-CP-11-00438. Los Alamos National Laboratory)
[9] Cassola V F, Lima V J D, Kramer R and Khoury H J 2010 *Physics in Medicine and Biology* **55** 133–62
[10] Cassola V F, Lima V J D, Kramer R and Khoury H J 2010 *Physics in Medicine and Biology* **55** 163–89
[11] ICRP Publication 103 2007 *The 2007 Recommendations of the International Commission on Radiological Protection* (Ann. 37 (24))
[12] LND INC 2017 Cross Reference Chart to Other Mfg. Products, 7311 URL `http://www.lndinc.com/products/geiger-mueller-tubes/7311/`

IOP Conf. Series: Journal of Physics: Conf. Series **975** (2018) 012051 doi:10.1088/1742-6596/975/1/012051

Geant4 simulation of ion chambers response to ^{60}Co spectrum of LNMRI/IRD Shepherd 81-14D Radiator

P P Queiroz Filho and C N M Da Silva

Instituto de Radioproteção e Dosimetria, IRD-CNEN, Rio de Janeiro, RJ, Brasil

queiroz@ird.gov.br, cosme@ird.gov.br

Abstract. The National Ionizing Radiation Metrology Laboratory of the Radioprotection and Dosimetry Institute (LNMRI / IRD) has recently acquired a Shepherd 81-14D Radiator. In this work we simulate, using Geant4, the behavior with the inverse square law radiation for 3 models of PTW spherical chambers used in radioprotection, a relevant information to planning the measurements. We did the corrections for the attenuation and scattering in the air for each distance, where we used the ^{60}Co spectrum simulated previously.

1. Introduction

Recently the LNMRI / IRD acquired a Shepherd 81-14D irradiator with ^{137}Cs sources with activity of 54 Ci and ^{60}Co with activity of 5 Ci, for the facilities of the Radioprotection Laboratory. The laboratory has been carrying out measurements and characterized the beams of ^{137}Cs and ^{60}Co in a previous work [1]. The spectra of the ^{137}Cs and ^{60}Co were simulated in another work [2] and used in our simulations.

It is necessary to carry out Monte Carlo simulations to know in more detail the behavior of the measurement system [3-5]. Therefore, we simulated the complete irradiator as well as three PTW spherical chambers with distinct sensitive volumes: 27.9 cm^3, 1 liter and 10 liters. In this work we simulate the behavior with the inverse square law radiation to the beam of the ^{60}Co spectrum generated after crossing the structure of the irradiator. The attenuation and scattering effects were obtained in the air to correct the results. The simulations, using the Monte Carlo method, were performed with the Geant4 Program version 10.02 [6].

2. Metodology

The source of ^{60}Co J. L. Shepherd & Associates Model No. 7810-150 is in the form of solid metal pellets occupying a cylindrical volume with double encapsulation of 304 stainless steel.

We extracted from the previous paper [2] the simulated spectrum of the ^{60}Co source after the exit of the collimation system. It will be used in the simulations and can be seen in figure 1.

IOP Conf. Series: Journal of Physics: Conf. Series **975** (2018) 012051 doi:10.1088/1742-6596/975/1/012051

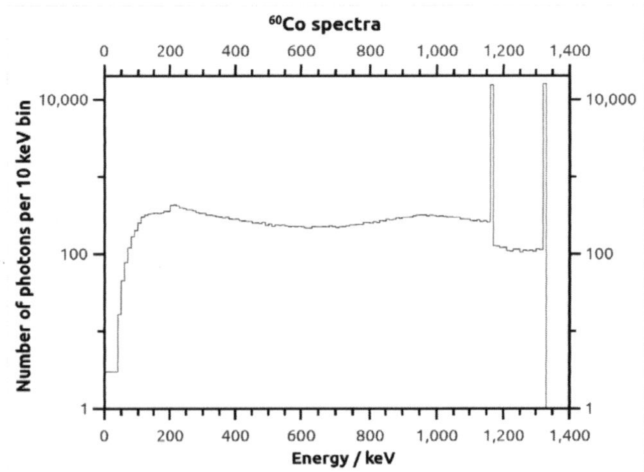

Figure 1. Distribution of the simulated spectrum of the ^{60}Co source.

The LNMRI / IRD ionization chambers are the PTW type 32005, 32002 and 32003 models shown in figures 2, 3 and 4 [7].

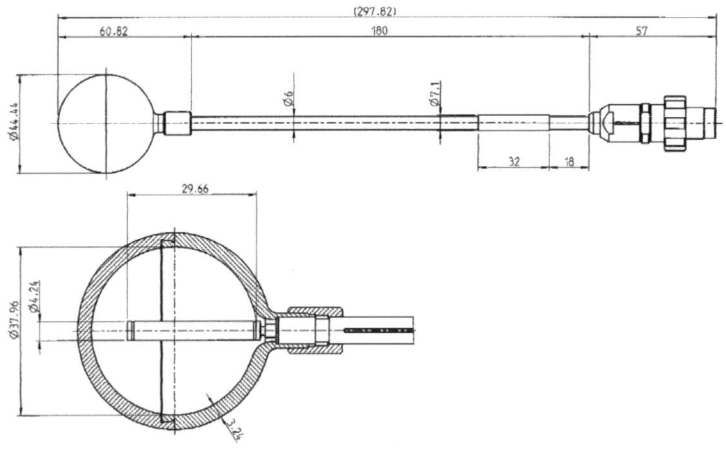

Figure 2. Spherical Ionization Chamber TK-30 32005 of PTW.

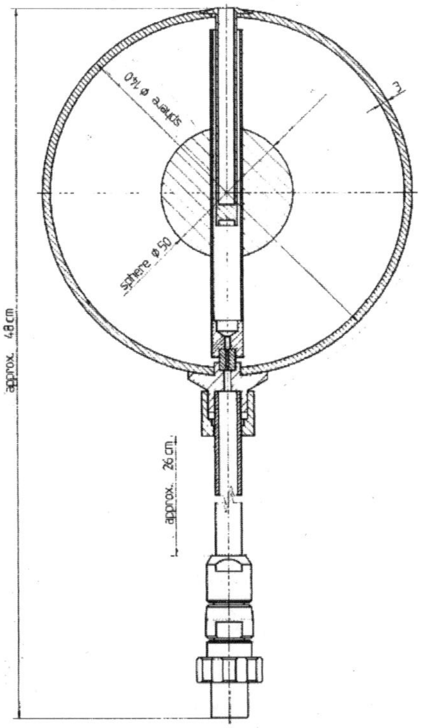

Figure 3. 1 Liter Spherical Ionization Chamber 32002 of PTW.

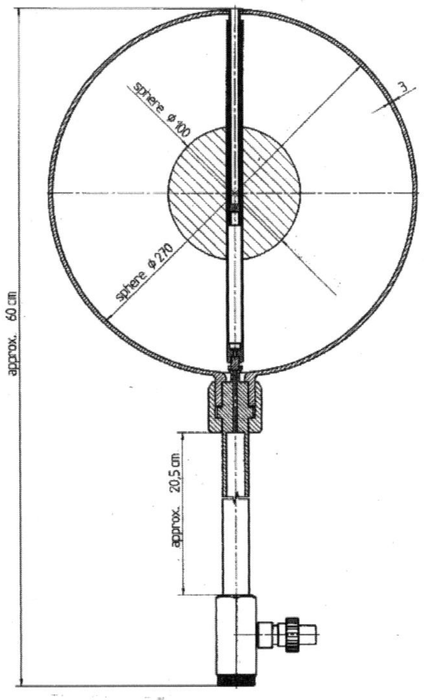

Figure 4. 10 Liter Spherical Ionization Chamber 32003 of PTW.

The technical information of the simulated cameras was taken from the Ionizing Radiation Detectors catalog of PTW [7]. The chambers are spherical with a long metal rod, suitable for use in radioprotection measurements. The spherical wall surrounding the sensitive volume of the 3 chambers has a thickness of 3 mm of graphite polyoxymethylene. The chamber of 27.9 cm^3 has a sensitive volume with a radius of 22 mm and a cylindrical central electrode attached to the rod made of graphite coated with PMMA with 4.2 mm in diameter. The 1 and 10 liter chambers have a spherical central electrode, made of PMMA coated graphite, with diameters of 50 mm and 100 mm, respectively, and a sensitive volume consisting of a spherical shell with a thickness between the diameter of the electrodes and the diameters of the 140 mm and 276 mm cameras, respectively.

The three camera models were simulated with their centers distant 1 m, 2 m, 3 m and 4 m from the center of the cobalt source.

Ideally, the calculated doses should decrease proportionally to the inverse square law radiation. However, we know that some effects contribute to change this result. In this work we eliminate, through simulation, the effects of the interaction of the ^{60}Co photon spectrum with the air layer between the irradiator and the chambers. Although we know that this spectrum interacts very little with air and that the attenuation and scattering factors compete with each other to further diminish its effect, it should not be neglected. For this we made a correction for the attenuation and scattering in the air. In addition, the larger the diameter of the chambers, the greater the effect of their geometry. 10 billion events were simulated for the smaller camera and 1 billion events for the 1liter and 10 liter cameras at the four distances.

3. Results

The factors obtained for the four distances and three ionization chamber geometries are shown in tables 1, 2 and 3.

Table 1. Attenuation and scattering effects for 27.9 cm^3 chamber.

Distance (m)	factor	Uncertainty (%)
1	1.004	0.18
2	1.019	0.54
3	1.014	0.92
4	1.01	0.98

Table 2. Attenuation and scattering effects for 1 liter chamber.

Distance (m)	factor	Uncertainty (%)
1	1.002	0.15
2	1.007	0.32
3	1.013	0.60
4	1.014	0.64

Table 3. Attenuation and scattering effects for 10 liter chamber.

Distance (m)	factor	Uncertainty (%)
1	1.005	0.11
2	1.009	0.16
3	1.010	0.29
4	1.024	0.27

We can observe that the factors obtained are very small since they combine the effects of attenuation and scattering. The factors of the chamber of 27.9 cm^3 are larger than those of the 1 liter chamber which undergoes a greater compensation of the scattering due to its larger volume. The same behavior can be seen for the 10 liter chamber relative to the 1liter chamber. Table 4 shows the values of the ratio between the doses in the chambers with the distance, corrected by the attenuation factor and the scattering factor.

Table 4. Comparison between simulation and the inverse square law radiation.

Distance ratio d_i/d_f	Dose ratio D_i/D_f	Uncertainty (%)	Percent deviation	Volume (cm^3)
1/2	3,99	0,44	0,27	27,9
1/3	9,02	0,85	0,24	27,9
1/4	16	0,81	0,23	27,9
2/4	4,02	0,90	0,50	27,9
1/2	4,03	0,25	0,76	1000
1/3	9,05	0,34	0,56	1000
1/4	16,1	0,47	0,63	1000

2/4	4,00	0,51	0,12	1000
1/2	3,635	0,12	9,12	10000
1/3	8,21	0,23	8,80	10000
1/4	14,59	0,22	8,79	10000
2/4	4,02	0,23	0,36	10000

We can observe that the inverse square law radiation was respected. As expected, the percentage deviation between the simulated result and the inverse square law radiation increased with the volume of the chambers. This is because the effect of the geometry of the chambers increases with the chamber diameter. Consequently, we also observe that as distances increase this effect decreases. This is clearly observed by comparing the values of the percentage deviation for each of the chambers, and we can particularly highlight the deviation of the ratio between doses at 1 m and 2 m for each chamber and between doses at 2 m and 4 m for the 10 liter chamber.

4. Conclusions

We simulated irradiation of the 3 PTW chamber models used in radioprotection with the simulated spectrum for the LNMRI / IRD Shepherd 81-14D Irradiator with the ^{60}Co source. The chambers were positioned at 4 distances where the doses were corrected for the air attenuation and scattering effects. We show the behavior of each camera in relation to the inverse square law radiation.

We observe, through the percentage deviation, that our results respect the inverse square law radiation. The percentage deviation to the mentioned law, as expected, increases with the increase of the dimensions of the chambers, as well as with the greater proximity between camera and source. The next step will be the calculation of wall chamber correction factors.

References

[1] Da Silva C N M and Cabral T S *2015 Proc. 8º Congresso Brasileiro de Metrologia*
[2] Queiroz Filho P P, Da Silva C N M and Cabral T S 2016 *Proc. 9º Congresso Brasileiro de Metrologia*
[3] Mora G M, Maio A and Rogers D W O 1999 *Med. Phys.* **26** 2494
Burns D T 2003 *CCRI(I)/03-40.*
[4] *ISO-4037-1* 1996- X and gama reference radiations for calibration dosemeters and doserate meter and for determining their response as a function of photon energy- Part1: Radiation characteristics and production methods.
[5] *ISO-4037-2* 1997 X and gama reference radiations for calibration dosemeters and doserate meter and for determining their response as a function of photon energy- Part 2: Dosimetry for radiation protection over the energy ranges 8 keV to 1.3 MeV and 4 MeV to 9 MeV.
[6] Geant4 - Available in < http://geant4.cern.ch/ > Access in 07/24/2017.
[7] Ionizing radiation detectors - Including codes of practice - Available in < www.ptw.de/fileadmin/data/download/catalogviewer/DETECTORS.../complete.pdf> Access in 07/24/2017.

IOP Conf. Series: Journal of Physics: Conf. Series **975** (2018) 012052 doi:10.1088/1742-6596/975/1/012052

Linearity and reproducibility response of Fricke dosimetry for low energy X-Ray beam

A Mantuano [1], G J de Amorim [1], M G David [1], P H G Rosado [1,4], C Salata [2], L A G Magalhães [1] and C E deAlmeida[3]

[1] Radiological Sciences Laboratory, State University of Rio de Janeiro, Rio de Janeiro, Brazil
[2] National Nuclear Energy Commission (CNEN)
[3] Fundação do Câncer, Rio de Janeiro, Brazil
[4] Institute of Radiation Protection and Dosimetry (IRD/CNEN), Rio de Janeiro, Brazil

E-mail: mantuanoandrea@gmail.com

Abstract. The Fricke dosimeter is the most used, liquid chemical dosimeter. It has been shown to be a feasible option for the absorbed dose standard. The present work aims to determinate a dose-response curve of Fricke solution using different doses and reproducibility test comparing the calculated dose to Fricke solution and Ionizing Chamber. Tests were performed using an X-ray irradiator for biological research at Radiological Science Laboratory (LCR/UERJ). The results showed a linear response to different doses of type A uncertainties from 0.08 to 1.2%. Reproducibility test showed type A uncertainties of 0.16% to the dosimeter.

1. Introduction

Chemical dosimetry using a standard FeSO4 solution has been shown potential to be a reliable standard of absorbed dose.

However, the linearity for this dosimeter and reproducibility tests are currently not available. Even though the measurements of Fricke dosimetry show a good linearity, its methodology requires a lot of attention and care. Organic or inorganic impurities into the dosimeter can change the response of the solution.

In order to obtain lower uncertainties in the dose measurement, it is necessary to eliminate all the possibilities of influence on the solution that are not related to the ionizing radiation. Repeatability and reproducibility tests can be used to find and discuss those influences and to provide greater security and accuracy for the results.

Controlled temperature, pressure and humidity are crucial to perform the tests. Few authors have shown a dose-response curve using different qualities comparing absorbance and dose or Fricke dose and the dose calculated through other methods [1-3].

This work aims to evaluate the Fricke solution stability and linearity through the dose-response curve and reproducibility. The obtained results will be compared with the ionizing chamber response, the most used dosimeter, according to the literature. We expected to develop and implement a standard for the absorbed dose in water based on the Fricke chemical dosimetry; and also the use this methodology for different energy dosimetry used for research.

2. Materials and methods

For the process of irradiation, the Fricke solution was disposed into a bag and it was put individually in an acrylic holder for their total irradiation. The irradiation was performed using an X-Ray irradiator for biological research, completely self-shielded (RS 2000X, Radsource, CA, USA), at the Radiological Science Laboratory (LCR/UERJ), Rio de Janeiro, Brazil. The figure 1 shows the acrylic holder to bag (holder 1), ionizing chamber (holder 2) and another holder that was not used in this work (holder 3). Figure 2 shows the complete setup into the X-Ray irradiator.

In this paper the range to dose to water used was according with the literature from 14 to 40 Gy [2,4,5].

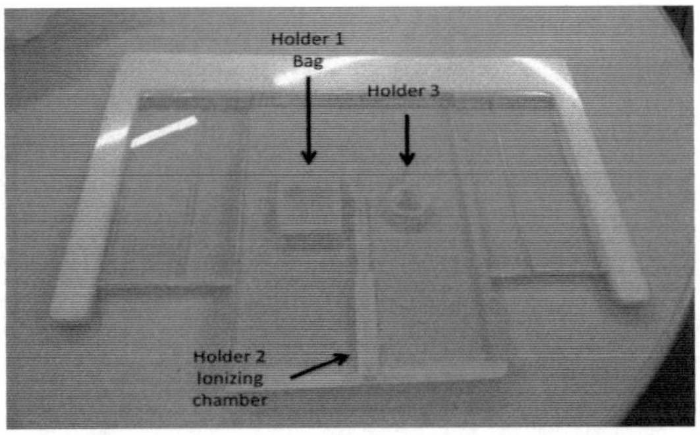

Figure 1. Acrylic experimental design.

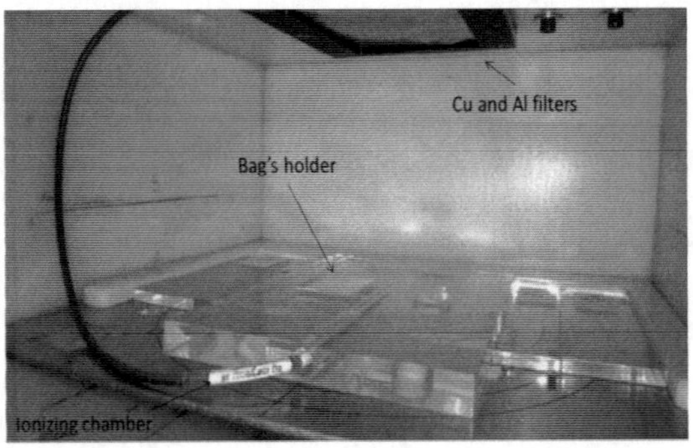

Figure 2. Irradiation Setup for Bags containing Fricke solution and ionizing chamber.

Cu (0,32 mm) and Al (1,75 mm) filters were used taking into account a half value layer HVL of 0,66 mmCu. The HVL was calculated using a X-Ray source (IRD-Brazil) considering an effective energy of 68 KeV.

2.1. Linearity study

The electron accelerating voltage used on the X-ray tube for the linearity study was 150 kV and different doses were tested.

The dose-response curves were obtained changing the current and maintaining the time of irradiation, and after, maintaining the current and changing the time, ranging from 7 to 46 Gy for both tests.

During the irradiation, another bag containing Fricke solution was kept at the same room of the irradiator to be used as a control solution. Control and irradiated solutions had their absorbance read using a Varian cary 50 Bio spectrophotometer.

The absorbed dose to Fricke solution (DF) was calculated through the equation (1) [6-7].

$$D = \frac{\Delta OD}{G(Fe^{3+}) \cdot l \cdot \rho \varepsilon} \qquad (1)$$

Where, ΔOD is defined as the optical density (OD) at 304 nm, taking into account the temperature effect, G(Fe3+) is the radiation chemical yield of the ferric ions for 150 kV, previously calculated by our group (1.447·10-6 mol·J-1) using the parameters described in the methodology, l is the optical path length of the cuvette (1 cm), ρ is the density of the Fricke solution (1.024 g·cm-3) at 25°C and ε is the value for the molar linear absorption coefficient for ferric ions (equal to 2174 M-1 cm-1 at 304 nm, which is numerically equal to the value that was used previously [4,7].

The Absorbed dose to water D_w measured with ionizing chamber NE2571 was calculated using the equation (2) according to TRS 277 [9].

$$D_W = M_u \cdot N_{D,W} \qquad (2)$$

M_u is the charge collected by ionizing chamber after the correction to pressure and temperature and $N_{D,w}$ is the calibration factor in terms of absorbed dose to water [8,9].

3. Results

3.1. Linearity

The figure 3 shows the linearity response of the chemical dosimeter when the current was increased. The figure 4 shows the same response when the time of exposure was increased. Linear regression showed a R-squared of 0.999 for both cases.

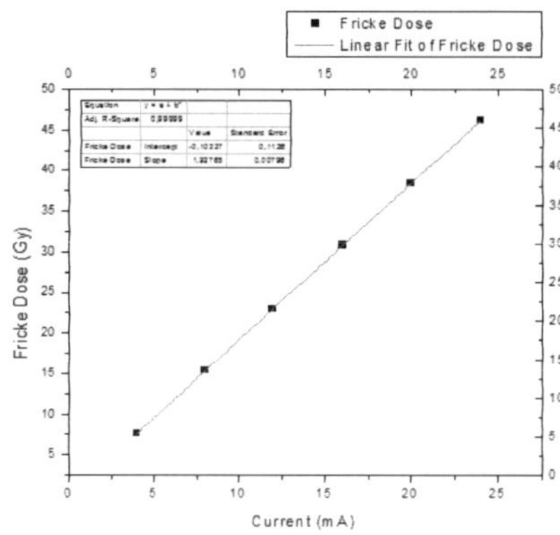

Figure 3. Dose response curve changing the current.

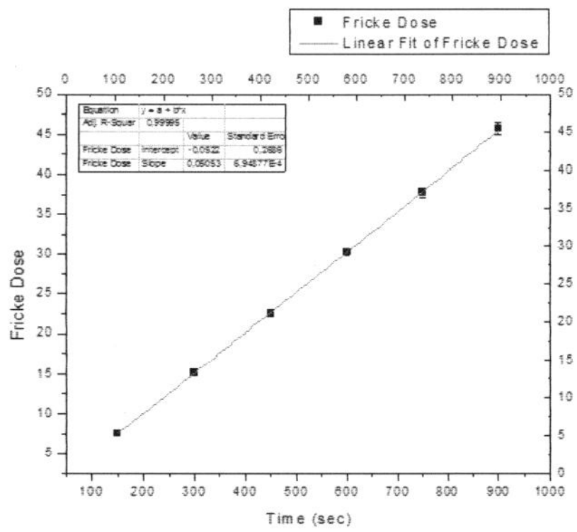

Figure 4. Dose response curve changing the time.

The range of the doses corresponding to different times and/or current was from 7 to 40 Gy. The doses to Fricke solution were calculated using the equation (1).

The Ionizing chamber was used as a monitor dosimeter and the dose to water was calculated by equation (2).

It is important to note that to the repeatability measures, the type A uncertainties were 1.2 % at most, to doses under 15 Gy. On the other hand, the doses from 15 to 46 Gy, showed type A uncertainties from 0.08 to 0.3%.

3.2. Reproducibility

The figure 5 shows a stability of the measurements during the days of repetition.

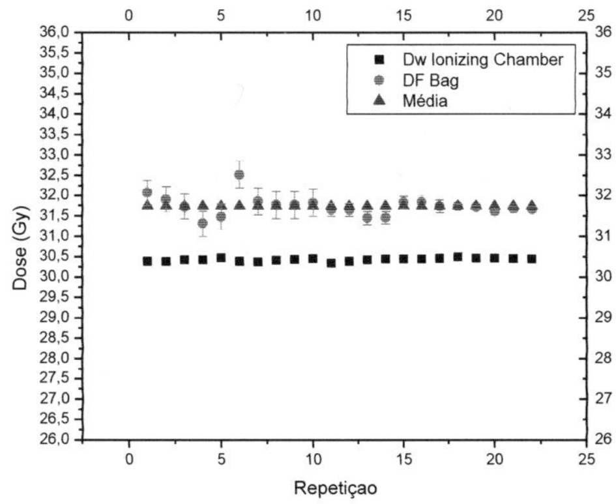

Figure 5. Graphic of stability during different days of measurements.

The stability measurements showed that on the first 5 repetitions the standard deviation was higher than the others. It could have happened due to the position of the ionizing chamber or other geometry factors that could have influenced the experiment, and this was corrected at the 6th measurement.

The dispersion of values in relation to the mean to dose of Fricke solution was 0.7% disregarding the first five measures. The results showed type A uncertainties to dose to Fricke solution of 0.16%.

The absorbed dose to water calculated through dose of Fricke solution can be obtained calculating some Monte Carlo correction factors. It was not considered in this paper, but open opportunities for the future.

4. Conclusions
The absorbed doses derived from Fricke Solution measurements show that the chemical dosimeter response is linear with R-squared of 0.999. Repeatability showed uncertainties lower than 0.3%.

This study can open new opportunities to use the Fricke dosimetry as an alternative method to be a feasible option for the absorbed dose to water.

Acknowledgements
The authors acknowledge the financial support by Brazilian Government organizations CAPES and the organization within the Research Support Foundation of Rio de Janeiro State FAPERJ.

References
[1] Pimpinella M, Guerra A S, La Civita S, and Laitano R 2007 *Absorbed Dose and Air Kerma Primary Standards* (Paris)

[2] deAlmeida C E, Ochoa R, deLima M C, David M G, Pires E J, Peixoto J G, Salata C, Bernal M A 2014 A feasibility study of Fricke dosimetry as an absorbed dose to water standard for 192Ir HDR sources *PlosOne* p 1-13

[3] Moussous O, Khoudri S, and Benguerba M. 2011 Characterization of a Fricke dosimeter at high energy photon and electron beams used in radiotherapy *Australas Phys Eng* 34 523-28.

[4] Franco L, Gavazzi S, Coelho M, and De Almeida C E 2011 *In Standards, Applications and Quality Assurance in Medical Radiation Dosimetry (IDOS). Proceedings of an International Symposium*

[5] Meesungnoen J, Benrahmoune M, Filali-Mouhim A, Mankhetkorn S and Jay-Gerin, J P 2001 Monte Carlo Calculation of the Primary Radical and Molecular Yields of Liquid Water Radiolysis in the Linear Energy Transfer Range 0.3–6.5 keV/μm: Application to 137Cs Gamma Rays *Radiat Res.* **155** 269-78.

[6] ICRU International Commission on Radiation Units and Measurements - *Report 35 Radiation Dosimetry: Electron Beams with energies between 1 and 50 MeV* 1984

[7] Klassen N V, Shortt K R, Seuntjens J, and Ross C K 1999 Fricke dosimetry: the difference between G($Fe3^+$) for 60Co -rays and high-energy x-rays *Phys Med Biol* **44** 1609

[8] International Atomic Energy Agency. Absorbed dose determination in external beam radiotherapy: an international code of practice for dosimetry based on standards of absorbed dose to water. *Technical Reports Series no. 398. Vienna: IAEA* 2006

[9] International Atomic Energy Agency. Absorbed dose determination in photon and electron beams: an international code of practice. *Technical Reports Series no. 277, 2nd ed. Vienna: IAEA* 1987

MicroCT parameters for multimaterial elements assessment

Olga M. O. de Araújo[1], Jaqueline Silva Bastos[2], Alessandra S. Machado [1], Thaís M. P. dos Santos[1], Cintia G. Ferreira[1], Ana Paula Rosifini Alves Claro [2] and Ricardo T. Lopes [1]

[1]Nuclear Engineering Program, UFRJ - Federal University of Rio de Janeiro, Rio de Janeiro, RJ, Brazil
[2]Paulista State University, UNESP, São Paulo, Brazil
E-mail: olgaufrjlin@gmail.com

Abstract. Microtomography is a non-destructive testing technique for quantitative and qualitative analysis. The investigation of multimaterial elements with great difference of density can result in artifacts that degrade image quality depending on combination of additional filter. The aim of this study is the selection of parameters most appropriate for analysis of bone tissue with metallic implant. The results show the simulation with MCNPX code for the distribution of energy without additional filter, with use of aluminum, copper and brass filters and their respective reconstructed images showing the importance of the choice of these parameters in image acquisition process on computed microtomography.

Keywords: Microtomography, additional filters, multimaterial elements.

1. Introduction

In non-destructive tests involving the use of X-rays, the choice of voltage, current and additional filters are determining factors in the image quality. The basic principle of image formation is the interaction of ionizing radiation with the sample. Image quality depends on several factors, such as contour sharpness, noise level and contrast between structures. Another factor that contributes to image quality is presence or absence of artifacts, taking into account their intensity and their influence in detection of relevant structures. The analysis of samples of materials with large difference of density, such as bone tissue with metallic material, can be a difficult task and provide an incorrect analysis if inappropriate parameters are selected.

The knowledge of the energy distribution of the photons produced by X-ray tube is an important information for different fields of research involving the use of ionizing radiation. In this work, to demonstrate the importance of the selection of these parameters, a spectra of X-ray tube with the code MCNPX were simulated and qualitative and quantitative parameters such as effective energy, half value layer and homogeneity coefficient were evaluated [1, 2]. The results show the acquisition of the images obtained by the microCT with the different combinations where it is possible to verify the final results on the reconstructed samples and the result when choosing a combination of additional filters that may generate unwanted artifacts in image such as metal scattering, a result hardly corrected by microCT reconstruction softwares.

2. Material and methods

2.1. MCNPX

To verify the distribution of the X-ray spectrum, was performed a simulation with MCNPX tool based on the Monte Carlo N-Particle Transport Code (MCNP), originally developed at the Los Alamos National Laboratory (LANL). The Monte Carlo method provides approximate solutions for a variety of physical and mathematical problems performing statistical samplings. Individual particles are simulated and the result of their histories are used to determine the average behavior of the particles carried [1]. Particle histories is determined by probability distributions of physical events such as interactions, scattering, capture, and so on.

In the proposed study, fundamental elements and processes in the operation of an X-ray tube were considered. The electron source was positioned in cathode towards anode where the electrons are accelerated to interact with a tungsten target. The selected voltage was 60kV compatible with this material analysis and the simulation was performed to estimate a spectrum without filters and for different types of additional filters. Some parameters are necessary to describe the beam of radiation and its power of penetration, to characterize the X-ray is necessary to determine quantitative and qualitative parameters which were estimated from energy distribution calculated with MCNPX code.

2.2. Average energy

Information about energy of the X-ray beam provides parameters on beam quality, such as semi-reducing layer and average energy [3, 4, 5]. The average energy of X-rays spectrum was obtained using the equation below:

$$E_{average} = \frac{\sum_{n=1}^{n}\left[\frac{d\Phi(E_n)}{dE}\right]E_n(\Delta E_n)}{\Phi} \tag{1}$$

Where $d\Phi(E_n)$ is the number of photons of energy spectrum distribution in energy interval ΔE, whose energy is E_n and Φ is the photon fluence [5, 6].

2.3. Half Value Layer (HVL) and Homogeneity Coefficient

HVL is defined as the thickness of a given absorber material required to reduce the intensity of the X-ray beam to 50% of the initial value under good geometry conditions [7]. This implies a configuration that minimizes the spreading that tends to increase its value. Is a parameter related to penetration power of X-ray beam. As more energetic the beam, greater the thickness required to reduce its intensity by half.

In some cases is possible to obtain the same HVL value for different configurations of additional filters and voltage applied to an X-ray tube, although the energy spectra in each case are not similar. This difference can be characterized by the homogeneity coefficient (CH) and can be defined as the ratio of first to second HVL[8]:

$$CH = \frac{1^{a}\text{HVL}}{2^{a}HVL} \tag{2}$$

Where, $CH = 1$ for monoenergetic beams and $CH < 1$ for heterogeneous beams.

The lower the CH, smaller the homogeneity of the beam. As more low energy photons are attenuated, more 'hardened' or homogeneous the beam becomes and consequently an higher CH.

The microCT system used in this work is Skyscan model 1173 (*Bruker*, Belgium). The equipment has X-ray source operating between 40kV and 130kV voltages and detection system a Flat-Panel sensor, operating with a matrix of 2240x2240 (Hamamatsu Photonics, 2012a). Each pixel in image corresponds to average absorption, in reconstruction process, where slices are assembled to produce 3D reconstructed image pixel is mathematically transformed into voxel, capable of representing the depth of the radiological image. The main parameters in image acquisition in this study were: 15μm pixel size, 60kV voltage, current of 133μA, detector array 2240x2240, rotation step 0.5° (degrees).

3. Results

The spectra obtained from simulation without additional filter and filters of Aluminum, Copper and Brass with different thicknesses and tension of 60kV are shown in figure 1, and the qualitative and quantitative parameters in table 1.

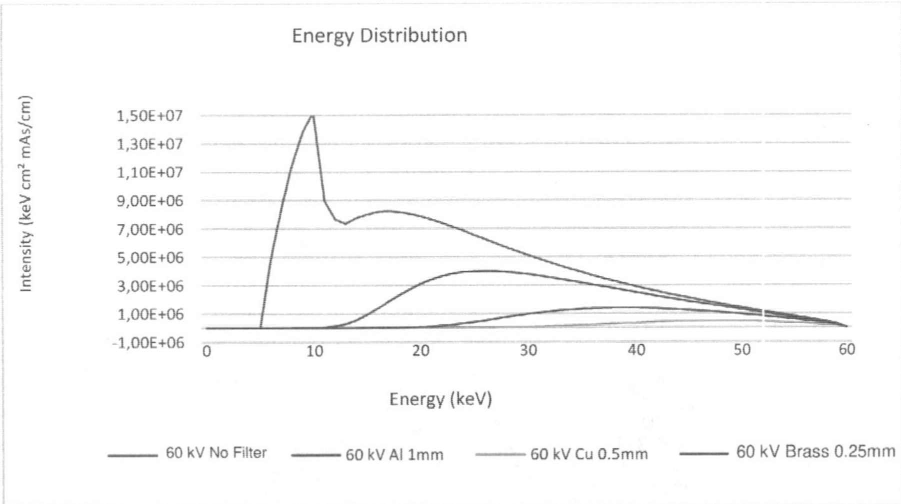

Figure 1. Spectrum of energy distribution for different additional filters and without the use of filters.

Table 1. Results of qualitative and quantitative parameters for different configurations.

60 kV Filter	No Filter	Al 1mm	Cu 0.5mm	Brass 0.25mm
Average Energy (keV)	22.536	32.322	46.236	40.436
HVL (cm)	0.008	0.118	0.020	0.0149
CH	0.471	0.624	0.858	0.795

Can be seen in graphs without additional filters a low energy photon peak which agrees with an X-ray beam without additional filtering of these low-energy beams [9, 10]. The 1 mm aluminium filter has a higher efficiency for lower voltages, filtering the low energy photons that do not contribute to formation of radiographic image, also not reducing considerably the intensity of beam, such as observed for copper and aluminium.

In spectrum for copper filter a shift of the maximum curve point to right is observed, which corresponds to an average upper energy. The additional filters increase beam quality by increasing its penetration power and average energy, as can be seen in table 1. On the other hand it reduces beam intensity and spectral counts for copper and brass.

Quantitative and qualitative parameters for copper filter agree with the results obtained in Figure 1, presenting the highest increase in average energy. Brass and copper filters presented the highest result for the homogeneity coefficient compared to other combinations used in the simulation. The higher density of copper and brass related to aluminium can be considered a determining factor for this behaviour.

In figure 2 the comparison of images was possible visually verify the difference in effects of additional filtration: a) without additional filters presents a sample without definition at the edges, due to low energy photons that cause image noise; b) aluminium filter proved to be more efficient for artifacts that impair the quality of acquisitions, such as metal artifacts, because it absorbs the low energy photons without losing the intensity of the x-ray spectrum; c) and d) with the copper and brass filters is lost definition with a grainy image and a lot of metal spreads, due to loss of intensity that we can observe in figure 1 and the effect of scattering due do increase on average energy, estimated in table 1.

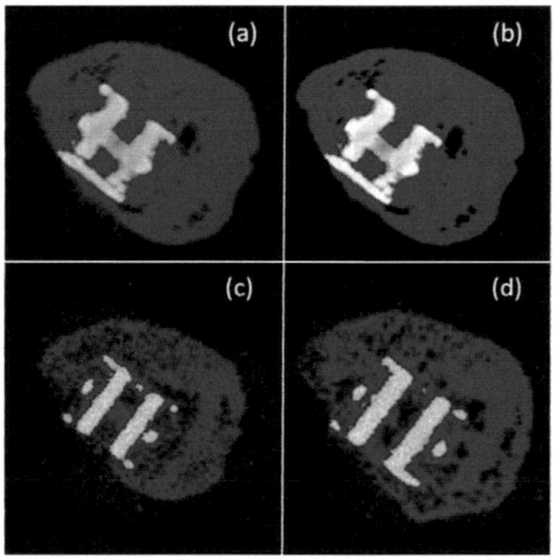

Figure 2. 2D segmented slices for configurations (a) without additional filter (b) aluminium filter (c) copper filter and (d) brass filter.

MicroCT is a non-destructive test technique that allows inspect materials with a definition of micrometers. In this example, studies involving analysis for evaluate bone growth by diversity of information that can be obtained. Images of figure 3 were obtained with Avizo Fire 9.01 software for projections with copper filter and shows the undesired effect of scattering caused by metal scattering in a sample containing bone and metal tissue.

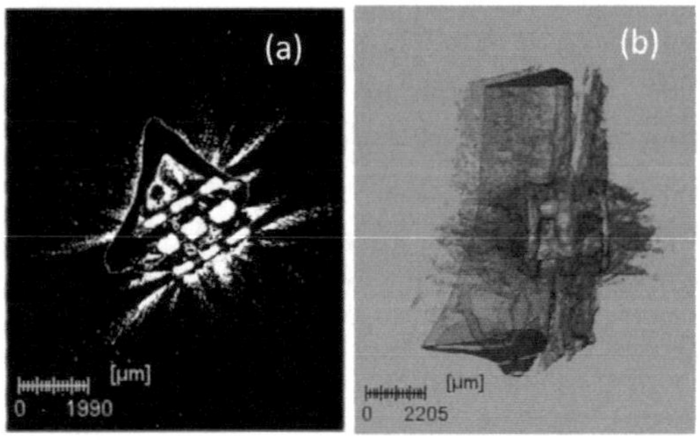

Figure 3. Images demonstrate spreading of implant metal without correction (a) 2D slice of metal implant (b) 3D reconstructed sample.

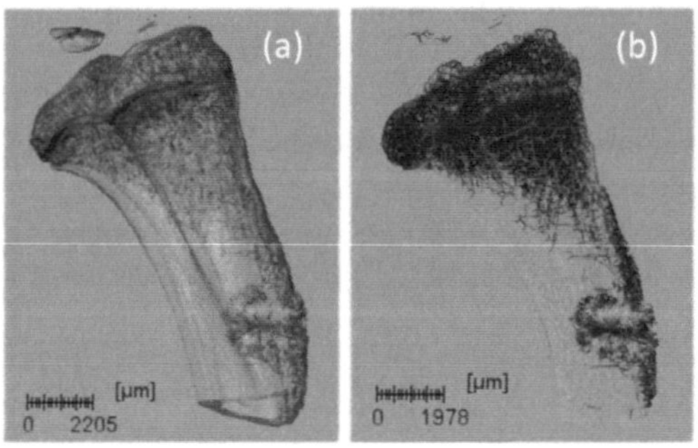

Figure 4. Reconstruction of the trabecular microarchitecture without the metallic implant (a) volume of trabecular connected (b) trabecular thickness. The degree of connection of the trabecular thickness present different colors, separated as warmer to colder colors.

4. Conclusion

Results obtained by microCT provide analysis of microarchitecture of the trabecular bone, which are used to evaluate the bone integrity. However, the choice of morphometric indexes to be studied depends mainly of the research objectives. The parameters of acquisition and reconstruction are of extreme importance in image quality as can be observed in comparative figures between different filters and this loss of image quality directly influences posterior analysis. Energy distribution simulation provides beam quality information with and without the use of additional filters. These data contributed to interpret the reconstructed images where it was possible to observe that use of aluminium filter for multimaterial sample of this study presents the best reconstruction results for analysis.

5. References

[1] Astuto, A., Salgado, A. P., Leite, S. P., Patrão, K. C. S., Fonseca, E. S., Pereira, W. W., & Lopes, R. T. (2014). "Thermal neutron calibration channel at LNMRI/IRD", Radiation Protection Dosimetry, 161(1-4), 185-189.

[2] Podgorsak, E., 2005. “Radiation Oncology Physics: A Handbook for Teachers and Students”, International Atomic Energy Agency, pp. 373-398, Vienna.

[3] Giarola, R. S., Coelho, T. S., Fernande S, M. A. R., ribeiro, V. A. B., Yoriyaz, H., 2013. “Quantification of the anode effect by Monte Carlo method (MCNP-5) simulation of a X-ray generator in diagnostic radiology”, Revista Brasileira de Física Médica, v. 7, n. 3, pp. 157-161.

[4] Carlsson, G.A., Carlsson, C.A., 1984. “Effective energy in diagnostic radiology – a critical review”, Physics in medicine and biology, v. 29, n. 8, pp. 953-958.

[5] Kharrati, H., Zarrad, B., 2003. “Computation of beam quality parameters for Mo/Mo, Mo/Rh, Rh/Rh, and W/Al target/filter combinations in mammography”, Medical physics, v. 30, n. 10, p. 2638-2642.

[6] Attix, F.H, 1986. “Introductionto Radiological Dosimetry”, John Wiley& Sons, 2ª Ed, 607 p.

[7] Kerur, B. R., Manjula, V. T., Lagare, M. T., AnilKumar, S., 2009. “Mass attenuation coefficient of saccharides for X-rays in the energy range from 8 keV to 32 keV”, Radiation Measurements, v.44, n.1, pp. 63-67.

[8] Lacerda, M. A. D. S., Silva, T. A. D., Oliveira, A. H. D., 2007. “Influência da metodologia de avaliação da camada semi-redutora em radiologia diagnóstica” Radiol. bras, v. 40, n. 5, p. 331-336.

[9] Podgorsak, E. B., “Review of radiation oncology physics: a handbook for teachers and students”, International Atomic Energy Agency, Educational reports series, Vienna 2003.

[10] Deslattes, R.D., Kessler Jr, E.G., Indelicato, P., Billy, L., Lindroth, E., Anton, J., Cours EY, J. S., Schwab, D. J., Chang, J., Sukuma R, R., Olsen, K., Dragoset R. A., 2005. “X-ray transition energies”, Nist Standard reference database. http://www.nist.gov/pml/data/xraytrans/index.cfm (acessado em 03.02.2017).

Acknowledgments

To UFRJ Nuclear Instrumentation Laboratory, CNPq and FAPERJ for their financial support and scholarships granted for this study.

Evaluation of the influence of acquisition parameters of microtomography in image quality applied by carbonate rocks

T M P Santos, A S Machado, O M O Araújo, C G Ferreira, R T Lopes

Nuclear Instrumentation Laboratory, Federal University of Rio de Janeiro, Rio de Janeiro, BR

thais@lin.ufrj.br

Abstract. X-ray computed microtomography is a powerful nondestructive technique for 2D and 3D structure analysis. However, parameters used in acquisition promote directs influence in qualitative and quantitative results in characterization of samples, due image resolution. The aim of this study is value the influence of theses parameters in results through of tests changing these parameters in different situations and system characterization. Results demonstrate those pixel size and detector matrixes are the main parameters that influence in resolution and image quality. Microtomography was considered an excellent technique for characterization using the best image resolution possible.

1. Introduction

Carbonate rocks are sedimentary rocks widely studied due the importance in petroleum extraction after pre-salt discovery. This type of rock represents half of the petroleum extraction in Brazil [1, 2].

X-ray Computed microtomography (MicroCT) is an important non-destructive technique for analysis of inner structures with high resolution in the order of micrometer. Its physical principle are based on attenuation of X ray by matter that result a gray scale image proportional to the material density [3].

This technique has been used to characterize carbonate rocks in relation to porosity and pore's size distribution in this work [4, 5]. However, the choice of acquisition parameters is essential to obtain the best resolution and image quality, ensuring the most accurate results [6, 7, 8].

The aim of this study is evaluate the influence of these acquisition parameters in image quality and through of obtained results optimize these parameters to obtain the better image quality for this application.

2. Materials and Methods

The system used was SKYSCAN/BRUKER High Energy model 1173, installed in nuclear instrumentation laboratory (UFRJ). The specimen was an Indiana Limestone and was acquired changing the acquisition parameters like pixel size, detector matrix, and frames number (case 01 – case 04). These cases can be visualized in table 1.

The reference case was using detector matrix 2240 x 2240 pixels, pixel size of 18.16 μm, 130 kV and 61 μA of voltage and current, additional filter of Copper of 0.5 mm, 5 frames and rotation of 360°, represented in table 1 like case 01. These parameters were chosen according many references of works in this area [6].

After the acquisition, the projections were reconstructed in the software NRecon with the best corrections parameters. The next step was segmentation that was realized in Avizo Fire 9.01 with the watershed's method [9].

The watershed's method is based in one or more choices of threshold by operator and the software does the segmentation through of the gradient image and provide the binary image. Also on Avizo Fire, it was obtained the porosity and the porous' size distribution for each cases, where the porosity can be calculate through of the equation 1.

$$Porosity = \frac{Pore's\ Volume}{Total\ Volume} \qquad (1)$$

To analyze the results, it was done a characterization of the system to obtain information about the spatial resolution. In order to characterize the system in relation the influence of detector matrix, pixel size and spatial resolution, it was realized a test of modular transfer function (MTF), where is possible calculate the resolution.

This test was realized using the inclined plate method through of a radiography of a Tungsten (W) plate inclined 5° with the system calibrated to work with voltage of 130kV and current of 61μA, and a additional Copper filter of 0.5mm [10]. The radiography was analyzed on software Isee, as figure 1, and spatial resolution was calculated by the equation 2.

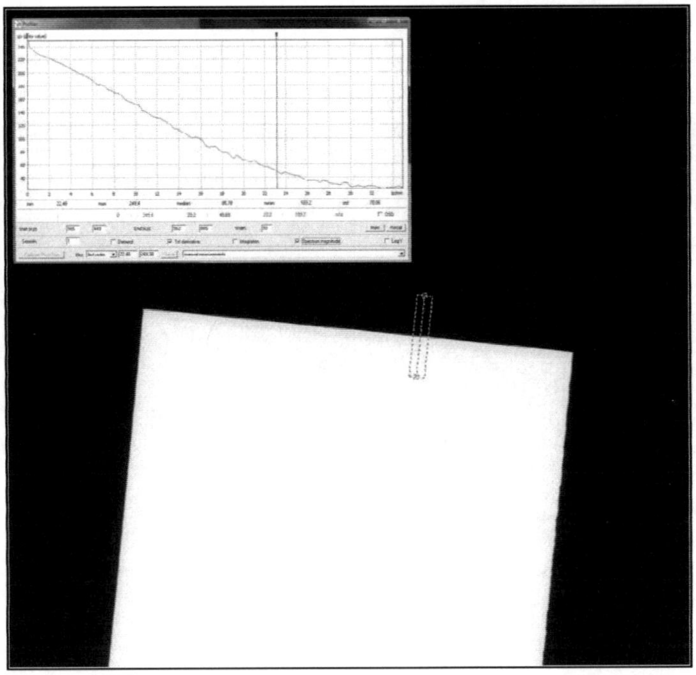

Figure 1 - Analyze of Tungsten plate radiography on software Isee.

The test was realized with two different detector's matrix, 1120x1120 pixels and 2240x2240 pixels, and different magnification and pixel size.

$$Resolution = \frac{1}{2x(MTF_{20\%})} \quad (2)$$

To obtain the signal-noise relation (SNR), it was obtained an empty projection, a projection without sample, using different detector's matrix and frames numbers. The projections were analyzed on software Isee in 25 different rectangle regions of interest of 20x55 pixels, as in figure 2.

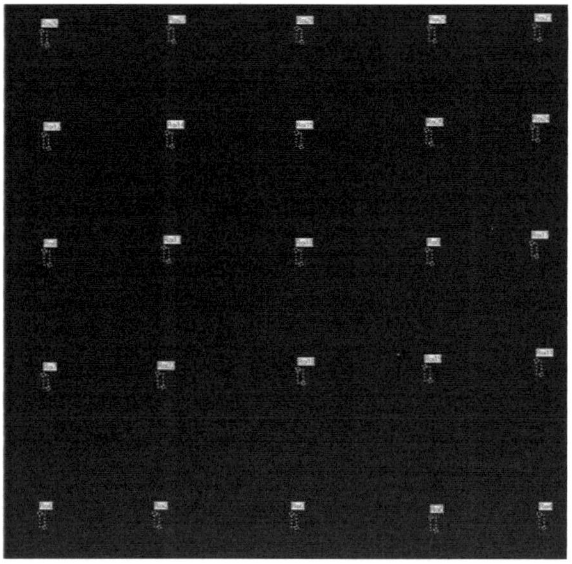

Figure 2 - This picture illustrate the localization of ROIs on empty projection

3. Results and Discussion

With the results, it was possible generate a graphic of spatial resolution vs pixel size in two detector matrix and a graphic of spatial resolution vs magnification also in two detector matrix, figure 3. Observe that for different detector matrix the same magnification is obtained with different pixel size, for instance, when you use detector matrix of 2240 x 2240 pixels you use a pixel size two times less than when you use detector matrix of 1120 x 1120 pixels.

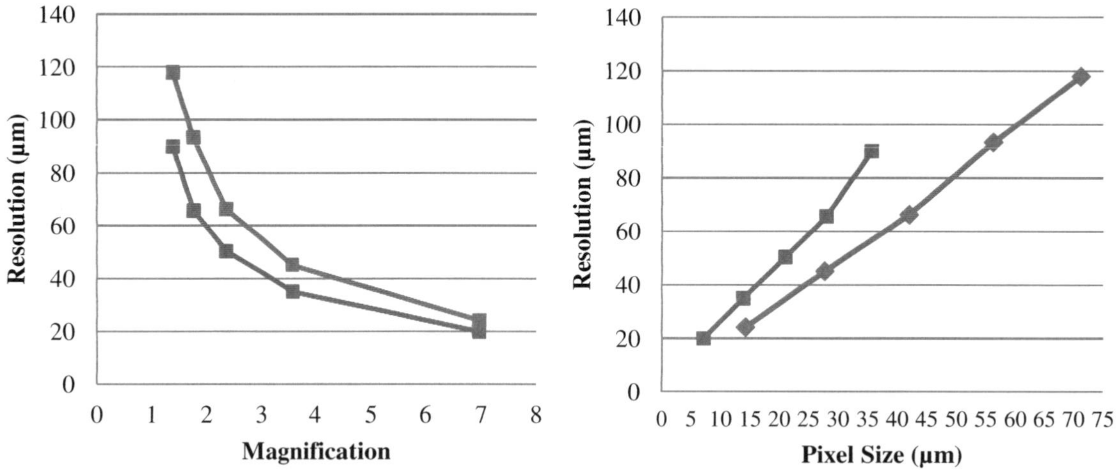

Figure 3 - Graphic of the relation between Spatial Resolution and Magnification, on left, and between Spatial Resolution and Pixel Size

It was possible also generate a graphic of SNR vs frames number, figure 4, through of the test previous cited and evaluate the best or approximated frames number ideal for the best image quality.

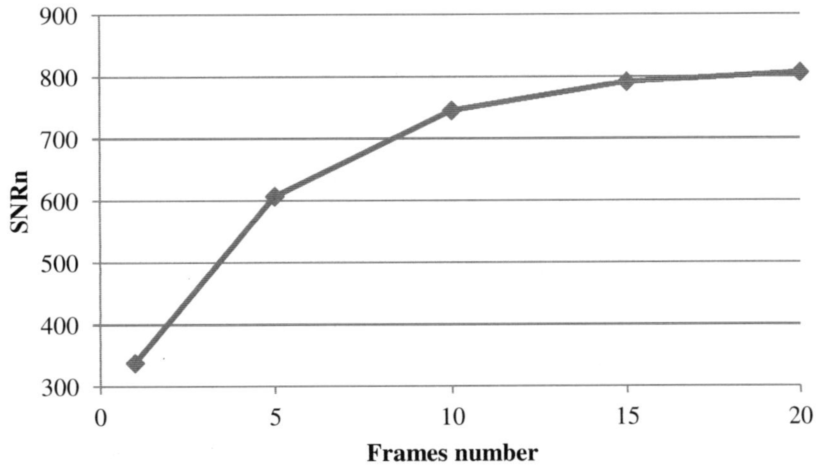

Figure 4 - Graphic of the relation between signal noise relation and frames number.

It is possible to observe in graphic on figure 4 that the curve begin to have a constant behavior from 15 frames. However, comparing the acquisition time of the case 1 and 4, when the frames number is 15, it is observed that the acquisition time in case 4 is three times great than case 1. It is necessary to balance the gain and the time acquisition to choice this parameter.

The porosity results obtained in each case can be visualized on table 1. It is possible to observe through of the results a strong influence of pixel size (case 3) and detector matrix (case 2) in resolution compared to reference case (case 1).

Table 1 - Table of the acquisition parameters and the porosity found in each case

Case	Detector matrix	Voltage and current	Pixel size	Frames	Add Filter	Rotation	Acquisition duration	Porosity (%)
1	2240 x2240	130kV e 61μA	18.16 μm	5	Cu 0.5 mm	360°	1:38:26	6.99
2	1120 x 1120	130kV e 61μA	36.32 μm	5	Cu 0.5 mm	360°	0:31:59	2.97
3	2240 x 2240	130kV e 61μA	29.91 μm	5	Cu 0.5 mm	360°	1:38:28	2.20
4	2240 x 2240	130kV e 61μA	18.16 μm	15	Cu 0.5 mm	360°	4:18:07	5.83

It was generate a graphic with results of case 02, figure 5 and it is observed that the spatial resolution for the second case, where the pixel size is 36.32μm and detector matrix of 1120x1120 pixels, is approximated 60μm. Then, pores less than this value will not be counted. It justified why don't have counting on the first interval of pore's size.

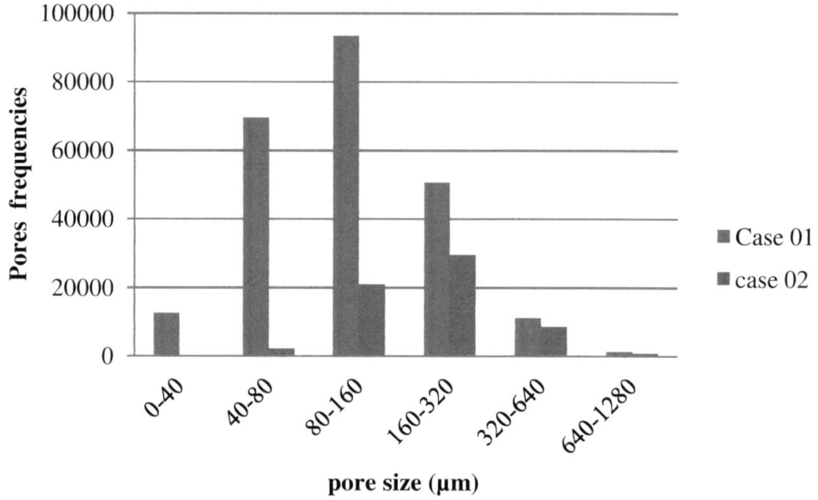

Figure 5 - Graphic of pore distribution of the case 01 and 02

On case 03, 29.91µm of pixel size and detector matrix of 2240x2240 pixels, the spatial resolution is little more than 70µm. This way, pores on the first interval of pore size and majority of second interval are no counting resulting in a low porosity, as on figure 6.

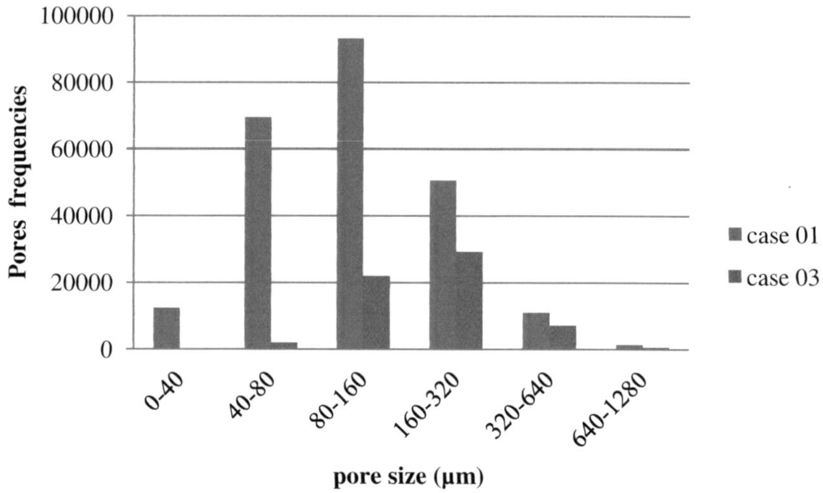

Figure 6 - Graphic of pore distribution of the case 01 and 03

It is believed that a bigger SNR result in a better image quality and the frames number can be associated with these improvements. In this case the pore's boundary is more defined than in the case 01, where the frames number is 5, and it is related with the sharpness.

With this improvement on the sharpness little pores were counting and the boundary was not super estimated on segmentation. Other point is the acquisition time, that in case 04 is three times greater than in case 01.

Besides that, on figure 7, was possible to observe the difference between each case in relation spatial resolution and sharpness. For this application, to characterize carbonate rocks, the technique provide porosity, pore's distribution, even as pore space and connected pores, like on figure 8.

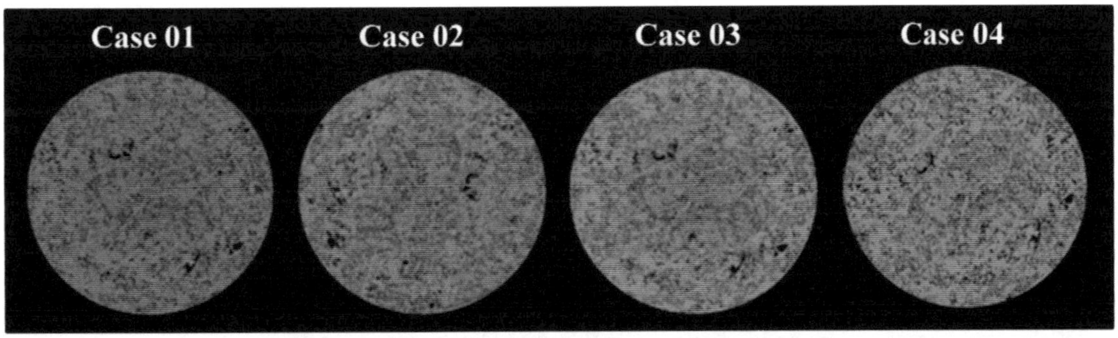

Figure 7 - Comparison of a slice of each case

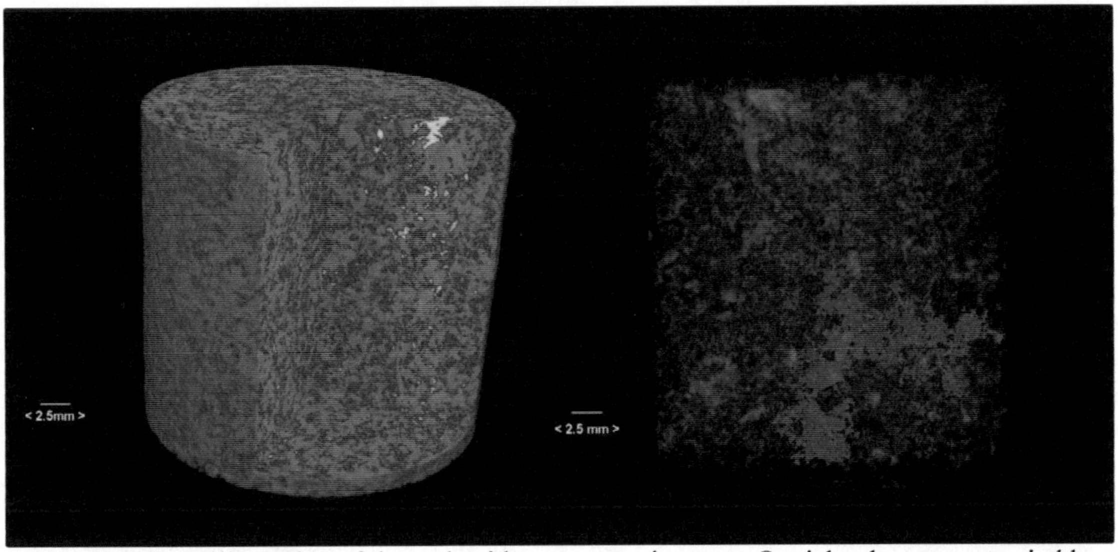

Figure 8 - On left, an illustration of the rock with pore space in green. On right, the pore space in blue with the connected pore in orange.

4. Conclusion

It is possible to conclude that all parameters influence in image quality with different intensity and it is necessary to optimize these parameters correctly to obtain the best image quality on the best acquisition time.

Detector matrix and pixel size were the main influencing parameters in relation to spatial resolution, and for this system and this type of sample the best detector matrix is 2240x2210 pixels and the best pixel size is the smallest provided for the system. However, it is also necessary to balance the pixel size and frames number with the time and the gain obtained with this improvement.

Microtomography is an excellent technique for image analyzes and with the best acquisition parameters is possible to obtain accurate results without damaging the samples.

Acknowledgements

The authors of the present work wish to thank the Brazilian supporting agency CNPq.

References

[1] Arh W M, 2008, Geology of carbonate Reservoirs the identification, description, and characterization of hydrocarbon reservoirs in carbonate rocks, John Wiley & Sons.

[2] Oliveira M F S, Lima I, Ferrucio P I, Abreu C J, Borghi L, Lopes R T, 2011, "Petrophysical

analysis of limestone rocks by nuclear logging and 3D high – resolution X-ray computed microtomography", Nuclear Instruments and Methods in Physics Research A, v. 652, pp. 905 – 910.

[3] Knoll G F, Radiation Detection and Measurement, 2000.

[4] Arzilli F, Cilona A, Mancini L, Tondi E, 2016, *Advances in Water Resources*, **95**, 254 – 263.

[5] Oliveira M F S, Lima I, Borghi L, Lopes, R. T., 2012, "X ray microtomography application in pore space reservoir rock", Applied Radiation and Isotopes, v. 70, pp. 1376 – 1378.

[6] Machado A C, Lima I, Lopes R T, 2014, *Radiation Physics and Chemistry*, **95**, 405 – 407.

[7] Buzug T M, Computed Tomography from Photon Statistics to Modern Cone-Beam CT, 2008.

[8] Marques F J M O, 2009, Controlo de qualidade em tomografia computadorizada. M.Sc, Dissertação, Universidade de Lisboa, Lisboa, Portugal.

[9] Barraud J, 2006, *J. Volcanol Geotherm*, **154**, 17–23

[10] EUROPEAN COMMITTEE FOR STANDARDIZATION, EN 14784 – 1, "Classification of systems". Non-destructive testing – Industrial computed radiography with storage phosphor imaging plates, Part 1, 2005.

Pressure sound level measurements at an educational environment in Goiânia, Goiás, Brazil

J J L Costa [1,2], E O do Nascimento[1], L N de Oliveira [1,3] and L V E Caldas[3]

[1]Instituto Federal de Educação, Ciência e Tecnologia de Goiás-IFG, Campus Goiânia, Goiânia-GO, Brazil
[2]Polícia Militar de Goiás, Goiânia-GO, Brazil
[3]Instituto de Pesquisas Energéticas e Nucleares-IPEN/CNEN, São Paulo-SP, Brazil

E-mail: contato@jjlopes.com.br, eriberto.on@gmail.com, lucas@ifg.edu.br, lcaldas@ipen.br

Abstract. In this work, 25 points located on the ground floor of the Federal Institute of Education, Science and Technology of Goias - IFG - Campus Goiânia, were analyzed in morning periods of two Saturdays. The pressure sound levels were measured at internal and external environments during routine activities seeking to perform an environmental monitoring at this institution. The initial hypothesis was that an amusement park (Mutirama Park) was responsible for originating noise pollution in the institute, but the results showed, within the campus environment, sound pressure levels in accordance with the Municipal legislation of Goiânia for all points.

1. Introduction

The noise pollution is the undesired noise released into the environment, without the perception of its damage and the negative results that it can cause to human health [1].

Many researchers [2, 3] showed that the Noise Pollution is constantly and repeatedly delivered, making people not aware of the harmful consequences of high Sound Pressure Levels (SPL) toward their health.

Therefore, in classroom environment, high levels of noise bring serious damages to the teachers as to the students. Consequently, in educational facilities, people can be subject to Noise Pollution and its hazards, but not noticing its complications. Thus, the environment monitoring of educational places is justified and has the objective to quantify through the Equivalent Sound Level (LAeq) the pollution levels to which the population is exposed to. Then, with the measurements performed with LAeq (the weighted continuous sound pressure level on the A scale), it is possible to establish new strategies seeking for adjustments into the cases where the sound levels do not agree with the standard limits proposed by the Regulatory Agencies. In addition, with the identification of irregularities it will be possible to trace actions against negative recurrences of noncompliance institutions.

Environmental monitoring consists in the accomplishment of specific measurements and/or observations, benchmarked in front of some indicators and parameters, in order to verify if certain environmental impacts are occurring. As a result, it will be possible to estimate its magnitude and to evaluate the efficiency of any possible preventive measures, and it can be used in the diagnosis of noise pollution.

In this sense, this work aims to carry out the environmental monitoring of the (LAeq) levels and to evaluate the measured sound levels accordingly with the current limits of Noise Pollution proposed by the Brazilian government. These measurements were taken, during the Saturday morning work routine activities developed at the Federal Institute of Education, Science and Technology of Goias (IFG) - Campus Goiânia, which consists of an educational environment.

2. Materials and Methods

For data acquisition and analysis, the authors followed the environmental monitoring techniques proposed by Zannin et al [3].

The research was carried out at the Federal Institute of Goiânia (IFG) - Campus Goiânia, during the period of normal activities, from 7:00 am to 9:00 am, in the last Saturday of February and in the first Saturday of March in 2015, trying to interfere as little as possible on the administrative and schooling activities during the monitoring time.

To perform the measurements, a sound level meter with data logger, model DEC-490 (INSTRUTHERM), and an acoustic calibrator CAL-4000 (INSTRUTERM), both with calibration certificates, were utilized.

Analyzing the Brazilian legislation, it was possible to extract the Level of Evaluation Criteria (LEC), shown in Table 1. The determination of the LAeq, Eq. 1, obeyed the standard NBR 10151/2000 [4].

$$LAeq = 10\log\frac{1}{n}\sum_{i=1}^{n}10^{L_i/10} \tag{1}$$

where LAeq [dB (A)] is the A-weighted equivalent sound pressure level, L_i is the sound pressure level in dB (A) taken in fast response every 5 s during the noise measurement, while n represents the total number of recorded measurements.

Table 1. Brazilian Levels of Evaluation Criteria (LEC) [4].

Environment	Level of Evaluation Criteria - LEC
External Environment	55 dB(A)
Internal environment with open windows	45 dB(A)
Indoor environment with closed windows	40 dB(A)

For each point, three measurements were taken (triplicate) at different times, the LAeq was calculated through equation (1) and, afterwards, the average LAeq obtained in each situation was determined. The corrected noise level (Lc) was applied in cases where there was a need for the results obtained experimentally compared with the Level of Evaluation Criteria of Table 1.

Eight points located in the external perimeter of the institution were chosen to monitor the LAeq outside the limits of the IFG, and 17 points within the institute internal perimeter, with 10 points located in external environments and 7 points located indoors (classrooms) as shown in Figure 1.

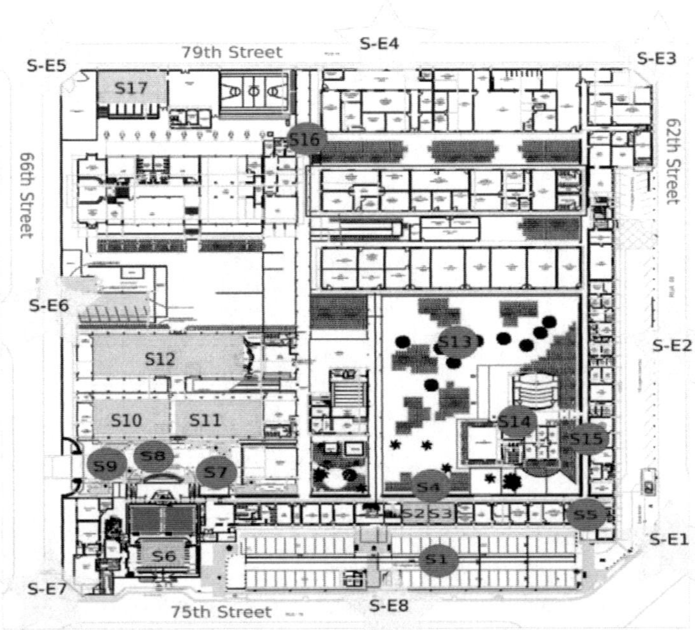

Figure 1. Ground floor of the IFG, with the places where measurements were taken. The stars represent the measurements in the outer perimeter, the red circles the measurements in external environments and the green rectangles the measurements indoors.

3. Results

Eight places S-E1, S-E2, S-E3, S-E4, S-E5, S-E6, S-E7 and S-E8 outside the IFG on the Saturday morning were monitored. The results are shown in Figure 2. The uncertainty obtained was lower than 3% in all results.

It can be noted that from the 8 points measured in the external area during that Saturday morning, only 3 presented values above the Evaluation Criteria Levels established in the Brazilian Law, by the NBR 10151/2000 [4].

At the ground level, 11 points S1, S2, S4, S5, S7, S8, S9, S13, S14, S15 and S16 were monitored, including external and internal environments. The results obtained are shown in Figures 3 and 4.

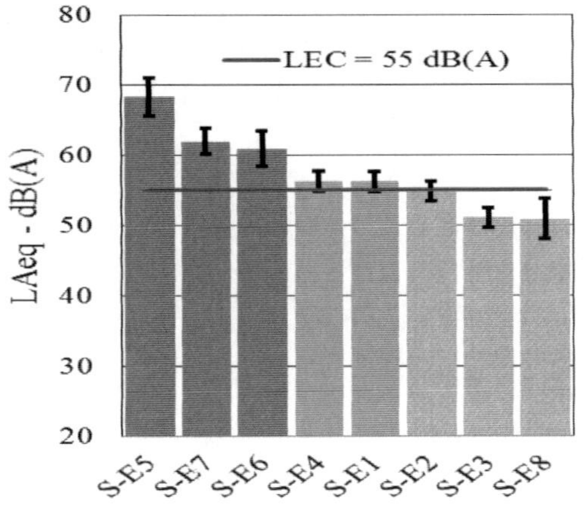

Figure 2. Equivalent Sound Pressure Level LAeq [dB (A)] for points at the external perimeter.

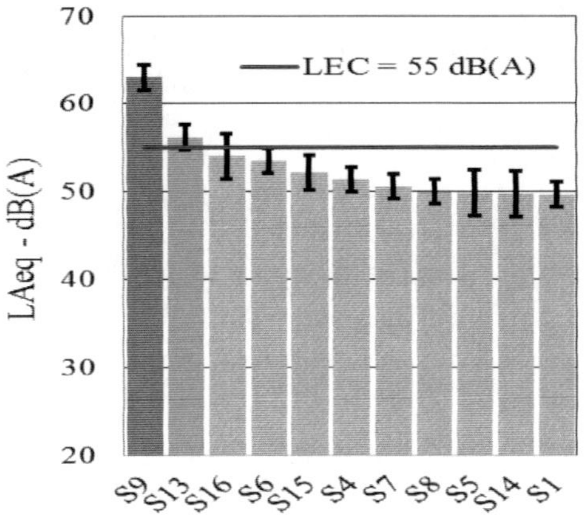

Figure 3. Equivalent Sound Pressure Level LAeq [dB (A)] for the institution ground points.

Measurements were taken in a classroom near the institution entrance, nearest to the Mutirama Park, with open windows; the measurements pointed out LAeq values within the legal limits.

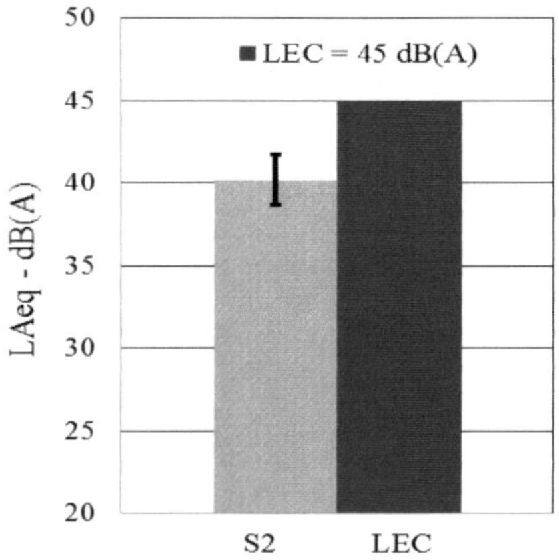

Figure 4. Equivalent Sound Pressure Level LAeq [dB (A)] for classroom on the ground floor with open windows.

4. Conclusions

This work presents partial results of the research carried out, opting for the survey on Saturdays because of complaints from teachers and students that an amusement park next to the institution (Mutirama Park) was responsible for producing the noise pollution.

During this work, measurements were taken on a Saturday morning, on a non-busy day and only one point (S9) presented LAeq above the LEC, at the external yard in front of the arts complex that, at the moment of the monitoring, presented sound coming from musical instruments. The other points monitored on the Saturday morning did not present LAeq in disagreement with the municipal legislation of Goiânia.

The situation differs from the rest of the research, when the environmental monitoring in different periods of the work days; at a few points no sound pollution was identified.

The initial hypothesis that the Mutirama Park would be responsible for generating noise pollution in the educational institution was not proven, and it is possible to conclude that points where there are excessive sound pressure levels are a consequence of the human activities at the place.

Acknowledgements

This work was supported in part by the Brazilian agencies: CNPq, under Grants No. 165466/2015-4, 151013/2014-4, and 301335/2016-8, FAPESP and MCTIC.

References

[1] Strite Tripathy D P 2008 *Noise Pollution* (New Delhi: APH Publishing Corporation) p 59
[2] Passchier-Vermeer W and Passchier W F 2000 *Environ. Health Perspect.* **108** 123-31
[3] Zannin P H T, Engel M, Fiedler P E K and Bunn 2013 F *Cities* **31** 317–27
[4] ABNT-NBR 2000 *10.151: Acoustics—Noise Assessment in Inhabited Areas Aiming at the Community's Comfort-Procedure* (Rio de Janeiro - In Portuguese)

IOP Conf. Series: Journal of Physics: Conf. Series **975** (2018) 012056 doi:10.1088/1742-6596/975/1/012056

The quantity time relation in the ionizing radiations

B O Jordão[1], D S Quaresma[1] and J G P Peixoto[2]

[1] Primary Laboratory of Time and Frequency (LPTF) of the National Observatory (ON); Rio de Janeiro, RJ 20921-400, Brazil
[2] National Laboratory of Metrology of Ionizing Radiation, Institute of Radioprotection and Dosimetry, Rio de Janeiro, RJ 22783-127, Brazil

E-mail: bjordan@on.br

Abstract. The metrology area has taken a step forward with regard to the uncertainty calculation. This mathematical tool used in laboratories is essential to ensure that the values resulting from a measurement are reliable. For this to be possible, all equipment used in a measurement process must be reliable and, above all, traceable to the international metrology system. We propose to present in this work: (i) the development and calibration of a microcontrolled time device with a resolution of $1x10^{-4}$ s, in order to characterize the time greatness and make it reproducible; (ii) the calibration of the quartz clock present in a computer present in the dosimetry laboratories; (iii) a more in-depth study of the influence of time quantity on calibrations of instruments used in the area of radiological protection, diagnostic radiology and radiotherapy, with measurements performed on the Kerma magnitude in air or its rate.

1. Introduction

The continuous quality of life improvement, the research for knowledge of the input quantity related to ionizing radiations used in radiodiagnosis, radiotherapy and radioprotection, determining the quantity specific and/or their ratios, as well as the reduction of unnecessary doses exposure of workers and public, safety respecting for ionizing sources handling which are: time of exposure, distance and shielding. In this way, exposures in the areas surrounding the dosimeter should be guaranteed as low as reasonably possible, taking into account social and economic factors known as the ALARA principle [1].

It The ionizing radiation reference dosimetry is based on the instruments calibration used in laboratories, hospitals, clinics, inspections, industries among many other applications.

Therefore, ionizing radiation quantities measurement used as reliable and reproducible way [2], maintaining, tracing and disseminating according to international standards and the international system units [3 – 7].

In order to realize these quantities measurements, with and without radiation and their ratios, it will be necessary to correlate the instrument or suitable material measurement, charge accumulated in a time interval, in seconds.

In order to uncertainty contribution evaluate of a time measurement device, defined as the elapsed time between two events, a microcontroller device, a universal counter, and a computer timer are compared to a traced cesium standard [8].

The time quantity measurement as ionizing radiation measurement system input quantity occurs when the device begins its measurement and after a time interval has elapsed, it ends.

2. Calculations details

The quantity integral or its ratio is correlated with the elapsed measurements at a time interval given, that way the quantity time and frequency must be taken into account, both in the correction coefficient and its associated uncertainties [9, 10].

Because the measurement, in the initial and final recording case on the device or counter, will follow an equipment itself probability distribution.

The uncertainties analyses increasing the need to arising from traceability automated systems takes place in the quantity time in the computer or operating system, at first without analyses integrated system and memories the use of the computer used [11]. Thus, the uncertainty comparison function between the quantity time instruments will follow equation (1).

$$U_c^2(d) = u^2(\Delta t) + u^2(disp1) + u^2(cont) + u^2(ls) \qquad (1)$$

Where $U_c^2(d)$ the comparison is expanded uncertainty, $u^2(\Delta t)$ is the elapsed time ty, $u^2(disp1)$ is the device 1 uncertainty, $u^2(cont)$ is the counter uncertainty and $u^2(ls)$ is the calibration standard uncertainty.

At this method, the evaluated uncertainty were considered all input quantities independent, between the quantity time and the measurement systems, but all probability distributions will be taken into account.

The computer time absolute accuracy was determined by elapsed time comparison between the computer's stopwatch and the reference counter.

Although, the absolute accuracy of measurement were useful, but at this results comparison were taken the standard time measurement as reference for relative time computer [8].

3. Material and method

The partnership development between the Primary Time and Frequency Laboratory (LPTF) of the Division of Time Service (DISHO / ON) and the National Laboratory of Ionizing Radiation Metrology (LNMRI) of the Institute of Radioprotection and Dosimetry (IRD / CNEN) could be possible the method development comparison using the equipment and components below:

- Standard Cesium – CSAC, made by Symmetricom, model AS.45s (Chip Scale Atomic Clock).
- Universal Counter, made by Fluke, model PM6681 (High Resolution Programmable Time/Counter Analyzer).
- Microcontroller, made by Microship, model ATMEGA328P.
- Personal Computer, made by HP, processor AMD Phenom (tm) IIx4 B95 3.00GHz, memory RAM de 4.00GB, operational system Windows 7 professional - 64 Bits.

3.1. Comparison

The 10 MHz frequency from Cesium standard output thought BNC - J3, plus BNC – type T, is connected to the Universal Counter input, using the BNC - IN Reference connector with also in channel A input has another BNC - type T connector.

This parallel connection series was to obtain the same frequency of 10 MHz at cesium standard reference, Microcontrollers devices (PPS) and Counter systems, figure 1.

The range measurements Microcontroller system were 600, 1200, 1800, 2400, 3000, and 3600 PPS, where an initial start pulse was sent to each PPS value for the Universal Counter BNT - EXT ARM connector on the back side to Microcontroller and to computer timer.

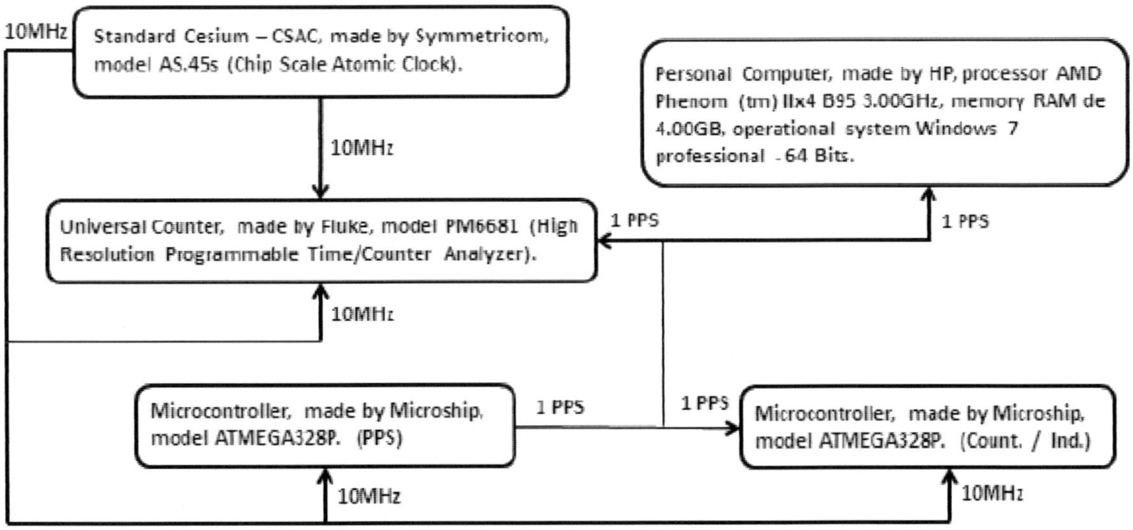

Figure 1. Connection diagram from Standard Cesiun at Microcontroller device and to computer timer.

Simultaneously start the time counting on the three time indicators. After select elapsed time desired, the final pulse - Stop, were sent to the systems, again simultaneously interrupting the time count, thus start the new measurement, by sending another pulse.

The quantity time mean and standard deviation from five data measurement was given by the range from 600, 1200, 1800, 2400, 3000 and 3600 s.

3.2. System device.

3.2.1. Microcontroller. The device uses as base of time an cesium atomic standard, reference standard for measurement of the time quantities, with $1x10^{-4}$ of resolution.

Made by Microchip Technology / Atmel from ATMEGA family 328P, model MCU AVR 32K FLSH 1K EE2K SRAM - 20MHz, with 8 - bit microcontrollers.

It had two operation stages; the first one is to provide the pulse per second, named PPS, in order to start counting simultaneously between the microcontroller, the universal counter and the personal computer. After the programmed elapsed time interval has another count end pulse.

3.2.2. Personal computer. The measurement of the time quantity relative to the personal computer was carried out through the LABVIEW platform, specifically developed for this comparison.

The third PPS pulse was clear the display, them at the LOG A field the program data was record.

3.3. Calibration: Microcontroller and personal computer.

The temperature and humidity ambient conditions range measurement was performed under with of (24 ± 2) °C and (40 to 70) %, respectively.

The cesium atomic standard, as reference for the calibration between the microcontroller device and the personal computer was obtained by comparing the final and initial measurements.

4. Results

Tables 1 and 2 present the results of the calibrations of the microcontroller device and the personal computer in relation to the cesium standard, respectively.

Table 1. Microcontroller device calibrations results considering a coverage factor k = 2.0

Nominal Value (s)	Correction (s)	Expanded Uncertainty (±s)
600	0.00012	0.00013
1200	0.00003	0.00012
1800	-0.00002	0.00014
2400	0.00012	0.00013
3000	-0.00003	0.00014
3600	-0.00004	0.00012

Table 2. Personal computer calibrations results considering a coverage factor k = 2.9.

Nominal Value (s)	Correction (s)	Expanded Uncertainty (±s)
600	0.01	0.00
1200	0.05	0.04
1800	0.04	0.01
2400	0.01	0.01
3000	0.04	0.01
3600	0.07	0.00

Table 3 shows the average relative accuracy of the personal computer obtained from the collection of three samples. One, two and three hours, respectively.

Table 3. Results table for relative computer accuracy (pc) in 3 hours.

Average ($s.s^{-1}$)	Expanded Uncertainty ($\pm s.s^{-1}$)	Coverage factor (k)
18.5E-6	0.60E-6	2.9

Figure 2 shows the plot of the correction - time difference between the standard meter and the microcontrolled device - with their respective uncertainties considering a coverage factor k = 2.9.

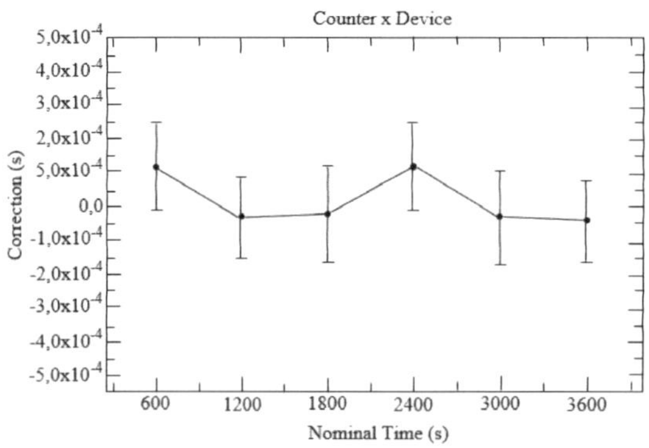

Figure 2. Correction chart x time referring to the time difference between the counter and the device.

5. Conclusions

It can be observed that, in both Table 1 and Figure 2, the microcontroller device presented a correction in the order of quantity of 10^{-4} and, therefore, showed good reproducibility and could therefore be used in laboratories as a standard device.

Regarding the computer calibration, we can observe that U = 40 ms, having as a discrepant point, the uncertainty referring to the interval of 1200 s.

Observing table 3, it is verified that the measurement of the time quantity relative to the personal computer, obtained an average relative precision of 18.5 μs and with a measurement uncertainty of 0,60 μs considering a coverage factor k = 2.9. To know how much the personal computer clock delays per day, multiplies the relative accuracy of the computer by 86400 which is the amount of seconds that a day has, arriving at a delay of 1.60 s / day on that personal computer.

References

[1] CNEN 2014 *Diretrizes Básicas de Proteção Radiológica* Norma CNEN NN 3.01
[2] Peixoto J G et al 2016 *Ionization Radiation Metrology* ed J G Peixoto (Rio de Janeiro: Instituto de Radioproteção e dosimetria)
[3] IAEA 2000 *Technical Reports Series 398 - Absorbed Dose Determination in External Beam Radiotherapy* STI/DOC/010/398 /(Vienna: IAEA)
[4] IAEA 2007 *Technical Reports Series 457 - Dosimetry in diagnostic radiology: An international code of practice* STI/ PUB/ 1294 (Vienna: IAEA)
[5] ICRP 1991 *1990 Recommendations of the International Commission on Radiological Protection* ICRP Publication 60. Ann. ICRP 21 (1-3)
[6] ICRP 2007 *The 2007 Recommendations of the International Commission on Radiological Protection* ICRP publication 103. Ann. ICRP 37 (2-4)
[7] IEC 2005 *IEC 61267 - Medical diagnostic X-ray equipment - Radiation conditions for use in the determination of characteristic*
[8] Gust J C and E A l 2009 NIST Recommended Practice Guide *Stopwatch and Timer Calibrations* Spec. Publ. 960-12
[9] BIPM 2008 *Evaluation of measurement data - Guide to the Expression of Uncertainty in Measurement (GUM)*
[10] Baratto A C, Damasceno J C, Alves J A P, Filho J T, Couto P R G and Oliveira S P 2012 *Avaliação de dados de medição: Guia para a expressão de incerteza de medição-GUM 2008* (Duque de Caxias: Instituto Nacional de Metrologia, Qualidade e Tecnologia)
[11] Portalaction 2017, "*Ambiente Virtual de Aprendizagem.,*". [Online]. Available: http://www.portalaction.com.br/incerteza-de-medicao/17-incerteza-expandida Access in 05/30/2017.

Uncertainty evaluation of a modified elimination weighing for source preparation

F L Cacais[1,2], J U Delgado[1] and V M Loayza[2]
[1] Laboratório Nacional de Metrologia das Radiações Ionizantes / Instituto de Radioproteção e Dosimetria – LNMRI/IRD – Brazil
[2] Instituto Nacional de Metrologia, Qualidade e Tecnologia – Inmetro – Brazil

E-mail: facacais@gmail.com

Abstract. Some modifications in elimination weighing method for radioactive source allowed correcting weighing results without non-linearity problems assign a uncertainty contribution for the correction of the same order of the mass of drop uncertainty and check weighing variability in series source preparation. This analysis has focused in knowing the achievable weighing accuracy and the uncertainty estimated by Monte Carlo method for a mass of a 20 mg drop was at maximum of 0.06%.

1. Introduction

In radionuclide metrology, radioactive sources preparation encompasses a weighing procedure able to achieve standard uncertainties below than 20 μg in the range from 10 mg to 100 mg [1]. The Elimination Weighing method meets this requirement for micro-drops deposition or dilution of a master solution using a plastic pycnometer [2]. In this weighing procedure three weighing steps are performed per source: the pycnometer is weighed before and after dispensing the drop of solution, and by knowing the mass of the drop (by the weighing difference), one or several standard weights [3] are added on the balance load receptor and the third weighing reading is recorded. The mass of the drop is obtained from the difference between first and third weighing thus avoiding non-linearity errors problems. To avoid correlations in series sources preparation this method suggests to weigh the pycnometer before each drop deposition. The absence of systematic errors between the weighing (evaporation, drying of a drop in capillary stem, zero drift of the balance) is checked by the criterion that the difference between third and second weighing readings should be in agreement with conventional mass [4] of standards within twice the uncertainty.

In despite of, by elimination weighing it is not possible to correct systematic errors without take into account non-linearity errors on the third and second weighing and its uncertainty contribution. Furthermore, this method does not provide any estimate to check variability in series sources preparation. Thus, in order to improve the reliability of the elimination method, this work proposes a modification that allows: correcting weighing results without non-linearity problems, assign a uncertainty contribution for correction of the same order of the mass of drop uncertainty and check weighing variability in series source preparation.

The modification consists in performer the second weighing in elimination method also with the same or different standard weights used in third weighing. In the original elimination method [5] the second weighing was performed by this way and it relies on the previously planned procedure. For an experienced staff the diameter of the stem tip of a laboratory-made pycnometer [6] is adjusted such the mass per drop is previously known with 2 mg of difference from the true value. This approximation could be improved by performing the weighing of some drops before to execute the weighing procedure. Thus, the suitable amount of mass standards can be available close to the balance at the moment of the weighing. By this way, the standard weights and the pycnometer are placed together on the balance receptor at the second weighing and its mass is adjusted until the difference to first reading is less than 3 mg (for 0.001 mg resolution balance) limiting non-linearity error [7].

By this proposal, one can obtain two weighing differences to form a restrained underdetermined equation system which can be solved including a restrain, the mass of standard weights. This solution provides a mass value of the standard weights in addition to the mass of the drop. The mass of the standards could be used to estimate errors difference in weighing on some assumptions about linear error structure (constant, increasing and different in one weighing from the two equal others). The values for mass standards obtained from weighing of a series source preparing could be used to set a long-run standard deviation [8] and to check weighing variability in source preparation serie, so providing information to variability studies [9].

In this study the measurement models for the mass measurement are showed and an estimate of the correction to linear errors in weighing is presented. A routinely assumed error structure on weighing is defined to evaluate the applicability of correction and its uncertainty. The uncertainty to the correction and mass values corrected are calculated by Monte Carlo method [10] to a simulated weighing condition in order to evaluate the achievable accuracy for this modified method in order to implement it.

2. Measurement model

On the assumption of the balance has been adjusted before weighing, the difference ΔI_{1i} between the first weighing, that includes the conventional mass of the drop mc_{d}, and the ith (2nd or 3rd) weighing with standard weights of conventional mass $mc_{\mathrm{p}i}$ can be written as in equation (1).

$$\Delta I_{1i} = mc_{\mathrm{d}} - mc_{\mathrm{p}i} \tag{1}$$

In equation (2), the difference ΔI_{1i} includes the balance readings for 1st weighing R_1 and the ith weighing R_i, sensitivity error ΔS and the buoyancy correction [11] which take in account the air density ρ_{a}, the conventional air density ρ_0 (1.2 kg m^{-3}), density of the master solution ρ_{S}, density of balance reference standards ρ_{R} and the density of standards weighed together with the solution ρ_i.

$$\Delta I_{1i} = \left[\left(R_1 - R_i \right) \left(1 - \frac{\left(\rho_{\mathrm{a}} - \rho_0 \right)}{\rho_{\mathrm{R}}} \right) + \left(\rho_{\mathrm{a}} - \rho_0 \right) \left(\frac{R_1}{\rho_{\mathrm{S}}} - \frac{R_i}{\rho_{\mathrm{i}}} \right) \right] \left(1 - \Delta S \right) \tag{2}$$

From the two weighing differences a system of equations in matrix form can be mounted, as in equation (3).

$$\begin{bmatrix} \Delta I_{12} \\ \Delta I_{13} \end{bmatrix} = \begin{bmatrix} 1 & -1 & 0 \\ 1 & 0 & -1 \end{bmatrix} \times \begin{bmatrix} mc_{\mathrm{d}} \\ mc_{\mathrm{p1}} \\ mc_{\mathrm{p2}} \end{bmatrix} \tag{3}$$

This system could be solved regarding one (p_1) or two (p_1 and p_2) reference mass standards. The solution of this system, on the assumption of one reference standard with conventional mass $\mathrm{mc}_{\mathrm{p1}}$ provides the measurement models in equations (4 – 5) for the conventional mass of the drop mc_{d1R} (1R means obtained by one reference), and for the estimated mass of the other standard mc_{p2}.

$$mc_{\text{d1R}} = \left[(R_1 - R_2)\left(1 - \frac{(\rho_a - \rho_0)}{\rho_R}\right) + (\rho_a - \rho_0)\left(\frac{R_1}{\rho_S} - \frac{R_2}{\rho_2}\right)\right](1 - \Delta S) + mc_{\text{p1}} - (e_1 - e_2) \tag{4}$$

$$mc_{p2} = \left[(R_3 - R_2)\left(1 - \frac{(\rho_a - \rho_0)}{\rho_R}\right) + (\rho_a - \rho_0)\left(\frac{R_3}{\rho_3} - \frac{R_2}{\rho_2}\right)\right](1 - \Delta S) + mc_{p1} - (e_3 - e_2) \tag{5}$$

When the two standards are set as reference just one solution for the conventional mass of drop mc_{d2R} is possible. This solution is showed in equation (6).

$$mc_{\text{d2R}} = \left[\left(R_1 - \frac{R_2 + R_3}{2}\right)\left(1 - \frac{(\rho_a - \rho_0)}{\rho_R}\right) + (\rho_a - \rho_0)\left(\frac{R_1}{\rho_S} - \frac{R_2}{2\rho_2} - \frac{R_3}{2\rho_3}\right)\right](1 - \Delta S)$$

$$+ \frac{mc_{\text{p1}} + mc_{\text{p2}}}{2} - e_1 + \frac{e_2 + e_3}{2} \tag{6}$$

In equations (4 – 6) , the terms e_1, e_2 and e_3 are the sum of readability and repeatability errors (zero mean with uncertainty) and it is not considered non-linearity error (avoided by method) and evaporation error (because it is corrected by this approach). If one performs weighing with two references the three equations can be used. However if the second weighing is carried out with the same standard of the first, only the two first equations should be used.

A conversion factor F should be multiplied to conventional mass of the drop to obtain its mass value. As in equation (7), this factor takes in account the conventional air density ρ_0, density of the master solution ρ_S and conventional standard weight density ρ_c (8000 kg m^{-3}).

$$F = \frac{\left(1 - \dfrac{\rho_0}{\rho_c}\right)}{\left(1 - \dfrac{\rho_0}{\rho_S}\right)} \tag{7}$$

3. Errors structure

The errors in weighing (δ_1, δ_2, δ_3) were considered as additional linear terms in the equations of conventional mass values because for two weighing differences it is not possible to take conclusion about additional higher order systematic error as in mass comparisons [12]. The redefined values for the solution of matrix system mc_{d1R}, mc_{p2} and mc_{d2R} are showed in equations (8 – 10):

$$\text{mc}_{\text{d1R}} = mc_{\text{d1R}} - (\delta_1 - \delta_2) \tag{8}$$

$$\text{mc}_{\text{p2}} = mc_{\text{p2}} - (\delta_3 - \delta_2) \tag{9}$$

$$\text{mc}_{\text{d2R}} = mc_{\text{d2R}} + \delta_1 - \left(\frac{\delta_2 + \delta_3}{2}\right) \tag{10}$$

For two references the errors difference ($\delta_3 - \delta_2$) should be estimated from the difference between mc_{p2} and the calibration result mc_{p2C}, however if only one reference was used ($\delta_3 - \delta_2$) is ($\text{mc}_{\text{p2}} - mc_{\text{p1}}$) because mc_{p1} is determined from its calibration result. It is emphasized that the underlying hypothesis for this approach is no standard weights mass drift. This way, the estimate $(\delta_3 - \delta_2)^*$ can be written as in equations (11 – 12).

$$(\delta_3 - \delta_2)^* = mc_{p2} - mc_{p2C} \quad \text{(2 references)} \tag{11}$$

$$(\delta_3 - \delta_2)^* = mc_{p2} - mc_{p1} \quad \text{(1 reference)} \tag{12}$$

Five assumptions about the difference of the systematic errors in weighing were studied:

a) Constant ($\delta_1 = \delta_2 = \delta_3$), in this case the value ($\delta_3 - \delta_2$) = 0 and ($\delta_1 - \delta_2$) = 0, so no correction should be applied to mc_{d1R} or mc_{d2R}. This case could occur in practice for zeroing balance before each weighing and evaporation and balance drift are constant but not necessarily negligible.

b) Linear increasing ($\delta_2 = \delta_1 + k$ and $\delta_3 = \delta_2 + k$), thus ($\delta_3 - \delta_2$) = k and ($\delta_1 - \delta_2$) = -k. Here $(\delta_3 - \delta_2)^*$ should be add with changed signal to mc_{d1R} and to mc_{d2R} added as 1.5 × $(\delta_3 - \delta_2)^*$. The linear increasing shows commonly for linear evaporation and/or balance drift when balance is left to drift.

c) ($\delta_1 = \delta_3 \neq \delta_2$) implies ($\delta_3 - \delta_2$) = ($\delta_1 - \delta_2$), by this way the estimate –1 × $(\delta_3 - \delta_2,)^*$ should be summed to mc_{d1R} and –0.5 × $(\delta_3 - \delta_2)^*$ to mc_{d2R}. When the balance is zeroing before weighing, this case will most likely occur in drying of a drop in capillary stem or improper handling of standard weights or pycnometer in put them on the balance receptor.

d) ($\delta_2 = \delta_1 \neq \delta_3$), in this case ($\delta_1 - \delta_2$) = 0 and the estimated value for $(\delta_3 - \delta_2)^*$ shows a value which should not correct mc_{d1R}. This case should be true for improper handling of weights or pycnometer in the third weighing.

e) ($\delta_2 = \delta_3 \neq \delta_1$), here ($\delta_3 - \delta_2$) = 0 implying that no correction should be applied when it should be. This case is the most likely to occur if the care previously to weighing as suggested in [2] was not taken.

From these assumptions, some conclusions can be taken about the applicability of the estimate $(\delta_3 - \delta_2)^*$ and its uncertainty $u(\delta_3 - \delta_2)^*$:

1. $(\delta_3 - \delta_2)^* < u(\delta_3 - \delta_2)^*$: Cases **a** ($\delta_1$=$\delta_2$=$\delta_3$=0), **b** ($\delta_1$=0 and k= 0) and **c** ($\delta_3 - \delta_2 = \delta_1 - \delta_2$ =0) are the fundamental hypothesis to a carefully weighing practice, for a trained staff, so $(\delta_3 - \delta_2)^* < u(\delta_3 - \delta_2)^*$ accomplishes it and no correction should be applied but the uncertainty should be applied. If there is some evidence that had occurred case **e**, uncertainty should be taken in the same way.

2. $(\delta_3 - \delta_2)^* > u(\delta_3 - \delta_2)^*$: Case **b** can occur even for a carefully weighing so $(\delta_3 - \delta_2)^*$ and the uncertainty should be applied to the mass of the drop. Case **c** and **d** relies on the technician judgment. For case **c** applies the correction and uncertainty, however for case **d** no correction is applied but the uncertainty will be.

Due to errors structure, $(\delta_3 - \delta_2)^*$ is correlated with mc_{d1R} and mc_{d2R} thus, in according to the multiplicative term for $(\delta_3 - \delta_2)^*$ to correct mass, the correlation effect could be higher or lower.

4. Uncertainty evaluation

In order to evaluate if the uncertainty of mass achievable by this method complies with weighing requirements, the uncertainty for estimate $(\delta_3 - \delta_2)^*$, mass corrected m_{d1R}, mass corrected m_{d2R} and conventional mass mc_{p2} were calculated to a 20 mg weight for a simulated weighing condition close to the real. The uncertainty was calculated by Monte Carlo method to 1×10^6 trials.

The environmental conditions taken in these calculations are: temperature variation within 21.5 °C ≤ T ≤ 22.5 °C, pressure variation is 995 hPa ≤ p ≤ 1005 hPa, relative humidity variation within 40% ≤ h ≤ 60%. By these values, the air density value is 1.181(12) kg m^{-3} [13]. The assumed density of master solution is 1000(3) kg m^{-3}, the density of balance reference weights is 8000(200) kg m^{-3} and the density of standard weights 8000(15) kg m^{-3}. The balance was adjusted before the weighing and loads are centred carefully.

The Table 1 shows the uncertainty components, probability distribution and parameters to be used in Monte Carlo simulation for the uncertainty of $(\delta_3 - \delta_2)^*$ and mass corrected md_{1R}, mc_{p2} and md_{2R}.

Table 1: Parameters for uncertainty simulation by Monte Carlo method.

Uncertainty component	Distribution	Parameters	Unit
Readability (zero)	R(-a,a)	a = 1	μg
Readability (load)	R(-a,a)	a = 1	μg
Repeatability	$N(\mu,\sigma^2)$	$\mu = 0$; $\sigma = 4$	μg
Sensitivity tolerance	R(-a,a)	$a = 1.5 \times 10^{-6}$	
Temperature sensitivity	R(-a,a)	$a = 5.8 \times 10^{-7}$	
Air density	$N(\mu,\sigma^2)$	$\mu = 1.181$; $\sigma = 0.012$	kg m^{-3}
Solution density	$N(\mu,\sigma^2)$	$\mu = 1000$; $\sigma = 3$	kg m^{-3}
Balance standard density	R(-a,a)	a = 200	kg m^{-3}
Standard weights 1 density	$N(\mu,\sigma^2)$	$\mu = 8000$; $\sigma = 15$	kg m^{-3}
Mass instability of weight 1	R(-a,a)	a = 6	μg
Mass weights 1 20 mg E_2	$N(\mu,\sigma^2)$	$\mu = 20 \times 10^3$; $\sigma = 1.5$	μg
Standard weights 2 density	$N(\mu,\sigma^2)$	$\mu = 8000$; $\sigma = 15$	kg m^{-3}
Mass instability of weight 2	R(-a,a)	a = 6	μg
Mass weights 2 20 mg E_2	$N(\mu,\sigma^2)$	$\mu = 20 \times 10^3$; $\sigma = 1.5$	μg

Additional parameters	Value	Unit
R1= R2= R3	2000	mg
$\rho_2=\rho_3$	1.095	kg m^{-3}
ρ_0	1.2	kg m^{-3}
ρ_c	8000	kg m^{-3}

Table 2 shows the uncertainty for $(\delta_3 - \delta_2)^*$ and mass corrected md_{1R}, mc_{p2} and md_{2R}. Additionally, the non-corrected mass uncertainty, relative standard uncertainties and errors are presented.

Table 2: Errors and uncertainties (μg).

Errors	$(\delta_3 - \delta_2)^*$	md_{1R}		mc_{p2}		md_{2R}	
Values	u	u	u_r(%)	u	u_r(%)	u	u_r(%)
$\delta_1=0$ $\delta_2=0$ $\delta_3=0$	6.1	10.8	0.05	6.5	0.03	8.1	0.04
$\delta_1 = 10$ $\delta_2 = 20$ $\delta_3 = 30$	6.1	10.8	0.05	6.5	0.03	11.6	0.06
$\delta_1 = 10$ $\delta_2 = 20$ $\delta_3 = 10$	6.1	6.5	0.03	6.5	0.03	6.3	0.03
No Error	----	6.5	0.03	6.5	0.03	5.4	0.03

By Table 2, the relative uncertainty to 20 mg is always lower than the limit for relative standard uncertainty, i.e., 0.1%. The higher than 6.5 μg uncertainty of md_{1R} and md_{2R} for cases with error zero means the correlation effect which arises from the sum of $(\delta_3 - \delta_2)^*$ to the mass. As previously, the higher uncertainty in error linearly increasing means the correlation effect from the added corrections $(\delta_3 - \delta_2)^*$ for md_{1R} and $1.5 \times (\delta_3 - \delta_2)^*$ for md_{2R}. Otherwise, for errors just in the second weighing the correlation in md_{1R} and md_{2R} is reduced due to correction, respectively, $-1 \times (\delta_3 - \delta_2)^*$ and $-0.5 \times (\delta_3 - \delta_2)^*$. If it was not regard the error, the lowest uncertainty for mass is obtained. Just in this case the usage of two mass standards would be justified.

5. Conclusion

The uncertainty calculation achievable for a modification in elimination weighing method was performed. This modification, in contrast to the standard elimination weighing, allows correcting the common errors in weighing of radioactive source and provide information for variability studies, thus improving the reliability of elimination weighing.

By the results, the relative standard uncertainty complies with the limit for relative standard uncertainty of the drop mass and the usage of additional standard weights is not fundamental.

Acknowledgments

One of the authors wishes to thank IRD/CNEN and Pronametro/Inmetro for the financial scholarship sponsor.

References

[1] Campion P J 1975 *Procedures for Accurately Diluting and Dispensing Radioactive Solutions (Monographie BIPM-1)* (Sèvres: BIPM) p 33
[2] Lourenço V and Bobin C 2015 Weighing uncertainties in quantitative source preparation for radionuclide metrology *Metrologia* **52** pp S18–29
[3] OIML TC 9/SC 3 Weights 2004 *Weights of Classes E_1, E_2, F_1, F_2, M_1, M_{1-2}, M_2, M_{2-3} and M_3, Part 1: Metrological and Technical Requirements (OIML R 111-1: 2004 (E))* (Paris: BIML) p 80
[4] OIML TC 9/SC 3 Weights 2004 *Conventional Value of the Result of Weighing in Air (OIML D 28: 2004 (E))* (Paris: BIML) p 12
[5] Gallic Y L 1973 Problems in microweighing *Nucl. Instr. and Meth* **112** pp 333–41
[6] Sibbens G and Altzitzoglou T 2007 Preparation of radioactive sources for radionuclide metrology *Metrologia* **44** pp S71–78
[7] Gläser M *et al* 2000 *Comprehensive Mass Metrology* (Berlin: WILEY-VCH) p 561
[8] Croarkin C 1989 An extended error model for comparison calibration *Metrologia* **26** pp 107–13
[9] Fitzgerald R *et al* 2015 Uncertainties in 4πβ-γ coincidence counting *Metrologia* **52** pp S86–96
[10] JCGM 2008 *Evaluation of Measurement Data—Supplement 1 to the "Guide to the Expression of Uncertainty in Measurement"—Propagation of Distributions Using a Monte Carlo Method: JCGM 101:2008* (Sevres: BIPM) p 134
[11] Malengo A 2014 Buoyancy effects and correlations in calibration and use of electronic balances *Metrologia* **51** pp 441–51
[12] Sutton C M and Clarkson M T 1993/1994 A general approach to comparisons in the presence of drift *Metrologia* **30** pp 487–93
[13] EURAMET e.V. 2015 *Guidelines on the Calibration of Non-Automatic Weighing Instruments (EURAMET Calibration Guide No. 18, Version 4.0 (11/2015))* (Braunschweig: EURAMET e.V) p 120

A portable detection system for *in vivo* monitoring of ^{131}I in routine and emergency situations

EA Lucena, ALA Dantas and BM Dantas
Instituto de Radioproteção e Dosimetria – CNEN -
Av Salvador Allende s/n - Rio de Janeiro – RJ – 22780-160

E-mail: bmdantas@ird.gov.br

Abstract: *In vivo* monitoring of ^{131}I in human thyroid is often used to evaluate occupational exposure in nuclear medicine facilities and in the case of accidental intakes in nuclear power plants for the monitoring of workers and population. The device presented in this work consists on a Pb-collimated NaI(Tl)3"x3" scintillation detector assembled on a tripod and connected to a portable PC. The evaluation of the applicability and limitations of the system is based on the estimation of the committed effective doses associated to the minimum detectable activities in different facilities. It has been demonstrated that the system is suitable for use in routine and accidental situations.

1. Introduction

^{131}I plays an important hole in radiation protection since it is widely used in Nuclear Medicine and is one of the radionuclides of interest in the case of accidents and incidents in nuclear power plants. It is highly volatile and radiotoxic, which may represent a radiological hazard, especially in places where significant activities are present in the workplace and in the environment. In such situations, inhalation should be the main pathway of internal exposure. According to the IAEA, a monitoring program for workers occupationally exposed to ^{131}I may include direct thyroid measurement, indirect bioassay of urine samples and workplace monitoring [1]. This work presents a methodology developed for *in vivo* monitoring of internal contamination using a NaI(Tl)3"x3" scintillation detector system installed in a lead collimator and assembled on a tripod. The system is fully portable and suitable for use either in routine monitoring in nuclear medicine clinics as well as in the field as a tool for the response to radiological and nuclear accidents.

2. Methodology

2.1 Calibration procedures

The calibration procedures were carried out in four medical facilities located in the city of Rio de Janeiro and at the IRD Whole Body Counter Laboratory. The NaI crystal is collimated with a 9 mm lead cylinder aimed to reduce ambient background and increase sensitivity up to a level suitable for internal monitoring of occupationally exposed workers. The detector signal output is connected to a compact electronic module which includes the whole electronics (HV power supplier, preamplifier, shaping amplifier, gain stabilizer, ADC and MCA). The whole system is powered through a USB port and spectrum acquisition and processing is controlled by Genie 2000 Basic Spectroscopy Software installed in a portable PC [2].

The neck-thyroid phantom used for the calibration of the detection system was designed and produced at the In Vivo Monitoring Laboratory of IRD [3]. It contains a known activity of ^{133}Ba (^{131}I simulator) standard solution, uniformly distributed on the surface of a filter paper previously cut in the form and shape of a human thyroid (Fig. 1a). Besides the emission characteristics of the radionuclide of interest, the design and production of the phantom must consider the biokinetic of the element in the human body [4] in order to simulate the real distribution in the whole body or in specific organs and tissues [5]. The calibration setup (Fig 1 b) was identical to the counting geometry established for *in vivo* monitoring as shown in Figure 1c.

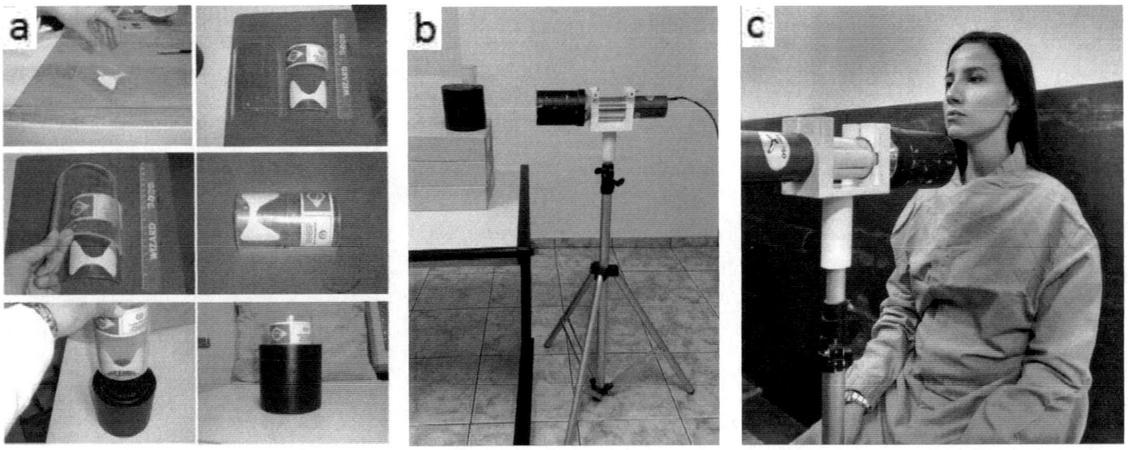

Figure 1. (a) Assembling of the IRD Neck-thyroid phantom; (b) Calibration setup; (c) In Vivo measurement of ^{131}I in the thyroid, using a Pb-collimated NaI(Tl)3x3 installed on a tripod in Lab 1.

The monitoring protocol (including count time and geometry) was optimized in each facility in order to achieve a balance between sensitivity and comfort for the individual. Thus, the neck-thyroid phantom was measured in a series of distances between 14.5 to 25 cm. Although the shorter the distance the higher the counting efficiency, *in vivo* measurements performed too close to the body should be avoided in order to overcome morphological differences between the phantom and an actual human organ. It should also be highlighted that the application of a monitoring protocol should always result in minimum disturbance in routine activities, especially in the case of medical facilities.

2.2 Evaluation of the sensitivity of the technique

The evaluation of the sensitivity is based on the calculation of the minimum detectable activity (MDA) and the corresponding minimum detectable intake (MDI) and the minimum detectable effective dose (MDED. The MDA is based on the Publication ISO12790-1 [6] and is calculated as shown on equation (1):

$$MDA = \frac{4.65\sqrt{N}}{CF \times T} + \frac{3}{CF \times T} \quad (1)$$

Where: N = Total counts of the background; CF = Calibration Factor (cps/Bq) and T = count time (s).

The Minimum Detectable Intake (MDI), is a function of the MDA and depends on the biokinetic of the element, the exposure scenario and time elapsed between intake and measurement. It is calculated as shown on Equation (2):

$$MDI = MDA / m(t)_{inh} \tag{2}$$

Where: MDA=Minimum Detectable Activity (Bq) and m(t) = Retention fraction by inhalation (Bq/Bq).

The last parameter to be calculated is the Minimum Detectable Effective Dose. It is based on the MDI and in the dose coefficient, e(g), associated to the corresponding intake scenario adopted in the simulation. It is calculated as shown on Equation (3):

$$MDED(Bq) = MDI \times e(g)_{inh} \tag{3}$$

Where: MDI = Minimum Detectable Intake (Bq) and $e(g)_{inh}$ = Dose coefficients (mSv/Bq).

In this work it has been used m(t) values corresponding to 1 day after the intake and the dose coefficient, $e(g)_{inh}$, corresponding to volatile iodine in the form of gas. The standard biokinetic and dosimetric values are available in the Publication 78 of the ICRP [4] and can also be generated for specific exposure scenarios and times through the software AIDE [7].

3. Results and Discussion

The calibration factors and sensitivity evaluation for the various facilities which participated in this project are shown in Table 1.

Table 1. Calibration Factors and Sensitivity of the portable NaI(Tl)3"x3" detector system assembled on a tripod (Values of Minimum Detectable Effective Doses were calculated for measurements performed 7 and 14 days after the intake).

Facility	Room Type	[1]D (cm)	[2]CF (cpm/Bq)	[3]MDA (Bq)	[4]MDED (x 10^{-2} mSv)	
					7 days	14 days
Hospital 1	Examination room	23.0	0.18 ± 0.01	296	4.2	8.2
Hospital 1	Office	23.0	0.18 ± 0.01	266	3.8	7.3
Hospital 2	Shadow-shield	14.5	0.33 ± 0.01	136	1.9	3.8
Hospital 3	Examination room	23.0	0.18 ± 0.01	326	4.7	9.0
Hospital 4	Office	23.0	0.18 ± 0.01	386	5.5	10.7
Laboratory 1	Lead-Shielded Room	20.0	0.24 ± 0.01	226	3.2	6.2
Laboratory 2	Mobile Lab	20.0	0.22 ± 0.01	163	2.3	4.5
Laboratory 2	Mobile Lab	25.0	0.15 ± 0.01	168	2.4	4.6
Laboratory 3	Whole Body Counter	20.0	0.15 ± 0.01	51	0.7	1.4

[1]D = Distance; [2]CF = Calibration factor; [3]MDA = Minimum Detectable Activity in 20 min count time;
[4]MDED = Minimum Detectable Effective Dose

The distance of 14.5 cm in Hospital B, where a shadow-shield is available, was defined due to detector support and shielding characteristics. In this case, the detector is positioned in a fixed distance in relation to the neck-thyroid phantom as compared with other facilities. The highest sensitivity was obtained in Laboratory 3, where a heavy shielded whole body counter facility is available, allowing to achieve a lower background count rate as compared to the other facilities where calibrations where performed in conventional rooms.

4. Conclusions

The Minimum Detectable Effective Doses (MDED) for the *in vivo* measurement of ^{131}I in the thyroid obtained at different facilities using a Pb-collimated NaI(Tl)3"x3" detector assembled on a tripod, in 20 minutes count time, is far below the recording level of 1 mSv. Such statement is valid for an exposure scenario considering a single inhalation of ^{131}I vapor and measurements performed up to 14 days after the intake. It should be noted that the count time can be reduced to a certain minimum value depending on local background, providing more comfort to the subject. However, such reduction is limited to the increase in the MDED up to a maximum value suitable for internal monitoring in a routine basis or in accident situations.

This work provides a useful, simple and cost-effective method to evaluate ^{131}I incorporation. The methodology can be implemented routinely in Nuclear Medicine Services as well as in emergency situation in order to estimate the committed effective doses associated to the internal exposed of workers and public.

5. REFERENCES

[1] International Atomic Energy Agency (IAEA) 1999. *Assessment of Occupational Exposure due to Intakes of Radionuclides*. Safety Standards Series. Safety Guide No. RS-G-1.2.

[2] Canberra 2017. http://www.canberra.com/products/radiochemistry_lab/digital-signal-processing.asp.

[3] Dantas BM, Dantas ALA, Cruz-Suarez R 2015. IAEA intercomparison exercises of thyroid measurement: Performance of Latin American and Caribbean laboratories. Radiation Protection Dosimetry. , **v.170**, p.455 - 69

[4] International Commission on Radiological Protection (ICRP) 1998. *Individual Monitoring for Internal Exposure of Workers*. Publication 78

[5] International Commission on Radiation Units (ICRU) 1992. *Phantoms and computational models in Therapy, Diagnosis and Protection*. ICRU Publication 48

[6] International Standard Organization. Radiation Protection (ISO) 2001. *Performance Criteria for Radiobioassay*. Part 1: General principals - ISO 12790-1.

[7] Bertelli L, Melo DR, Lipsztein J, Cruz-Suarez R 2008. AIDE: Internal Dosimetry Software. Radiation Protection Dosimetry **v. 130**, n. 3: 358–367

Acknowledgments

This project was received financial support from IRD-CNEN, IAEA and FAPERJ. The authors are especially grateful to the fellows from the Mechanic Workshop of IRD for their permanent and competent support for the fabrication of essential parts used in the course of this work.

A study to assess the long-term stability of the ionization chamber reference system in the LNMRI

O L Trindade Filho, D A Conceição, C J da Silva, J U Delgado, A E de Oliveira, A Iwahara, L Tauhata

Instituto de Radioproteção e Dosimetria (IRD/CNEN), Avenida Salvador Allende, s/n, Barra, RJ, Brazil, CEP 22783-127

E-mail: octavio@ird.gov.br

Abstract. Ionization chambers are used as secondary standard in order to maintain the calibration factors of radionuclides in the activity measurements in metrology laboratories. Used as radionuclide calibrator in nuclear medicine clinics to control dose in patients, its long-term performance is not evaluated systematically. A methodology for long-term evaluation for its stability is monitored and checked. Historical data produced monthly of 2012 until 2017, by an ionization chamber, electrometer and ^{226}Ra, were analyzed via control chart, aiming to follow the long-term performance. Monitoring systematic errors were consistent within the limits of control, demonstrating the quality of measurements in compliance with ISO17025.

1. Introduction

National Metrology Laboratory on Ionizing Radiation LNMRI/IRD acts in two main areas, radionuclide metrology and radiation dosimetry.

Created in 1973, the IRD has the function of guard, maintenance, dissemination of Brazilian standards of radioactivity for calibrating this quantity. In this way, LNMRI provides traceability for calibrated instruments for all nuclear area. To ensure the accuracy of the measurements, it is necessary that the instrument is calibrated to standards traceable to the LNMRI. To this, the LNMRI must be traceable internationally to other national laboratories through key- comparisons organized by the International Bureau of Weights and Measures (BIPM) or Regional Metrological Organizations (RMO). To keep the results of measurement of radionuclide activity, from the primary standardization until several subsequent applications of the traceability, a well-type ionization chamber in geometry $4\pi\gamma$ was used [1]. This chamber is an integral part of the secondary reference system of measurements in LMNRI. However, to ensure the accuracy of the measurements obtained must be carried out quality control tests to check the performance of the measurement system over time, evaluated by means of a control chart [2]. Quality assurance of measurements of activity is also a requirement of the Norm CNEN 3.05, from the point of view of radiation protection and quality control of radiopharmaceuticals used in nuclear medicine services [3]. Quality assurance is also a requirement of the Norm ABNT NBR ISO/IEC 17025, for calibration and essay laboratories [4]. The LNMRI, while reference laboratory for calibration of the radioactive sources, develops, coordinates and performs the interlaboratorial comparison programs of the radionuclide calibrators used not only in nuclear medicine clinics as well as for the certification of the producer centers of radiopharmaceuticals in different institutions in the country. Radionuclide metrology laboratory monitors the statistical fluctuations in the reference system through source of ^{226}Ra, but does not use the control chart. In 2017, at the meeting of the Consultative Committee for ionizing radiation, radionuclide area, of the BIPM, was forwarded a proposal demanding that each system of ionization chamber that has been calibrated for a period more than 2 years, to validate their calibrations keep a control chart in order to demonstrate that there is no trend or systematic error.

2. Objective

The aim of this paper is to apply the method of statistical control chart showing the variation of the activity of a reference source of ^{226}Ra, long-term, to assess the stability of the measurement system of the secondary standard ionization chamber. In this case, it is proposed to monitor whether the values of the reference source activity are kept within the limits of the control as a stable system.

3. Materials and methods

The construction of a control chart requires the following planning to define: which processes will be controlled (validated essays methods or instruments/measuring equipment); what characteristics of the process will be monitored; which materials will be monitored (certified reference materials-MRCs, reference material, standard solutions, etc.); equipment/measuring instruments and supplies required; and choose the type of control.

Control charts can be classified into two types: control charts for variables or control charts for attributes. If the feature to be controlled is expressed as a number in a continuous scale of measurement is called variable. In this case, it is appropriate to describe the characteristic as a measure of central tendency and variability. These graphs are called control charts for variables. Many features are not measurements on a continuous scale or even on a quantitative scale. In these situations, it is possible to judge each process unit as or based if she has or not certain attributes, or else, you can count the number of "defects" that appear in one unit of the product. Control charts for such characteristics are called control charts for attributes [2].

Control charts for variables, which a characteristic that is measured on a numeric scale is called variable. In this article will address three charts for variables: charts $\bar{X}$ and R (mean and range); charts $\bar{X}$ and s (mean and standard deviation); and charts I and MR (individual values and amplitude), according figure 1[2].

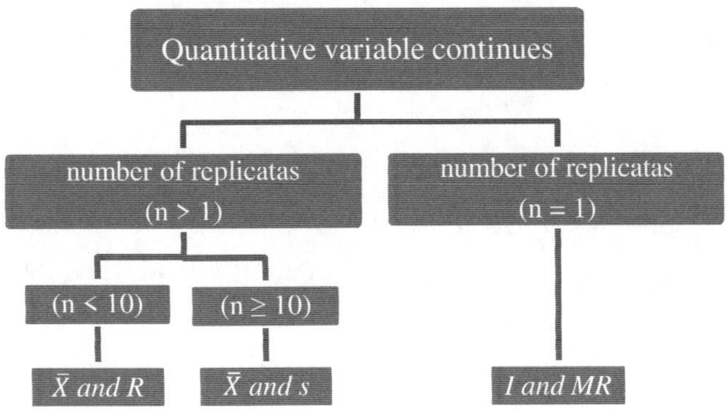

Figure 1. Construction scheme of control charts for variables.

To determine the activity of radionuclides in the ionization chamber (CENTRONIC IG11) with electrometer coupled, it were here performed measurements of the background and of a reference source of ^{226}Ra, as can be seen in figure 2 [5]. For more than 5 years, the activity of different radionuclides had been measured in the LNMRI by an ionization chamber type $4\pi\gamma$. The stability of the chamber has been systematically verified by measurement of the reference source of ^{226}Ra of long half-life ($T_{1/2}$ = 1600(7) a) and the results of the measurements were obtained with their uncertainties

[6]. This reference source was measured in regular interval of at least once per month, and each measurement has been corrected for background radiation. Thus, 71 values were generated over the period between 2012-2017, and each value is an average of 10 measurements. The activities of historical measurements stored in a database were all corrected to a reference date.

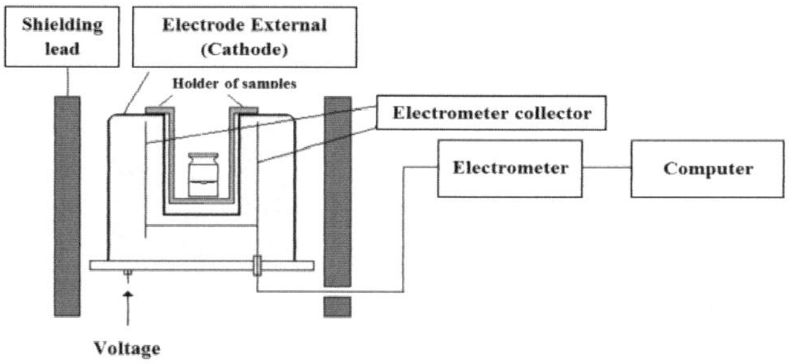

Figure 2. Ionization Chamber CENTRONIC IG11 with electrometer KEITHLEY 6517A of LNMRI.

In this way, were here adopted the methods of Control Chart for Individual Values (I) and Control Chart of Mobile Amplitudes (MR), used for continuous data, individual, without references, considering only their own measurements and the date chosen, including individual value graphs and amplitude mobile graphs, as follow [2].

For individual values:

$$\mathrm{LSC} = \overline{\mathrm{X}} + \mathrm{E}_2\,\overline{\mathrm{MR}} \quad (1)$$
$$\mathrm{LC} = \overline{\mathrm{X}} \quad (2)$$
$$\mathrm{LIC} = \overline{\mathrm{X}} - \mathrm{E}_2\overline{\mathrm{MR}} \quad (3)$$

For individual mobile:

$$\mathrm{LSC} = \mathrm{D}_4\overline{\mathrm{MR}} \quad (4)$$
$$\mathrm{LC} = \overline{\mathrm{MR}} \quad (5)$$
$$\mathrm{LIC} = \mathrm{D}_3\overline{\mathrm{MR}} \quad (6)$$

Where:
LSC: Upper Control Limit;
LC: Control Limit;
LIC: Lower Control Limit;
$\bar{X}$: Average of the Activities;
$\overline{MR}$: Average Amplitude Mobile (difference of consecutive values);
d_2, D_3, D_4: Values of Tabulated Constant;
$E_2 = \frac{3}{d_2}$, in that 3 is the number of standard deviation for the outer rows, the Upper Control (LSC) and Lower (LIC) Limits, and d_2 is a given value [7].

The points that remain between LSC and LIC shall be considered to be "under control". However, if one or more points are outside these limits, there is a need for appropriate corrective actions.

4. Results and discussion

For this study were used the data obtained by the laboratory of radionuclides in June 2012 until June 2017, given a total of 71 current values, converted into net activity, produced by ^{226}Ra source. All data were referenced to June 6, 2017. Table 1 lists the measurements and the calculated MR values, with both in unit kBq. Figure 3 shows the results of the individual values of the control chart, with standard uncertainty of 0.04%. Figure 4 shows the mobile amplitude control chart. It can be checked in these control charts that no point was outside the limits of control (outlier). I.e., not showing trends or special causes despite of the narrow difference between the upper and lower limits, of 0.19% of the average (Chart I). Therefore, in general, is not necessary to apply corrective actions, because the secondary standard measurement system which uses ionization chamber is under statistical control for the evaluated period, demonstrating the high long-term stability, which is a characteristic of ionization chambers.

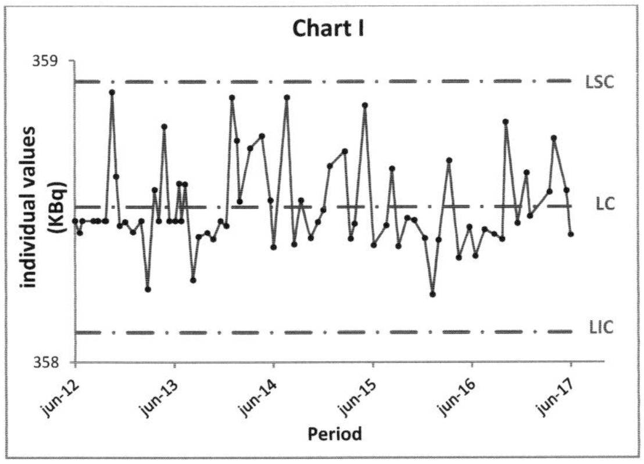

Figure 3. Individual value control chart, with standard uncertainty of 0.04%.

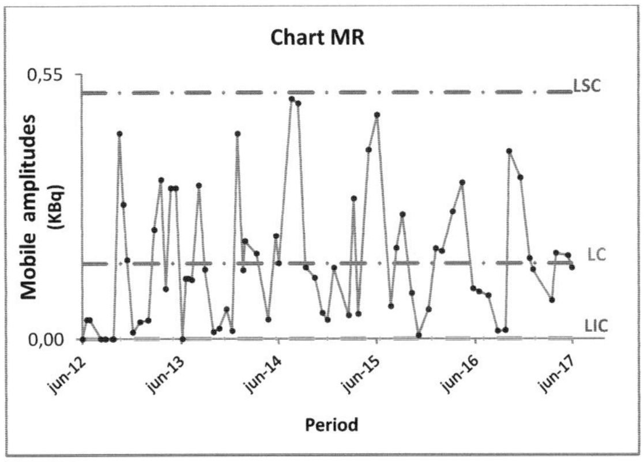

Figure 4. Mobile amplitude control chart.

5. Conclusion

The performance of the CENTRONIC IG11 ionization chamber, in accordance with the method adopted here remains within the limits that the showing as a stable system, despite the stability evaluation criteria have been limited (not using replica of each measurement). Additionally, this stability comes to demonstrate the quality of the measurements carried out in compliance with the requirements of the Norm ABNT NBR ISO/IEC 17025 and in compliance with the recent recommendations of the Consultative Committee for Ionizing Radiation of BIPM, CCRI (II), which requires the adoption of control charts for measurement systems with ionization chambers calibrated for over two years.

On the other hand, according to the Norm CNEN (NN 3.05), all the radionuclide calibrators used in Nuclear Medicine Services (SMN) in the country should have their results produced in ionization chamber. This suggests that control chart-based methods can be applied to SMN, taking advantage of that already use standardized sources of intermediary half-lives, for example, ^{133}Ba or ^{137}Cs, for essaying of equipment. It should be emphasized that there are other methods of control chart in literature which may fit better in stability evaluation, according to the case.

6. References

[1] Siegert H, Schrader H and Schoetzig U 1998 Half-life Measurements of Europium Radionuclides and the Long-term Stability of Detectors. *Appl. Rad. and Isto.* **49**, No. 9-11, pp. 1397-1401.

[2] Oliveira C C , Granato D, Caruso M S F, Sakuma A M 2013 Manual para elaboração de cartas de controle para monitoramento de processos de medição quantitativos em laboratórios de ensaio. 1ª edition São Paulo - SES/SP

[3] CNEN.NN 3.05 Requisitos de segurança e proteção radiológica para serviços de medicina nuclear Resolução CNEN 159/13.

[4] ABNT NBR ISO / IEC 17025 2005 Requisitos Gerais para Competência de Laboratórios de Ensaio e Calibração.

[5] Da Silva C J, Iwahara A, Moreira D S, Delgado J U, Gomes R S 2012 ^{57}Co half-life determination *Appl. Rad. and Isot..* **70**, 1924-1926.

[6] Delgado J U, Morel J, Etcheverry M 2002 Measurements of photon emission probabilities from the decay ^{226}Ra . *Appl. Rad. and Isot.* **56**, 137- 140.

[7] Mullins, E 1994, *Analyst,* **119**, 369-375.

Activity measurements of ^{55}Fe by two different methods

Paulo A. L da Cruz[1], Akira Iwahara[1], Carlos J. da Silva[1], Roberto Poledna[1], Jamir S. Loureiro[1], Monica A. L. da Silva[1], Anelise Ruzzarin[2]

[1]Laboratório Nacional de Metrologia das Radiações Ionizantes/Instituto de Radioproteção e Dosimetria/Comissão Nacional de Energia Nuclear (LNMRI/IRD/CNEN) Av. Salvador Allende n° 9, Barra da Tijuca, Rio de Janeiro, Brasil, CEP 22783-127;

[2]Programa de Engenharia Nuclear/Coordenação de Programas de Pós-Graduação em Engenharia/Universidade Federal do Rio de Janeiro (PEN/COPPE/UFRJ) Rua Moniz de Aragão, 360, Bloco 1, Ilha do Fundão – Cidade Universitária, Rio de Janeiro, Brasil, CEP 21941-594

palcruz@ird.gov.br

Abstract. A calibrated germanium detector and CIEMAT/NIST liquid scintillation method were used in the standardization of solution of ^{55}Fe coming from a key-comparison BIPM. Commercial cocktails were used in source preparation for activity measurements in CIEMAT/NIST method. Measurements were performed in Liquid Scintillation Counter. In the germanium counting method standard point sources were prepared for obtaining atomic number versus efficiency curve of the detector in order to obtain the efficiency of 5.9 keV KX-ray of ^{55}Fe by interpolation. The activity concentrations obtained were 508.17 ± 3.56 and 509.95 ± 16.20 kBq/g for CIEMAT/NIST and germanium methods, respectively.

1. INTRODUCTION

^{55}Fe disintegrates by electron capture to ^{55}Mn with emission of electrons and X-ray of low energy. A gamma transition of 125.95 keV with a very low emission probability of 1.3 x 10^{-7} % has been observed [1]. It is a very important radionuclide in the calibration of low energy photon spectrometers, proportional and scintillation detectors due to its long half-life of 1001.1(21) d and emission of 5.9 keV characteristic X-ray. This work describes the results of the standardization of a ^{55}Fe solution by using a planar type germanium detector and the CIEMAT/NIST liquid scintillation counting methods [2, 3]. The measured solution originates from an international key-comparison organized by Bureau International des Poids et Mesures (BIPM) in 2005.

For germanium counting standard point sources of ^{51}Cr, ^{54}Mn, ^{57}Co and ^{65}Zn with standard combined uncertainties lower than 1 % were used for plotting the atomic number versus efficiency curve of the detector in order to get the efficiency of 5.9 keV energy of ^{55}Fe by interpolation. Correction for the attenuation in source support, air and beryllium window of the detector were determined for the X-rays energies of the radionuclides used to obtain the efficiency curve.

In the CIEMAT/NIST method with ^{3}H-tracer, a Liquid Scintillation Counter model Wallac 1414 was used and five samples made by scintillation cocktails with ^{55}Fe and $FeCl_3$ inactive carrier were prepared for measurements. Several commercial cocktails were tested for checking the stability of the sources in a period of 12 days, evaluated by means of the counts normalized to the value of the first measurement and the spectra obtained over the period. The results showed that only the samples in 'Hisafe'3 with and without carrier showed to be stable.

2. MEASUREMENT BY PLANAR GERMANIUM DETECTOR

The planar germanium detector used was a Canberra model GL2020R with 400 eV FWHM energy resolution at 5.9 keV with Beryllium window of 0.5 mm thick. Standard point sources of ^{51}Cr, ^{54}Mn, ^{57}Co and ^{65}Zn, previously standardized by 4πβ(PC)-γ(NaI) coincidence counting method were used in the calibration of efficiency versus nuclide atomic number for the activity determination of ^{55}Fe solution. Standard uncertainties (k = 1) of these sources were 0.29, 0.41, 0.29 and 0.40 %, respectively for ^{51}Cr, ^{54}Mn, ^{57}Co and ^{65}Zn. The detection efficiency versus atomic number at 5 cm distance between the detector window and the source was obtained with these sources.

The uncertainty in determining the K X-ray net peak area of each standard source were of the order of 0.1 to 0.3 % for each measurement. The output pulses were used to a multi-channel analyzer and the peak areas were analyzed by commercial Maestro software [4].

Each measurement was corrected for the attenuation effects of source support (polystyrene), of air and of Beryllium window of the detector. The data for calculating the attenuation effects were taken from NIST X-ray Mass Attenuation coefficient [5]. The detection efficiency at 5.90 keV ^{55}Fe K X-ray was evaluated by interpolating the efficiency versus atomic number curve obtained from the standard sources. The activity of ^{55}Fe sources were obtained by dividing the X-ray count rate by the interpolated efficiency and emission intensity of 5.9 keV X-ray of ^{55}Fe.

The sources were prepared by dropping quantitative masses of radioactive solution onto the center of a polystyrene disk of 25.4 mm diameter and 0.05 mm thick. A drop of TWEEN 20 was diluted by 1000 times and added to each source and dried in a desiccator with silica gel. After drying, the sources were covered by the same polystyrene film and placed at 5 cm from the detector for measurements.

Figure 1 shows the efficiency versus atomic number curve obtained with ^{51}Cr, ^{54}Mn, ^{57}Co and ^{65}Zn and sources placed at 5 cm from de detector window. Table 1 gives the relevant data of mass attenuation corrections to correct count rates and to obtain the efficiency versus atomic number curve. Table 2 shows the atomic data of this radionuclides.

The measured activity obtained was 509.95 ± 16.20 (3.2 %) kBq/g at reference date of 1 December 1st, 2005, 00:00 UTC. The uncertainty budget is presented in table 3.

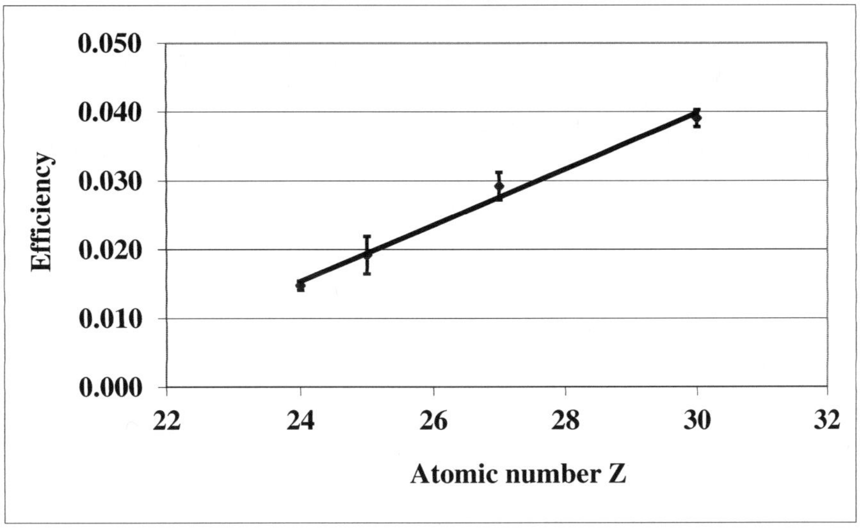

Figure 1 Efficiency versus atomic number curve.

Table 1 Mass attenuation correction for X-ray of ^{51}Cr, ^{54}Mn, ^{57}Co and ^{65}Zn used to obtain the detection efficiency of 5.9 keV X-ray of ^{55}Fe.

Nuclide (Z)	Mass attenuation correction		
	Polystyrene	Air	Beryllium
^{51}Cr (24)	0.9109	0.7852	0.6688
^{54}Mn (25)	0.9282	0.8244	0.7267
^{57}Co (27)	0.9550	0.8863	0.8183
^{24}Zn (30)	0.9044	0.9441	0.9044
^{55}Fe (26)	0.9466	0.8659	0.7879

Table 2 X-ray energy and efficiency of ^{51}Cr, ^{54}Mn, ^{57}Co and ^{65}Zn used to obtain the detection efficiency of 5.9 keV X-ray of ^{55}Fe.

Nuclide (Z)	Energy K X-ray (keV)	Efficiency	Standard uncertainty
^{51}Cr(24)	4.95	0.014773	0.000338
^{54}Mn (25)	5.40	0.019174	0.001366
^{57}Co(27)	6.40	0.029212	0.000999
^{65}Zn(30)	8.04	0.039044	0.000611
^{55}Fe (26)	5.90	0.023077	0.000641

Table 3 Uncertainty budget evaluated in the determination of ^{55}Fe activity concentration by germanium method.

Component due to	Relative standard uncertainty
Counting statistics	0.30
Decay data	1.63
Detection efficiency	2.8
Weighing	0.05
Pile-up	0.01
Half-life	0.006
Combined standard uncertainty	3.2

3. MEASUREMENT BY THE LIQUID SCINTILLATION COUNTER

The Liquid Scintillation Counter Wallac 1414 was used in the CIEMAT/NIST counting method with ^{3}H as tracer and efficiency calculation CN2003 code [6].

Samples made by commercial scintillation cocktails Ultima Gold, Instagel Plus and Optiphase 'Hisafe'3, manufactured by Perkin Elmer were measured by a period of 12 days to stability checking of the samples. The measurement counts were normalized to the first measurement and the results are presented in the figure 2.

The results showed that only the samples prepared in Optiphase 'Hisafe' 3 with and without carrier showed to be stable. The spectra obtained from this cocktail in the initial data and after 12 d are presented in the figure 2. Five samples were prepared with this cocktail and the activity concentration obtained was 508.17 ± 3.56 (0.70 %) kBq/g at reference date. The uncertainty budget is presented in Table 4.

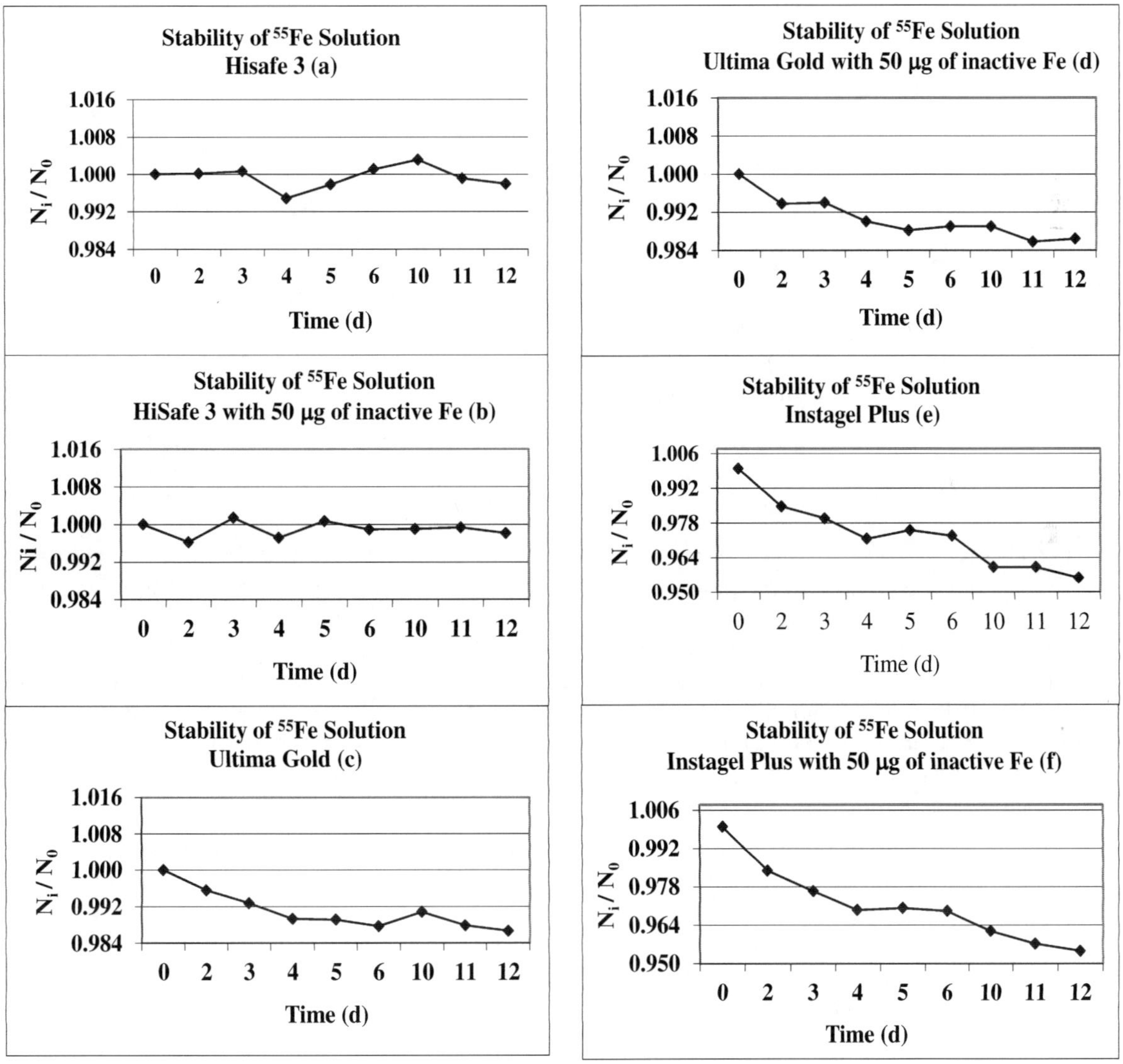

Figure 2 (a) to (f) Test of source stability checking for samples prepared in several cocktails.

Table 4 Uncertainty budget evaluated in the determination of activity concentration of ^{55}Fe by CIEMAT/NIST method.

Component due to	Relative standard uncertainty
Counting statistics	0.03
Decay data	0.30
Weighing	0.01
Half-life	0.006
Quench determination of ^{3}H	0.363
^{3}H activity standardization	0.307
^{3}H counting	0.07
PMT efficiency lost	0.33
Quenching ^{55}Fe	0.20
PMT asymmetry	0.10
Ionization quenching	0.10
Combined standard uncertainty*	0.70

4. DISCUSSION

The two major contributions to the total uncertainty in the germanium method are the determination of the efficiency of the 5.9 keV K X-ray of ^{55}Fe (2.8 %) by interpolation of the efficiency curve and the atomic decay parameters of the radionuclides found in literature used to obtain this curve (1.63 %). Standard sources with smaller uncertainties should be used to reduce these uncertainties. Much lower uncertainty (0.70 %) was found in the CIEMAT/NIST method, so the value of this method was adopted as a reference value for comparison with the results of the BIPM key-comparison. However, the activities of ^{55}Fe obtained by planar germanium and liquid scintillation methods agreed well within the evaluated uncertainties.

The values of activity concentration by germanium and liquid scintillation methods agree also reasonably well with the published reported value (without outliers) of the international key-comparison, which was 516.8 ± 2.9 kBq/g [7], as depicted in figure 3. The standardization of ^{55}Fe is not trivial due to its decay scheme (it decays mainly by electron capture with nearly 100 % emission probability to the ^{55}Mn ground state), emitting a very low energy characteristic X-ray (5.9 keV). This has been shown in the international comparison between the results reported by participants where the spread of the results relating to the arithmetic mean ranged between -1.7 % and 3.7 % [7].

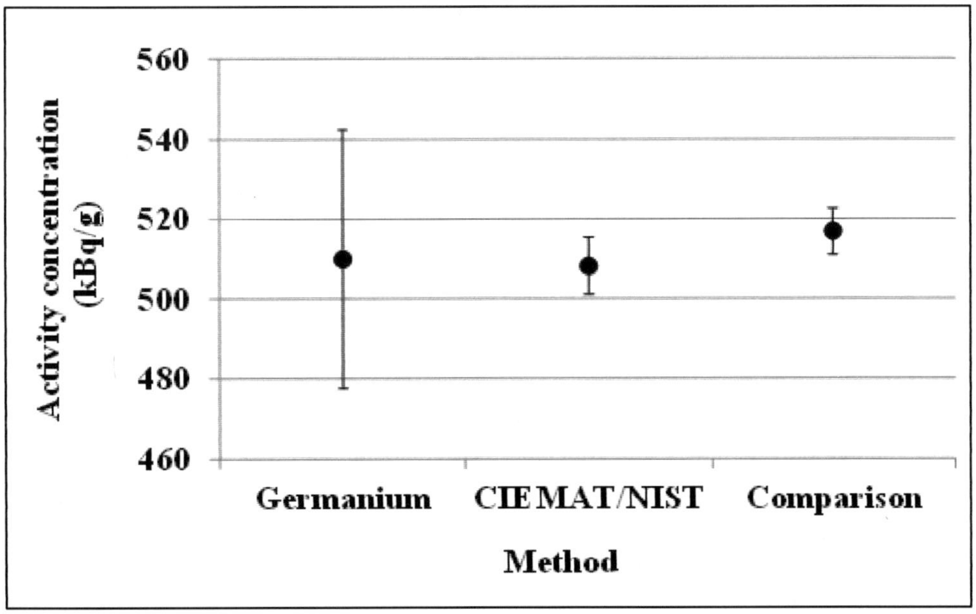

Figure 3 Activity concentration of ^{55}Fe solution (k = 2).

5. CONCLUSIONS

Despite large uncertainty, the indirect measurement method using a planar germanium detector was satisfactory for activity determination of ^{55}Fe with results comparable to the liquid scintillation method. Of the two methods the most simple with less work is the liquid scintillation counting using CIEMAT/NIST method which gave uncertainty compatible with the needs of the end users as laboratories of metrology and medical application. Most cocktails tested proved to be unstable over time requiring more research in finding a suitable scintillation cocktail for ^{55}Fe radioactive solution. The results obtained in LNMRI presented good agreement with BIPM key-comparison reference value (k = 2).

REFERENCES

[1] Bé M M and Chisté V 2006 Table de Radionucléides, *LNE-LNHB/CEA* France.
[2] Delgado, J U 2000 Probabilidade de Emissão X e Gama Determinadas por Espectrometria em Regiões Complexas, Tese de Doutorado *COPPE/UFRJ* Brasil.
[3] Malonda A G 1999. Free Parameter Models in Liquid Scintillation Counting, *Editorial CIEMAT* Spain.
[4] Maestro II for Windows 1991 Version 1.40 model A64-BI*EG&G ORTEC.*
[5] Hubbell J H and Seltzer S M 2004 Tables of X-Ray Mass Attenuation Coefficients and Mass Energy Absorption Coefficients, *National Institute of Standards and Technology* Gaithersburg MD/USA.
[6] Gunther E 2003 Program CN2003: A program to calculate the LC efficiency of a nuclide vs. efficiency the tracer H-3 (CNET) *PTB* Germany.
[7] Ratel G 2008 Analysis of the results of the international comparison of activity measurements of a solution of ^{55}Fe, Appl. Radiat Isot. **66** pp. 729-732.

IOP Conf. Series: Journal of Physics: Conf. Series **975** (2018) 012061 doi:10.1088/1742-6596/975/1/012061

General requirements to implement the personal dose equivalent Hp(10) in Brazil

Amanda Gomes Lopes and Francisco Cesar Augusto Da Silva

Institute of Radiation Protection and Dosimetry (IRD/CNEN), Rio de Janeiro, Brazil
E-mail: amandagl@bolsista.ird.gov.br; dasilva@ird.gov.br

Abstract. To update the dosimetry quantity with the international community, Brazil is changing the Individual Dose (Hx) to the Personal Dose Equivalent Hp(10). A bibliographical survey on the technical and administrative requirements of nine countries that use Hp(10) was carried out to obtain the most relevant ones. All of them follow IEC and ISO guidelines for technical requirements, but administrative requirements change from country to country. Based on countries experiences, this paper presents a list of important general requirements to implement Hp(10) and to prepare the Brazilian requirements according to the international scientific community.

1. Introduction

In Brazil occupationally exposed workers can only use personal dosimeters provided by services approved by the Regulatory Authority, the National Commission of Nuclear Energy (CNEN-Brazil). The responsibility to provide the formal approval of operation for External Individual Monitoring Services has been delegated to the Institute of Radiation Protection and Dosimetry (IRD/CNEN) since 1995 by CNEN. A formal Committee named "Committee for the Evaluation of Services of Essays and Calibration" (CASEC/IRD) has the responsibility of defining the requirements for approval, auditing the laboratories, organizing inter-comparison exercises, certifying the technical officers of the laboratories and recommending the approval or cessation of laboratories activities to the Director of the Institute.

The External Individual Monitoring Services (SMIE) must comply with a series of management and specific requirements to obtain the approval. Ten laboratories in Brazil are approved to provide dosimetry services using Film Badge, TLD and OSLD, with around 183,000 radiation workers.

The problem is that the dosimetric quantity used in Brazil is the Individual Dose (Hx), which is an operational quantity for photons obtained by multiplying the value obtained in the detector, calibrated in kerma in the air, by the factor f=1.14 Sv/Gy [1]. Internationally, the main operational quantity, defined by International Commission on Radiation Units and Measurements (ICRU) and adopted by International Commission on Radiological Protection (ICRP), is the Personal Dose Equivalent Hp(d). This is the dose equivalent in soft tissue below a specified point on the body, at an appropriate depth d (10 millimeters) [2].

In order to be updated with the international community, Brazil is changing the old quantity H(x) to the useful quantity Hp(10) for Effective Dose and making a review of general requirements to implement the Hp(10). As the technical requirements are generally based on and follow the International Electrotechnical Commission (IEC) and/or International Organization for Standardization (ISO) guidelines [3],[4], the main objective of this paper is to suggest a list of the most important general requirements to implement the personal dose equivalent Hp(10) in Brazil, based on the Countries experiences. A bibliographical survey was made on technical regulations of nine (9) Countries, that use Hp(10), highlighting the required technical and administrative criteria.

IOP Conf. Series: Journal of Physics: Conf. Series **975** (2018) 012061 doi:10.1088/1742-6596/975/1/012061

Administrative requirements are usually developed in accordance with the regulations and laws of each Country and have differences from one to the other.

2. Methodology

2.1. International Technical Requirements

There are two most important international technical standard that provides requirements applied to dosimetry system: IEC 62387:2012 "Radiation protection instrumentation – Passive integrating dosimetry systems for personal and environmental monitoring of photon and beta radiation" [3] and ISO 4037:1999 "X and gamma reference radiation for calibrating dosimeters and doserate meters and for determining their response as a function of photon energy" [4].

The IEC 62387 is the leading global organization that prepares and publishes International Standards for all electrical, electronic and related technologies. Many Countries conduct testing for system evaluation following guidelines recommended in this standard. Two tests can be highlighted: the radiation energy and angle of incidence for Hp(10) and the non-linearity test.

The ISO 4037 - Part 3 describes procedures for calibrating and determining the response of dosimeters and doserate meters in terms of the ICRU operational quantities for radiation protection purposes. It also describes the reference angles and quality of energies used in the evaluation tests described in IEC.

2.2. International Management Requirements

Based on technical regulations of nine (9) Countries, that use Hp(10) in the dosimetry system, a selection of some management requirements was done to support a new Brazilian regulation.

In the Spanish technical regulation the highlight is about dosimeter reading that must have the following automatically operations: dosimetric identification, reading, recording and data storage for elaboration of thermoluminescent emission curves, application of correction factors, verification of system stability, estimation and allocation of doses [5].

According to Canadian Requirements, for routine performance tests, the dosimetry service shall comply with the following requirements: irradiate test dosimeters to known doses, usually under standard exposure conditions (e.g., at normal incidence with the calibration radiation); treat test dosimeters in the same way as routine dosimeters; if processing is required, provide test dosimeters without identifying them to the processing laboratory [6].

Metrological tests are carried out in Germany for investigate whether the response or the indication of the dosimeter varies by no more than the permitted maximum values if an influence quantity is changed from its reference value to any other value within its rated range of use [7].

The Portuguese accreditation body audits the services, outside the evaluation program, to verify compliance with the accreditation decision and inter-laboratory comparisons are mandatory to demonstrate the competence and performance for some accredited evaluations [8].

European Commission [9] and Ireland Environmental Protection Agency [10] have in their technical recommendations and guidelines for monitoring individuals occupationally exposed to external radiation that is necessary the payment of fees for Approval of Dosimetry Services.

Approval of Dosimetry Services (ADS) in United Kingdom will be granted for an indefinite period of time. Approved Body (HSE) will, however, carry out a reassessment of the dosimetry service, usually at intervals of 5 or 7 years depending on the nature of the dosimetry service. Fees are payable for all applications, whether or not they are successful [11],[12],[13].

In France, laboratories that provide dosimetry services, private or public sector, must be both accredited and approved. Performance tests are done "announced" with dosimeters irradiated under defined conditions. The laboratory carries out in "blind" analysis and the result is compared with international standards, either ISO or IEC (depending on the techniques) [11].

In the USA there are two dosimetry service assessment programmes: The National Voluntary Laboratory Accreditation Program (NVLAP) and the Department of Energy Laboratory Accreditation Program (DOELAP). They perform a proficiency test and an audit visit in the personal dosimetry laboratory to grant accreditation. The validation period of accreditation is two years in both programmes [11].

At present, in European Commission (EU) there is increasing pressure for accreditation/certification of approved dosimetry services (ADS), and in particular, demonstration of conformity with ISO/IEC 17025:2005 [9],[14]. Dosimetry services in different EU Member States do not have to comply with the same legal or approval requirements, and these requirements are not always based to the same degree on standards. Nevertheless, many dosimetry services in the EU are accredited according to ISO/IEC 17025:2005 and this provides a certain uniformity of quality in individual monitoring services in Europe [9].

2.3. Brazilian Management Requirements

The External Individual Monitoring Services (SMIE), to be approved by the CASEC/IRD, must comply with one general requirement and three specific requirements. Some items of the "General Requirements to Approval Individual Monitoring Services" - RT N° 001/1995 [15] to be followed by the SMIE are: dosimetry procedures and QA program must be evaluated and approved by CASEC/IRD; a specialized staff with a technical officer, a quality manager and dosimetry technicians must be organized; technical officers must be submitted a qualify certification made by CASEC/IRD. After approved process the SMIE receives a Three Years Certification and must participate in a monthly CASEC/IRD Follow-up Performance Program with 5 dosimeters to be irradiated by Secondary Standards Dosimetry Laboratory (SSDL/IAEA/IRD) for a blind test. In this Program the SMIE are evaluated in the Trumpet Curves and 90% of the dosimeters have to be within it. After three years new audit is done for renovation.

3. Results and Discussion

Based on the Countries experiences to approval dosimetry services, a list of ten relevant general requirements are presented, as result of this bibliographic survey. This list shows some requirements to be added in the new Brazilian requirement and others that are the same. The Brazilian requirements showed in item 4 are not in this list.

The top 10 general requirements to implement the personal dose equivalent Hp(10) in Brazil are:

(1) Private or public laboratories, which provide dosimetry services, are accredited according to ISO/IEC 17025:2005 and approved by the National Requirement and these processes are generally paid.

(2) Performance dosimetry tests are carried out for several irradiation condition to confirm the accuracy system.

(3) Independent inter-comparison test can be done by the dosimetry service at least once every two years.

(4) The results of performance tests and of inter-comparisons are evaluated using the Trumpet Curves.

(5) The performance of the approved dosimetry services are evaluated by an "announced test" made by the approval body.

(6) The accreditation and approval are granted for a period of two years maximum.

(7) The reference calibration of the dosimetry system should be repeated at regular intervals of two years.

(8) All reported doses are stored for an appropriate period, in conformity with ISO/IEC 17025.

(9) Annual doses are sent to a National Dose Register by the dosimetry service.

(10) Review, annually, the dosimetry process to optimize, to have under control and produce accurate results to be comply with the requirements.

4. Conclusion

The main objective of this paper is to suggest a list of the most important general requirements to implement the personal dose equivalent Hp(10) in Brazil, based on experiences of nine Countries. It was observed that some Countries requirements must be taking into account in the new Brazilian

IOP Conf. Series: Journal of Physics: Conf. Series **975** (2018) 012061 doi:10.1088/1742-6596/975/1/012061

requirement. These requirements must be analysed if they can be applied in Brazil. For other side, it was found some Countries requirements that are the same of the Brazilian requirements. It shows that the Brazilian processes are updated on international one. The main conclusion is that with the requirements from the Countries the new Brazilian general requirement will be in accord to the international scientific community.

References

[1] Comissão Nacional de Energia Nuclear 2011 Critérios para Cálculo de Dose Efetiva a partir da Monitoração Individual *Posição Regulatória 3.01/005* (Brasil: CNEN)

[2] International Atomic Energy Agency 1999 Assessment of Occupational Exposure due to External Sources of Radiation *Safety Standards Series RS-G-1.3* (Austria: IAEA)

[3] International Electrotechnical Commission 2012 Radiation Protection Instrumentation – Passive Integrating Dosimetry Systems for Personal and Environmental Monitoring of Photon and Beta Radiation *International Standard IEC 62387* (Switzerland: IEC)

[4] International Organization for Standardization 1999 X and Gamma Reference Radiation for Calibrating Dosemeters and Doserate Meters and for determining their response as a function of Photon Energy – Part 3: Calibration of area and Personal Dosemeters and the Measurements of their response as a function of energy and angle of incidence *International Standard ISO 4037-3* (Switzerland: ISO)

[5] Consejo de Seguridad Nuclear 2006 Requisitos Técnicos-Administrativos para los Servicios de Dosimetría Personal *Guia de Seguridad 7.1* (Spain: CSN)

[6] Canadian Nuclear Safety Commission 2006 Technical and Quality Assurance Requirements for Dosimetry Services *Regulatory Standard S-106 Revision 1* (Canada: CNSC)

[7] Physikalisch-Technische Bundesanstalt 2013 Personal Dosimeters for Measuring the Personal Dose Equivalent at 10 mm depth, Hp(10) and the Personal Dose Equivalent at 0.07 mm depth, Hp (0.07) *Radiation Protection Instruments* (Germany: PTB)

[8] Instituto Português de Acreditação 2016 Regulamento Geral de Acreditação *DRC001* (Portugal: IPAC)

[9] European Commission 2009 Technical Recommendations for Monitoring Individuals Occupationally Exposed to External Radiation *Radiation Protection No 160* (Luxembourg: EC)

[10] Environmental Protection Agency 2015 Approval of Dosimetry Services in Ireland *Guidelines for Applicants* (Ireland: EPA)

[11] Health Protection Agency 2006 Review of the level of Accuracy required and means of demonstrating that accuracy for approval of Dosimetry Services by the Health and Safety Executive *Research Report 477* (UK: HPA)

[12] Health and Safety Executive 2008 Requirements for the Approval of Dosimetry Services under the Ionising Radiations Regulations 1999 Part 1 – External Radiations (UK: HSE)

[13] Health and Safety Executive 2010 Statement on the Approval of Dosimetry Services *Ionising Radiations IRR Regulation 1999 REPPIR* (UK: HSE)

[14] International Organization for Standardization 2005 General requirements for the competence of testing and calibration laboratories *International Standard ISO 17025 (*Switzerland: ISO)

[15] Comitê de Avaliação de Serviços de Monitoração Individual Externa 1995 Critérios Gerais para Certificação de um Serviço de Monitoração Individual Externa *Regulamento Técnico IRD-RT No 001.01/95* (Brasil: CASMIE/IRD)

Acknowledgments

The authors are grateful to Brazilian development agency (*Conselho Nacional de Desenvolvimento Científico e Tecnológico - CNPq)* to grant a scholarship project (*Projeto de Capacitação Individual - PCI*) that made this work possible.

IOP Conf. Series: Journal of Physics: Conf. Series **975** (2018) 012062 doi:10.1088/1742-6596/975/1/012062

Primary Standardization of ^{152}Eu by 4πβ(LS) – γ (NaI) coincidence counting and CIEMAT-NIST method

A Ruzzarin [1], P A L da Cruz [2], A L Ferreira Filho [2], A Iwahara [2]

[1] Laboratório de Instrumentação Nuclear/Programa de Engenharia Nuclear/Instituto Alberto Luiz Coimbra de Pós-Graduação em Pesquisa de Engenharia /Universidade Federal do Rio de Janeiro (LIN/PEN/COPPE/UFRJ), Rio de Janeiro, Brasil – CEP 21945-972
[2] Laboratório Nacional de Metrologia das Radiações Ionizantes/Instituto de Radioproteção e Dosimetria/Comissão Nacional de Energia Nuclear (LNMRI/IRD/CNEN), Av. Salvador Allende s/no, Barra da Tijuca, Rio de Janeiro, Brasil – CEP 22783-127

E-mail: aruzzarin@nuclear.ufrj.br

Abstract. The 4πβ-γ coincidence counting and CIEMAT/NIST liquid scintillation method were used in the standardization of a solution of ^{152}Eu. In CIEMAT/NIST method, measurements were performed in a Liquid Scintillation Counter model Wallac 1414. In the 4πβ-γ coincidence counting, the solution was standardized using a coincidence method with ''beta–efficiency extrapolation''. A simple 4πβ-γ coincidence system was used, with acrylic scintillation cell coupled to two coincident photomultipliers at 180° each other and NaI(Tl) detector. The activity concentrations obtained were 156.934 ± 0.722 and 157.403 ± 0.113 kBq/g, respectively, for CIEMAT/NIST and 4πβ-γ coincidence counting measurement methods.

1. Introduction

^{152}Eu is widely used as gamma standard spectrometry sources for the calibration of solid-state detectors. The main method used for the standardization of radionuclide activity by most National Metrology Institutes worldwide is the 4πβ-γ coincidence counting method, a primary method that has great versatility as may be used for those radionuclides which decay by alpha-gamma and beta-gamma emission, electron capture and pure beta emitters [1]. Another method currently used for determination of activity is the CIEMAT/NIST liquid scintillation method, consisting of three or two photomultipliers based on statistical Free Parameter model of the scintillation photons distribution and their detection probabilities [2]. The decay of ^{152}Eu is divided in branches of β- decay (27.8%), followed by photon emission, and electron capture (72.2%), with subsequent gamma emission. The gamma-ray spectrum ranges from 121 to 1769 keV, with 85 gamma emissions. The half-life of ^{152}Eu is (13.53 ± 003) years [3, 4]. Due the complex decay of ^{152}Eu, the activity determination is complicated, creating difficulties in the absolute standardization of this nuclide. The main features of the complex decay are shown in figure 1.

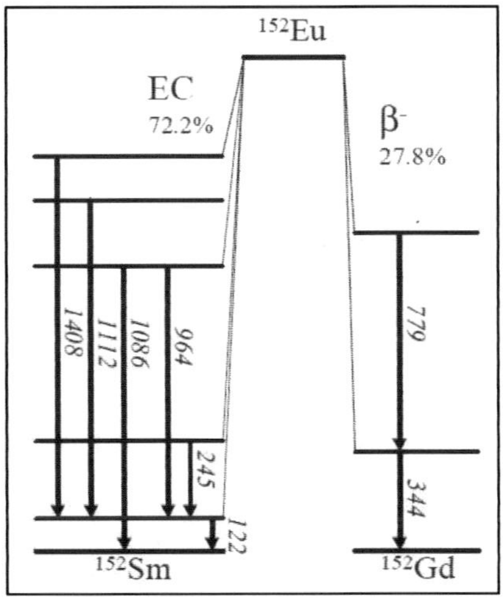

Figure 1. Simplified decay scheme of ^{152}Eu

2. Methodology

2.1. $4\pi\beta$-γ coincidence counting method

This well-known counting method for absolute activity measurement is commonly used for decay modes including β-emission, electron capture or a mixture of the two [6, 7].

For the determination of the activity, 14 sources were prepared by adding aliquots of ^{152}Eu solution in a glass vial with 10 mL of commercial cocktail Ultima Gold, manufactured by Perkin Elmer/USA. Sources were measured in 4πβ(LS)-γ (NaI) coincidence system, consisting of an acrylic scintillation cell coupled to two coincident photomultipliers at 180° each other, in which the scintillation vial is placed, and the generated signals analyzed and processed in a conventional electronic chain, forming the beta channel.

The signals come mainly from the Auger electrons and characteristic X-rays generated by the electronic capture events. The counting rate of these events is called N_β. A 10 x 10 cm^2 NaI(Tl) crystal is placed at the bottom of the acrylic scintillation cell for detection of gamma rays whose signals are analyzed and processed in a conventional electronic chain, forming the gamma channel. The counting rate of these events is denominated N_γ. The counting rate of pulses from beta and gamma channels arriving simultaneously in a coincidence circuit with a fixed resolution time of 1.27 μs is denominated N_c. The activity was determined by means of a variation of the mean beta efficiency, $\varepsilon_\beta = N_c/N_\gamma$, by electronic discrimination and applying the extrapolation method to the measurement results [8, 9]. The coincidence equation is expressed as

$$\frac{N_\beta N_\gamma}{N_c} = N_0\left[1+\frac{1-\frac{N_c}{N_\gamma}}{\frac{N_c}{N\gamma}}c\right] \quad (1)$$

where N_o is the activity of the source, N_β, N_γ and N_c are beta, gamma and coincidence count rates, and *c* is a constant dependent of decay scheme of the measured radionuclide. In (1), $N_\beta N_\gamma/N_c \rightarrow N_o$, the activity of the source, when $N_c/N_\gamma \rightarrow 1$ [10].

2.2. CIEMAT/NIST measurement method

This counting system consists of two photomultipliers placed in coincidence at 180° to each other. The CIEMAT/NIST method uses a set of standard tracer tritium (^{3}H) samples for the commercial scintillator characterizing in terms of the quenching parameter by the introduction of the chemical quenching agent in increasing degree. Figure 2 shows the correlation between Free Parameter, quenching and efficiencies for standard tracer and radionuclide in analysis by CIEMAT/NIST method [2, 5]. The equation 2 gives the detection efficiency for two photomultipliers obtained from Poisson model.

$$\varepsilon = \left(1 - e^{-\frac{\nu m}{2}}\right)^2 \tag{2}$$

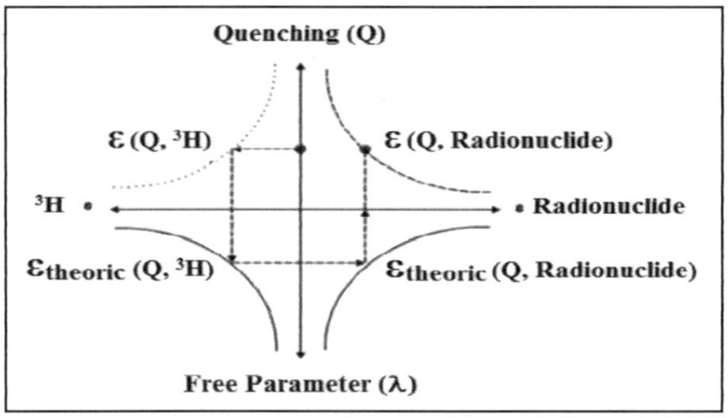

Figure 2. CIEMAT/NIST model for theradionuclide standardization.

3. Results and discussion

3.1. 4πβ-γ coincidence counting method

The activity was determined through the variation of the beta efficiency by electronic discrimination, and applying an extrapolation technique to the measurement data. Two levels of gamma discrimination were adjusted for the counts in the gamma channel: 100 and 180 keV.

To obtain the extrapolation curve, the expression ($N_\beta N_\gamma$) /N_c is plotted as a function of (1 - ε_β) / ε_β and extrapolated to ε_β = 1. The function is linear in the region of high efficiencies [6]. ε_β efficiencies ranged from 0.78 to 0.55 and 0.81 to 0.72 for the gamma window range of 100 and 180 keV, respectively. The extrapolated value to ($1-N_c/N_\gamma$)/(N_c/N_γ), where ε_β=1, provides the activity of the radioactive solution from which the sources were prepared. Figure 3 presents the extrapolation curves on the coincidence measurements of all 14 sources.

The consistency of the results of value of activities from all sources was verified using the statistical test called "Birge ratio" [11]. The calculated value with the experimental results was 1.17. The value adopted for the activity concentration was the weighted average of the 14 results: 157.403 ± 0.113 kBq/g, where the uncertainty refers to the standard deviation of the mean of the results. Table 1 presents the components of the uncertainties in determining the activity of ^{152}E by the coincidence method.

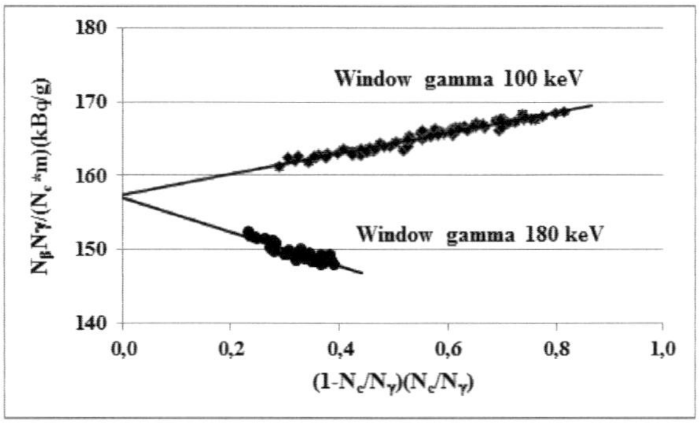

Figure 3. Extrapolation applied to the measurements by 4πβ-γ coincidence counting on all sources, measurements at gamma channel of 100 and 180 keV.

Table 1. ^{152}Eu uncertainty components by coincidence method.

Components	Type	Relative standard uncertainty (%)
Counting statistics	A	0.50
Weighing	B	0.050
Background	A	0.030
Dead-time	B	0.033
Resolving time	B	0.20
Gandy effect	B	0.093
Fitting procedure	A	0.072
Decay correction	B	0.00033
Combined		0.60
Expanded (k=2)		1.2

3.2. CIEMAT/NIST measurement method

Eight sources were prepared by adding aliquots of ^{152}Eu solution in a glass vial with 10 mL of commercial cocktail Ultima Gold, manufactured by Perkin Elmer/USA. During the measurement period the samples remained stable, showing no tendency. Measurements were performed in a Liquid Scintillation Counter model Wallac 1414.

The activity was calculated using the CN2003 computational code [12]. The value adopted for the activity concentration was the weighted average of the 8 results: 156,934 ± 0,722 kBq/g, where the uncertainty refers to the standard deviation of the mean of the results. Table 2 presents the components of the uncertainties in determining the activity of ^{152}E by the CIEMAT / NIST method.

Table 2. ^{152}Eu uncertainty components by CIEMAT/NIST method.

Components	Type	Relative standard uncertainty (%)
^{3}H activity	B	0.20
Weighing	B	0.05
^{3}H Quenching (SQPE)	A	0.05
Nuclear and Atomic data	B	0.13
Photomultiplier asymmetry	B	0.23
kB value	A	0.10
^{152}Eu quenching (SQPE)	A	0.09
Statistic counts	A	0.28
Combined		0.46
Expanded (k=2)		0.92

Figure 4 summarizes the results of two absolute measurements methods applied in the determination of the activity concentration of the ^{152}Eu solution. The uncertainties bars refer to combined uncertainty (coverage factor k = 1).

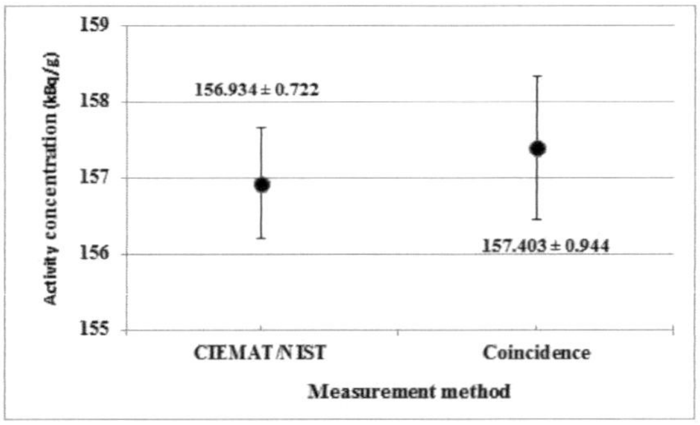

Figure 4. Activity concentration of the ^{152}Eu solution determined by two absolute methods.

4. Conclusion

The results of the standardization of ^{152}Eu solution by the absolute CIEMAT / NIST and 4πβ(LS)-γ(NaI) methods were consistent within the uncertainties evaluated at a confidence level of ≈ 68% (k = 1). The largest contributions to assessing the combined uncertainty came from the counting statistics for both methods indicating the need to increase the counting time or the mass of the sources to minimize these uncertainties. As a reference activity for calibration of other secondary calibration systems in operation at LNMRI, the value of the CIEMAT/NIST method activity was adopted because it presented a lower combined uncertainty (0.92%) than that of coincidence (1.2%).

5. References

[1] Kawada Y 1972 Extended application and improvement of the 4πβ-γ coincidence method in the standardization of radionuclide. Researches of the Electrotechnical Laboratory No. 730
[2] da Cruz PAL *et al* 2016 *J. Phys. Conf. Ser.* **733** p.1-9
[3] Vanin V R, de Castro R M, Browne E 2004 *BNM-LNHB/CEA Table de Radionucléides* USP, LNHB

[4] Lagoutine F, Coursol N, Legrand J 1987 *Table de Radionucleides Laboratoire Primaire des Rayonnements Ionisants* Commissariat À L'Energie Atomique
[5] Malonda AG 1999 *Free Parameter Models in Liquid Scintillation Counting* Editorial CIEMAT/Spain
[6] Campion P J 1959 *Appl. Radiat. Isot.* **4** p.232
[7] Baerg AP 1973 *Nucl. Instr. and Meth.* **112** p.43
[8] Houtermans H, Miguel M 1962 *Appl. Radiat. Isot.* **13** p.137
[9] Funck E, Larsen A N 1983 *Appl. Radiat. Isot.* **34** p.565
[10] Johansson L *et al* 2003 *Nucl. Instr. Meth. Phys. Res.* A **508** p.378–387
[11] Birge R T 1932 *Phys. Rev.* **40** p.207-227
[12] Gunther E 2003 *Program CN2003: A program to calculate the LC efficiency of a nuclide vs. efficiency the tracer H-3* (CIEMAT/NIST) PTB/Germany

^{134}Cs emission probabilities determination by gamma spectrometry

M C M de Almeida [1,2*], R Poledna [2], J U Delgado [2], R L Silva [2], M T F Araujo[2,3], C J da Silva [2]

[1] Comissão Nacional de Energia Nuclear (DISEN) [2] Instituto de Radioproteção e Dosimetria (LNMRI) [3] COPPE (Eng. Nuclear)

[*] candida@ird.gov.br

Abstract. The National Laboratory for Ionizing Radiation Metrology (LNMRI/IRD/CNEN) of Rio de Janeiro performed primary and secondary standardization of different radionuclides reaching satisfactory uncertainties. A solution of ^{134}Cs radionuclide was purchased from commercial supplier to emission probabilities determination of some of its energies. ^{134}Cs is a beta gamma emitter with 754 days of half-life. This radionuclide is used as standard in environmental, water and food control. It is also important to germanium detector calibration. The gamma emission probabilities (Pγ) were determined mainly for some energies of the ^{134}Cs by efficiency curve method and the Pγ absolute uncertainties obtained were below 1% (k=1).

1. Introduction

^{134}Cs is a radioisotope of extreme importance for the calibration of HPGe spectrometers, among others. The determination of nuclear parameters beyond the absolute supply activity should be considered as a factor of relevance.

The determination of the gamma emission probabilities, for example, allows improvements in the characterization of this radioisotope depending on the accuracy of the measured values and serves as quality indicator of the spectrometry system and the methodology used in the determination of this parameter. Another factor that highlights the use of ^{134}Cs is that through the determination of efficiency one can use this radioisotope as a plotter to measure ^{137}Cs.

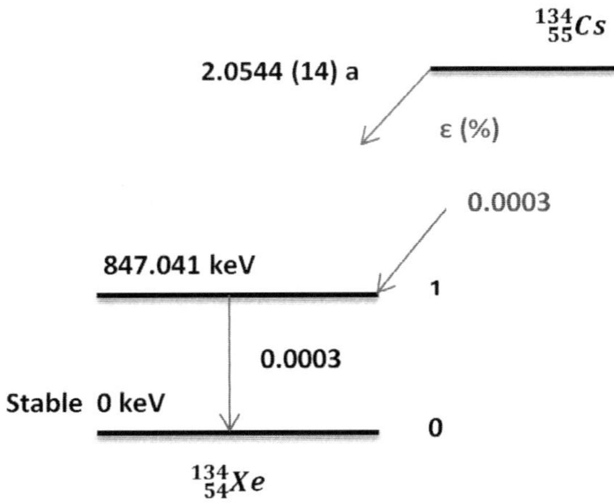

Figure 1. ^{134}Cs decay scheme

^{134}Cs has a radioactive half-life of 2.06 years and it has two mechanisms by which it can disintegrate showed on figure 1. One mode is by β- emission which is the most likely occurrence option being responsible for 99.9997% of the disintegrations. Only 0.0003% of the disintegrations occur by electronic capture and positron emission (β +).

In the event of a nuclear accident, such as the leakage or explosion of a reactor as happened in 1986 in Chernobyl, Ukraine, some radioisotopes produced in the ^{235}U nuclear fission reaction are released into the atmosphere, such as ^{131}I, ^{137}Cs, ^{134}Cs and ^{90}Sr [1]. After this accident, Brazil worried about the importation of foods that could have the presence of radioisotopes ^{137}Cs, ^{134}Cs and ^{90}Sr. These radioisotopes are absorbed by the plants that are consumed by the animals resulting in the possible contamination of imported beef and milk [2].

2. Methodology

2.1 Relative Efficiency Curve Method

The efficiency curve was obtained using various radionuclidic standardized sources such as: ^{166m}Ho, ^{152}Eu, ^{137}Cs,^{241}Am, ^{65}Zn totaling 59 energy points and the graphic can be observed in figure 2. The HPGe spectrometric system has been calibrated in efficiency through the use of standard point sources [3, 4, 5]. The range of energy was established between 48 keV and 1427 keV originally. Then a cut was made considering only the energies above 300 keV since the low energy region was not necessary for the purpose of the calibration. The efficiency curve fitting was performed by a 5th degree polynomial.

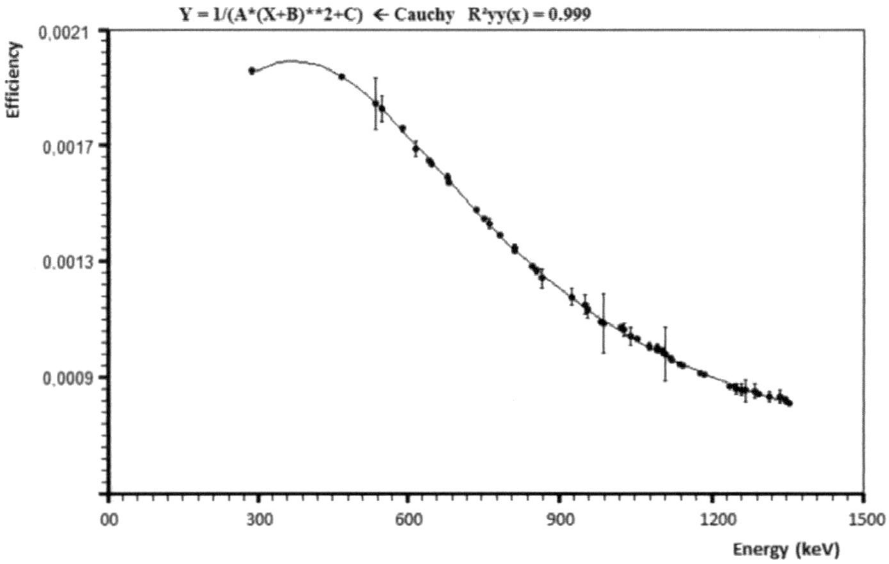

Figure 2. Efficiency curve with ^{166m}Ho, ^{152}Eu, ^{133}Ba, ^{137}Cs, ^{241}Am, ^{54}Mn and ^{65}Zn radionuclidic standards

2.2. *Gamma-ray emission probability measurements*

In order to associate the main peaks of the spectrum to the radionuclide, the energy-channel relation of a spectrometer needs to be obtained. Afterwards, the total absorption efficiency curve is determined, as a function of energy, and the radionuclide activity may be calculated from region of interest. The expression that represents an activity area is:

$$\frac{CPS_{corrected}}{P_\gamma \varepsilon_\gamma} \qquad (1)$$

where: $CPS_{corrected}$ is the count rate of full energy peak; ε_γ is the full energy peak efficiency for specific gamma energy; and P_γ is the emission probability for specific gamma energy.

However, as the source activity is known , P_γ was calculated for ranges of 475 keV, 563 keV, 569 keV, 604 keV, 796 keV, 802 keV, 1039 keV, 1168 keV and 1365 keV to ^{134}Cs by means of the following expression, taking into account the corrections as decay, background and positioning:

$$P_\gamma = \frac{CPS_{corrected}}{\varepsilon_\gamma . A} \qquad (2)$$

where A is the absolute activity. The P_γ determination depends on the precision achieved in efficiency curve.

2.3. *Source preparation and measurements*

^{134}Cs point source was prepared by means of the pycnometer technique, depositing drops of radionuclide solution in a polystyrene film, with a thickness of 0.05 mm, set in one acrylic ring. The ring has an external diameter of 25 mm, inner diameter of 4 mm and a thickness of 1 mm. Once dried, the source was covered with the same polystyrene film .

The spectrometry system used consist of HPGe detector type a planar with beryllium window with 20% of relative efficiency (d3gem). Some appropriate electronics composed basically of the elements: High voltage supply, signal amplifier and a multichannel analyzer. The multichannel analyzer associated with the data acquisition program is responsible for subtracting background beyond managing dead time.

The conditions of ^{134}Cs source measurements are :HPGe detector position- d3gemp4 (20cm); source activity : 2820.71Bq at 12:00h of date 20170601. This activity was determined in the LNMRI laboratory by 4πβ(LS)-γ(Na(Tl)) anticoincidence method with 0.38 % of uncertainty (k=1) [6].

3. Results

The components of the uncertainties considered for the determination of gamma emission probability were uncertainty of the photopic area, uncertainty in the activity, uncertainty in efficiency and uncertainty in ^{134}Cs decay.

The gamma spectrometry system was stable throughout the measurement. During ^{134}Cs spectrum acquisition some problems can occur when considering the peaks of 563 keV and 569 keV that overlap over the 605 keV peak when using a NaI (Tl) type detector but in this case the use of the HPGe detector stands out for such use because it has a much greater resolution power and facilitates the separation of these peaks with greater ease.

Table 1. Counts per second to the different energies of ^{134}Cs

Energy (keV)	Area	Uncertainty u_a (area)	$CPS_{corrected}$ (Bq/s)
475	16538	322	0.0476
563	82224	421	0.2366
569	149331	499	0.4297
604	854624	965	2.4591
796	630766	811	1.8149
802	62050	268	0.1785
1039	5796	92	0.0166
1168	9681	105	0.0279
1365	14951	123	0.0430

Table 2. Results of gamma emission probabilities to ^{134}Cs energies

Energy (keV)	Gamma Efficiency (ε_γ)	Pγ (%)
475	0.001278	1.509(7)
563	0.001135	8.446(44)
569	0.001115	15.625(39)
604	0.001034	96.41(85)
796	0.000861	85.48(77)
802	0.000857	8.440(24)
1039	0.000673	1.0039 (19)
1168	0.000618	1.827(18)
1365	0.000579	3.013(21)

The results of gamma emission probabilities (Pγ) to ^{134}Cs energies are near with the LNHB results [7].

The absolute uncertainties results, table 2, for the Pγ of each energy are within the expected although the values were higher than those of the LNHB, table 3. This fact is also due mainly to the uncertainties of the efficiency curve and the source activity.

Table 3. LNHB [7] data of the gamma emission probabilities (P_γ) to ^{134}Cs energies

Energy (keV)	**Pγ (%)**
475	1.479 (7)
563	8.342 (15)
569	15.368 (21)
604	97.63 (8)
796	85.47 (9)
802	8.694 (16)
1039	0.9909 (33)
1168	1.791 (5)
1365	3.019 (8)

The calculation of the deviations of Pγ results, table 4, from each energy was:

$$\Delta P\gamma\ (\%) = [(P\gamma_{WORK} - P\gamma_{LNHB})/ P\gamma_{LNHB}]*10 \qquad (3)$$

Table 4. Deviation between this work results and LNHB [7] data of the gamma emission probabilities (P_γ) to ^{134}Cs energies

Energy (keV)	ΔPγ (%)
475	+ 2.03
563	+ 1.25
569	+ 1.67
604	- 1.25
796	+ 0.01
802	- 2.92
1039	+ 1.31
1168	+ 2.01
1365	- 0.20

The experimental results of Pγ showed a tendency of higher values than those of the literature [7], anyway the results were close.

4. Conclusion

The efficiency curve method is widely used for radioisotopes whose half-life is not long as in the case of the ^{134}Cs. Satisfactory results were obtained for the gamma emission probabilities in which a mean deviation was obtained compared to the literature of 0.43 %. The gamma emission probabilities were determined mainly for some energies of the ^{134}Cs by efficiency curve method and the absolute uncertainties obtained were below 1 % (k=1).

5. References

[1] Gazal S and Prange H 2005 *Radioprotection, Suppl.1.* **40 S** 747

[2] Pawel J and Kalita S J 2010 *Nukleonika* **55 (2)** 143

[3] Dabb H M 2015 *Arab Journal of Nuclear Science and Applications* **48 (4)** 53

[4] Silva, R L , Poledna, R , Delgado, J U, de Almeida M C M, 2011 , *6o Congresso Brasileiro de Metrologia*, Natal,. **1** 20.

[5] Araujo, M T F ; de Almeida, M C M , Poledna, R. , Delgado, J U , Silva, R L, Lopes, R T , Cagido, A C F 2016 *Journal of Physics. Conference Series (Online)*, **1** 6

[6] da Silva C J *The ^{134}Cs activity determination by $4\pi\beta(LS)$-$\gamma(Na(Tl))$ anticoincidence method* will be published.

[7] LNE-LNHB/CEA 2012 Table de Radionucleides : www.nucleide.or/Cs-134-tables

IOP Conf. Series: Journal of Physics: Conf. Series **975** (2018) 012064 doi:10.1088/1742-6596/975/1/012064

Implementation of the qualities of radiodiagnostic: mammography

L C Pacífico [1], L A G Magalhães [1], J G P Peixoto[1,2], E Fernandes[1]

[1] Radiological Science Laboratory, University of the State of Rio de Janeiro, Rio de Janeiro, RJ 20550-900, Brazil
[2] National Laboratory of Metrology of Ionizing Radiation, Institute of Radioprotection and Dosimetry, Rio de Janeiro, RJ 22783-127, Brazil

E-mail: leonardocpacifico@gmail.com

Abstract. The objective of the present study was to evaluate the expanded uncertainty of the mammographic calibration process and present the result of the internal audit performed at the Laboratory of Radiological Sciences (LCR). The qualities of the mammographic beans that are references in the LCR, comprises two irradiation conditions: no-attenuated beam and attenuated beam. Both had satisfactory results, with an expanded uncertainty equals 2,1%. The internal audit was performed, and the degree of accordance with the ISO/IEC 17025 was evaluated. The result of the internal audit was satisfactory. We conclude that LCR can perform calibrations on mammography qualities for end users.

1. Introduction

The Laboratory of Radiological Sciences (LCR), belonging the university of the state of Rio de Janeiro (UERJ), is already accredited to perform calibrations on area and individual monitors. With the objective of attending to society providing calibration services as a laboratory tracked to the National Laboratory of Metrology of Ionizing Radiations (LNMRI / IRD / CNEN), the LCR is increasing its scope of calibration and implementing the calibration service for the mammography qualities, in order to fill a gap that exists between ionizing radiation users and the Primary Standard Dosimetry Laboratory (PSDL) [1], in addition to providing metrology training of ionizing radiation, to improve calibration techniques and to keep its traceable standard to a primary laboratory with recalibration every 3 years [2, 3]. To do so, the LCR will be treated as a Secondary Standard Dosimetry Laboratory (SSDL).

It is desirable that the influence quantities of a SSDL have a relative standard uncertainty less than 0.1% [4]. The influence of the magnitudes relative to the user's detector is not considered here. The objective of the present study was to evaluate the expanded uncertainty in this scope increase and present the result of the internal audit performed.

2. Material and method

For the quality of radiation in no-attenuated and attenuated mammography beam, the Comet MXR-160/22 industrial tube with tungsten target and additional 0.060 mm nominal molybdenum filtration was used to obtain a similar energetic spectrum as a tube with Mo target and 0.030 mm Mo filtration [5], like those used in clinic.

For the mammographic qualities under attenuated condition, 2 mm of nominal Al are inserted, under conditions already obtained previously. Tables 1 and 2 shows the mammographic qualities that are references in LCR calibrations for end users. To calculate the relative uncertainty, we used equation (1), with 16 input quantities, discriminated in Table 3.

$$\left[\left(\frac{u_G}{G}\right)\right]^2 = \sum_{i=1}^{16}\left(\frac{u_i}{M_i}.100\right)^2 \tag{1}$$

where u_i is the standard uncertainty of the i-th quantity and M_i is the value of the i-th quantity of influence.

Table 1. Mammographic qualities for non-attenuated condition (0.060 mm Mo)

Condition no-attenuated				
Quality	Voltage (kV)	HVL (mm Al)	Filtration (mm Mo)	HC
W23_60Mo	23	0,3318	0,060	0,81
W25_60Mo	25	0,3499	0,060	0,82
W28_60Mo	28	0,3520	0,060	0,79
W30_60Mo	30	0,3626	0,060	0,80
W35_60Mo	35	0,3905	0,060	0,76

Table 2. Mammographic qualities for attenuated condition (0.060 mm Mo + 2.0 mm Al)

Condition attenuated				
Quality	Voltage (kV)	HVL (mm Al)	Filtration (mm Mo)	HC
W23_60Mo_2Al	25	0,5246	0,060+2,00	0,93
W25_60Mo_2Al	28	0,5780	0,060+2,00	0,90
W28_60Mo_2Al	30	0,6359	0,060+2,00	0,89
W30_60Mo_2Al	35	0,7133	0,060+2,00	0,87

If $u_i(\%)=\left(\frac{u_i}{M_i}\right).100$, we can replace in (1):

$$\left(\frac{u_G}{G}\right) = \sqrt{\sum_{i=1}^{i=16} u_i(\%)^2} \tag{2}$$

where $\boldsymbol{u_G}$ is the combined uncertainty concerning the ionization chamber reading unit, Coulomb (C), and G is the reference value of the charge. Doing $\mathbf{u_r} = \left(\frac{u_G}{G}\right)$ the relative combined standard uncertainty, we have:

$$u_r = \sqrt{\sum_{i=1}^{i=16} u_i(\%)^2} \tag{3}$$

To find the expanded uncertainty, with a confidence level of approximately 95% and a coverage factor equals to 2, simply multiply (3) by 2.

3. Results

Table 3 presents the results referring to the sources of uncertainties considered, as well as the expanded uncertainty.

Table 3. Considered influence quantities, relative combined standard uncertainty and expanded uncertainty, for a coverage factor k = 2, for the non-attenuated CMC.

Condition: No-Attenuated / Attenuated

	Input Quantity	Unity	Relative Uncertainty
u1	Charge	*nC*	0,06%
u2	Temperature	*°C*	0,07%
u3	Pressure	*kPa*	0,00%
u4	Distance	*nC*	0,23%
u5	Time	nC	0,14%
u6	Calibrated standard	nC	0,56%
u7	Electrometer resolution	nC	0,00%
u8	Leakage	nC	0,05%
u9	Energy Dependence	nC	0,00%
u10	Loss by recombination	nC	0,10%
u11	Positioning	nC	0,00%
u12	Homogeneity	nC	0,20%
u13	Spectral Difference	nC	0,00%
u14	Kerma Rate on Air	nC	0,13%
u15	Stability	nC	0,79%
u16	Polarity	nC	0,04%
	Relative combined uncertainty (%)		1,0%
	Expanded Uncertainty (%)		2,1%
	K		2,0

3.1 Expanded uncertainty

Calibration and Measurement Capacity (CMC), defined as the smallest uncertainty that a laboratory can present within its scope of accreditation [6], is the smallest expanded uncertainty itself. In the LCR, the value was 2,1% for both no-attenuated condition and attenuated condition. Table 3 shows the relation of the influence quantities considered in the CMC calculation.

3.2. Increase of Scope

Considering the increasing demand for mammography examinations, it is necessary to trace the mammographic beam qualities practiced in the clinics, to focus on patient safety. Given the scientific importance of the LCR, an increase in scope was requested from the National Institute of Metrology, Quality and Technology (INMETRO) so that calibrations could be carried out.

3.3 Audit
The internal audit was carried out with the purpose of verifying the conformity of the operations referring to the request of the increase of scope with those required in ISO/IEC 17025, General Requirements for Testing and Calibration Laboratories Competence [7]. The no-conformities were corrected, thus allowing a CMC of 2,1% for both conditions.

4. Discussion
The uncertainty regarding the calibration of the LCR standard was obtained by dividing the value of the relative uncertainty given in the certificate issued by the secondary laboratory LNMRI, by the coverage factor 2, corresponding to a confidence level of approximately 95%.

With the addition of 0.060 mm Mo in a tungsten anode tube, we are simulating the X-ray spectrum of a conventional Mo-target mammograph and 0.030 mm Mo filtration. Since both laboratories used the same irradiation conditions, the uncertainty of the spectral difference is negligible. However, it will become significant when performing the calibration of the user's detector, assigning the value of 0.46% in the uncertainty of the two spectra a [8].

The kerma rate in the air can be influenced by possible oscillations in the tube current as well as in the voltage. The uncertainty value associated with this oscillation was not considered to be 0.13%, following the literature [8].

For the calculation of the leakage current, measurements were made in the W28_60Mo quality, an electric charge accumulated in 1 minute. Thereafter, the tube was disconnected, and the charge variation within 5 minutes was evaluated. The uncertainty associated with the leakage current, given by the variation of such charge in time, was negligible.

The ion recombination loss in an ionization chamber is small, since its electric field is sufficient to collect almost all the charges released in the ionization process [8]. The value of this type B uncertainty was 0.03%.

The uncertainty associated with the focus-detector distance, which is 1000 mm, was 0.14%, as it was considered that it could vary by ± 5 mm. This value was considered because the measurement of the focus-detector distance was not performed with a laser but with a tape measure.

The positioning was analyzed by displacing the ionization chamber ± 2 mm perpendicular to the focus-detector axis, with no significant contribution, with an uncertainty of 0.00%.

The resolution of the electrometer had no influence on the result, being negligible. The same for energy dependence, because both LCR and LNMRI spectrum are same. The field size, or homogeneity of the field, has an assigned uncertainty of 0.20 %, where the anodic effect of the tube is the major contributor, since it has a 20 ° target angulation.

The stability of the system, i.e., its response over time, was evaluated in six months, and we obtained an uncertainty of 0.79%.

When the chamber polarity is changed of 300 V to -300V, the associated uncertainty is 0,04%.

5. Conclusion
As discussed above, we have concluded that the LCR is able to perform calibrations intended for users in the mammographic qualities presented in tables 1 and 2, since the expanded uncertainty, with a coverage factor equal to 2, is 2,1 %, which allows to deliver to the end user an uncertainty of less than 5%. [10],

6. Reference

[1] Cerqueira JP, Martins MC. Auditorias de sistemas de gestão: ISO 9001, ISO 14001, OHSAS 18001, ISO/IEC 17025, SA 8000, ISO 19011. Rio de Janeiro, Brazil: Qualitymark; 2005.

[2] Technical Reports Series No. 457: Dosimetry In Diagnostic Radiology: An International Code Of Practice. INTERNATIONAL ATOMIC ENERGY AGENCY VIENNA, 2007

[3] Peixoto JG, et al. Ionization Radiation Metrology. 1º Ed. IRD, Rio de Janeiro, 2016

[4] International Atomic Energy Agency (IAEA). Measurement Uncertainty: A Practical Guide For Secondary Standards Dosimetry Laboratories. Vienna, 2008.

[5] Corrêa, E.L., Lucena, R.F., Potiens, M.P. A. e Vivolo,V. Comparação de espectros de raios x gerados por um alvo de W e filtração adicional de Mo e Al para aplicações em metrologia na mamografia. Associação Brasileira de Física Médica. XIV Congresso Brasileiro de Física Médica, São Paulo.

[6] Instituto Nacional de Qualidade e Tecnologia (Brasil). Norma Nº Nit-Dicla-021, de 03/2013. Expressão Da Incerteza De Medição Por Laboratórios De Calibração

[7] Associação Brasileira de Normas Técnicas (Brasil). Norma Nº ISO/IEC 17025, de 25/09/2016. Requisitos Gerais Para a Competência de Laboratórios de Ensaio e Calibração.

[8] Peixoto JG, Almeida CE. The radiation metrology network related to the field of mammography: implementation and uncertainty analysis of the calibration system. Meas. Sci. Technol. 12 (2001) 1586–1593.

[9]. International Atomic Energy Agency (IAEA). Dosimetry and Medical Radiation Physics: Guidelines for Member States on the Designation of SSDLs. Disponível em: < http://www-naweb.iaea.org/nahu/dmrp/SSDL/guidelines.asp>. Accessed: may 8, 2017.

[10] International Electrotechnical Commission (Brasil). Norma IEC-61674 de 29/11/2012. Medical Electrical Equipment – Dosimeters with ionization chamber and/or semiconductor detectors as used in x-ray diagnostic imaging.

IOP Conf. Series: Journal of Physics: Conf. Series **975** (2018) 012065 doi:10.1088/1742-6596/975/1/012065

Status of ionization chambers calibration for radiation therapy in Brazil

M. Gonçalves, G. Joana, P. Leal, R. Vasconcelos, N. do Couto, F. C. Teixeira, A. D. Soares, E. S. Santini and C. Salata

Comissão Nacional de Energia Nuclear, R. General Severiano, 90, 22290-151, Rio de Janeiro, RJ, Brazil.

E-mail: mgoncalves@cnen.gov.br.

Abstract. CNEN makes a constant effort to keep updated with international standards and national needs to strengthen the radiological protection status of the country. The guidelines related to radiation treatment facilities have been revised in the last five years in order to take in consideration the most relevant aspects of the growing technology as well as to mitigate the accidents or incidents observed in practice. Hence, clinical dosimeters have gained special importance as significant items in Brazilian regulation. In the present work we discuss the importance of inspections from the point of view of equipment dosimetry and instruments quality control. The dosimeter sets based on thimble and well ionization chambers need periodic calibration, and this calibration becomes a fundamental task in order to guarantee the dose prescribed-delivered to patients.

Thus Brazilian guidelines enforce the need of at least two sets of clinical dosimeters with thimble chambers calibrated and one set of electrometer with well ionization chamber for hdr equipment. We call attention to the fact that inspections are a very valuable tool in order to enforce the application of guidelines around the country both by enlightening the weaker aspects of facilities concerning radiological protection and by stating in loco that reasons which lead the regulatory body to enforce such guidelines items.

pacs: 87.62.+n, 87.90.+y, 87.52.Tr, 87.53.Dq, 87.58.Sp

Keywords: Therapeutic Beam, Ionization Chambers, Regulatory Body, Inspections, Calibration Laboratory

1. Introduction

According to Brazilian law, all radiation therapy facilities must be regulated in terms of taxes and licensing processes. This regulation takes place through document analysis and in loco inspections performed by a small group of inspectors of Brazilian regulatory body (CNEN) in the last 20 years. Besides conforming all facilities to a required standard of radiation protection, this regulation also provides national global data which could elucidate the importance of the normatization (guides, norms and laws) to the overall process. Thus, these data offer a national panorama and help on deciding whether the current guidelines have been taken properly or not.

CNEN makes a constant effort to keep updated with international standards and national needs to strengthen the radiological protection status of the country. The guidelines related to radiation treatment facilities have been revised in the last five years in order to take in consideration the most relevant aspects of the growing technology as well as to mitigate the accidents or incidents observed in practice.

Hence, clinical dosimeters have gained special importance as significant items in Brazilian regulation. Three years ago the guidelines changed to include as an obligation, for every radiation therapy facility, the need of two calibrated sets of thimble ion chamber and electrometer for clinical dosimeter for photon beams dosimetry and one calibrated well-chamber for high dose rate brachytherapy equipment.

In the present work we show the status of the calibration of these instruments observed during inspections, as an attempt to evaluate the effectiveness of the guidelines and to investigated whether other actions are needed to guarantee the precision in the dose delivered to patients. This investigatory aspect of CNENs inspection program has already shown effective, allowing actions by the regulatory body in terms of local and national perspectives [1, 2, 3]. With these data, guidelines may be updated for ionization Chambers Calibration for Radiation Therapy in Brazil conforming to the real national panorama. Results are sufficient to represent the global status and help in making the profile of Brazilian facilities.

2. Methodology

In the last five years inspectors from CNEN have evaluated dosimetric systems of 168 radiation therapy facilities all over the country. We performed a retrospective analysis of the 168 inspection reports, focusing on the following items: a) if the facility had two sets of thimble ion chamber plus electrometer; b) if both sets have a calibration certificate; and c) if the facility treating patients with HDR had a well-chamber calibration certificate. Some aspects were considered: the calibration certificates must be emitted from an accredited Brazilian Laboratory, and it is valid for two years.

At the end of the inspection, inspectors have a form listing instruments which presented a valid calibration certificate or not. This form is shown in Figure 1 and the whole set of forms gathered by inspectors in a period constitutes our database here discussed. With this database we can classify facilities by year, region, province/state, number of equipment, treatment modality and so on. Here we concentrate our attention on the validity of dosimetric instruments and only evaluate whether the guidelines are being accomplished by facilities in the Brazilian territory.

3. Results and Discussion

In order to understand the Brazilian radiation therapy profile, we firstly display results dependent on the region: North, North-East, South, South-East and Center-West. North and

<table>
<tr><td rowspan="10">ELECTROMETER + CHAMBER</td><td colspan="8">DOSIMETRIC SET</td></tr>
<tr><td colspan="8">THIMBLE CHAMBER</td></tr>
<tr><td rowspan="4">1</td><td colspan="3">MANUFACTURER</td><td colspan="2">MODEL</td><td colspan="2">SERIAL</td></tr>
<tr><td colspan="3"></td><td colspan="2"></td><td colspan="2"></td></tr>
<tr><td>MANUFACTURER</td><td>MODEL</td><td>SERIAL</td><td>TYPE</td><td>LABORATORY</td><td>DATE</td><td>STATUS</td></tr>
<tr><td></td><td></td><td></td><td></td><td></td><td></td><td>Ok/Not Ok</td></tr>
<tr><td rowspan="4">2</td><td colspan="3">MANUFACTURER</td><td colspan="2">MODEL</td><td colspan="2">SERIAL</td></tr>
<tr><td colspan="3"></td><td colspan="2"></td><td colspan="2"></td></tr>
<tr><td>MANUFACTURER</td><td>MODEL</td><td>SERIAL</td><td>TYPE</td><td>LABORATORY</td><td>DATE</td><td>STATUS</td></tr>
<tr><td></td><td></td><td></td><td></td><td></td><td></td><td>Ok/Not Ok</td></tr>
</table>

Figure 1. Form employed by inspectors to gather information about clinical dosimeter set. For the sake of simplicity, the most valuable information presented here is the name of the calibration laboratory (IRD or IPEN in Brazil) and the date of calibration.

Center-West are less populated and have just a few facilities. South and North-East are the intermediary cases between North and South-East.

Considering the total number of inspections analyzed in the current study, we display their proportion by region in Figure 2. We notice that the great majority was located in the South-East region, the most developed region in Brazil. This region counts with roughly 60% of all analyzed facilities and, since laboratories are also located in this region, we expect that these facilities reach laboratories easier than more distant facilities, such as those in the North and Center-West regions.

This distribution has two causes to be mentioned here. Firstly, it represents the real panorama in Brazil, i. e., most of facilities are located in South-East region and the inspection program shall be homogeneous covering all facilities in a period of two years. Secondly, due to this concentration in the South-East region, and since some of these inspections have also been motivated by other reasons such as re-reinforcements, we expect a kind of double counting of

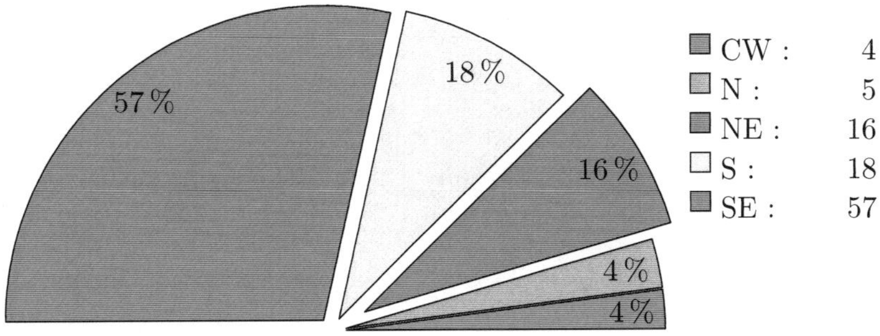

Figure 2. Distribution of facilities in the present analysis. In a total of 168 inspections, 96 were performed in the South-East region, 30 in the South region, 27 in the North-East region, 8 in the North region and 7 in the Center-West region.

some facilities.

Concerning thimble ionization chambers, in Figure 3 we show the sum of all facilities with one or none calibrated dosimeter and facilities which meet the regulation with at least two clinical dosimeters calibrated. We expected that for the year of 2015 we could get an inversion of bars Yes/No, since the guidelines changed in 2014 and facilities had two years to comply with the regulation. However, since that was not the case, all those facilities received a notification by CNEN and some of them had their activities discontinued for a while until they achieve the minimum requirements for operation.

Some facilities argued that both laboratories of calibration presented no schedule for their clinical dosimeters. Thus in 2016 CNEN started an approximation to laboratories in an attempt to improve or enlarge the schedule window. This turned out to be a controversial theme since laboratory directors kindly presented their Ionization Chambers Calibration for Radiation Therapy schedules and we observe some room for calibration of several instruments. We finally understood this issue seemed to rely on some kind of miscommunication between facilities and laboratories. Also, in 2016 CNEN started a new strategy during inspections enforcing that facilities contact laboratories to schedule their calibration. The effects of that action are expected

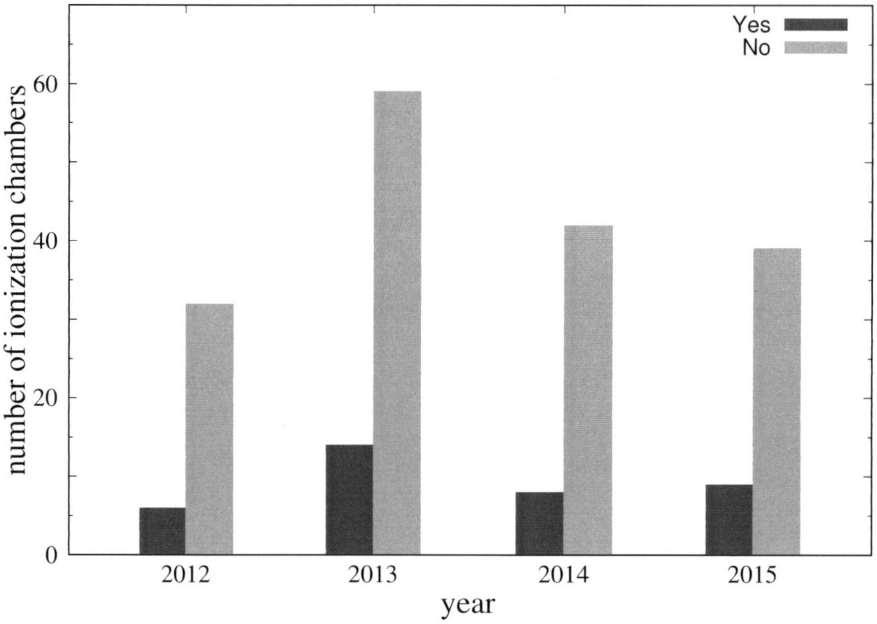

Figure 3. Number of clinical dosimeters with valid (Yes) and out-of-date (No) calibration certificates distributed by region for the period 2012–2015. The number of instruments with out-of-date certificates is large even after the guidelines have changed.

to be seen in 2017 and 2018.

Another important issue observed in inspection relates to well ionization chambers used for dose control of equipment of high dose rate brachytherapy [4, 5]. Differently of thimble ionization chambers, the sole calibration laboratory for that kind of chamber only started its activities in 2014, roughly at the same time the regulatory body changed the guidelines to include the need of calibration of this kind of chamber.

Before 2014 CNEN accepted that foreign laboratories could perform the calibration of well chamber and most of facilities in the country bought their set with a first calibration certificate endorsed by the manufacturer. However, after that no other calibrations were performed and

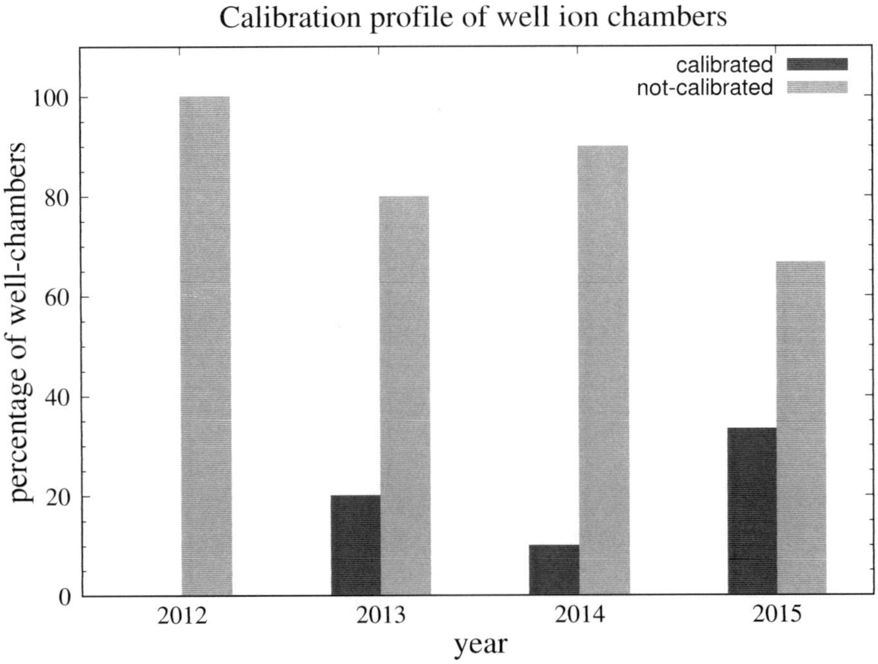

Figure 4. Total number of well chambers calibrated or not in the 2012–2015 period. The regulatory need to present a valid certificate for these chambers was only introduced in 2014.

inspectors did not point that as a non-conformity since it was not a regulation item, resulting in a lack of data before 2014. After that, inspectors began to observe the well chamber calibration as an inspection aspect and data became more reliable. Figure 4 shows results for the period from 2012 up to 2015 in terms of the presence of a valid calibration certificate for the well chambers in facilities.

We clearly note there was an increment in the number of calibrated equipment after 2014 as a consequence of the regulatory need for this calibration and the appearing of the first calibration laboratory. However, this number is still not sufficient since the majority of the facilities did not enroll with the national calibration process. It is a clear fact that the regulatory body, mainly through inspections, must act together with those facilities to strengthen the dosimetry for HDR equipment. Unlike the situation observed for thimble chambers, we did not observe any complaint in respect to the well chamber calibration laboratory and it seemed to the regulatory body that facilities had not understood the need of calibration of their HDR dosimetry system.

4. Summary and Conclusions

In the present work we discuss the importance of inspections from the point of view of equipment dosimetry and instruments quality control. The dosimeter sets based on thimble and well chambers need periodic calibration, and this calibration becomes a fundamental task in order to guarantee the accuracy of prescribed-delivered dose to patients. Brazilian guidelines changed in 2014 to enforce the need of at least two sets of calibrated clinical dosimeters with thimble chambers and one set of calibrated electrometer with well ionization chamber for HDR equipment.

Data analysis up to 2015 showed that Brazilian radiation therapy facilities are far from accomplishing the regulatory needs, in spite of there being two calibration laboratories for thimble ionization chambers and one calibration laboratory for well chambers available in the national territory. New efforts should be made by the regulatory body in order to enhance the strength of dosimetric levels in this aspect and also in accelerating the communication between facilities and laboratories.

Concerning well chambers, results have shown a different and better perspective than that for thimble chambers. Up to now there is only a calibration laboratory in the country (in Rio de Janeiro, south-east region) available to perform calibration of more than a hundred well chambers. However we observed that the number of calibrated well chambers is increasing from 2014 to the present moment, not just to accomplish the regulation but also as an awareness of the need of a precise determination of the dose delivered to patients during brachytherapy treatments.

As a final conclusion we call attention to the fact that inspections are a very valuable tool in order to enforce the application of guidelines around the country both by enlightening the weaker aspects of facilities concerning radiological protection and by stating *in loco* those reasons which lead the regulatory body to enforce such guidelines items. Thus, although results have not shown a perfect effectiveness of the inspection program, they may be interpreted as an improvement in the status of radiological protection in Brazil. For those facilities which still do not comply strongly with the regulation, the regulatory body shall find new ways of bringing them to attention, clarifying the necessity of guidelines to ensure that all patient treatments are performed in the country with the highest level of precision, accuracy, quality and safety.

References

[1] da Rosa L A, Brito R R, Gonçalves M, de Paiva E, Dovales A C, Freire B L, Giannoni R A, Di Prinzio R, Velasco A F, Castelo L H and Oliveira R V 2008 Radiation Protection Dosimetry **131** 272

[2] da Rosa L A R, de Paiva E, Gonçalves M, Velasco A F, Di Prinzio R, Dovales A C M, Freire B L V, Brito R R A, Giannoni R A, Castelo L H R and Marechal M H H 2005 Revista Brasileira de Pesquisa e Desenvolvimento **7** 127–132

[3] Alves C E G R, Azevedo E M, de Sá L V, da Rosa L A R, Mendes L C G, França W F L, Gutterres R F and Gonçalves M 2009 Journal of Radiological Protection **29** 507

[4] Saminathan S, Godson H F, Ponmalar R, Manickam R and Mazarello J 2016 Journal of Medical Physics **41** 234–239

[5] Smith B R, Micka J A, Aima M, DeWerd L A and Culberson W S 2017 Medical Physics **44** 311–320

Testing an extrapolation chamber in computed tomography standard beams

M C Castro, N F Silva and L V E Caldas

Instituto de Pesquisas Energéticas e Nucleares (IPEN / CNEN - SP) - Av. Professor Lineu Prestes, 2242 – CEP:05508-000 São Paulo, SP.

E-mail: maysadecastro@gmail.com

Abstract: The computed tomography (CT) is responsible for the highest dose values to the patients. Therefore, the radiation doses in this procedure must be accurate. However, there is no primary standard system for this kind of radiation beam yet. In order to search for a CT primary standard, an extrapolation ionization chamber built at the Calibration Laboratory (LCI) of the Instituto de Pesquisas Energéticas e Nucleares (IPEN), was tested in this work. The results showed to be within the international recommended limits.

1. Introduction

The computed tomography (CT) diagnostic exams are responsible for the highest dose values to the patients. The radiation doses in this procedure have to be determined with the highest possible precision and accuracy. However, there is no primary standard system for this kind of radiation beam. So, it was decided to use an extrapolation ionization chamber, built at the Calibration Laboratory (LCI) of the Instituto de Pesquisas Energéticas e Nucleares (IPEN) to establish a CT primary standard [1,2].

An extrapolation chamber is a parallel-plate ionization chamber that allows the variation of its sensitive air volume. Usually, this ionization chamber is utilized in beta radiation dosimetry [1], but this chamber was already used for low-energy radiation beams too, and it showed results within the international recommended limits [2-4].

The aim of this work was to performe some of the main characterization tests (stabilization time, leakage current and linearity of response) in two different chamber depths (interelectrode distances) of the extrapolation chamber in standard X-ray beams established for computed tomography at the LCI. When the depth is varied there are different sensitive volumes; therefore it is possible to decide which chamber depth is the best for standard CT beams.

2. Materials and Methods

The measuring system used in this work was an extrapolation chamber with a collecting electrode of 30 mm in diameter, entrance window made of aluminized polyethylene terephthalate and a graphited guard ring. This ionization chamber was developed by Dias and Caldas [1,2] at the LCI.

The characterization tests performed in this work were stabilization time, leakage current and linearity of response for the extrapolation chamber in two different depths (0.75 mm and 1.00 mm). For the stabilization time and leakage current, a ^{90}Sr + ^{90}Y control source (33 MBq) was utilized. For the linearity of response test the X-ray system Pantak/Seifert (ISOVOLT model 160 HS) operating up to 160 kV was utilized. Table 1 shows the CT radiation qualities established at the LCI.

Table1. Characteristics of the CT standard beams based on the IEC [5].

Radiation Quality	Tube Voltage (kV)	Tube Current (mA)	Air Kerma Rate (mGy/min)
RQT 8	100	10	22.0
RQT 9	120	10	34.0
RQT 10	150	10	57.0

For the stabilization test of the ionization chamber, the measurement standard deviation must not exceed ± 2% when comparing the response ionization current obtained in 15 min and 60 min [5].

For the leakage current test of the ionization chamber, the measurement standard deviation obtained before and after irradiation must not exceed 5 % [5].

The linearity of response of the extrapolation chamber must present a linear curve, so the correlation factor needs to be close of 1.0 (R^2= 1.00).

The uncertainties of type A and type B were determined, with the combined uncertainty of factor k = 2 when necessary.

3. Results and Discussion

Initially are presented the results obtained for the stabilization test in both depths. These results are shown in Table 2.

Table 2. Stabilization time of the extrapolation chamber response in two different depths.

Depth (mm)	Stabilization Time (%)
0.75	0.17
1.00	0.15

The results shown for the stabilization time of the extrapolation chamber in both depths are within the internationally acceptable limits [5].

For the leakage current tests the results obtained before and after irradiation for both depths can be seen in Table 3.

Table 3. Leakage current of the extrapolation chamber response in two different depths.

Depth (mm)	Before Irradiation (%)	After Irradiation (%)
0.75	2.39	3.30
1.00	3.51	0.94

The results obtained for the leakage current of the extrapolation chamber response in both depths are within the recommended limits [5].

For the linearity of response, the results obtained can be seen in Figures 1 and 2: the linearity of response for the extrapolation chamber in the respective chamber depths of 0.75 mm and 1.00 mm.

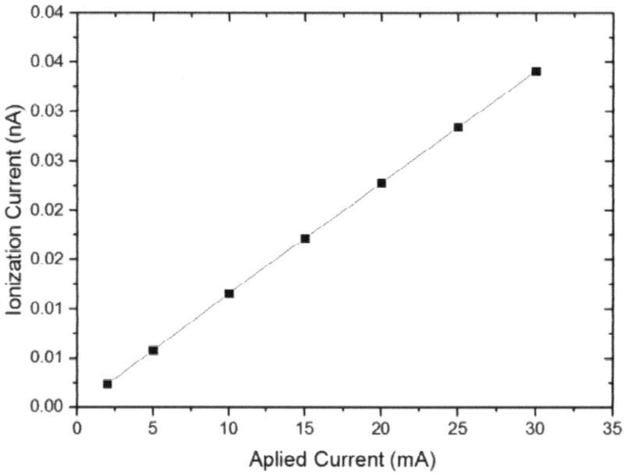

Figure 1. Linearity of response curve of the extrapolation chamber in the reference radiation quality (RQT 9) in the chamber depth of 0.75 mm. The maximum measurement uncertainty was 0.0004 %.

As can be observed in Figure 1, the correlation factor for the linearity of response curve in depth of 0.75 mm was 0.9999.

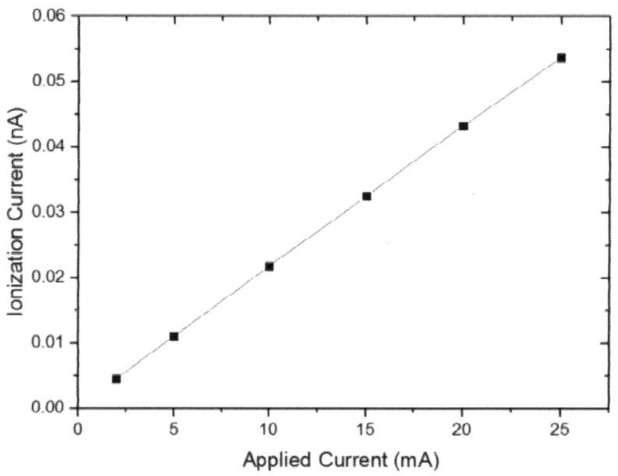

Figure 2. Linearity of response curve of the extrapolation chamber in reference radiation quality (RQT 9) in depth of 1.00 mm. The maximum measurement uncertainty was 0.0004 %.

As observed in Figure 2, the correlation factor for the linearity of response curve in profundity of 1.00 mm was 0.9999.

4. Conclusion

The results obtained for the characterization tests of the homemade extrapolation chamber in computed tomography standard beams were in agreement with the international recommendations.

When analyzing the results obtained in the two different chamber depths it is possible to observe that for the computed tomography beams the depth of 1.00 mm presented the best results.

Acknowledgments

The authors acknowledge the partial financial support from the Brazilian agencies: CNEN, CNPq, CAPES and MCTIC (Project: INCT – Radiation Metrology in Medicine).

References

[1] Dias S K and Caldas L V E 1998 Development of an extrapolation chamber for the calibration of beta-ray applicators *IEEE Trans. Nucl. Sci.* **45** 1666-1669

[2] Dias S K and Caldas L V E 2001 Extrapolation chamber response in low-energy radiation standard beams *J. Appl. Phys.* **89** 669-671

[3] Figueiredo M T T and Silva T A 2015 Determination of the effective volume of an extrapolation chamber for X-ray dosimetry *Congress Procedings: X Congreso Regional Latinoamericano IRPA de Protéction y Seguridad Radiológica* Buenos Aires

[4] Neves L P, Silva E A B, Perini A P, Maiadana N L and Caldas L V E 2012 Characterization of an extrapolation chamber for low-energy X-rays: experimental and Monte Carlo preliminary results *Appl. Radiat. Isot.* **70** 1388-1391

[5] IEC, INTERNATIONAL ELECTROTECHNICAL COMMISSION 2005 Medical diagnostic X-ray equipment. Radiation conditions for use in the determination of characteristics. 2nd ed. IEC, Genève, (IEC 61267)

IOP Conf. Series: Journal of Physics: Conf. Series **975** (2018) 012067 doi:10.1088/1742-6596/975/1/012067

AC-DC Transfer Difference Measurements using a Frequency Output Thermal Converter

G M Geronymo, W G K Ihlenfeld and R P Landim

Instituto Nacional de Metrologia, Qualidade e Tecnologia - Inmetro, Duque de Caxias, Brazil.

E-mail: `gmgeronymo@inmetro.gov.br`

Abstract. This paper presents the design of a Frequency Output Thermal Converter (FOTC), built using commercially available discrete components. The FOTC structure consists of a discrete resistor as heater element with a thermistor connected to an oscillator as temperature sensor. The AC-DC transfer difference of the FOTC was measured by direct comparison with calibrated TCs.

1. Introduction

Despite the recent developments in Josephson-based AC voltage calibration systems [1], the accurate characterization of the root-mean-square (rms) values of time-varying (ac) signals is still most widely done by the use of thermal converters (TCs). A basic TC consists of a heater structure and a temperature sensor, which allows the comparison of ac and dc quantities [2]. We are working on an alternative TC design, using a discrete surface-mount device (SMD) resistor as heater element and a thermistor connected to an oscillator as temperature sensor [3]. In this work we present the results obtained with an improved structure.

2. Frequency Output Thermal Converter

The basic structure of the FOTC (figure 1) consists of two main blocks: the thermal converter and the oscillator. The heater element is a SMD resistor and the temperature sensor is a NTC (negative temperature coefficient) thermistor. The output of the thermistor is connected to an oscillator. The output frequency of the oscillator is proportional to $1/(R_{ntc}\,C)$, where C is a fixed capacitor and R_{ntc} is the output resistance of the thermistor, which is inversely proportional (but not linearly) to the temperature of the heater. Thus, this composition of proportionality constants results in an output frequency proportional (but not linearly) to the input power (i.e., the RMS value of the input voltage).

2.1. *Thermal Converter*

The layout of the thermal converter is presented in figure 2. The thermal converter was built using a printed circuit board (PCB) substrate, rated for frequencies up to 40 GHz. A microstrip transmission line was designed for characteristic impedance (Z_0) of 50 Ω, matching with the nominal value of the SMD resistor used as heater element.

The heater element (a SMD 0402 form factor resistor, 50 Ω) was mounted like a line terminating resistor, perpendicular to the substrate. A layer (of approximately 1.5 mm) of

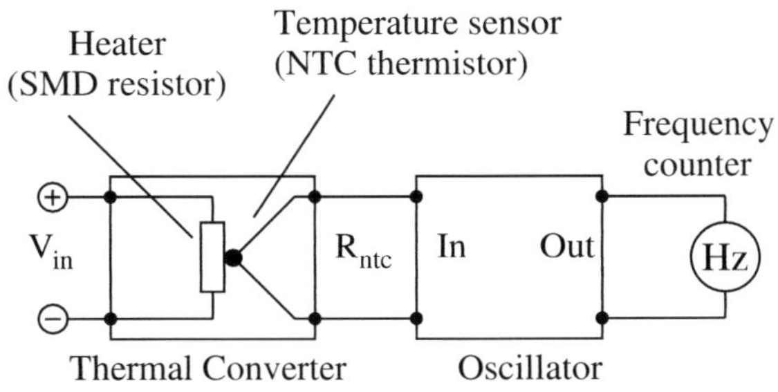

Figure 1. FOTC basic structure.

thermal insulating material (expanded polystyrene) was used to improve the thermal resistance. The temperature sensor is a micro-bead NTC thermistor (0.33 mm bead diameter), fixed next to the low potential terminal of the heater resistor using epoxy glue mixed with alumina, in order to improve the thermal contact between heater and temperature sensor.

A chip carrier structure was built using the same PCB substrate used for the main structure. With the chip carrier closed, only the input terminals of the heater and the output terminals of the temperature sensor are accessible, as shown on figure 2.

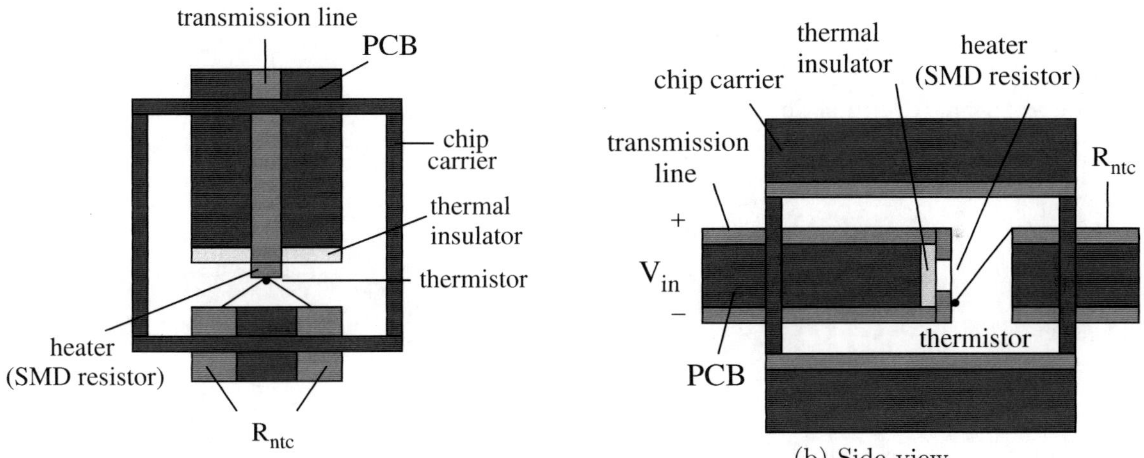

(a) Top view. (b) Side view.

Figure 2. Thermal converter layout.

2.2. Oscillator

A comparison of several oscillator structures showed that the circuit of figure 3 provides the best stability [4]. The basic topology of the oscillator is composed by an integrator block, a comparator block and a double pole double throw (DPDT) switch connected to stable voltage references ($+V_{ref}$ and V_{ref}).

The fractional frequency ($\Delta f/f_0$) of the oscillator was measured using a frequency counter, with averaging time (τ_0) set to 0.1 s. Figure 4 shows the fractional frequency variations over a period of 15 minutes (9000 samples).

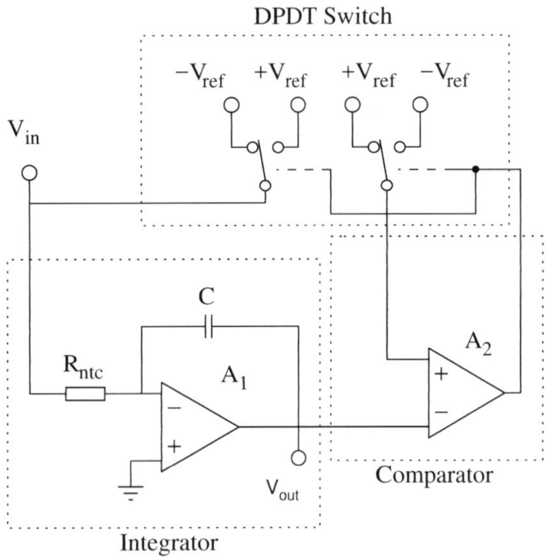

Figure 3. Oscillator topology.

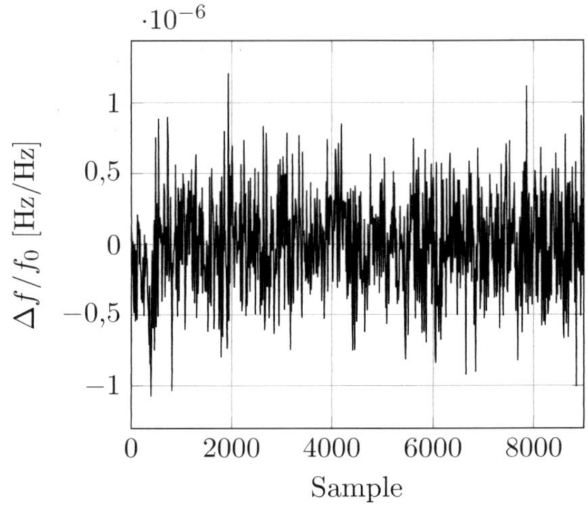

Figure 4. Frequency stability (15 minutes).

3. Results

3.1. Measurement setup

The AC-DC transfer difference of the FOTC was determined by direct comparison with calibrated Planar Multijunction Thermal Converters (PMTCs) of our reference system [5]. The measurement setup is presented on figure 5. The system is fully-automated, controlled by locally developed control software.

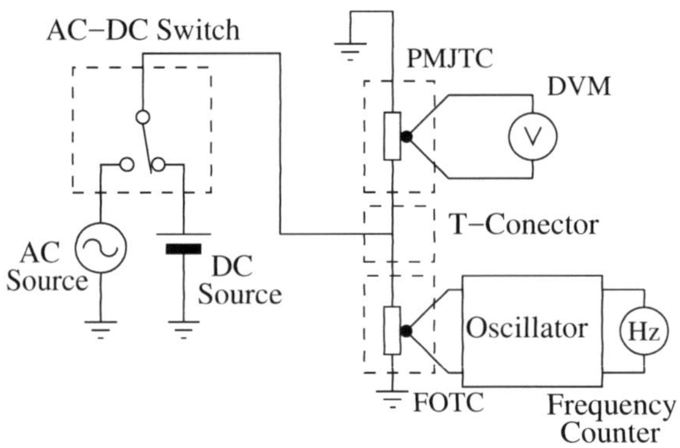

Figure 5. Measurement setup.

One major drawback of using a thermistor instead a thermocouple as temperature sensor is an undesirable sensitivity to fluctuations of the ambient temperature (the differential nature of the thermocouple minimizes this issue). To overcome this, a temperature controlled air bath was used. In future works we plan to develop a compensation circuit or an active temperature control to avoid this limitation.

IOP Conf. Series: Journal of Physics: Conf. Series **975** (2018) 012067 doi:10.1088/1742-6596/975/1/012067

3.2. Measurement results

The AC-DC transfer difference of the FOTC δ_{FOTC} was measured at 1 V level using a calibrated 90 Ω PMJTC as standard. The results are presented in table 1, where δ_{FOTC} is the AC-DC transfer difference of the FOTC, δ_{PMJTC} is the AC-DC transfer difference of the PMJTC, and $U(\delta_{FOTC})$ and $U(\delta PMJTC)$ are the corresponding measurement uncertainties. δ_M is the measured AC-DC transfer difference (relative), and σ is the standard deviation of the measurement.

Table 1. AC-DC transfer difference in μV/V (at 1 V level).

f [kHz]	0.065	0.12	1	10	100	1000
δ_{PMJTC}	0.4	0.0	0.0	1.0	12.0	58
$U(\delta_{PMJTC})$	1.3	1.3	1.5	1.5	3.5	19
δ_M	0.1	-2.1	-1.4	-1.3	7.6	86.6
σ	1.4	0.4	1.9	2.7	0.5	1.0
δ_{FOTC}	0.5	-2.1	-1.4	-0.3	19.6	144
$U(\delta_{FOTC})$	1.6	1.6	1.9	2.4	3.9	34

At 1.5 V level, the FOTC AC-DC transfer difference was measured using a 180 Ω PMJTC as standard. The results are presented in table 2, following the same naming convention.

Table 2. AC-DC transfer difference in μV/V (at 1.5 V level).

f [kHz]	0.065	0.12	1	10	100	1000
δ_{PMJTC}	0.5	0.1	0.1	0.8	3.3	-11.1
$U(\delta_{PMJTC})$	1.3	1.3	1.5	1.5	3.5	19
δ_M	-0.2	-0.9	-0.3	1.4	20.9	181.2
σ	1.4	0.4	1.9	2.7	0.5	1.0
δ_{FOTC}	0.3	-0.8	-0.2	2.2	24.2	170
$U(\delta_{FOTC})$	2.5	2.5	2.6	2.6	4.3	22

The AC-DC transfer difference of the FOTC does not show a significant level dependence, as we can see comparing the results obtained at 1 V and at 1.5 V, considering the uncertainties. For frequencies up to 10 kHz the AC-DC transfer difference is small, typically smaller than 3 μV/V.

3.3. Validation

In order to validate the results, we did measurements of a second PMJTC, following the scheme of figure 6.

In the first step we have determined the AC-DC difference of the PMJTC 2 directly against the PMJTC 1 (δ_{Y1}). Then, the AC-DC difference of the FOTC was measured using the PMJTC 1 as standard, and finally the AC-DC difference of the PMJTC 2 was obtained using the FOTC as standard (δ_{Y2}).

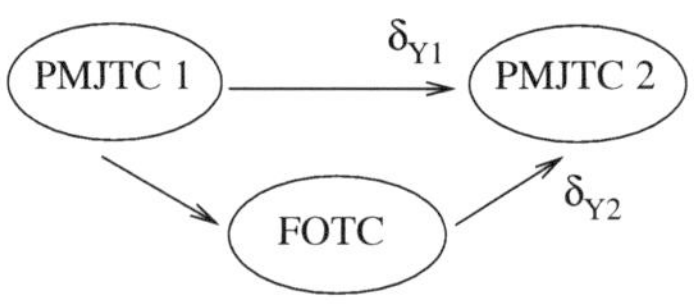

Figure 6. Validation scheme.

To validate the results, we have calculated the normalized error (E_n) using the result δ_{Y1} as reference:

$$E_n = \frac{|\delta_{Y2} - \delta_{Y1}|}{\sqrt{U(\delta_{Y2})^2 + U(\delta_{Y1})^2}} \tag{1}$$

When $E_n < 1$, the result is considered validated. The results are presented in table 3.

Table 3. Validation of the results at 1 V (AC-DC transfer differences in μV/V).

f [kHz]	0.065	0.12	1	10	100	1000
δ_{Y1}	0	0	0	0	7	32
$U(\delta_{Y1})$	1	1	2	2	4	34
δ_{Y2}	1	0	0	1	9	45
$U(\delta_{Y2})$	1	2	2	2	4	44
E_n	0.1	0.1	0.2	0.2	0.4	0.2

4. Conclusion

We have presented a Frequency Output Thermal Converter, built using commercially available discrete components, that can be be used as an AC-DC voltage transfer standard.

Future works:

- Improve the TC immunity to ambient temperature fluctuations;
- Improve the design and build a high frequency optimized TC to extend the traceability to frequencies higher than 1 MHz.

References

[1] Ihlenfeld W G K and Landim R P 2014 An automated josephson based ac-voltage calibration system *29th Conference on Precision Electromagnetic Measurements (CPEM 2014)* pp 468–469 ISSN 0589-1485

[2] Inglis B D 1992 *Metrologia* **29** 191–199

[3] Geronymo G M, Ihlenfeld W G K and Landim R P 2016 Design of a thermal converter with frequency output *2016 Conference on Precision Electromagnetic Measurements (CPEM 2016)* (Ottawa: IEEE)

[4] Schulz P 2011 *AC-DC Transfermessungen mittels thermisch gekoppelter Relaxationsoszillatoren* Master's thesis Ostfalia Hochschule fr angewandte Wissenschaften, Fakultt Elektrotechnik

[5] Geronymo G M, Jr R A, de Barros e Vasconcellos R T and Poletaeff A 2014 *NCSL International Measure: The Journal of Measurement Science* **9** 66–73

IOP Conf. Series: Journal of Physics: Conf. Series **975** (2018) 012068 doi:10.1088/1742-6596/975/1/012068

Behavior of voltage transformers under distorted waveforms

Daniel Slomovitz[1], Gonzalo Aristoy[1], Alejandro Santos[1], Marcelo Brehm[1], Leonardo Trigo[1]

[1] Administración Nacional de Usinas y Transmisiones Eléctricas (UTE)

dslomovitz@ute.com.uy

Abstract. In this work, we present testing results of voltage transformers (VT) errors, under distorted waveforms. Different types of VTs used in high voltage power networks were analysed. This includes inductive as well as capacitive VTs. The main tests were done at nominal voltages, but a comparison with another low-voltage testing method was included.

1. Introduction
The continuous proliferation of non-linear loads in power networks leads to increase the harmonic content of the voltage. To mitigate this problem, filters and other devices are used. Anyway, measurements of voltage harmonic distortion are necessary to demonstrate that they are under regulated limits. In high-voltage networks, the voltage measurement is done through voltage transformers (VT) with primary voltages in the range of 6 kV up to 500 kV, or even more. There are different types of VTs. Inductive type (IVT) is based on the same principle than power transformers. They have a magnetic core and primary and secondary windings. Other type called capacitive voltage transformer (CVT) has a high-voltage capacitive voltage divider followed by a conventional medium-voltage transformer. The behavior under distorted waveform is very different for one type or the other. International most used standards only propose to test these transformers at power frequency [1, 2]. In this case, the errors at higher frequencies are not known. Additional requirements for harmonic response would be find in other standard [3], but most VT in service were not manufactured to fulfil this standard. Table II of that standard shows the limit ranges for harmonic measurement. They go from 1% and 1° for 1st and 2nd harmonic, and 5% and 5° for harmonics from 3th to 50th. For this work, a high-voltage source with programmable harmonic content was used, as well as a measuring bridge that gives the error in ratio and phase for each harmonic order. They are described in [4].

2. Tested units
Units of both types, IVT and CVT, were tested. Their behaviours are very different, as will be shown in the following paragraphs.

2.1. Inductive type
A low frequency model for IVTs is shown in figure 1.

IOP Conf. Series: Journal of Physics: Conf. Series **975** (2018) 012068 doi:10.1088/1742-6596/975/1/012068

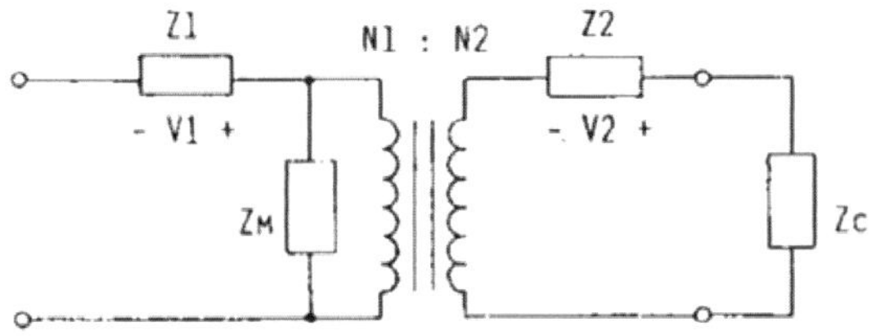

Figure 1. Model for inductive voltage transformers.

Z1 and *Z2* are the primary and the secondary winding impedances. Z_M is the magnetizing branch, modelled by a resistor in parallel with a capacitor and a non-linear inductor. The behaviour of the inductor is taken into account modifying its value in function of the applied voltage. At low frequencies, errors in IVTs are due to voltage drops in the series impedance *Z1* and *Z2*. We performed two different tests: low-voltage tests up to 20 kHz, and nominal-voltage tests up to 3 kHz.

2.1.1 Low voltage tests

Tests were done with a FRA equipment [5], with voltages around 10 V. The ratio between the output and input voltages was computed, expressed in dB. Figures 2 and 3 show the behaviour of the frequency response of different IVT used in high voltage power networks. Their nominal voltages go from 6 kV up to 150 kV. All curves show a similar behaviour up to 2 kHz, corresponding to the 40th harmonic. They remain approximately constant up to 2 kHz, with variations lower than 2 dB. At higher frequencies, the 6-kV IVT reduces the ratio voltage, but the 60-kV and 150-kV ones, increase it. This is due to resonances between series inductance and parallel capacitances. The ratio variation at 2 kHz reaches values up to 25% for the 150-kV IVT. The others have lower variations.

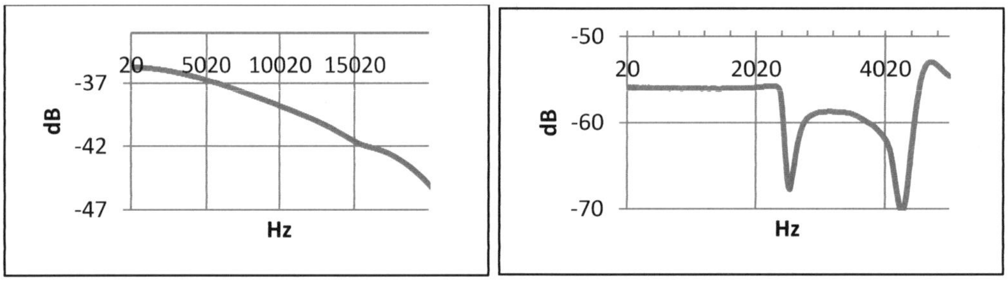

Figure 2. Frequency response of 6 kV and 60 kV IVTs.

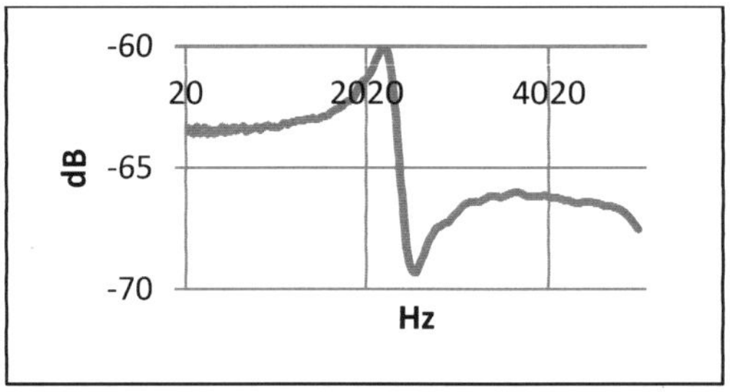

Figure 3. Frequency response of a 150 kV IVT.

2.1.2 Nominal voltage tests

These tests were done at 53 Hz of fundamental frequency, to prevent noise influence from the power network. The voltage at fundamental frequency was set between 80% and 100% of the nominal value of the VTs. A first test was done adding single tones to the fundamental frequency, from 3th (159 Hz) component to 49th (2597 Hz) one. The amplitude of the harmonic components was set around 10% of the fundamental one. In this condition, transformer errors were measured for fundamental and for each harmonic. A second test was done applying the fundamental component and all harmonics together. Again, the transformer errors were measured at fundamental and at each harmonic component. Only results on units of 6 kV and 60 kV are shown, but they are similar for other IVT. Figures 4 and 5 show the ratio error and phase shift with single tones and with high distortion (all harmonic components at the same time). No significant difference between both tests was detected, which corresponds to linear behavior of the tested IVTs.

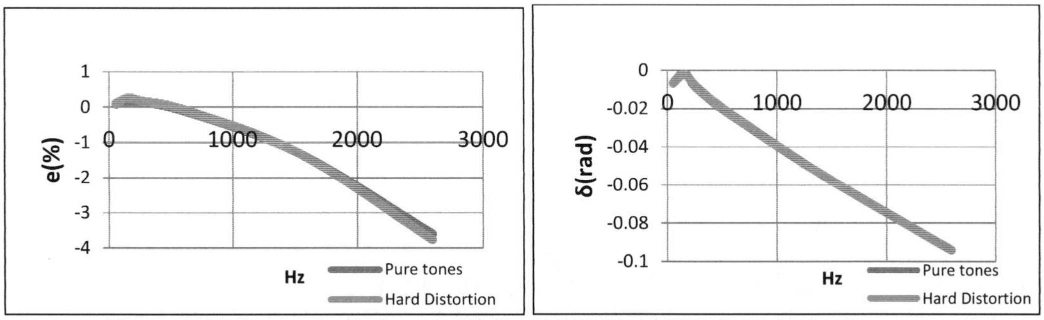

Figure 4. Ratio error and phase shift of a 6 kV IVT at 80% of nominal voltage.

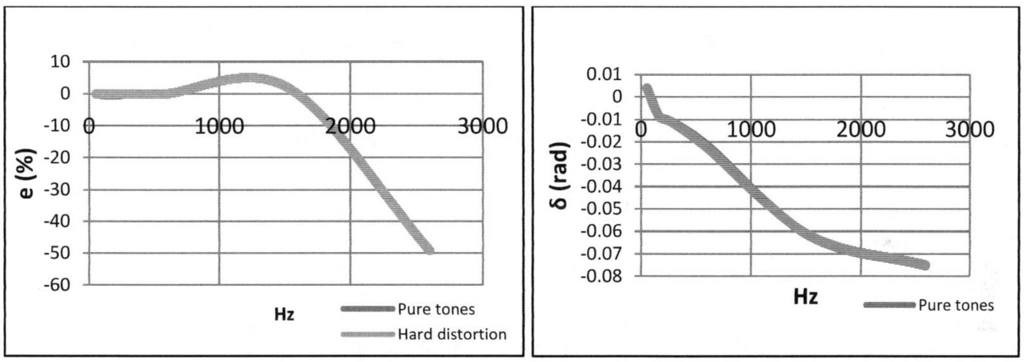

Figure 5. Ratio error and phase shift of a 60 kV IVT at 100% of nominal voltage.

Ratio errors at 2 kHz reaches -2.2% for the 6-kV IVT and -18% for the 60-kV unit. Although they are much larger than the accuracy class of these transformers (0.5), they can be compensated applying a correction table.

The shapes of these curves are close to those got from low-voltage tests, but the differences between both tests do not allow the use of low-voltage tests for error correction. Low-voltage test is good for having a quick view of the frequency response, but if corrective factors are intended to use, tests at nominal-voltage must be done.

2.2. Capacitive type

The simplest model for this type of VT is shown in figure 6.

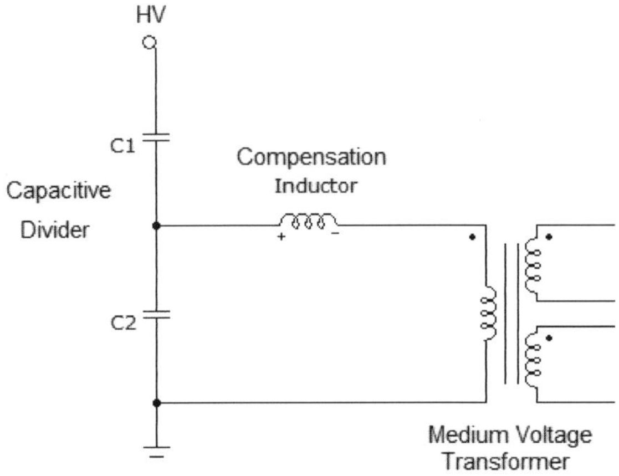

Figure 6. Simplified model of a capacitive voltage transformer

C1 and *C2* form a capacitive divider that is followed by a compensation inductor and a conventional medium-voltage transformer. The inductor is designed to resonate with the capacitors at power frequency to null the capacitance impedance. But, at other frequencies, the series impedance increases, so that, the errors. For the test, it was used the equivalent circuit, shown in [2]. In this configuration, the low terminal of *C2* is disconnected from ground, and connected to the upper terminal of *C1*. In this way, both capacitors are in parallel, in a Thevenin configuration. The advantage of this testing method is that the value of the applied voltage must be the rated voltage of the medium voltage transformer, around 10 kV, much lower than the nominal voltage of the CVT under test. Many tests were done on different units from 60 kV up to 500 kV. As an example, figure 7 shows the ratio error and phase shift of a 150-kV CVT, 200 VA, class 0.5, at two different voltages: 80% and 120% of the nominal one. The applied voltage had harmonic contents that include many orders from the 3th up to the 49th. The amplitude of each one was around 10%.

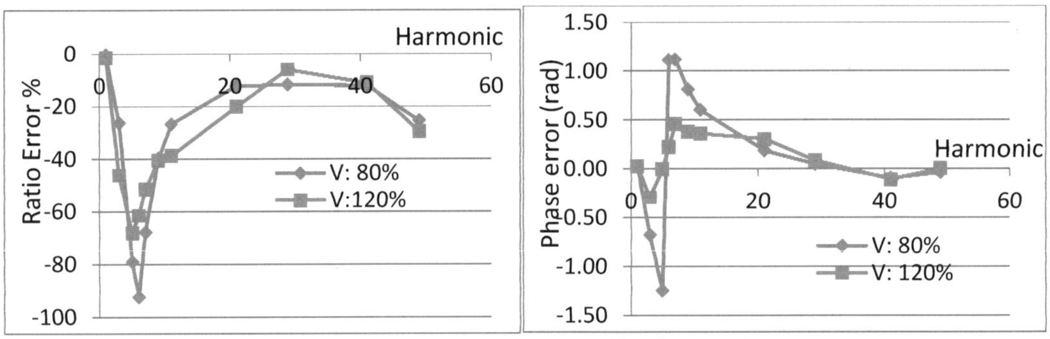

Figure 7. Ratio error and phase shift of a 150-kV VCT.

The results show that this transformer has very large errors at any harmonic frequency. Even, these errors depend on the voltage. Other tests were done varying the burden, and again, a behaviour that depends on the burden was detected. This means that in practice it is not possible to use correction factors for compensating those errors. This kind of VT is not suitable for harmonic measurement unless compensating devices were added. There are proposals [6] for that, but they need of special internal connections and add external devices. Unfortunately, most of the installed CVTs do not have the possibility to be compensated using that technique, because internal points, were currents must be measured, are not accessible.

It is undergoing a project for compensating conventional CVTs, avoiding to change them for other types that allow harmonic measurement.

3. Conclusions

Results of tests on VTs under distorted waveform were presented. Low-voltage tests method (FRA) is useful to get a quick view of the frequency response, but tests at nominal-voltage must be performed to know precisely the value of the errors. Many IVT can be compensated using corrective factors for each harmonic, but this cannot be done for CVT. Their errors are so large that other compensation techniques are necessary.

4. References

[1] International Electrotechnical Commission 2011 Instruments transformes Part 3 Additional requirements for inductive voltage transformers IEC Standard 61869-3.
[2] International Electrotechnical Comission 2011 Instruments transformes Part 5 Additional requirements for capacitor voltage transformers IEC Standard. 61869-5.
[3] International Electrotechnical Comission 2012 Instrument transformers The use of instrument transformers for power quality measurement IEC/TR 61869-103.
[4] Aristoy G, Trigo L, Brehm M, Santos A, Slomovitz D 2016 Measuring system for calibrating high voltage instrument transformers at distorted waveforms Conference on Precision Electromagnetic Measurements Canada.
[5] Doble M4100 High Voltage Apparatus Tester http://www.doble.com/ product/m4100/
[6] Ghassemi F, Gale P, Cumming T, Coutts C. 2005, Harmonic voltage measurements using CVTs, IEEE Transactions on Power Delivery 20 pp 443 – 449.

Acknowledgments
This work has been partially financed by ANII, Project FSE_1_2014_1_102482.

IOP Conf. Series: Journal of Physics: Conf. Series **975** (2018) 012069 doi:10.1088/1742-6596/975/1/012069

Development of an ultrastable laser at 1550 nm

J De Martin Júnior [1,3], S K N Valappil [1], S T Müller [1], P W Courteille [2], V S Bagnato [2], D V Magalhães [1]

[1] São Carlos School of Engineering, São Carlos, Brazil
[2] São Carlos Institute of Physics, São Carlos, Brazil
[3] Federal Institute of Education, Science and Technology of São Paulo, Catanduva, Brazil

E-mail: jairdemartinjr@ifsp.edu.br

Abstract. This work deals with the development of a 1550 nm diode laser stabilized against a high finesse cavity (>250,000). It reports the components, apparatus and some preliminary results for the system under construction.

1. Introduction

Lasers are applied in a wide range of sectors, from industry to high precision physics. In order to be used in some physics experiments, like high resolution spectroscopy, there is a need for very accurate and stable systems. In this sense is fundamental to reduce the linewidth of the laser system, to be able to resolve certain atomic transitions. Our Group has been developing a system of this kind, with linewidth enough to deal with atomic reference standards that we have in other experiments [1].

One of the main aspects concerning the quality of a laser is its linewidth. For a free-running laser, this characteristic can be on the order of megahertz. There are several techniques to reduce the linewidth, for example, the use of an extended cavity for a diode laser (ECDL) [2] or servo the laser to a high finesse Fabry-Perot cavity (PDH technique) [3]. The use of external Fabry-Perot cavities to reference laser systems is one of the most common techniques for stabilizing ultra stable lasers [4].

On the other hand, special attention should be devoted when preparing the mechanical structure to isolate the cavity, since the deformations in the cavity degrade the stabilization system [5], and a temperature control in the mK range is crucial, in order to maintain the cavity around the most stable region of operation. Also, the quality of the controllers has a fundamental whole in the laser stabilization, where a reflection from the optical cavity is used to lock on the desired frequency. The aim of this work is to develop a system able to stabilize a diode laser at 1550 nm in a sub-hertz level linewidth. Figure 1 summarizes the system.

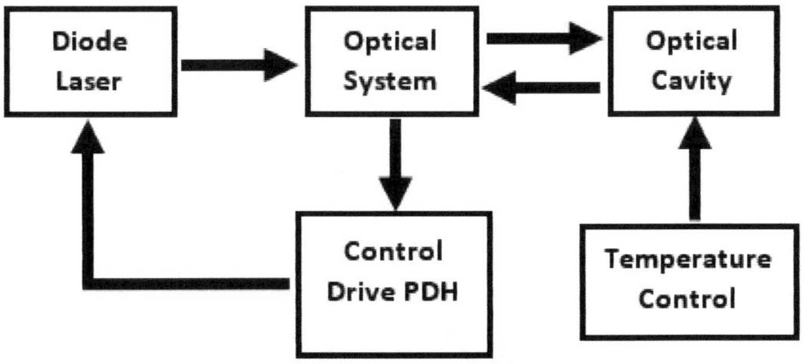

Figure 1. General scheme of the parts for the ultrastable laser system at 1550 nm.

2. System description
Following are presented the components for the laser stabilization system.

2.1. Diode Laser
The system has a diode laser of 1550 nm with linewidth in free-running regime of 1 kHz (RIO Orion Laser). The laser parameter settings (temperature and current) are set through a SPI / USB interface with the computer. In addition the laser has a modulation input used as corrective feedback.

2.2. Optical System
In order to stabilize the laser in the sub-hertz level is essential to use an optical cavity for reference. The cavity used has a finesse 250,000 and length of 10 cm.

The big issue for the laser stabilization is the injection of laser light into the optical cavity and the optical system should be done to achieve the best possible impedance and mode matching for the cavity.

2.3. Vacuum chamber
In order to provide a shielding for temperature stabilization, a vacuum chamber is used to protect the optical cavity. The chamber used is also involved with an acrylic blanket to increase their isolation with the variation of ambient temperature. In addition there are four Peltier elements coupled to the chamber for temperature control, figure 2 shows the vacuum chamber used in this project.

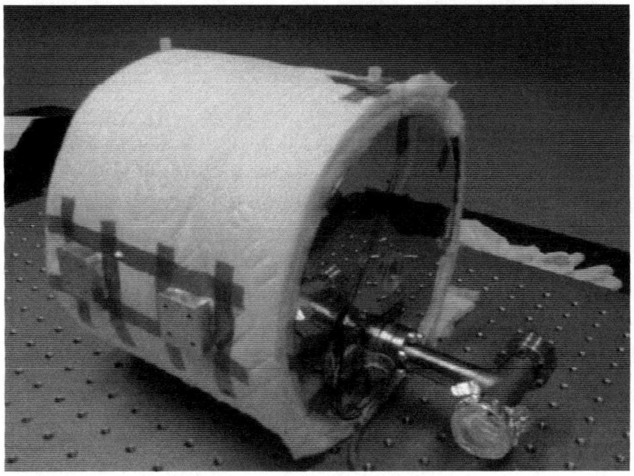

Figure 2. Vacuum chamber involved with the acrylic blanket for thermal insulation.

2.4. Mechanical Insulation System

The vacuum chamber, diode laser and the optical system are being mounted on a breadboard attached to a high isolation passive platform (Minus K). This platform performs isolation through a negative constant spring system, which provides excellent isolation down to 0.5 Hz. In addition the entire system is placed in an acoustic box to prevent coupling of noise. Figure 3 illustrates the pre mounted system.

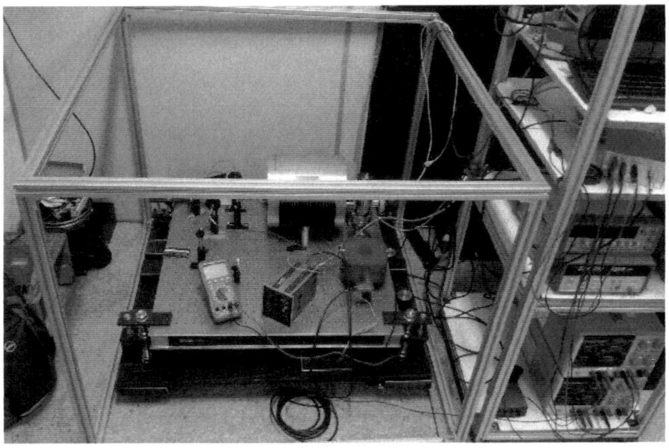

Figure 3. Preliminarily assembled system.

It is noteworthy that the platform used has an acceptable weight range of 164 - 239 kg i.e. the optical table over the vacuum chamber is not sufficient to achieve the minimum weight limit, therefore, steel bars were added to the system to work in nominal mass of the platform. This provides a good margin for additional components and, at the same time, keeps the mass center of the system very low.

2.5. Laser locking system

To lock the laser, we will use the Pound-Drever-Hall technique, shown schematically in figure 4.

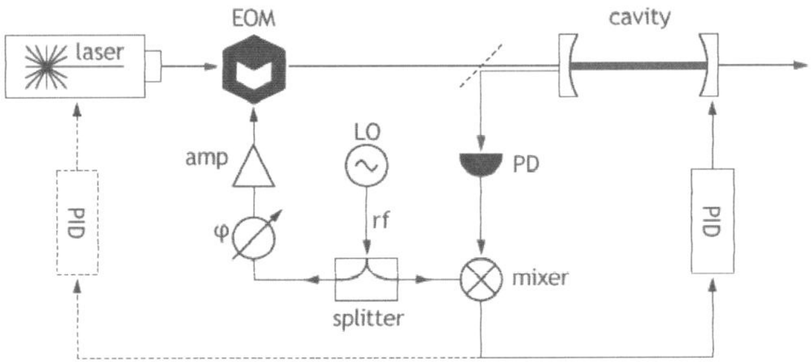

Figure 4. Pound Drever Hall technique [6].

Basically, this technique is achieved through modulation and demodulation of a signal in the megahertz order in the laser light. A radio frequency signal is injected through an electro-optic modulator, which induces sidebands to the laser light. The light of the optical cavity resonance and the sidebands are collected by a photo-detector and demodulated with a mixer.

2.6. Temperature control of the vacuum chamber

To implement the temperature control is essential to know the thermal behavior of the vacuum chamber. So it is necessary to take into account the plant parameters. The technique used is the relay

IOP Conf. Series: Journal of Physics: Conf. Series **975** (2018) 012069 doi:10.1088/1742-6596/975/1/012069

method, which induces a relay-type non-linearity in the plant, founded by descriptive function [7], the output is basically a sinusoidal signal, with time and amplitude depending on the input signal (relay signal). Figure 5 shows a control loop using this technique and figure 6 illustrates the relay signal and the plant response.

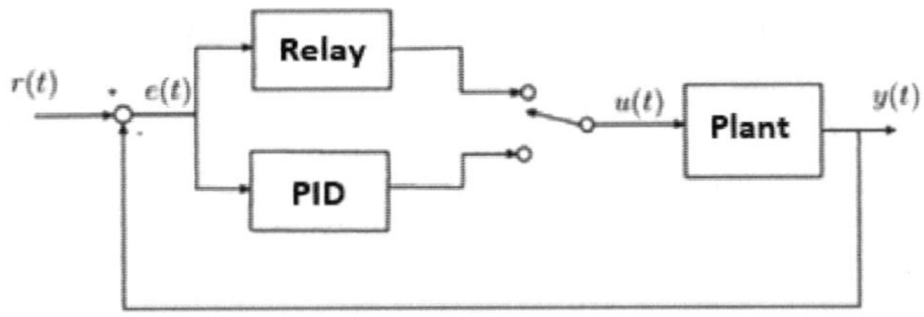

Figure 5. Loop Control used to determine the plant characteristics using a relay-type non-linearity.

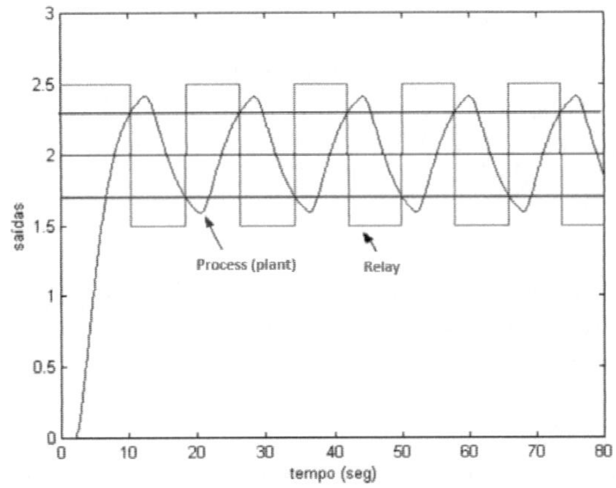

Figure 6. Relay signal and plant response.

The control was first implemented using LabView platform with a data acquisition board PCI-6259. The initial focus was to understand system behavior as well as their parameters, critical gain and critical period, with these two parameters the gains Proportional, Integral and Derivative are calculated using the method of Ziegler-Nichols [8]. Thermal stability found is already under 10 mK variation.

3. Current results

The optical table assembly has been completed, as shown in figure 7.

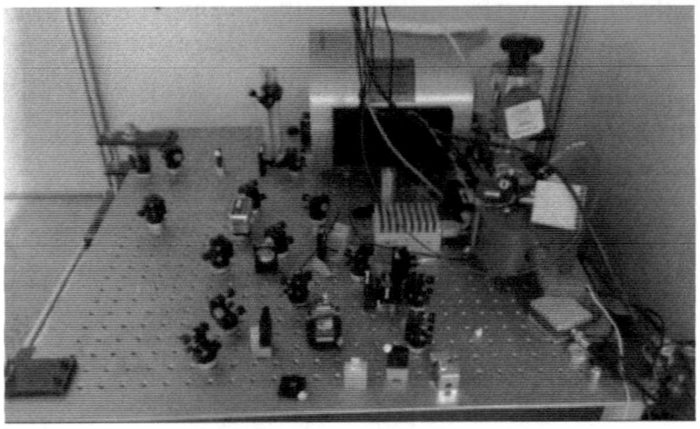

Figure 7. Optical table.

Adjusting the alignment of the laser beam at the entrance of the cavity and modulating the laser into current, we can visualize the transmission signal of the optical cavity represented in figure 8.

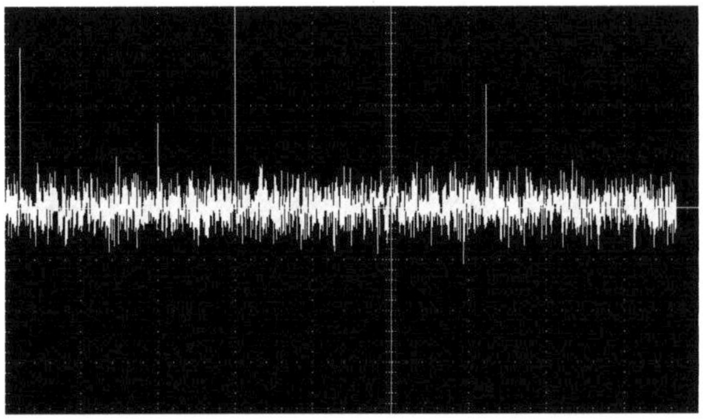

Figure 8. Transmission signal of the optical cavity.

4. Conclusion

We present the current status of an ultra stabilized laser at 1550 nm. The mechanical assembly and temperature control are being finalized. The next steps are design the control driver for laser locking, characterize the acoustic and mechanical vibration insulation system and characterize the ultra stable laser against frequency comb.

Acknowledgments

Grant 2013/04162-5, São Paulo Research Foundation (FAPESP), grant 2013/07276-1, São Paulo Research Foundation (FAPESP), CAPES and CNPq for financial support.

References

[1] De Martin Jr J, Pechoneri R D, Otoboni F A, Müller S T, Bagnato V S and Magalhães D V 2012 Development of a Mobile Atomic Frequency Standard based on Cold Atoms *Latin America Optics and Photonics Conference*, paper **LM2A.4.**

[2] Fleming M and Mooradian A 1981 Spectral characteristics of external-cavity controlled semiconductor lasers in Quantum Electronics *IEEE Journal of Quantum Electronics* **17** 44-59.

[3] Black E. D 2001 An introduction to Pound–Drever–Hall laser frequency stabilization *American Journal of Physics* **69** 79-87.

[4] Saleh B E A and Teich M C 1991 Fundamentals of Photonics *John Wiley & Sons* **1** 947.

[5] Millo J, Magalhães D V, Mandache C, Le Coq Y, English E M L, Westergaard P G, Lodewyck J, Bize S, Lemonde P and Santarelli G 2009 Ultrastable lasers based on vibration insensitive cavities *Physical Review A* **3** 8.

[6] Image available in: http://www.qubig.com/pound-drever-hall.htm.

[7] Almeida M A and Prado M L M 2011 Identificação de uma planta térmica através do método do relé *Sbmac* 351-354.

[8] Ziegler J B and Nichols N B 1942 Optimum settings for automatic controllers *ASME Transactions* **64** 759-768.

IOP Conf. Series: Journal of Physics: Conf. Series **975** (2018) 012070 doi:10.1088/1742-6596/975/1/012070

GMR Sensors and Neural Networks Applied to the Contactless Measurement of Direct Electrical Currents

C Schuina[1], D P Magalhães[2] and C R Hall Barbosa[1]

[1]Postgraduate Program in Metrology, [2] Department of Electrical Engineering, Pontifical Catholic University of Rio de Janeiro, Rio de Janeiro, 22541-900, Brazil

E-mail: hall@puc-rio.br

Abstract. Clamp ammeters based on coils are restricted to the measurement of alternating currents, through the associated magnetic field. There are also commercial versions able of measuring direct currents, based on Hall effect sensors. This manuscript presents improvements of a previously presented prototype of a contactless ammeter based on commercial giant magnetoresistance magnetometers, associated with neural networks for signal processing, able to measure direct electrical currents and to infer the distance between the sensor and the electrical current conductor.

1. Introduction

There are several types of ammeters applicable for measurement of electric currents in conductors, depending on their physical principle and on the need to interrupt the circuit for the current to be measured [1]. The goal of the present development is to perform measurements of direct currents with high resolution, in relation to clamp ammeters based on Hall effect sensors. The system, initially presented in [2-4], is able to also estimate the distance between the sensor and the conductor. The present work enhances the previous prototypes, by replacing the permanent magnet used to polarize the GMR sensors by a solenoid, and by employing artificial neural networks to estimate currents and distances.

Section 2 presents the design of the ammeter and its electronic circuit. Section 3 presents the measurement and processing of experimental data. The inverse problem solution based on neural networks is described in Section 4. At last, Section 5 presents the conclusions.

2. GMR Based Ammeter

2.1. GMR Sensors

A GMR sensor is characterized by the large variation of its resistance as a function of an external magnetic field [5], typically between 10 and 20%. This project is based on the GMR AA005-02, manufactured by NVE Corporation. The topology of the sensor is a Wheatstone half-bridge, having two magnetically shielded GMRs and two GMRs that are sensitive to the magnetic field, all in the same SOIC integrated circuit.

Without the presence of an external magnetic field the bridge is balanced and the output of the bridge is zero. In the presence of an external magnetic field, the field-sensitive GMRs change their values, unbalancing the bridge and generating a differential voltage output.

The output signal of the bridge is directly proportional to the supply voltage/current, with a typical sensitivity of 0.45 mV/V/Oe and a linear region ranging from 10 Oe to 70 Oe.

2.2. Ammeter Requirements

The goal of the ammeter under development is to measure electric currents in the range of -20 A to 20 A, passing through electrical conductors at close distance, from 1 cm up to 4 cm. An electric current passing through a wire generates a circular magnetic field around it that varies with the intensity of the current and with the distance from the conductor to the measuring point. By applying the Biot-Savart law to a straight conductor it is observed that the magnetic field H, generated is given by

$$H = \frac{I}{500r_1} \tag{1}$$

where I is the electric current in amperes, r1 is the distance between the conductor and the sensor in meters and H is the magnetic field in oersteds (1 Oe = 1000/4π A/m). Considering the values for current and distance above indicated, the maximum magnetic field to be measured by the GMR sensors is 4 Oe. For comparison purposes, the Earth's magnetic field is about 0.5 Oe.

2.3. Ammeter Design

As the GMR sensor behavior around H = 0 Oe is highly non-linear, it is necessary to apply a DC biasing magnetic field so that the sensors operate in their linear region. In the previous versions of the ammeter, this biasing field was generated by a permanent magnet, but it was quite difficult to control precisely its value. The present design uses a solenoid with 10 turns and 1.5 A to generate a biasing field of 16 Oe, so as to provide a dynamic range of ±6 Oe in the linear region with high accuracy and stability.

Also, the use of only one GMR sensor is not sufficient, as it is necessary to know the distance r1 to estimate the electric current I. Thus, the ammeter is based on two GMR sensors separated by a fixed distance (D = 3 cm), in addition to the solenoid above described that generates the biasing magnetic field, as shown in Figure 1.

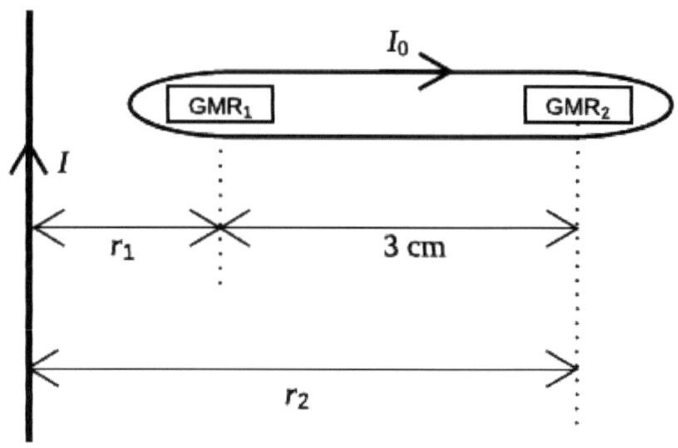

Figure 1. Schematic diagram of the ammeter.

By measuring the voltage generated by both GMR sensors it is possible to estimate the current I and also estimate the distance r1 between the sensor GMR1 and the conductor.

2.4. Electronic Circuit

The electronic conditioning and reading circuit shown in Figure 2 was designed and implemented. The electronic circuit has the functions of feeding the GMR sensors with a DC current of approximately 3 mA (current source based on the LM318 operational amplifier), as well as reading the differential output voltage of the bridge and amplifying it by 50 times (instrumentation amplifiers INA129, with gain defined by the 1 kΩ resistor).

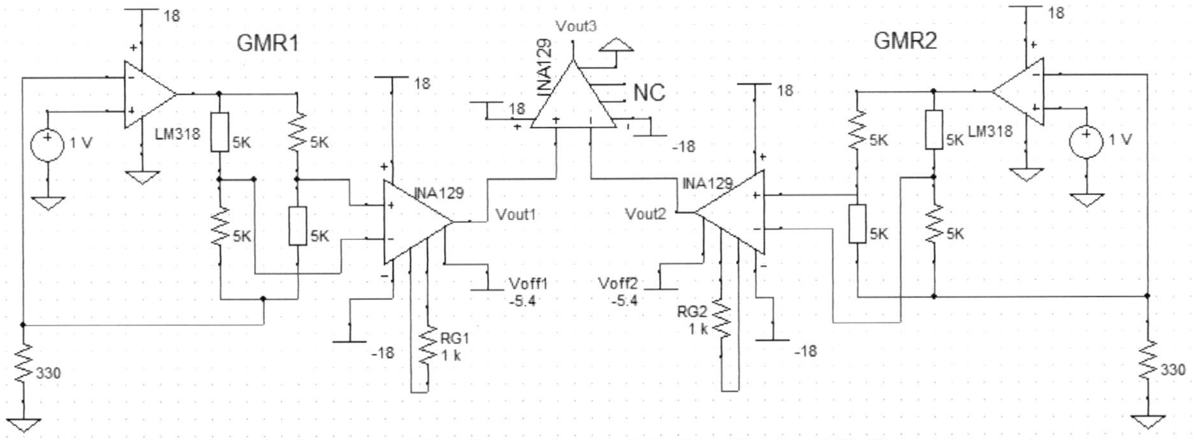

Figure 2. Schematic diagram of the electronic conditioning circuit of GMR magnetometers.

The offset voltage level of -5.4 V allows a zero output voltage, Vout, to be obtained when H = 16 Oe (polarization field). Thus, for a current of 0 A in the wire, there will be an output of 0 V. To minimize the effect of external interferences and improve the ANN performance, a differential reading of the sensors outputs is also made, by the third INA129 indicated in Figure 2.

3. Experimental Data

To observe the effect of the hysteresis and any other fluctuations, 15 tests with currents varying between -3 A and 3 A were performed, in steps of 0.2 A. The tests were performed with five different distances r_1, from 1.0 to 2.1 cm, with three repetitions for each distance.

The data were acquired using an A/D converter, model NI-USB 6229, with 16 bits resolution and measurements taken every 2 s with 5 kHz acquisition frequency.

It was observed a variation on the offset of the signals, probably caused by external interferences. To correct this problem, the linear coefficient of the estimated trend lines for each test was excluded. The results for two distances (1.0 and 2.1 cm) can be seen in figure 3, containing the electrical current values and the respective output voltages.

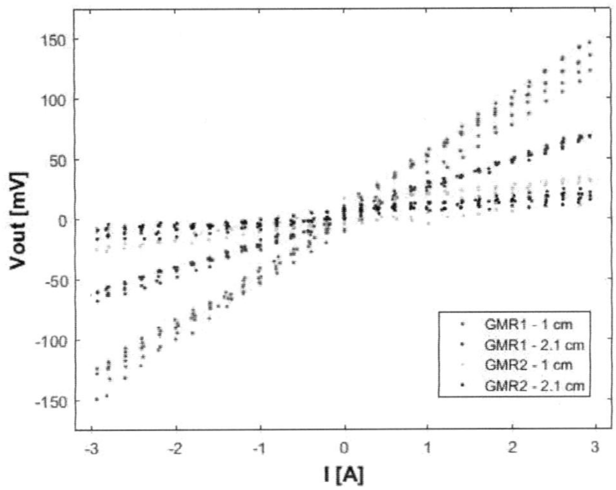

Figure 3. Data set for two distances.

IOP Conf. Series: Journal of Physics: Conf. Series **975** (2018) 012070 doi:10.1088/1742-6596/975/1/012070

4. Inverse Problem

As the objective is the development of an ammeter, it is necessary to solve the inverse problem, which is, given the output voltages of the GMR sensors, measured by the data acquisition system, to estimate the electric current flowing through the conductors and the distance r1. For that, two artificial neural networks (ANN) were used. The following subsections present the details of each one.

4.1. Estimation of Electric Current

A two-layer feed-forward network, with sigmoidal transfer function in the hidden layer and linear transfer function in the output layer was created using Matlab's Neural Network Fitting Tool. The final data set described in section 3, plus the differential reading, was randomly divided into training, validation and testing, respectively 70%, 20% and 10% of the data. The network was trained with Levenberg-Marquardt algorithm. Training process stops when generalization stops improving, as indicated by an increase in the mean squared error (MSE) of the validation samples [6]. The tests were performed using 20 neurons in the hidden layer. The ANN was trained 50 times, and the one with best validation Root Mean Squared Error (RMSE) was chosen for testing.

The regression analysis in figure 4 shows that the data estimated by the neural networks are strongly correlated with the experimental data with R (correlation coefficient) higher than 0.96. At last, the test RMSE was calculated. The network can predict the electric current value, given the GMRs voltage values, with reasonable accuracy, with an RMSE of 0.327 A.

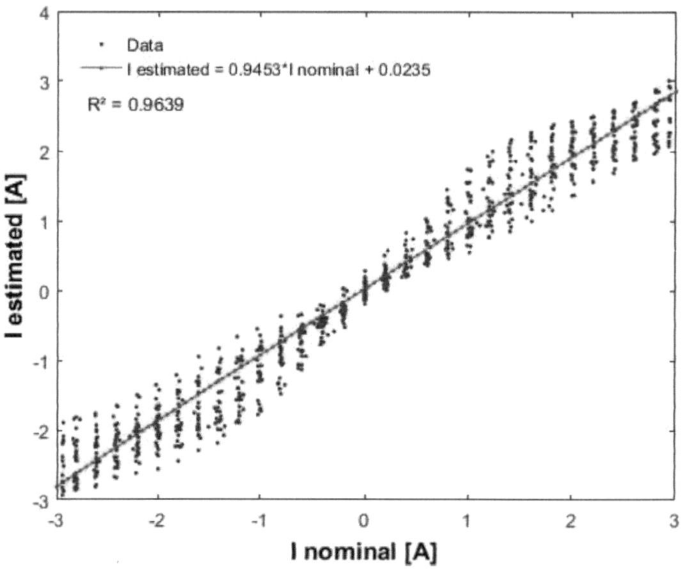

Figure 4. Regression analysis.

4.2. Estimation of Distance

A two-layer feed-forward network, with sigmoidal transfer function in the hidden layer and softmax transfer function in the output layer was created using Matlab's Neural Network Pattern Recognition and Classification Tool. The input variables are the same described in section 3, plus the electric current values, in a total of 3 inputs. However, as the distance can't be estimated when I = 0, these values were excluded from the final data set. After that, the new data set was randomly divided into training, validation and testing, respectively 70%, 20% and 10% of the data. The network was trained with scaled conjugate gradient algorithm. Training process automatically stops when generalization stops improving [6]. The tests were performed using 20 neurons in the hidden layer. The ANNs were trained 10 times, and the one with best percent of correctly classified cases was chosen.

IOP Conf. Series: Journal of Physics: Conf. Series **975** (2018) 012070 doi:10.1088/1742-6596/975/1/012070

Even though this ANN originally performs a classification between 5 measured distances, it is still possible to calculate the error between the actual distances and the classified distances. When using the actual current values as inputs the RMSE for the distance was 0.29 cm and, when using the current values estimated by the first ANN, this RMSE error increases slightly to 0.39 cm.

5. Discussion and Conclusions

The prototype described in this paper, as expected, presents a better performance than the one in [3], but it is less accurate than the one in [4]. This behavior can be explained by the inclusion of different distances and the interferences observed in the GMR2 signal.

In future works it is predicted to magnetically isolate the sensor, as well as improve the conditioning circuit, in order to eliminate the offset variations and interferences observed and obtain more consistent data, so as to improve the obtained uncertainty and RMSE.

Acknowledgements

The authors thank for the financial support provided by CNPq, FINEP and FAPERJ.

References

[1] Ripka P 2010 *Meas. Sci. Technology* **21** 1-23.
[2] Stefani Filho C L and Barbeta V B 2001 *Revista Pesquisa & Tecnologia FEI* **21** 14-18.
[3] Carvalho M C, Schuina S, Barbosa C R H, Silva E C and Gusmão L A P, "Contactless Ammeter Based on GMR Sensors", *Proc. CBM 2015* Bento Gonçalves.
[4] Schuina C, Magalhães D P, Barbosa C R H, Oliveira E C, "A contactless ammeter based on GMR magnetometers", *Proc. IMEKO 2017*, Rio de Janeiro.
[5] Càndid R, María-Dolores C and Diego R M 2010 *Sensors* **9** 24.
[6] MATLAB, "MathWorks," [Online]. Available: https://www.mathworks.com/help/nnet/gs/fit-data-with-a-neural-network.html . [Accessed 15 01 2017].

IOP Conf. Series: Journal of Physics: Conf. Series **975** (2018) 012071 doi:10.1088/1742-6596/975/1/012071

Progress toward Brazilian cesium fountain second generation

Caio Bueno [1], Andrés Rodriguez Salas [2], Stella Torres Müller[2], Vanderlei Salvador Bagnato [2], Daniel Varela Magalhães [1]

[1] Departamento de Engenharia Mecânica, Escola de Engenharia de São Carlos, Universidade de São Paulo, São Carlos, Brazil

[2] Departamento de Física e Ciência dos Materiais, Instituto de Física de São Carlos,Universidade de São Paulo, São Carlos, Brazil

E-mail: caiobuenofisica@gmail.com

Abstract. The operation of a Cesium fountain primary frequency standard is strongly influenced by the characteristics of two important subsystems. The first is a stable frequency reference and the second is the frequency-transfer system. A stable standard frequency reference is key factor for experiments that require high accuracy and precision. The frequency stability of this reference has a significant impact on the procedures for evaluating certain systematic biases in frequency standards. This paper presents the second generation of the Brazilian Cesium Fountain (Br-CsF) through the opto-mechanical assembly and vacuum chamber to trap atoms. We used a squared section glass profile to build the region where the atoms are trapped and colled by magneto-optical technique. The opto-mechanical system was reduced to increase stability and robustness. This newest Atomic Fountain is essential to contribute with time and frequency development in metrology systems.

1. Introduction

Since 1967 the time unit second of the international system of units (SI) has been defined using the ground state hyperfine transition of the Cesium atom 133 Cs. The International Atomic Time (TAI) is a time scale calculated with a weighted average algorithm (Algos) by BIPM. The timescale traceable to TAI is kept by laboratories (about two hundred atomic clocks) which maintain the best primary cesium standards.

Primary frequency standards are commonly used for calibration of secondary time and frequency standards. Atomic clocks also play a vital role in navigation systems, like the Global Positioning Systems (GPS) and other Global Navigation Satellite Systems (GNSS) (GLONASS, Galileo, Beidou), relying on the accuracy and the stability of atomic clocks. Atomic frequency standards are also used for network synchronization in telecommunication. They are also extremely useful in fundamental research fields, such as: relativity, atomic physics, astronomy and dark matter [1-6].

The goal of this work is to present some progress in the Brazilian Cesium fountain Second Generation under development in our laboratory at the University of São Paulo in São Carlos campus. This follows previous experiment that keep going in course for the first atomic fountain in our laboratories [7].

2. Optical setup and vacuum system

The optical system use three lasers: a master laser used to cool and to trap atoms; repumping laser used to the repump transition; and detection laser used to detect atomic transition after the interrogation of the atomic cloud. For the master laser we use a commercial laser made (TOPTICA TA100), which produces 400 mW of collimated laser beam. The detection laser is commercially made by Radiant Dyes and for repumping one of used our homemade laser. Both detection and repump lasers were built with an intracavity very narrow frequency filter in extended cavity. This filter ensures more frequency stability.

The Master laser is tuned to the transition $6^2S_{1/2}|F=4, m_f=0> \rightarrow 6^2P_{3/2}|F^{'}=5, m_f=0>$. Its light is divided in six beams, each of them coupled in optical fiber and taken to the beam expanders. This expanders have fixed alignment that provide more robustness.

The repump laser is tuned in the transition $6^2S_{1/2}|F=3, m_f=0> \rightarrow 6^2P_{3/2}|F^{'}=4, m_f=0>$. This laser beam is divided in two other beams which feed two optical fibers.

Using a pair of coils that generate a quadrupole magnetic field and six laser beams, we can trap and cold an atomic cloud, using a mogneto-optical technique [8]. The trap region is shown in Figure 1.

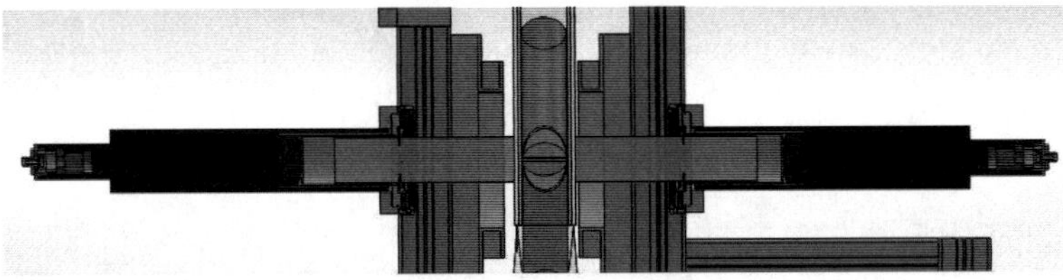

Figure 1. Trapping atoms region, using two coils and six beam expanders. In figure is shown a pair of transverse beams.

We used a square glass profile as a cavity to trap atoms. This provides a great optical access to the preparation chamber and gives more flexibility concerning alignment. The use of glass qalso prevents the generation of spurious fields by residual magnetization of traditional stainless steel chambers. The alternative aluminum or titanium chambers are a bit tricky to work or expensive. The vacuum system is composed by an ionic pump of 60 l/s.

The Cesium reservoir uses three getters of cesium. These getters are capable of liberating cesium when a current up to I = 3 A is applied. The assembly of getters is shown in Figure 2.

Figure 2. Three Cesium Getters used as a cesium supply.

3. Results

Up to now, we were concerned with the general assembly. Two coils for the magnetic field were made and put on anti-Helmholtz configuration, that produces a quadrupole magnetic field. Using a Hall probe, we characterized the magnetic field. The coils can produce 20 gauss/cm of field divergence, using only 3 A of current. This data is shown in Figure 3. This field is enough for a magneto-optical trap [7].

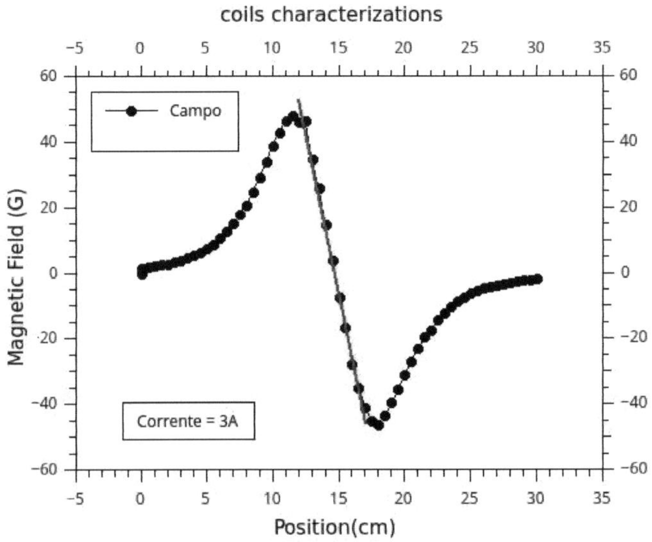

Figure 3. Characterization of the magnetic field produced by two coils in anti-Helmholtz configuration. The measurement was performed using a Hall sensor and coils supplied with 3 A of current.

For the vacuum system, we use an ionic pump controlled by Driver Multivac, made by Varian. Using its electrical current meter, we can estimate the pressure This measurement was made for many days and shown in Figure 4. The vacuum evolves during this time. By the date of this text, the pressure is 9,1x10^{-7}Pa. We are finalizing the optical alignment of the MOT beams and the trapping of atoms is

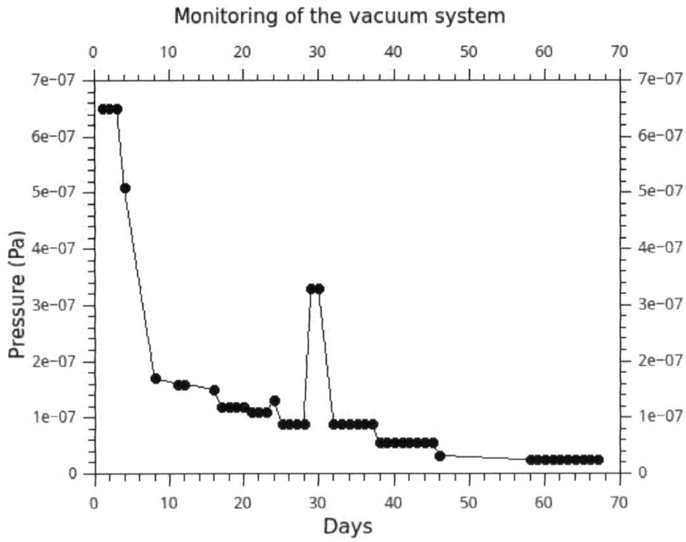

Figure 4. Monitoring of the vacuum system, using Multivac driver controller.

expected to happens in a short timescale. This should validate the system as it is and provide more data concerning the reliability of using the getters supply with the glass chamber. This new kind of system should provide a new manner of producing atomic fountains, opportunity of simpler systems.

References

[1] K. Gibble, S. Chang and R. Legere. Direct observation of s-wave atomic collisions Phys. Rev. Lett. 75, 2666 (1995).

[2] Marion, F. Pereira dos santos, m. Abgrall, s. Zhang, y. Sortais, s. Bize, i. Maksimovic, d. Calonico, j. Gruenert, c. Mandache, p. Lemonde, g. Santarelli, ph. Laurent, a. Clairon, c. Salomon. Search for variations of Fundamental constants using atomic fountain clocks. Physical Review Letters, v. 90, n. 15, p.150801-1 - 150801-4, abr. 2003.

[3] Bebeachibuli, A.; Santos, m. S.; Magalhães, d. V.; Müller, s. T.; Bagnato, v. S. Characterization of the main frequency shifts for the brazilian 133Cs atomic frequency standard. Brazilian Journal of Physics, v. 35, n. 4A, p. 1010-1015, 2005.

[4] P. Wolf. Proposed satellite test of special relativity Phys. Rev. A. 51,5016 (1995).

[5] N. F. Ramsey. The method of successive oscillatory fields Phys. Today.33(7), 25 (1980).

[6] A. Hees, J. Guèna, M. Abgrall, S. Bize, and P. Wolf .Searching for an Oscillating Massive Scalar Field as a Dark Matter Candidate Using Atomic Hyperfine Frequency Comparisons. Physical Review Letters, p.177,1016

[7] Metcalf, H.; Straten, p. Van der. Laser cooling and trapping of atoms. Journal of Optical Society of America B, v. 20, n. 5, p. 887-908, 2003. 30 RAMSEY, N. F. A molecular beam resonance method with separated oscillating fields. Physical Review, v. 78, n. 6, p. 695699, June 1950.

[8] S. A. Rodriguez ; V. S. Bagnato ; C. Bueno ; R. D. Pechoneri ; S. T. Müller ; D. V. Magalhaães: Progress in Brazilian Cesium atomic fountain- BrCsF,Frequency Control Symposium (IFCS), 2016 IEEE International.

Acknowledgements

Grant 2013/07276-1, São Paulo Research Foundation (FAPESP), CAPES and CNPq for financial support

Trilateral South-American project: a reference system for measuring electric power up to 100 kHz – progress report II

G A Kyriazis[1], **L Di Lillo**[2], **D Slomovitz**[3], **R Iuzzolino**[2], **E Yasuda**[2], **L Trigo**[3], **R M de Souza**[1], **H Laiz**[2], **R M Debatin**[1], **E Afonso**[1]

[1] Instituto Nacional de Metrologia, Qualidade e Tecnologia, Av Nossa Senhora das Graças 50, Duque de Caxias, RJ 25250-020, Brazil
[2] Instituto Nacional de Tecnología Industrial, Av Gral Paz 5445, B1650KNA, San Martín, Buenos Aires, Argentina
[3] Administración Nacional de Usinas y Transmisiones Eléctricas, Paraguay 2385, Montevideo – 11800, Uruguay

gakyriazis@inmetro.gov.br

Abstract. Three countries in South America are jointly developing a reference system for measuring electric power up to 100 kHz. The objective is the construction of three similar measuring systems, one for each institute. This project will contribute to provide calibration services in measuring ranges still not covered by the three institutes. The status of its development by the end of 2017 is described here.

1. Introduction

As reported in [1], the Instituto Nacional de Metrologia, Qualidade e Tecnologia (Inmetro), in Brazil, the Instituto Nacional de Tecnología Industrial (INTI), in Argentina, and the Administración Nacional de Usinas y Transmisiones Eléctricas (UTE), in Uruguay, are jointly developing a reference system for measuring electric power up to 100 kHz. The objective is the construction of three similar measuring systems, one for each institute. The project is supported in part by the Conselho Nacional de Desenvolvimento Científico e Tecnológico (CNPq), in Brazil.

Apart from coordinating this project, Inmetro is also responsible for the development of the power amplifiers and the transconductance amplifiers. INTI is responsible for the development of the arbitrary waveform function generators, the dual-channel digitizers and the current shunts (and their calibration). UTE is responsible for the development of the resistive voltage dividers (and their calibration). The integration and testing of the measuring systems are to be performed by the three institutes.

The current status of this development is described in this article.

2. Measuring system

The reader should consult [1] for a description of the measuring system. It comprises: an arbitrary waveform function generator with two synchronized channels, a power amplifier, a transconductance amplifier, a resistive voltage divider, a current shunt and a digitizer with two synchronized channels.

2.1 Arbitrary waveform function generator

The design requirements for such generator were presented in [1]. A two-channel version of the design described in [2]-[4] will be implemented. The circuit changes are not significant. The input buffer and analog-to-digital converter (ADC) are doubled and two power supplies for the ADC modules need to be used.

The generator circuits and printed circuit boards (PCBs) layout were designed by INTI. The circuit components were purchased by Inmetro and the PCBs fabricated in Brazil, all being transported to INTI. The prototype integration and testing is now being carried out by INTI.

Electronic components and product enclosures, for assembling 3 (three) two-channel arbitrary waveform function generators, and PCBs for assembling the dual-channel digitizers, weighing 39.85 kg and amounting US$ 12,052.25 FOB embarked to INTI on August 29, 2017. The PCBs for the function generators amounting US$ 1,890.00 embarked to INTI on November 29, 2017.

2.2 Power amplifier

The design requirements for such amplifier were presented in [1]. The amplifier topology was selected by Inmetro. A total of 2 (two) amplifiers will be assembled: one using pin-through-hole (PTH) components based on [5] and another using surface-mount-device (SMD) technology based on a one-stage version of [6]. The National Institute of Standards and Technology (NIST) is collaborating with the development of the latter. The behavior of the two amplifiers will be compared. The circuit components were purchased by Inmetro and the PCBs of the PTH power amplifiers fabricated in Brazil. Four units of each amplifier were assembled at Inmetro and are now under test.

Inmetro purchased electronic components for assembling 4 (four) PTH power amplifiers weighing 2.80 kg and amounting US$ 1,744.91 FOB on September 03, 2016 and electronic components for assembling 4 (four) SMD power amplifiers weighing 8.55 kg and amounting US$ 4,321.78 FOB on March 21, 2017. They arrived at Inmetro only on June 27, 2017 as CNPq had problems with the renewal of her transportation security contract since January 2017. Inmetro also purchased on December 05, 2016, 12 units of 500-V and 6 units of +/−15-V linear power supplies for assembling the abovementioned amplifiers weighing 33 kg and amounting US$ 5,671.20 FOB (plus US$ 953.35 for transportation). They also arrived at Inmetro only on June 27, 2017. The PCBs for assembling the PTH power amplifiers were fabricated in Brazil for US$ 620.

2.3 Tranconductance amplifier

A total of 4 (four) transconductance amplifiers are being constructed each covering the following ranges: 20 mA, 200 mA, 5 A, and 20 A. The nominal output voltage of the shunts is always 0.8 V. Budget restrictions prevented the construction of the 100 A shunt as had been planned in [1].

The design requirements for such amplifiers were presented in [1]. The amplifiers contain only analog circuits to reduce electrical noise. Linear power supplies are utilized in all amplifiers except the 20 A module also to reduce electrical noise. Vishay Z-Foil resistors (VTA) with low TCR are employed in critical circuit locations to increase the output stability of the amplifier. For 20 A capability, the amplifier needs adequate means to dissipate power while having a small physical geometry to maintain a low inductance in the output current circuit.

The amplifier circuits and product integration were designed by Inmetro based on [7]-[9]. The prototype integration and testing is now being carried out by Inmetro.

Inmetro purchased on June 16, 2016, electronic components for assembling 12 transconductance amplifiers – 3 (three) units at each nominal range - weighing 15.55 kg and amounting US$ 9,936.77 FOB (plus US$ 227 for transportation). They arrived at Inmetro on November 14, 2016. Inmetro also purchased from May 11 to July 16, 2015, Vishay resistors for critical circuits of the abovementioned amplifiers amounting US$ 14,379.83 FOB. They arrived at Inmetro on January 15, 2016. Inmetro also purchased on July 29, 2015, 12 linear and 6 (six) switching-mode +/−12 V power supplies for assembling the abovementioned amplifiers weighing 51 kg and amounting US$ 7,659 FOB (plus US$ 953.35 for transportation). They arrived at Inmetro on January 20, 2016.

Inmetro developed a Brazilian supplier and purchased for US$ 174.86 a total of 24 units of 45-cm long flat flexible circuit material with 9-mm wide copper strips separated by a thin dielectric. Those cables have low impedance and are used as the output cables of the transconductance amplifiers. Inmetro also fabricated 11 cooling tunnel assemblies with 4 (four) black-anodized aluminum heat sinks each to dissipate up to 240 W from the amplifier output transistors. They are going to be used in the abovementioned 5 A and 20 A transconductance amplifiers (and in future ones). A total of 18 PCBs of the 5 A amplifier, 12 PCBs of the buffer amplifier and 12 PCBs of the +/−15 V linear power supply were fabricated in Brazil for US$ 1,300.

2.4 Resistive voltage divider

There are 9 (nine) resistive voltage dividers to be used one at a time whose (binary) nominal value (from 4 V to 1024 V) depends on the test voltage selected [10][11]. Each divider comprises two cascaded, independently shielded sections: a range resistor and a shunt resistor. The dividers have input resistances ranging from 1 kΩ @ 4 V to 256 kΩ @ 1024 V. Thus the power amplifier output current is always 0.004 A. A 200-Ω shunt resistor is used so that the nominal output voltage of the dividers is always 0.8 V. The two sections can be calibrated together as a voltage divider or separately as resistors. This provides flexibility in testing.

Vishay Z-Foil audio resistors (VAR) with low dissipation factor whose dissipated power in each resistor does not exceed 100 mW are employed. It was confirmed by UTE that nonlinear effects are reduced considerably by soldering such resistors on PTFE boards (see [1] and references therein). Such boards have a very low dissipation factor. Once nonlinear effects have been minimized, stray capacitances are the most relevant linear parasitic effect. A special shielding technique for nulling radial electric fields is employed in this work [12].

The voltage divider circuits, the special shielding and the product integration were all designed by UTE. The circuit components and PTFE boards were purchased by Inmetro and transported to UTE. The PCBs layout was carried out and a few prototypes were assembled and tested by UTE. Divider labels containing the project partners logos were fabricated in Brazil. All resistive voltage dividers are now being assembled by UTE. UTE is also developing a system [13] for comparing the divider ratio errors and phase displacements.

Vishay resistors, special connectors and PTFE boards, for assembling 27 resistive voltage dividers – 3 (three) units at each nominal value –, weighing 5.23 kg and amounting US$ 10.045.42 FOB (plus US$ 660 for transportation) embarked to UTE on April 08, 2016. Additional Vishay resistors and connectors weighing 2.80 kg and amounting US$ 2,602.60 FOB (plus US$ 460 for transportation) embarked to UTE on September 28, 2017. Further special connectors amounting US$ 1,161.12 (plus US$ 460 for transportation) to finalize the assembly of the dividers embarked to UTE on October 20, 2017.

2.5 Current shunt

There are 12 current shunts to be used one at a time whose nominal value (20 mA, 50 mA, 100 mA, 200 mA, 0.5 A, 1 A, 2 A, 5 A, 10 A, 20 A, 50 A and 100 A) depends on the test current selected [14]. Their input resistances range from 40 Ω @ 20 mA to 0.008 ohm @ 100 A. Thus the nominal output voltage of the shunts is always 0.8 V. Vishay Z-Foil resistors (Z201) with low TCR whose power dissipated in each resistor does not exceed 0.33 W are employed to satisfy stability requirements (see [1] and references therein).

The current shunt circuits and PCBs layout were designed by INTI and Inmetro based on [15]. The circuit components were purchased by Inmetro and the PCBs fabricated in Brazil, all being transported to INTI. All current shunts are now being assembled by INTI [16].

Special connectors and Vishay resistors for assembling 48 shunts – 4 (four) units at each nominal value – weighing 11.25 kg and amounting US$ 47,232.04 FOB, embarked to INTI on September 20, 2016. The PCBs were ready in July, 2016. The PCBs, weighing 26.70 kg and amounting US$ 4,080.88 FOB, embarked to INTI on September 12, 2016.

Two additional units of each nominal value were independently constructed by Inmetro for ac-dc transfer measurements and instrument calibration purposes. Inmetro already tested the 0.5 A, 1 A, 2 A, 5 A, 10 A and 20 A shunts [17]. Inmetro elaborated software to model the current shunts behavior and estimate their ac-dc transfers and phase displacements [18]. An M. Sc. dissertation will be produced under Inmetro´s metrology postgraduate program to document the modeling and experimental results obtained.

Inmetro purchased (May 22, 2015) special connectors, (May 11, 2015) Vishay resistors, and (April 26, 2016) PCBs for assembling 24 current shunts – 2 (two) units at each nominal value – amounting US$ 24,759.40 FOB (plus US$ 4,015.46 for transportation).The special connectors arrived at Inmetro on December 11, 2015 and the Vishay resistors on January 15, 2016.

A total of 24 units of 70x70x1.5-mm, 48 units of 130x130x1.5-mm, 24 units of 250x250x1.5-mm and 48 units of 46x18x0.5-mm metallic panels for assembling 72 current shunts – 6 (six) units at each nominal value – were fabricated in Brazil. All aluminum panels are anodized, matted (with small porous diameter for aesthetic purposes) and contains the colored logos of the project partners. Such service cost US$ 2,850. Two-thirds of these panels were transported to INTI and the remaining will stay at Inmetro.

2.6 Dual-channel digitizer

A high-resolution, digital sampling system based on a Sigma-Delta analog-to-digital converter (Σ–Δ ADC) has been developed and characterized [19]. The system has been modified to accommodate two synchronized channels.

The design requirements for such digitizer were presented in [1]. The digitizer circuits and PCBs layout were designed by INTI. The circuit components were purchased by Inmetro and the PCBs fabricated in Brazil, all being transported to INTI. The prototype integration and testing is now being carried out by INTI.

Electronic components and product enclosures for assembling 3 (three) dual-channel digitizers, weighing 68.53 kg and amounting US$ 16,609.47 FOB (plus US$ 1,000 for transportation) embarked to INTI on May 18, 2016. Information on the PCBs for the digitizers was given in section 2.1.

3. Project management

The project is coordinated by Inmetro, who is responsible for the purchase and transportation of the components and parts needed to the construction of the modules, and for the transportation of the modules needed to the assembly of the measuring systems, in each partner country.

The project activities include the design, layout and documentation of the PCBs, the design and documentation of the electronic packaging of the modules, the calibration certificates of resistive voltage dividers and current shunts, the design and documentation of the software and firmware, and the measuring system integration, testing and documentation.

Additional small scattered spending is not reported here. Nearly 95% of the funding allocated for this project has already been spent by December 2017. The project is a little delayed due to customs problems.

4. References

[1] Kyriazis G A, Di Lillo L, Slomovitz D, Iuzzolino R, Yasuda E, Trigo L, Laiz H, Debatin R, Franco A M R, Afonso E 2016 Trilateral South American project: a reference system for measuring electric power up to 100 kHz – progress report", *Journal of Physics, Conference Series (Online)*, v. 733, p. 012068

[2] Adad W 2010 *Investigación sobre la generación de señales arbitrarias por el método de síntesis digital directa para su aplicación a la metrología eléctrica*, Tese de Ingeniería, Facultad de Ingeniería, Universidad de Buenos Aires

[3] Adad W and Iuzzolino R 2012 Arbitrary function generator using direct digital synthesis *CPEM Digest (Washington)* 622 ISBN 978-1-4673-0440-5

[4] Adad W and Iuzzolino R 2012 Low distortion signal generator based on direct digital synthesis for ADC characterization *ACTA IMEKO* **1** 59
[5] Laug O B 1987 A precision power amplifier for power/energy applications *IEEE Trans. Instrum. Meas.* **36** 994
[6] Waltrip B C, Laug O B and Nelson T L 2015 A 600 V voltage amplifier for power measurements *IEEE Trans. Instrum. Meas.* **64** 1373
[7] Laug O B 1985 A wide-band transconductance amplifier for current calibrations *IEEE Trans. Instrum. Meas.* **34** 639
[8] Laug O B 1996 A 100 A, 100 kHz transconductance amplifier *IEEE Trans. Instrum. Meas.* **45** 440
[9] Hess D T and Clarke K K 1998 Evaluation of 100 Ampere, 100 kHz transconductance amplifiers *CPEM Digest (Washington)* 611
[10] Slomovitz D, Trigo L, Aristoy G, Brehm M and Kyriazis G A 2016 Wideband resistive voltage divider for a standard wattmeter, *Journal of Physics, Conference Series (Online)*, v. 733, p. 012072
[11] Slomovitz D, Trigo L, Aristoy G and Brehm M 2016 Resistor dividers for power quality meters *CPEM Digest (Ottawa)* ISBN 978-1-4673-9134-4
[12] Slomovitz D, Trigo L, Aristoy G, Brehm M and Kyriazis G A 2017 Wideband resistive voltage divider for a standard wattmeter. Progress report. *XII SEMETRO (Fortaleza)* Nov 26-29
[13] Trigo L, Slomovitz D and Kyriazis G A 2017 Voltage-ratio calibration system up to 50 kHz *XII SEMETRO (Fortaleza)* Nov 26-29
[14] Yasuda E, Di Lillo L and Kyriazis G A 2015 Design and construction of new shunts for a wideband sampling wattmeter *XI SEMETRO (Bento Gonçalves)* Nov 30 - Dec 02 ISBN 978-85-86768-10-1
[15] Rydler K -E and Tarasso V 2008 Extending ac-dc current transfer measurement to 100 A, 100 kHz *CPEM Digest (Boulder)* 28 ISBN 978-1-4244-2399-6
[16] Yasuda E, Di Lillo and Kyriazis G A 2017 Construction of new shunts for a wideband sampling wattmeter *XII SEMETRO (Fortaleza)* Nov 26-29
[17] De Souza R M, Kyriazis G A and Landim R P 2017 AC-DC current shunts: assembling and characterization at Inmetro *XII SEMETRO (Fortaleza)* Nov 26-29
[18] Kyriazis G A, De Souza R M, Yasuda E and Di Lillo L 2017 Modeling wideband cage-type shunts *XII SEMETRO (Fortaleza)* Nov 26-29
[19] Iuzzolino R 2011 *Josephson waveforms characterization of a Sigma-Delta Analog-to-digital converter for data acquisition in metrology*, Dr.-Ing. Dissertation, Technische Universität Braunschweig TU-BS Diss. ISBN 978-3-86387-095-9

Acknowledgments
This work is supported in part by the Conselho Nacional de Desenvolvimento Científico e Tecnológico (CNPq), of the Ministry of Science, Technology and Innovation of Brazil, under Grant CNPq/Prosul Processo Nº 490271/2011-1.

IOP Conf. Series: Journal of Physics: Conf. Series **975** (2018) 012073 doi:10.1088/1742-6596/975/1/012073

Wideband resistive voltage dividers. Progress report

D. Slomovitz[1], L. Trigo[1], G. Aristoy[1], M. Brehm[1], G. A. Kyriazis[2]

[1] Administración Nacional de Usinas y Transmisiones Eléctricas (UTE)

[2] Instituto Nacional de Metrologia, Qualidade e Tecnologia (INMETRO)

dslomovitz@ute.com.uy

Abstract. In this work, we describe the progress of the design of resistive voltage dividers for use in a standard wattmeter. It covers the frequency range from 50 Hz to 100 kHz and voltages from 4 V to 1024 V.

1. Introduction

Nonlinear loads in power networks have been increasing during the last decades. These include switched-mode power supplies used in many appliances that can produce harmonics beyond 1 MHz [1], [2]. This leads to distortion in currents and voltages which can disturb communication and control systems, as well as other electronic devices connected to the network. To calibrate voltage, current and power at high frequencies, it is necessary a standard meter. We are running a project [3], [4] for designing and constructing a reference system that can measure voltage, current and power up to 100 kHz and 1000 V. This project is being jointly developed by the National Metrology Institutes of Brazil, Argentina and Uruguay (Instituto Nacional de Metrologia, Qualidade e Tecnologia (Inmetro) in Brazil, Instituto Nacional de Tecnología Industrial (INTI) in Argentina, and Administración Nacional de Usinas y Transmisiones Eléctricas (UTE) in Uruguay). The objective is the construction of three measuring systems, one for each institute. This project will contribute to provide calibration services in measuring ranges still not covered by these institutes.

The wattmeter uses nine voltage dividers at its input to scale the input voltages from 4 V to 1024 V, to the voltage value required by the analogue-to-digital converter (0.8 V) [5], [6]. In this paper the progress of the development of a set of resistive dividers is described.

This divider design has two arms, an input one whose value depends on the nominal voltage of the divider, and an output arm that has a fixed value, 200 Ω, for all dividers. The nominal current is the same for all of them, 4 mA, so that the nominal output voltage is 0.8 V. Each arm is enclosed in separate cases, to allow the calibration together as a divider, or as single resistors.

It is not easy to design such dividers with very low error at high frequencies. There are parasitic capacitances and inductances that produce changes in the voltage ratio in amplitude and phase, which have also been studied in other electrical fields [7], [8]. Among these, the resistors and isolating materials used in the divider have nonlinear losses, which lead to nonlinear behaviors that cannot be compensated under distorted waveforms. To test the dividers, many methods have been proposed. One of them is the step-up calibration [9] and other is the ac-dc transfer test. In the latter, it is necessary to split both arms, disconnecting them, and calibrating each one separately.

At the moment, all designs have been concluded and a set of prototypes were manufactured. Preliminary results are in accordance with the specifications. The following sections show the progress in the development of these dividers.

2. Error sources

In previous papers [5], the influence of stray capacitances and nonlinearity were analyzed. A shielding technique was used for nulling radial electric fields. Two symmetric cone-shaped electrostatic-shields were installed at each end of the resistor to get null radial electric field at the whole surface of the resistor (see figure 1). The shape of the shield was designed using an electric-field simulation software.

Figure 1. Electrostatic shield made by two metallic cones.

Regarding non-linear behavior, the principal cause is dielectric losses in the printed circuit board (PCB) and in the isolation covers of the resistors. This non-linear behavior changes the resistance when varying the frequency. Note that this change cannot be compensated under distorted waveforms. To reduce this error factor, we use special Vishay naked Z-foil non-cover resistors without encapsulation, mounted on PTFE PCB.

Another error source comes from thermal effects. The higher nominal voltage of the divider set is 1024 V, with 4 W of dissipated power. Although the thermal coefficient of these resistors is low, it is necessary to take into account this error source because the temperature rise can be high.

3. Cooling design

The power dissipated in the input arm of the dividers increases the temperature, changing their values. The output arm has a fixed value of 200 Ω, so its temperature rise due to power can be neglected. The thermal coefficients of this type of resistor is lower than 1 (μΩ/Ω)/K, but the temperature rise for the 1024-V input arm can reach 60 K. This value causes resistance variations over the maximum project tolerance. To reduce the heating, a cooling mechanism was added for dividers from 256 V to 1024 V. It has a small fan installed at the output side of the external cylinder. The air flux is conducted to the resistor though small holes in both cones, as shown in figure 2. Experimental results show that the resistance variation was reduced from more than 60 μΩ/Ω to 35 μΩ/Ω in the 1024-V divider. The resistance variation of the input arm, applying step voltages of 250 V (blue), 500 V (red) and 1000 V (green) is shown in figure 3. Each step was applied immediately after the previous one. The temperature rise was reduced with this cooling system. The thermal constant is around 10 min, so that with 15 min of preheating, the difference with the asymptotic value is under 2 μΩ/Ω.

Figure 2. Cooling system for the input arm of dividers from 256 V to 1024 V.

However, the size and position of the holes was, at first, empirically determined, then, a better hole design would improve the thermal behavior. For that, a simulation software is being developed to determine the best number and position of the holes, to get lower temperature rises.

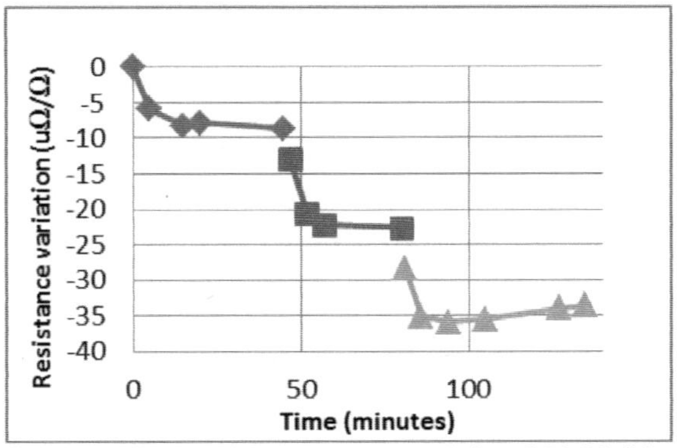

Figure 3. Resistance variation of the 1024-V input arm, applying 250 V (blue), 500 V (red) and 1000 V (green).

As the size of the holes increases, the electric field moves away from its ideal shape, so it is necessary to balance the temperature rise with the distortion in the electric field.

4. Conclusions

At the moment, in this project all designs have been concluded and a set of prototypes were manufactured. Preliminary results are in accordance with the specifications, although better results are expected using a computer-assisted design for the cooling system of the dividers with highest voltages.

5. References

[1] Kyriazis G A, 1997, Analysis of discrepancies between measured and predicted conducted emissions in switched-mode power supplies, *IEEE Internat. Symp. on Electromagnetic Compatibility (Austin)* 504.

[2] Kyriazis G A 1995 Prediction of conducted emissions in switched-mode power supplies, *EMC Test and Design,* Aug. 1995, p. 10.

[3] Kyriazis G A, Di Lillo L, Slomovitz D, Iuzzolino R, Yasuda E, Trigo L, Laiz H, Debatin R, Franco A M R, Afonso E, 2016, Trilateral South American project: a reference system for

IOP Conf. Series: Journal of Physics: Conf. Series **975** (2018) 012073 doi:10.1088/1742-6596/975/1/012073

measuring electric power up to 100 kHz – progress report, *Journal of Physics, Conference Series*, **733**, p. 012068.

[4] Kyriazis G A, Di Lillo L, Slomovitz D, Iuzzolino R, Yasuda E, Trigo L, de Souza R M, Laiz H, Debatin R M, Afonso E, 2017, Trilateral South American project: a reference system for measuring electric power up to 100 kHz – progress report II, *XII SEMETRO*, *this issue*.

[5] Slomovitz D, Trigo L, Aristoy G, Brehm M and Kyriazis G A, 2016, Wideband resistive voltage divider for a standard wattmeter, *Journal of Physics, Conference Series*, **733**, p. 012072.

[6] Slomovitz D, Trigo L, Aristoy G and Brehm M, 2016, Resistor dividers for power quality meters *CPEM Digest,* ISBN 978-1-4673-9134-4.

[7] Rungis J, Schon K, 1988, The evaluation of impulse divider response parameters, *IEEE Trans. Pow. Del.*, **3**, No 1, p. 88.

[8] Slomovitz D, Bergalli P, Daher J F, 1998, Evaluation of low-frequency voltage dividers, using the step response, *Conference on Precision Electromagnetic Measurements, CPEM 1998 Digest*, p. 319.

[9] Trigo L, Slomovitz D and Kyriazis G, 2017, Voltage-ratio calibration system up to 50 kHz, *this issue*.

Acknowledgments

This work is supported in part by Conselho Nacional de Desenvolvimento Científico e Tecnológico (CNPq), of the Ministry of Science, Technology and Innovation of Brazil, under Grant CNPq/Prosul Processo Nº 490271/2011-1.

Calibration system for voltage transformers under distorted waveforms

Gonzalo Aristoy [1], Leonardo Trigo[1], Marcelo Brehm[1], Alejandro Santos[1], Daniel Slomovitz[1]

[1] Administración Nacional de Usinas y Transmisiones Eléctricas (UTE)

dslomovitz@ute.com.uy

Abstract. In this work, we describe a programmable high-voltage source to test the behaviour of voltage transformers under distorted waveforms at nominal voltages, as well as the measurement system for computing their errors at each harmonic component.

1. Introduction

Measurements of voltage harmonic distortion in power networks require of voltage transformers (VT) to reduce the high-voltage level of the lines to a low-voltage level according to the measurement instrument specifications, around 100 V. Their primary voltages are in the range of 6 kV to 500 kV, or even more. Conventional tests are done only at power frequency, but if they are intended to use under distorted waveforms, it is necessary to test them under this condition. Even more, as they are not linear devices, the errors they have under distorted waveform must be tested at nominal voltages [1] [2]. For this reason, a programmable high voltage source was developed. It has the possibility to program any harmonic content, up to the 50th order of the power network frequency. Additionally, the same system allows to measure the errors of the VT under test, for each harmonic component. The following sections describe the developed system.

2. High voltage source

The voltage source comprises three main parts: an arbitrary waveform generator, a power amplifier and a step-up transformer, which are described in the following paragraphs.

2.1. Waveform generator

The waveform generator is based on a digital-to-analogue converter (DAC), with a resolution of 16 bits [3], and a PC running a LABVIEW [4] based software developed for this application. The software has six screens: *Generation, Measurements, Secondary, Primary, Ratio* and *Configuration*. The first and the last ones are used to configure the waveform and generate it. In these screens (see figure 1), the fundamental frequency (50 Hz to 60 Hz) and the percentage and angle of each harmonic, up to the 50th order, are entered. The software calculates the values of the data vector and finally loads this vector into the DAC. The output signal of the DAC has a range of ±10 V. A voltage divider attenuates this signal to values compatibles with the input of the power amplifier (0.5 V). There are buttons to record and load distorted waveforms, buttons to load waveforms into the card and to null the output voltage.

IOP Conf. Series: Journal of Physics: Conf. Series **975** (2018) 012074 doi:10.1088/1742-6596/975/1/012074

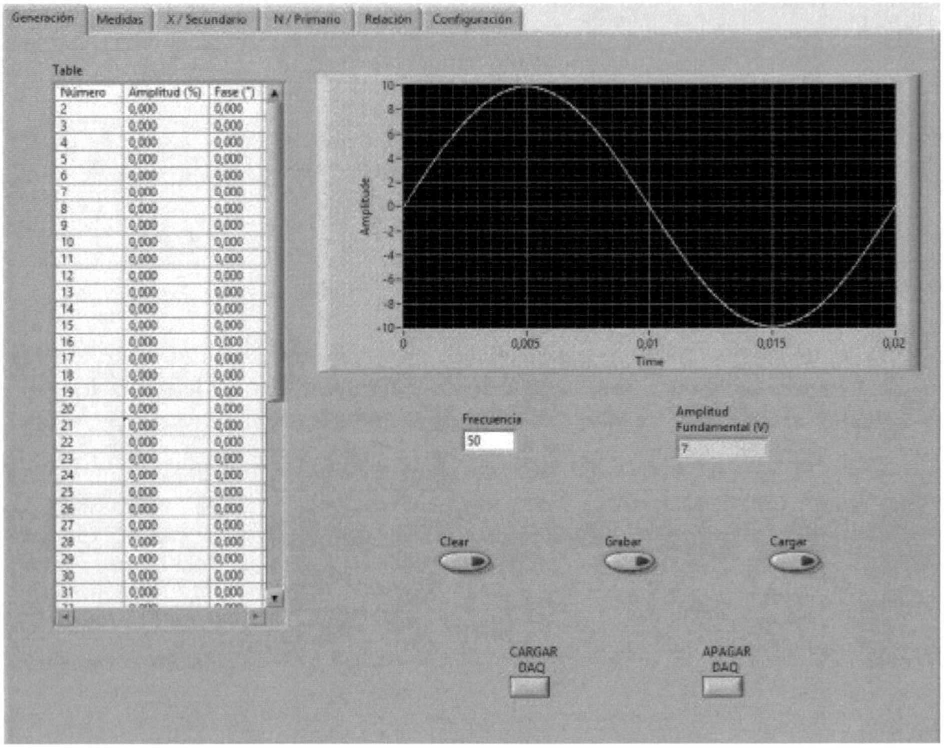

Figure 1. Waveform generation screen

2.2 Power voltage source

The power voltage source has two main blocks. The first block is the dc power supply (see figure 2). It uses a 20 kVA, 220 V /110 V, 50 Hz three-phase transformer with two secondary windings, one in delta and the other in star connections. A 12 pulse rectifier converts ac to dc. It has two three–phase full wave rectifiers each. At its output, there is a bank of capacitors (15 mF in total) for filtering.

The second block includes a 50-kW class-D audio power amplifier and auxiliary devices, as a 13-V-dc voltage source for the internal signal stage of the amplifier, multi-function meter for monitoring the power factor, current transformer, etc. The amplifier is energized with 300 V generated by the dc power supply. This amplifier can manage signals of 0.5 V, from 50 Hz to 3 kHz. These signals are amplified to values near 200 V rms, with currents of tens of amperes. Additionally, the system uses a series resistor at the output to increase the power factor (not shown in figure 2). Figure 3 shows the arrangement of the different parts of the system in the rack.

2.3 Step-up transformer

The third part of the source comprises a set of step-up transformers with secondary nominal voltages of 6 kV, 15 kV, 30 kV, 60 kV and 100 kV. Their bandwidths depend on the transformer. For the lowest voltage, it covers the range 50 Hz to 3 kHz. As the nominal voltage increases, the bandwidth decreases, attenuating the highest harmonics. Then, to get the required harmonic amplitude at the VT under test, larger values must be set in the waveform generator. This is not a problem, because the actual harmonic content is measured by standard voltage dividers and a harmonic meter. Although the maximum voltage is limited to 100 kV, even up to 500 kV capacitive-VTs were tested, as it is detailed in [1].

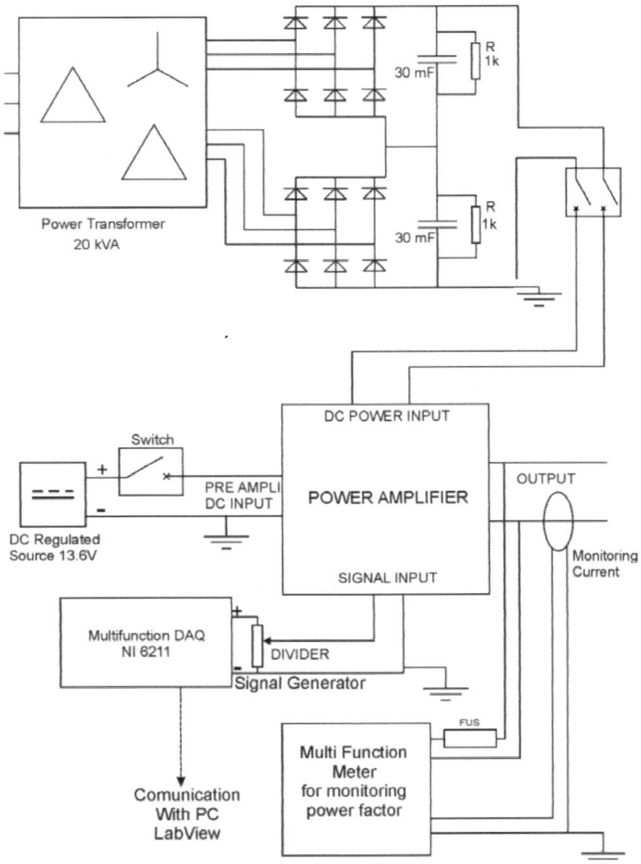

Figure 2. Power voltage source block-diagram.

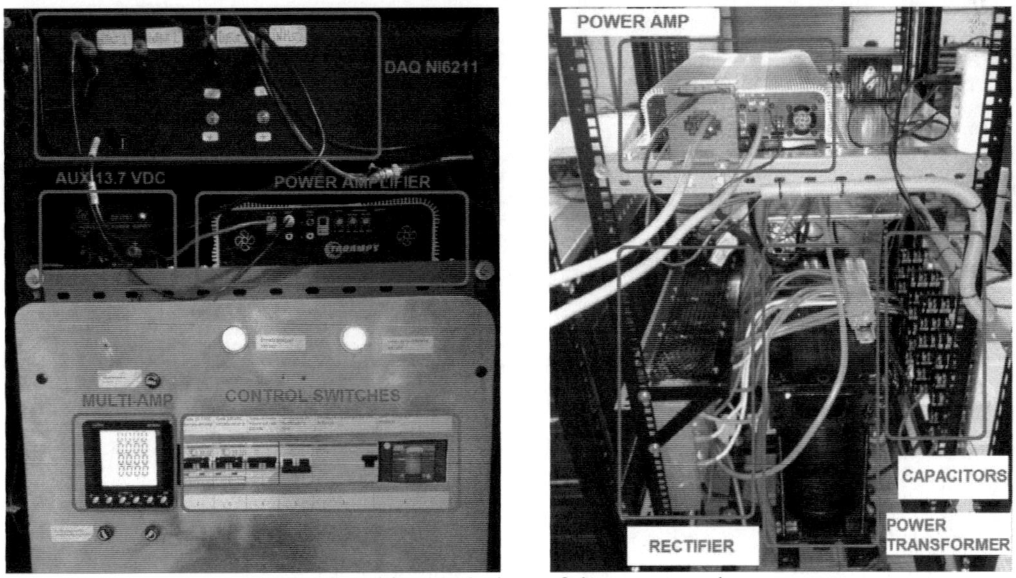

Figure 3. Panel and internal view of the power voltage source.

IOP Conf. Series: Journal of Physics: Conf. Series **975** (2018) 012074 doi:10.1088/1742-6596/975/1/012074

Figure 4. Step-up transformer-set used to elevate the voltage at the VT-under-test requirement.

3. Measurement system

The measuring system comprises high and low voltage dividers, a harmonic measurement standard meter and software to process the data. In this way, the primary and secondary voltages are measured, and their harmonic contents are compared. Figure 5 shows a general diagram.

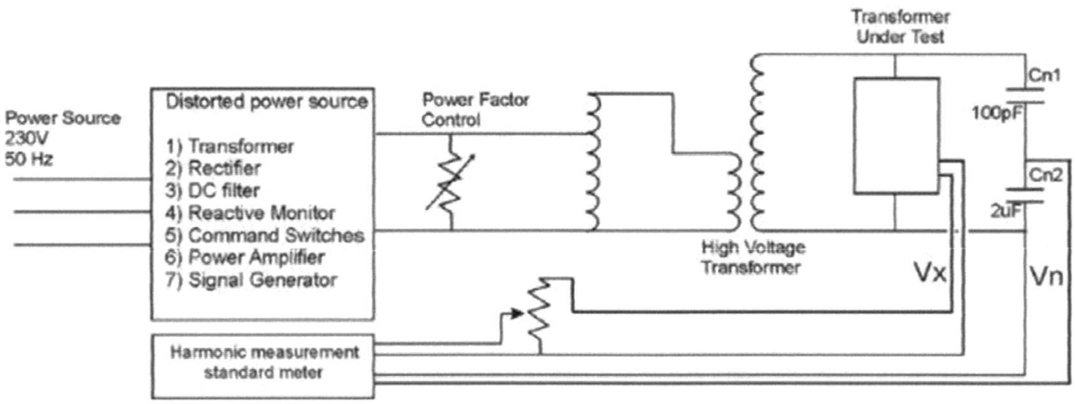

Figure 5. General diagram of the measurement system.

3.1 Voltage dividers

The primary and secondary voltages of the transformer under test are scaled down by voltage dividers (see figure 6). The primary voltage is scaled down by a capacitive-type-divider made by a 100-pF-SF6 high voltage capacitor in series with an array of low voltage capacitors of 2 μF. The divider constant has a value of 20 000. The secondary voltage is scaled down by a resistive-type-divider made by a set of high thermal stable resistors (Vishay, Z201). Its input impedance is 20 kΩ. The arrangement of these resistors has been constructed in such a way that it is possible to obtain two output values. The first one has a divider constant value of about 20 and the second one, a divider constant value of about 40. These two ratios are enough to cover a large range of VTs.

The outputs of the voltage dividers are sent to the harmonic measurement standard meter block.

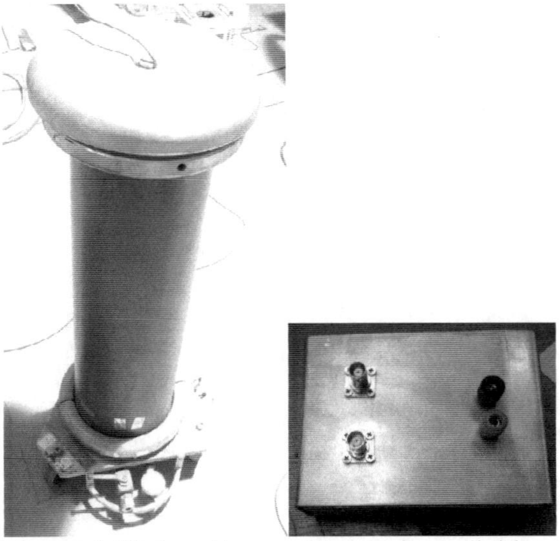

Figure 6. High-voltage and low-voltage dividers.

3.2 Harmonic measurement standard meter

This block has two analogue-to-digital converters (ADC) that are part of the same card of the DAC [3]. These ADCs have a resolution of 16 bits in 8 differential inputs with ranges: 0.2 V, 1 V, 5 V and 10 V. Our system only uses two of those inputs. Both input channels are digitized synchronously, to allow the phase measurement between them, and the data is sent to the PC.

An Interpolated Fast Fourier Transform algorithm reconstructs the two signals obtaining information about the harmonic content of each one. It includes the amplitude and the phase for each harmonic component. There is a screen of the software for entering the actual ratio values of the voltage dividers (module and phase), and the nominal ratio of the transformer to be tested.

The resulting ratio and phase shifts errors for each harmonic are computed and shown in the Ratio Screen (figure 7). It includes type-A uncertainties. Type B uncertainty was estimated based on the accuracy of the voltage dividers, the ADCs and the harmonic estimation software, around 0.05% in phase and quadrature.

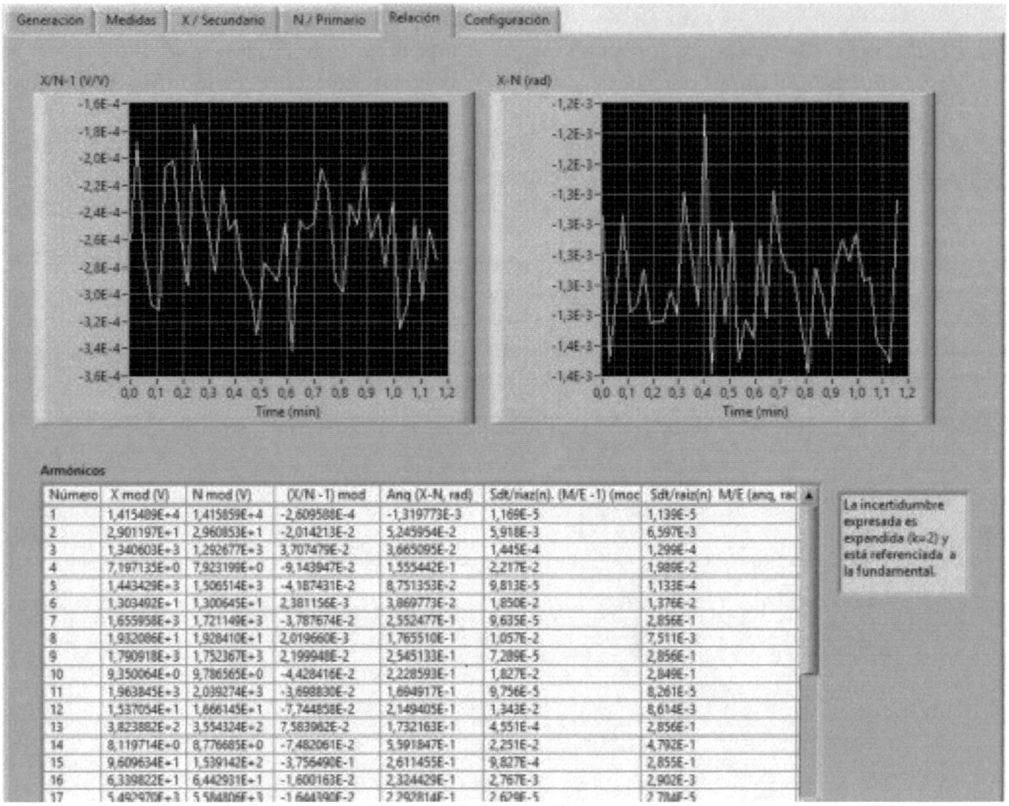

Figure 7. Ratio Screen.

Figure 8 reproduces the Primary Screen that shows the measured waveform of the primary voltage of the transformer under test and its harmonic content (bar graph and table). A similar screen exists for the secondary voltage.

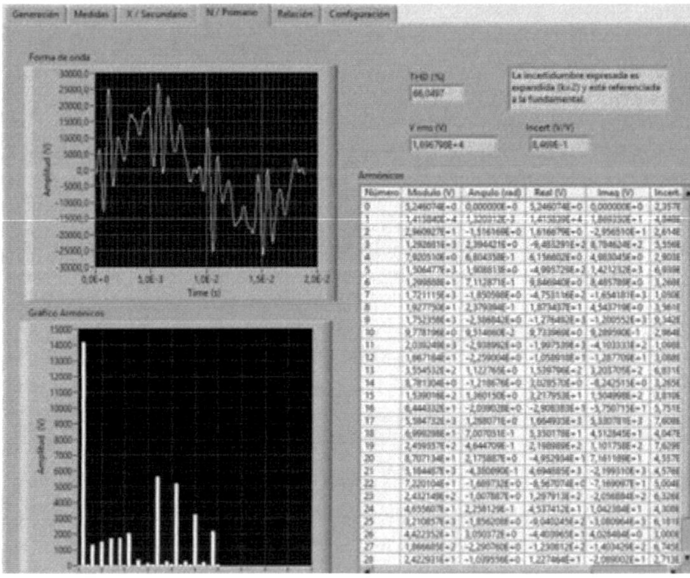

Figure 8. Primary Screen.

IOP Conf. Series: Journal of Physics: Conf. Series **975** (2018) 012074 doi:10.1088/1742-6596/975/1/012074

4. Conclusions

A system that generates and measures distorted voltage waveforms was described. It can calibrate voltage transformers determining the errors in amplitude and phase for each harmonic, up to the 50th component. It was used for voltage transformers from 6 kV up to 500 kV.

References

[1] Slomovitz D, Aristoy G, Santos A, Brehm M, Trigo L, Behavior of voltage transformers under distorted waveforms, this issue.
[2] Faifer M, Ottoboni R, Toscani S, Cherbaucich C, Gentili M, Mazza P, 2013, A measurement system for the calibration of MV instrument transformers under non sinusoidal conditions Semetro X, Buenos Aires, pp 25-27.
[3] DAQ 6211, National Instrument, www.ni.com/pdf/manuals/371931f.pdf
[4] LabVIEW, National Instrument, www.ni.com/en-us/shop/labview.html

Acknowledgments

This work has been partially financed by ANII, Project FSE_1_2014_1_102482.

IOP Conf. Series: Journal of Physics: Conf. Series **975** (2018) 012075 doi:10.1088/1742-6596/975/1/012075

Voltage-Ratio Calibration System up to 50 kHz

Leonardo Trigo[1], Daniel Slomovitz[1], Carlos Faverio[1], Gregory Kyriazis[2]

[1] Administración Nacional de Usinas y Transmisiones Eléctricas (UTE)

[2] Instituto Nacional de Metrologia, Qualidade e Tecnologia (INMETRO)

ltrigo@ute.com.uy

Abstract. This paper describes a voltage-ratio measuring system that can be used up to 50 kHz with high accuracy. It is based on two Keysight 3458A digital multimeters, working in DCV sampling mode. An external trigger based on a rubidium clock, drives both digital multimeters. It is applied to the calibration of a set of voltage dividers with primary voltages from 4 V to 1024 V.

1. Introduction

A project to design and construct a reference system for measuring voltage, current and power, up to 100 kHz, is been running [1], [2]. This project is being jointly developed by the National Metrology Institutes of Argentina (Instituto Nacional de Tecnología Industrial (INTI)), Brazil (Instituto Nacional de Metrologia, Qualidade e Tecnologia (INMETRO)) and Uruguay (Administración Nacional de Usinas y Transmisiones Eléctricas (UTE)). The objective is the construction of three measuring systems, one for each institute. This project will contribute to provide calibration services in measuring ranges still not covered by these institutes. The project will also contribute to improve the traceability not only of electric power but also of related quantities like ac-dc transfer, voltage ratio, phase angle, ac voltage and ac current.

For the voltage input, this standard system requires voltage dividers to scale the input values (4 V to 1024 V) to the value allowed by the analogue-to-digital converter (0.8 V) [3], [4]. The motivation for the present work comes from the need of calibration of these dividers.

Different methods can be used to calibrate this type of dividers. One of them is to compare the divider under test against an ac-dc thermal converter, other to measure the step response [5], [6], and third, to perform a step-up test, as described in the next section. For this last one, it is required a measuring system that can compare the output voltages of the two dividers under test.

Previous developments of a low-frequency power-quality meter [7], [8] based on two Keysight 3458A digital multimeters (DMM) [9] allowed to perform step-up calibrations for voltage dividers up to 5 kHz. In this work, an extension of that method, that reaches 50 kHz, is described. This method measures the amplitude and the phase-displacement differences between the two voltages to be compared.

2. Step-up method

The set of dividers includes the following nominal input voltages: 4 V, 8 V, 16 V, 32 V, 64 V, 128 V, 256 V, 512 V and 1024 V. The nominal output voltage is 0.8 V and the output impedance is 200 Ω, for all dividers.

The step-up method consists in comparing two adjacent dividers up to the voltage of the lower one. As the ratio between them is 2:1, the output voltages to be compared will be up to 0.8 V and 0.4 V. One of the dividers reaches its nominal voltage, but the other reaches only half of its nominal voltage. Therefore, this method requires the characterization of the voltage linearity of each divider. As in this case, the non-linear behaviour is mainly due to temperature variation, a low-frequency or dc test is enough for that. This ratio variation against the applied voltage is considered for correcting the step-up data.

If dividers with equal nominal voltage are compared, the ratio between them is 1:1 and equal voltages are measured by the calibration system. Then, this particular application of ratio calibration of this set of dividers implies only two ratios, 1:1 and 2:1.

3. Calibration system

The voltage ratio calibration system comprises two DMMs [9], whose inputs are connected to the voltages to be compared. They work in DC Direct Sampling mode. To be able to digitize signals up to 50 kHz, the DMMs are configured with a small aperture time of 1.4 μs. With this aperture, the resolution is limited to 16 bits [9]. Both DMMs are synchronized in parallel by a function generator [10] using a TTL signal of 100 kHz. The function generator is driven by a rubidium clock to reduce jitter [11]. A block diagram of the voltage-ratio calibration system is shown in figure 1.

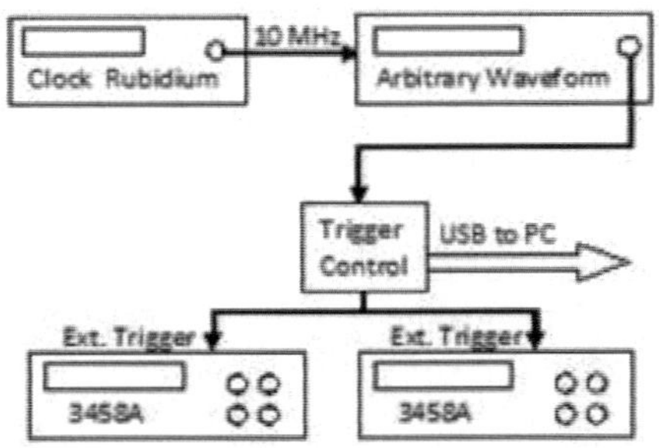

Figure 1. Block diagram of the voltage-ratio calibration system.

The 1-V range is used in both DMMs. The program runs the configuration software and reconstructs the digitized signals using Interpolated-Fast-Fourier-Transform (IFFT) techniques to get the amplitude ratio and phase difference. A Hanning window of the length of the original vector to be analysed was used. Then, Discrete-Fourier-Transform is performed, analysing the spectral leakage. With this information, the frequency of the signal is calculated allowing to compute modulus and angle of the original vector.

There is no synchronization between the input signals of the DMM and the sampling frequency. Each DMM runs at its internal 10 MHz clock. The acquisition times of the samples are around 0.7 s. Even very small differences between clocks, of few parts in 10^6, ensure asynchronous sampling, so that random points are sampled in each burst.

The configurations of commands of both DMMs are sent sequentially, so that one DMM is ready to measure before the other. The trigger control module enables the arrival of the pulses to the external trigger inputs of both DMM only when both DMMs are ready. The enabling and disabling of pulses is done by the software.

Before starting a ratio measurement, the system is adjusted to the ratio value that will be used. If the ratio is 1:1, the same voltage signal is applied to both DMMs. If the ratio is 2:1, a standard inductive divider 2:1 ratio is used to compensate for scale errors. In this way, most error sources are compensated.

3.1 Multimeter error sources

We consider five different influence factors on DMM errors when operating in DCV sampling mode at 1-V range [12], [13]: linearity, aperture time, dissipation factor of the input divider, input low-pass filter, and time differences of external trigger.

Linearity error in the 1-V range was studied in dc voltage, and this value is around some parts in 10^6. It is necessary to make a correction for the aperture time in amplitude and phase. This is done using equations (1) and (2).

$$Aperture\ Amplitude\ Correction = \frac{\sin(\pi \times f \times Taper)}{\pi \times f \times Taper} \quad (1)$$

$$Aperture\ Phase\ Correction = \frac{1}{2} Taper \times f \quad (2)$$

where *Taper* is the value of the aperture time, and *f* the frequency. The errors that these factors compensate are low at low frequencies, but increase at high frequencies. For 50 Hz, the amplitude error is lower than 10^{-6}, but it is about 8000×10^{-6} at 50 kHz. Phase error is linear with frequency, and reaches 35 000 μrad at 50 kHz. This means that even applying the correction factors, significant uncertainties can exist due to the uncertainties in *Taper* and *f*. Nevertheless, as only the difference between both DMMs is relevant, the influence of this error source can be significantly reduced performing the previous adjustment with the standard divider.

The DMM input impedance cause some errors in the voltage measurements. For the 1-V dc range, the nominal value is 1 MΩ with 140 pF in parallel. The influence of the input resistance is constant with the frequency, but the dissipation factor of the capacitance reduces the effective DMM input resistance at high frequencies. Both resistances result in parallel with the output resistance of the divider, attenuating the signal. The input impedances of both DMMs were measured with a RLC bridge. Figure 2 shows the values of the parallel capacitance (Cp) and parallel resistance (Rp) against frequency.

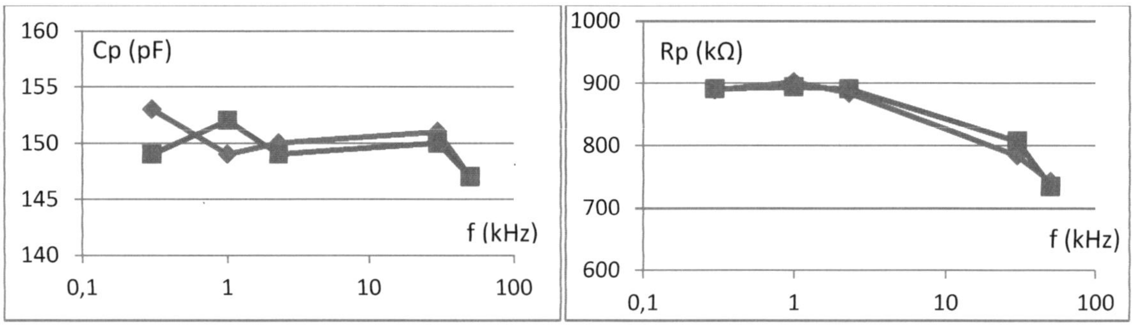

Figure 2. Parallel capacitance (Cp) and parallel resistance (Rp) against frequency of both DMMs.

The influence of this factor depends on the impedance of the voltage sources to be compared. In case of the voltage dividers previously mentioned, all of them have output resistance of 200 Ω. With this impedance the error caused by the input resistance of the DMM is -200 μV/V at low frequency, and it increases with frequency, up to -250 μV/V at 50 kHz because of the dissipation factor of the

input DMM capacitance. However, taking into account that the input impedances of both DMM are nearly equal, the ratio error for this effect is practically compensated
The input capacitance with the output divider resistance produce a linear phase shift with frequency around 10 000 µrad at 50 kHz, but as both DMMs have similar capacitance, this influence in ratio measurements is reduced to less than 600 µrad.

This DMM type has a low-pass filter at its input with nominal cut-off frequency of 150 kHz. The amplitude and phase error generated by this filter is evaluated by equations (3) and (4).

$$Bandwidth\ Amplitude\ Error = \frac{1}{\sqrt{1+(\frac{f}{bandwidth})^2}} \quad (3)$$

$$Bandwidth\ Phase\ Error = \tan^{-1}(\frac{f}{bandwidth}) \quad (4)$$

Corrections using these equations work very well at low frequencies, but near the cut-off frequency, the amplitude and phase errors are very large. They reach around 15% y 20° at 50 kHz. Even taken into account that in ratio measurements errors are similar for both DMMs, the compensation is not perfect and an excessive residual error remains. It is necessary to measure the actual bandwidth of each DMM for correcting the results. We found 129.854 kHz and 130.132 kHz for the tested units. Additionally, all these error sources, as trigger systematic differences, can be evaluated in the system calibration test with the inductive voltage divider, and discounted from the measurement results. Anyway, drifts and random variations after calibration change the correction factors generating errors that cannot be compensated. They were evaluated by analysis on experimental results, as shown in the following section.

3.2 Experimental results
A preliminary test was done applying to both DMM inputs the same signal (0.7 V rms). Table 1 shows the results. The uncertainty columns refer only to type-A uncertainty (expanded, k=2), averaging 20 measurements.

Table 1. Voltage ratio errors with 0.7 V applied to both DMMs.

Frequency (kHz)	Modulus ($\times10^{-6}$)	Phase (µrad)	Uncertainty (modulus) ($\times10^{-6}$)	Uncertainty (phase) (µrad)
1	-22.4	0.3	0.3	0.8
10	-33.3	10.9	0.6	3.8
20	-35.8	21	1.5	6.8
30	-35.2	59.6	2.1	8.7
40	-18.0	92.8	3.5	15.3
50	34.0	179.5	3.5	13.8

These values show that the DMMs are adequate stables and the method has acceptable error and dispersion for the objective of this work. Additional tests on ratio 2:1are being performed.

4. Conclusions
A measuring system for voltage ratio measurements has been presented, reaching frequencies up to 50 kHz. It is based on two high precision multimeters which are synchronized by a high stable external

trigger. Experimental results at ratio 1:1 show errors around 40×10^{-6} in modulus and 20 µrad in phase displacement, at 20 kHz. At 50 kHz, phase error increase, up to 180 µrad, but still the system is suitable for many ratio measurements applications.

5. References

[1] Kyriazis G A, Di Lillo L, Slomovitz D, Iuzzolino R, E. Yasuda E, Trigo L, de Souza R M, Laiz H, Debatin R and Afonso E, Trilateral South American project: a reference system for measuring electric power up to 100 kHz – progress report II, *this issue.*

[2] Kyriazis G A, Di Lillo L, Slomovitz D, Iuzzolino R, E. Yasuda E, Trigo L, de Souza R M, Laiz H, Debatin R, Afonso E and Franco A M, 2016, Trilateral South American project: a reference system for measuring electric power up to 100 kHz – progress report, *Journal of Physics, Conference Series*, **733**, p. 012068.

[3] Slomovitz D, Trigo L, Aristoy G, Brehm M and Kyriazis G A, 2016, Wideband resistive voltage divider for a standard wattmeter, *Journal of Physics, Conference Series*, **733**, p. 012072,

[4] Slomovitz D, Trigo L, Aristoy G and Brehm M, 2016, Resistor dividers for power quality meters, *CPEM Digest (Ottawa).*

[5] Fernandez J and Slomovitz D, Development of a fast step generator for testing impulse dividers, 1995, *ELECTROTECNICA,* **81**, pp. 233-240.

[6] Trigo L and Slomovitz D, 1999, Desarrollo de programas para la adquisición de señales y su evaluación, *IEEE, V EPIM, Montevideo, Uruguay.*

[7] Cardozo A, Camacho M I, Slomovitz D, Aristoy G and Trigo L, 2012, Progress in the standard system for power and energy measurements of UTE, *Sixth IEEE/PES Transmission and Distribution: Latin America Conference and Exposition (T&D-LA),* Montevideo.

[8] Trigo L, Camacho M I and Slomovitz D, 2014, Standard for Electric Distorted Waveforms, *IEEE, I2MTC 2014.*

[9] Keysight 3458A Digital Multimeter, https://www.keysight.com/en/pd-1000001297%3Aepsg%3Apro-pn-3458A/digital-multimeter-8-digit?nid=-536902435.536881781.00&cc=UY&lc=eng

[10] Tektronix, AFG3000C Arbitrary / Function Generator, http://www.tek.com/sites/tek.com/files/media/media/resources/AFG3000_Series_Arbitrary-Function_Generators_Datasheet_76W-18656-5.pdf

[11] Pokatilov A, Kübarsepp T and Vabson V, 2016, Effect of Keysight 3458A Jitter on Precision of Phase Difference Measurement, *IEEE Trans. Instrum. and Meas.*, **65**, pp. 2595-2600.

[12] Lapuh R, Voljc B, and Lindic M, . 2014, Accurate measurement of AC voltage in audio band using Agilent 3458 sampling capability, *CPEM 2014 Conf. Digest*, pp. 740-741.

[13] Kyriazis G. A. and Swerlein R., 2002, Evaluation of uncertainty in ac voltage measurement using a digital voltmeter and Swerlein´s algorithm, *CPEM Digest*, 24-25, Ottawa, Canada, Unpublished extended version available online at https://www.researchgate.net/publication/3968220_Evaluation_of_uncertainty_in_ac_voltage_measurement_using_a_digital_voltmeter_and_Swerlein´s_algorithm.

Acknowledgments

This work was partly supported by the Conselho Nacional de Desenvolvimento Científico e Tecnológico (CNPq), of the Ministry of Science, Technology and Innovation of Brazil, under Grant CNPq/Prosul Processo Nº 490271/2011-1.

Robotic tool positioning process using a multi-line off-axis laser triangulation sensor

T. C. Pinto, G. Matos

Metrology and Automation Lab. Labmetro-EMC-UFSC, Florianópolis-SC Brazil.

tiago.pinto@ufsc.br

Abstract. Proper positioning of a friction stir welding head for pin insertion, driven by a closed chain robot, is important to ensure quality repair of cracks. A multi-line off-axis laser triangulation sensor was designed to be integrated to the robot, allowing relative measurements of the surface to be repaired. This work describes the sensor characteristics, its evaluation and the measurement process for tool positioning to a surface point of interest. The developed process uses a point of interest image and a measured point cloud to define the translation and rotation for tool positioning. Sensor evaluation and tests are described. Keywords: laser triangulation, 3D measurement, tool positioning, robotics.

1. Introduction

Laser triangulation sensors integrated to robots have been used to measure three-dimensional shapes of complex objects [1] and following joints in welding [2-4], usually with the projection of a single laser line [1-3]. A measurement model for this type of system extends the classic pinhole camera model associated with a mathematical plane defined by the laser "light sheet"[1,8-10]. Multi-line sensors [4,5] can determine further information about the relative angle of the surface of interest in a single acquisition, as well as, to acquire a larger number of points.

Correct positioning and alignment of a friction stir welding head, driven by a developed closed chain robot, is essential to repair cracks appropriately on a damaged surface. Figure 1 shows the closed chain robot. Surface positioning and orientation at pin insertion point must be known and it motivates the development of a special measurement process and sensor that can measure a point cloud that represents the surface and the 3D position of the pin insertion point. Special characteristics of the developed sensor includes (1) measuring position and orientation in a single acquisition, (2) measurement coordinate system aligned to tool center point (TCP), (3) inability to remove the sensor during repairs and (4) spatial restrictions to avoid self-collision with the robot.

2. Laser Triangulation Sensor

2.1. Configuration

The mechanical and optical configuration was designed to position the laser emitters and the camera in an off-axis configuration regarding the robot tool axis. The sensor can detect surface position and orientation in a single acquisition, thus the measurement of the relative surface normal direction do not depend directly on the robot absolute accuracy. Figure 2 shows the actual developed sensor integrated with the robot. The optical configuration was defined for maximum sensitivity with 50 mm range in Z direction (tool axis) using a camera with 1280 x 960 pixels and a 12 mm focal length lens.

Figure 1. Closed chain robot with a friction hydro pillar processing (FHPP) head.

Figure 2. Sensor integrated with the robot.

2.2. Sensor model and calibration

The sensor measurement model and its calibration was based on [8-10]. The measurement model uses the pinhole camera model and a fitted mathematical plane for each projected laser light plane. For each laser peak detected in the image a line can be defined through the lens center using the pinhole projection matrix. The intersection of this line with the laser mathematical plane leads to a 3D measured point. In [9] a special part is used to calibrate the sensor and the sensor TCP simultaneously. In this work, a planar reference grid with controlled displacement is used for calibration.

2.3. Sensor TCP calibration

For a complete surface scan, each posture of the robot and consequently the sensor, the 3D measured points in the sensor coordinate system (SCS) must be transformed to a fixed reference, e.g. the robot base coordinate system (RCS). For this, the transformation between the SCS and the robot mechanical interface coordinate system (ICS) must be determined through a sensor TCP calibration using a structured pattern similarly as described in [6] but with only three planes and only one measurement acquisition.

3. TCP positioning process

After an initial sensor approximation two acquisitions are made, a point cloud in SCS and a point of interest image with lasers off and leds on. A plane is defined with the point cloud by a least-squares orthogonal distance fitting, as described in [7]. As a result, the plane normal direction $\boldsymbol{n_p}$ is determined. The goal of the alignment is to bring the TCP Z axis parallel to np and the TCP origin to a surface point of interest $\boldsymbol{M}$. The rotation axis ê is defined by a cross product between $\boldsymbol{n_p}$ and Z axis $\boldsymbol{Za} = [0\ 0\ 1]$, and the angle of rotation θ is defined as arccosine of the dot product between the same vectors, defining the axis-angle rotation. To define the 3D point of interest $\boldsymbol{M}$, the point cloud fitted plane coefficients (A, B, C, D) are used. Image processing define the point of interest $\boldsymbol{m} = (u,v)$ in image coordinates. The intersection of the line, starting at m and passing through lens center $\boldsymbol{c}$, with the point cloud fitted plane define the 3D position of the point of interest $\boldsymbol{M} = (X, Y, Z)$, as seen in Figure 3. A homogeneous transformation matrix, with $\boldsymbol{R}$ as the 3x3 rotation matrix notation of the axis-angle rotation, and translation as $-\boldsymbol{M}$ is defined to offset the TCP to align it with the surface normal and positioned at point of interest $\boldsymbol{M}$, as in (1).

$$Offset = \begin{bmatrix} R & -M \\ 0 & 1 \end{bmatrix}^{-1} \quad (1)$$

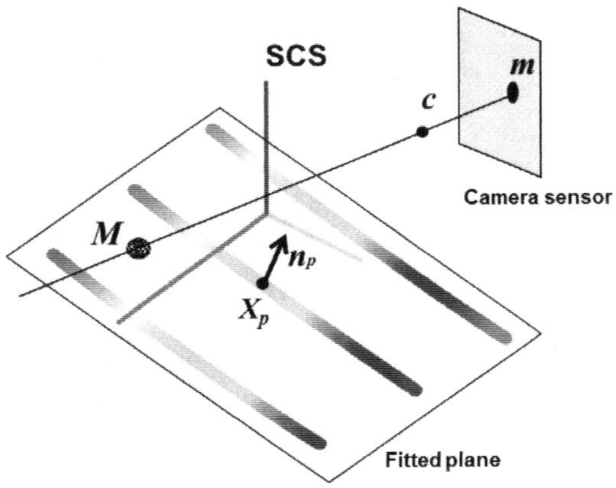

Figure 3. 3D point ***M*** defined at point cloud fitted plane.

Thus, the sensor can be used with two different modes, point cloud measurement and 3D point of interest measurement, the main contribution of this work.

4. Experiments

Several experiments were done to evaluate the laser sensor and the procedure described in this work. Tests include sensor evaluation, TCP positioning tests and final integration and tests with the developed closed chain robot.

4.1. Sensor evaluation

After calibration, the laser triangulation sensor was evaluated. For each reference plane position, the acquired data was evaluated for errors in the Z direction. The systematic errors was negligible and a random error was less than ± 0.15 mm for 95,45% confidence. These random errors are mainly cause by laser speckle.

4.2. Angle measurement

A reference sine table was used to evaluate the sensor angle measurement error. For each table angle, several acquisitions were performed with the laser sensor in the same posture. The table was positioned for rotation evaluation in X and Y axis in the SCS until 15°. Angle measurement errors are greater for greater angles, and standard deviations of the measurements were negligible. In every case, the error is below 0.35°.

4.3. TCP positioning at reference target

The TCP alignment and positioning was evaluated using a steel part measured in a smooth region with relative low form variation. For each acquisition the offset was determined with (1), and the industrial ABB IRB140 robot was moved.The images acquired by the sensor for an initial posture of the robot can be seen on figure 4. Left image is used for point cloud calculation and right image for target position M calculation. Figure 5 shows the projected laser lines for initial (left) and final (right) posture for positioning procedure. Note the central laser line aligned with the reference target on final robot posture (right).

4.4. Closed chain robot tests

Final tests include the robotic tool positioning process applied to the developed closed chain robot. The positioning of the friction stir welding head was performed to a plane metal part with reference marks for pin insertion. Figure 6 show the robot TCP, consequently the central laser line and the SCS, aligned with the first mark (left) and the second mark (right). The process was repeated for all marks. Figure 7 shows the friction stir welding head positioned for pin insertion process (left) and the milling cutter tool aligned at first mark (right).

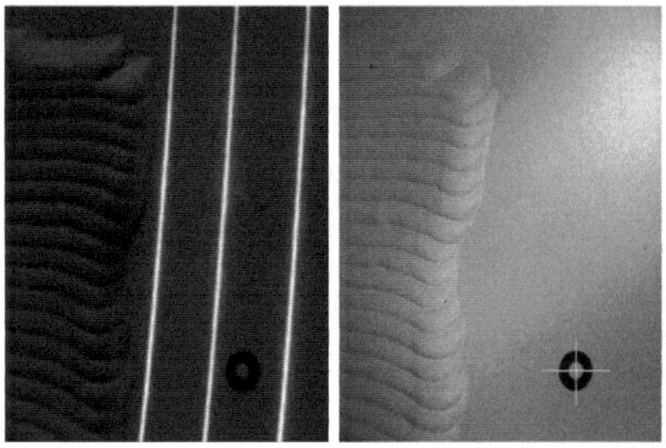

Figure 4. Laser image (left) and reference target image for ***M*** calculation (right).

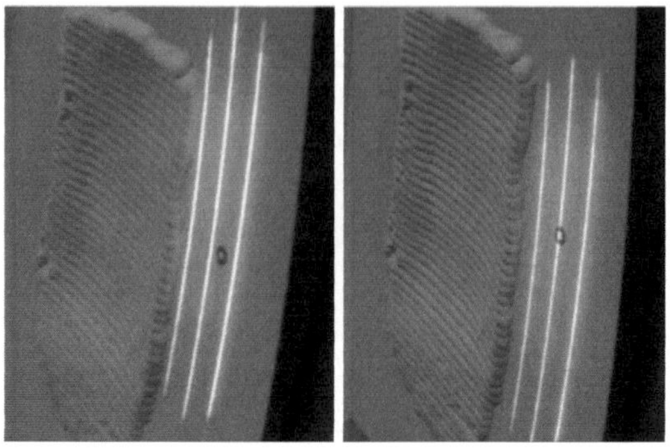

Figure 5. Laser lines for initial (left) and final (right) posture.

Figure 6. Central laser line aligned with the first mark (left) and with the second mark (right).

Figure 7 FHPP head positioned for pin inserction (left) and a tool aligned at first mark (right).

5. Conclusions

This work presents robotic tool positioning process using a developed off-axis multi-line laser triangulation sensor. The developed sensor has an off-axis configuration with coordinate system coincident with the TCP. The developed process uses the measured point cloud and a point of interest image to define the translation and rotation offsets to align the TCP to a reference point and normal to surface of interest, a significant contribution of this work. Practical results were described and include laser triangulation sensor errors evaluation, angle measurement errors evaluation and actual TCP alignment and positioning at reference targets. The proposed process is much easier and with better accuracy than the manual positioning and alignment process.

6. References

[1] Brosed F.J et al 2012 Laser triangulation sensor and six axes anthropomorphic robot manipulator modelling for the measurement of complex geometry products *Robotics and Computer – Integrated Manufacturing* 28 660–671

[2] Haug, K.; Pritschow, G. Robust laser-stripe sensor for automated weld-seam-tracking in the shipbuilding industry *IECON '98. Proceedings of the 24th Annual Conference of the IEEE*, doi: 10.1109/IECON.1998.724281

[3] Zou Y. et al 2008 Development of laser stripe sensor for automatic seam tracking in robotic tailored blank welding *7th World Congress on Intelligent Control and Automation* doi: 10.1109/WCICA.2008.4593410

[4] Sung K., et al 2009 Development of a Multiline Laser Vision Sensor for Joint Tracking in Welding *Welding Journal*, Vol. 88, ISSN: 0043-2296

[5] Bonser G., Parker G. A. 1999 Robotic gas metal arc welding of small diameter saddle type joints using multistripe structured light *Optical Engineering* Volume 38, Issue 11:1943-1949

[6] Zhongxue G., Qing T. 2011 *Visual Sensing and its Applications. Integration of Laser Sensors to Industrial Robots* Springer-Verlag Berlin Heidelberg, DOI: 10.1007/978-3-642-18287.

[7] Ahn, S. J, 2004 *Least Squares Orthogonal Distance Fitting of Curves and Surfaces in Space* Springer-Verlag Berlin Heidelberg, DOI: 10.1007/b104017.

[8] J. Santolaria, J. J. Pastor, F. J. Brosed and J. J. Aguilar 2009 A one-step intrinsic and extrinsic calibration method for laser line scanner operation in coordinate measuring machines *Measurement Science and Technology*, Volume 20, Number 4 IOP Publishing Ltd.

[9] J. Santolaria, J.J. Aguilar, D. Guillomía and C. Cajal 2011 A crenellated-target-based calibration method for laser triangulation sensors integration in articulated measurement arms *Robotics and Computer Integrated Manufacturing* 27,282–291.

[10] Brosed FJ, Aguilar JJ, Guillomía D and Santolaria J 2011 3D Geometrical Inspection of Complex Geometry Parts Using a Novel Laser Triangulation Sensor and a Robot *Sensors* 11, 90-110.

Acknowledgments

Authors thanks to Petrobras, FINEP and EngeMovi for founding this research project and the partners UFU, UFTPR and EngeMovi. Authors thanks to CERTI foundation for CMM measurements.

IOP Conf. Series: Journal of Physics: Conf. Series **975** (2018) 012077 doi:10.1088/1742-6596/975/1/012077

Standards fabrication to providing metrological traceability in micromass and nanoforce measurements results

Thaiane Vargas Pereira[1], Anderson Beatrici[1]

[1] National Institute of Metrology, Quality and Technology (Inmetro) - Directorate of Scientific Metrology and Technology (Dimci).
E-mail: tvpereira@inmetro.gov.br

Abstract: Some of the more sensitive weighing equipment available nowadays has its repeatability close to tenth of microgram. OIML characterize mass standards bigger them 1 mg, so in this range doesn't exist direct traceability to the kg prototype. The ASTM has a characterization of mass standard 50, 100, 200 e 500 micrograms. This work have a purpose of providing traceability to mass measurement in microgram scale (nanonewton scale in force) with the confection and calibration of a standard weights collection. At this time were studied two materials, Tungsten and MetGlass2705M (MetGlass), and produced 12 mass standards.

1. Introduction

The American Society for Testing and Materials (ASTM) recognized the advances in the area of weighing equipments that currently can achieve less than one microgram in the repeatability. The ASTM has specifications for precision mass standards (ASTM E617-13 revised in 2013 for including definition of weight specifications sub milligrams with maximum permissible error.

> "Weight manufacturers must be able to provide evidence that all new weights comply with specifications in this standard (material, density, magnetism, mass values, uncertainties)" [1].

In Brazil, specialized metrology regulations for the manufacturing and use of standard weights N°. 233/1994 [2] don't exist in 50 µg up to 500 µg range (500 nN up to 5 µN in the force case). The OIML R-111, specifies the format, dimensions, nominal values, nature of the material used in mass standards construction and assigns accuracy classes, has international classification and definitions related to mass standards and in this recommendation mass standards have their nominal value starting in 1 mg.

The US government is the most invests in micro and nanotechnology. In 2000, the Institute of Technology of the National Initiative of Nanotechnology was created, with initial investments of United State of American $ 495 million and continues being the country that invests more in this area. The science current state increasingly requires traceability in the microgram range, being relevant for balances sensitivity determination, in nanotechnology, in biotechnology, in fine chemistry, and others.

Researches in the Chemical, Biological, Biotechnology, Pharmaceutical and Materials areas often involve weighing-based measurement processes whose results must be traceable to mass units in order to ensure, in micro scale, the quality of their measurement results. Usually gravimetric measurements have the small uncertainties involved in this process.

IOP Conf. Series: Journal of Physics: Conf. Series **975** (2018) 012077 doi:10.1088/1742-6596/975/1/012077

This paper shows the process of choosing the adequate material for mass standards and stability monitoring. Was used aluminum to make the mass standards, as it is recommended in OIML to nominal values in milligram range, and testing the manufacturing with other materials, MetGlass and tungsten, which showed good stability during the monitoring tests.

2. Material characteristics

The manufacture of microgram standards requires proper material and precise production. It's hard to manipulate these mass standards, the size and shape are critical characteristics of these standards. Was used Aluminum (Al), MetGlass2705M and Tungsten, but we will describe just the MetGlass (MG) and Tungsten (W) characteristics because Aluminum is already recommended for use in standard weights.

2.1. Tungsten characteristics

The Tungsten was confirmed [3] by determining its specific mass. The value of its mass was obtained from weighing, and its dimensions determined by an optical measurement system [4] and tungsten wire diameter is 0.0767 mm obtained by optical device (figure 1.a), as shown in table 1.

Table 1. Tungsten characterization.

Tungsten	Tabulated value	Measured value
Density	19.2 g/cm³	19.1 g/cm³
Elastic Modulus	400 GPa	--

2.2. MetGlass characteristics

Amorphous metal alloys, MetGlass, are a new class of material, are composed of atoms, molecules or ions that don't have a long range ordering. The MetGlass properties are direct related with its homogeneous structure, thus allowing a different behavior of the amorphous alloys in relation to the crystalline alloys.

These alloys are produced by the fast cooling of their liquid phase (glasses), therefore metals without crystalline structure are called "metallic glasses" [5, 6, 7]. MetGlass used was the MetGlass2705M and have a chemical composition of 69% Co, 12% B, 12% Si, 5% Ni and 2% Mo [8]. And it has the following properties [9] as shown in table 2:

Table 2. MetGlass2705M characterization.

Density (g/cm³)	Vicker's hardness (50 g load)	Tensil strength (GPa)	Elastic Modulus (GPa)
7.80	900	1 – 2	100 – 110

2.3. Stability

An important step in this research is mass monitoring and stability after six months the weighs collection has the following distribution, table 3.

Table 3. Nominal value standard mass number.

Amount	Mass value (µg)
4	50
4	100
2	200
2	500

IOP Conf. Series: Journal of Physics: Conf. Series **975** (2018) 012077 doi:10.1088/1742-6596/975/1/012077

We currently manufacture a weights collection of nominal values outside the OIML recommendation, but which can be very useful for calibrating scales in the micromass range because it preserves the regularity of the intervals between each calibration point and also because it is easier to apply the subdivision calibration method to these weights. This weighs collection has the following distribution, table 4.

Table 4. Standards mass in microgram.

Amount	Mass value (µg)
3	250
6	500
2	750
1	1000

Was obtained from weighing of first standards the following results, show in table 5.

Table 5. Standards mass in microgram.

Standards Identification	Mean Value (µg)	Standard deviation (µg)
1 (MG)	47.00	0.87
2 (Al)	47.50	0.58
3 (MG)	50.00	1.04
4 (MG)	52.50	2.18
5 (MG)	96.50	1.60
6 (MG)	97.50	0.76
7 (MG)	99.00	0.76
8 (MG)	102.50	1.29
9 (MG)	201.00	1.92
10 (MG)	201.00	3.21
11 (W)	497.50	1.80
12 (MG)	499.50	2.32

3. Traceability

The traceability to the SI unit mass, the kilogram, is widely used and established in all countries. In Brazil, the mass measurements results are traceable of 1 mg up to greater nominal values. A lot of the modern weighing equipment has resolution much less than 1 milligram, in fact 0.1 microgram. In this case the traceability can be reached by interpolation (or extrapolation) [10] to determine the corrections and the respective measurement uncertainties [11]. Thus to provide a better reliability and reach lower uncertainties (Type A) is necessary to provide traceable standards weights in the microgram range.

3.1 Equipments and materials

The following materials were used in the standards mass production:

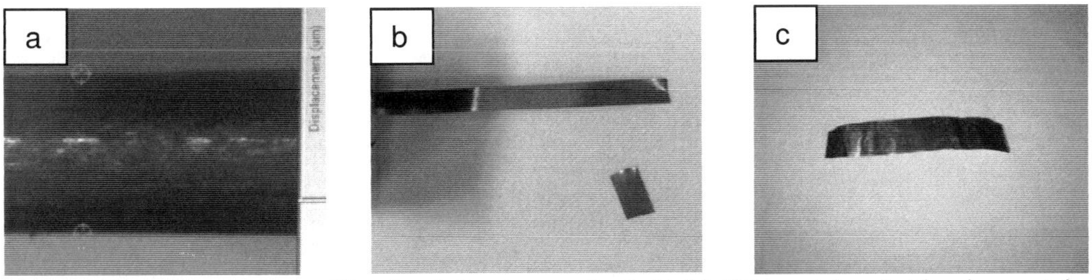

Figure 1. a) Wire tungsten magnification in system optical. b) MetGlass strip. c) Aluminum strip.

The equipments used are:

i) Mass comparator (ultramicro). Used in the samples mass determination and stability monitoring, resolution 0.1 µg.
ii) Standard weight collection PP016, 1 mg up to 20 mg [12]. To perform the mass comparator linearity e sensitivity.
iii) Digital caliper with 0.01 mm resolution. Just to samples length measurements.
iv) Optical dimensional system. 0.001 mm of resolution [3]. To determine the volume of tungsten used in calculating density.

4. Measurements and monitoring

To improve the measurements results reliability was made a mass comparator linearization in nominal value below 20 mg. The measurements points (table 6) were verified using the E_1 accuracy class standard weight collection PP016.

Table 6. Measurement results of balance.

	1 mg	**2 mg**	**5 mg**	**10 mg**	**20 mg**
1	0.9885	2.0003	4.9987	10.0087	19.9890
2	0.9887	2.0008	4.9988	10.0109	19.9909
3	0.9887	2.0006	4.9985	10.0086	19.9897
4	0.9889	2.0014	4.9984	10.0092	19.9901
5	0.9895	2.0005	4.9981	10.0098	19.9896
$\bar{m}$	0.9888	2.0007	4.9985	10.0094	19.9898

Where, $\bar{m}$ is the mean mass value.

Using the Calibration Certificate [12] we obtain the corrected values, table 7.

Table 7. Mass corrected measurement results.

Nominal value (mg)	Mean value (mg)	Corrected values (mg)
1	0.9888	0.9904
2	2.0007	2.0006
5	4.9985	4.9995
10	10.0094	10.0083
20	19.9898	19.9932

In figure 2 are shows the plotted data of table 7 (corrected values) and calculate the mass comparator angular a coefficients (sensitivity).

$$a = 0.9998 \text{ mg/mg}$$

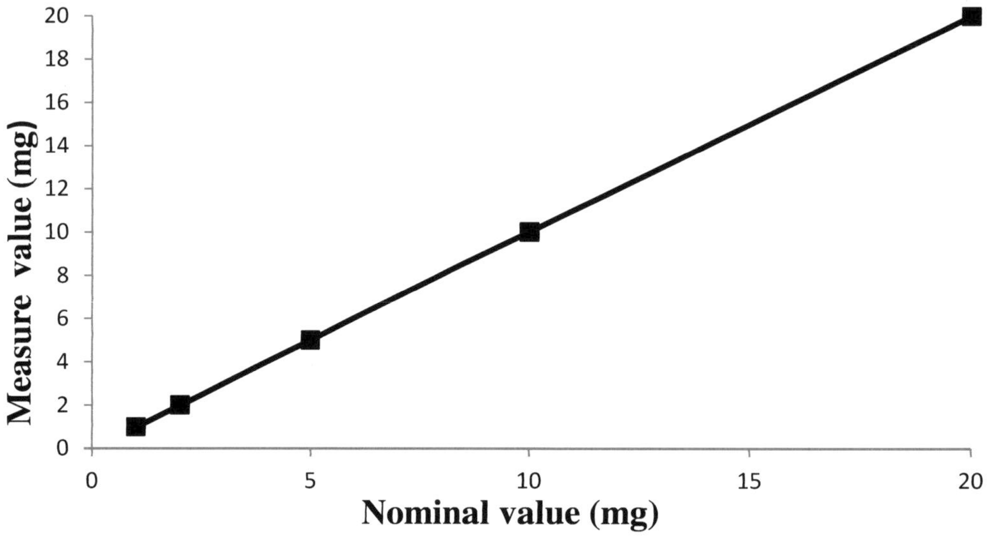

Figure 2. Nominal value *vs.* measurement result.

5. Conclusion

Nowadays, scientific researchers have greater requirements for traceability at lower mass values, lower than OIML R-111 recommendation provides. This lack accuracy in all fields that requires measurements results in micro or nanoscale in mass or force measurements.

This research has the purpose of determining the methodology of mass standards production, including the material choice and its metrological characterization.

The initial results point out that in this range of mass all of materials (Tungsten, MetGlass and Aluminum) are stable at least in these months of monitoring. Aluminum is already recommended by OILM as a material for mass standards, especially for its low density which allows better handling of the standards. Tungsten has very high density, creating a difficulty for its handling despite its advantages of homogeneity and chemical stability. The MetGlass alloy has an intermediate density, but it is very versatile and can be manufactured in very thin strips which allows for better standards manipulation, the disadvantage are its magnetic properties, in many cases the cutting process already cause magnetization, what is a factor against its use. The next steps Calibrate standards by subdivision method and continuous evaluation of the long term stability.

Acknowledgments

This work was funded by the National Program to Support the Development of Metrology, Quality and Technology (Pronametro) and Inmetro Mass Laboratory (Lamas).

6. References

[1] NCSLI 2013 *Standard Specifications for Laboratory Weights and Precision Mass Standards* Nashville
[2] Inmetro 1994 Portaria 233 - Regulamentação Metrológica Instrumentos de Medição e Medidas Materializadas (Brasil: Inmetro)
[3] AZoM 2012 Tungsten - Mechanical Properties and Material Applications Azo Materials
[4] Optical Grade Calibration Certificate DIMCI 0926/2016 Lamed/Inmetro
[5] HOLT, M. L. 1956 *Less-common* Metals and Alloys Metal Finish pp 48-55
[6] NAGEL, S, R. 1982 Advanced chemical physics Wiley pp. 227-275 New York
[7] SEARSON, P. C.; NAGASKAR, P. V.; LATANISION, R. M. In: WHITE, R. E.; BOCKRIS, J O. M.; CONWAY, R. B. 1990 Modern aspects of electrochemistry Plenum Press **Nº 20** pp 121-161 New York
[8] MATERIALS 1996 Conference proceedings. Technology & Engineering. **Volume II** SPE/ANTEC pp 1769
[9] Technical Bulletin Ref: 2705M08022014 - MAGNETIC ALLOY 2705M Metglass
[10] EURAMET 2015 *Guidelines on the calibration of non-automatic weighing instruments*: *Version 4.0 (11/2015)* (Braunschweig: EURAMET)
[11] BIPM, IEC, IFCC, ILAC, ISSO, IUPAC, IUPAP and OIML 2008 *Guide to Expression of Uncertainty in Measurement* (GUM: 1995 with Minor Corrections) JCGM 100:2008 (Sèvres: Bureau International dês Poids et Mesures)
[12] Standards Weighs Calibration Certificate (PP016) DIMCI 1247/2015 Lamas/Inmetro

Surface roughness influences on the behaviour of flow inside microchannels

M H Farias[1], **C S Castro**[1], **D A Garcia**[1,2] **and J S Henrique**[1]

[1] Fluid Dynamics Metrology Division (Dinam), Directorate of Scientific Metrology and Technology (Dimci), National Institute of Metrology, Quality and Technology (Inmetro), Av. Nossa Senhora das Graças, 50, Xerém, Duque de Caxias/RJ, 25250-020, Brazil
[2] Laboratory of Nano and Microfluidics and Microsystems (Labmems), Mechanical Engineering Program (PEM/COPPE), Federal University of Rio de Janeiro (UFRJ), Cx. Postal 68503, Cidade Universitária, Rio de Janeiro/RJ, 21945-970, Brazil

E-mail: mhfarias@inmetro.gov.br

Abstract. This work discusses influence of the surface roughness on the behavior of liquids flowing inside microchannels. By measuring the flow profile using the micro-PIV technique, the flow of water inside two rectangular microchannels of different wall roughness and in a circular smooth microchannel was studied. Comparisons were made among the experimental results, showing that a metrological approach concerning surface characteristics of microdevices is required to ensure reliability of the measurements for flow analyses in microfluidic processes.

1. Introduction

A reliable measurement for describing details of the fluid dynamics in microscale has been very challenger for scientists and for manufacturers of microfluidic devices. In very tiny geometries, the flow behavior is not well represented by the equations and hypotheses which are commonly applied to flow in macrodomain, for instance, Navier-Stokes equations and boundary conditions. So, it can affect, strongly, the performance and flow process control analyses. As metrological approach in microfluidic studies plays very important role for theoretical, experimental and computational developments devoted to provide reliability on achievements about such theme, technology advances in this field also has been driven by the need of ensuring confidence on the measurement for characterizing microflows. Besides, discussions, investigations and developments aiming metrological traceability chain establishment for flow measure in microscale have received attention from NMIs [1].

1.1. Aim of this work

When fluid flows in microscale, multiphysics effects [2] take place, and a process analysis can be misinterpreted. One parameter which clearly influences the microflow behavior is the roughness of the surface on which the flow is in contact. However, for flow inside microchannels is very difficult to eliminate these influences, since the surface quality depends on the techniques for microchannel fabrication, as well as the material, since each technique has limitations. In the literature, it is known that surface effects in microflow are important, nevertheless, researchers do not have focused the

IOP Conf. Series: Journal of Physics: Conf. Series **975** (2018) 012078 doi:10.1088/1742-6596/975/1/012078

discussions on this subject. Thus, aiming to contribute to call attention to the relevance of the metrological criteria in microflow characterization measurements, this work presents an introductory discussion about influences of the surface roughness of the channel walls for flow behavior in microscale. In this work, the flow of liquid occurs inside rectangular and circular channels, in which the resulting scale length characterizes the microfluidic behavior.

2. Material and methods

The microflow measurements were carried out in three different straight channels: two of square cross section and one of circular cross section, as shown on table 1.

Table 1. Channels characteristics.

Geometry	Size (mm)	Material
Circular	I.D: 1 L: 75	borosilicate glass
Rectangular: One with polished side walls and one with unpolished side walls	H: 0.8 W: 0.8 L: 75	borosilicate glass (side walls) and PET (bottom and ceiling)

The circular channel was a glass capillary (with smooth internal surface). The rectangular channels were constructed by carefully bonding two glass slides (microscope slides, plain) and two translucent and colorless polyethylene terephthalate (PET) strips. Care was taken to avoid the channel clogging with glue. In the rectangular channels, the mean roughness of the polished (ground glass) side wall is around 0.8 μm, while the unpolished side wall is bright and smooth (with roughness less than 0.05 μm), but it is slightly waviness.

The micro particle image velocimetry technique, micro-PIV [4], from Dantec Dynamics™, was employed to obtain the flow velocity field. A high-resolution CCD sensor (FlowSense 4M Mk2 with a resolution of 2048 x 2048 pixels) and a Nd:YAG double-pulsed laser light source, producing a short pulse of 5 ns of green light (532 nm), were adapted to the inverted microscope (Leica Microsystems™, model 090-135.003) to illuminate and visualize the flow using a 10x magnification lens. PMMA Rhodamine B-Particles of mean diameter 1 μm were seeded in the flow, as tracers. For each liquid flowrate, 100 pairs of images were captured at a frequency of 7.4 Hz, with time between pairs of images varying between 500 μs and 20000 μs (according to the fluid flowrate). A syringe pump (Havard Apparatus® Pump 11 Elite) promoted the displacement of a 10 mL syringe containing the working fluid (deionized water and tracer particles) according to the set flowrate. In previous work, the metrological reliability of this syringe pump was evaluated under the basis of totalized volume upon different liquid flowrate [3].

3. Experimental procedure

Each channel was connected to the syringe pump, separately, through a hose. Deionized water was seeded with PMMA Rhodamine B-Particles, and after setting the flowrate at the pump control panel, the water was sent to the channel and the resulting field of velocity was measured with the micro-PIV system. In this measurement, the center of the flow visualization window was placed at the middle length of the channels (L/2). The water flowrates in the experiments were the following:

Table 2. Water flowrate.

Reference Flowrate (μL/min)		
Circular channel	Rectangular channel (polished walls)	Rectangular channel (unpolished walls)
50	50, 100, 200, 500	10, 50, 100, 200, 500

The flow velocity field was mapped after a mean of 100 pair of the flow images were compared. It is important highlight here that the flow field which is shown in the next section is related to a horizontal plan which is placed around the middle height of the channel. It is said "around" because when microchannels are studied with the use of the micro-PIV technique, the correct position of the plan of measurement cannot be determined with accuracy. This problem means a limitation in the micro-PIV technique use, and this lack shows a need of metrological investigation and the need of procedures and standards development for positioning the focused optical plan when aiming microflow dynamics analysis. It is a challenge. At Inmetro, this issue has received attention by the Fluid Dynamics Metrology Division.

4. Results and discussion

The pictures below show the velocity fields for minimum and maximum flowrates. The graphs were plotted using the same size and color scales to allow a quick comparison.

4.1. Water flowrate 50 μL/min

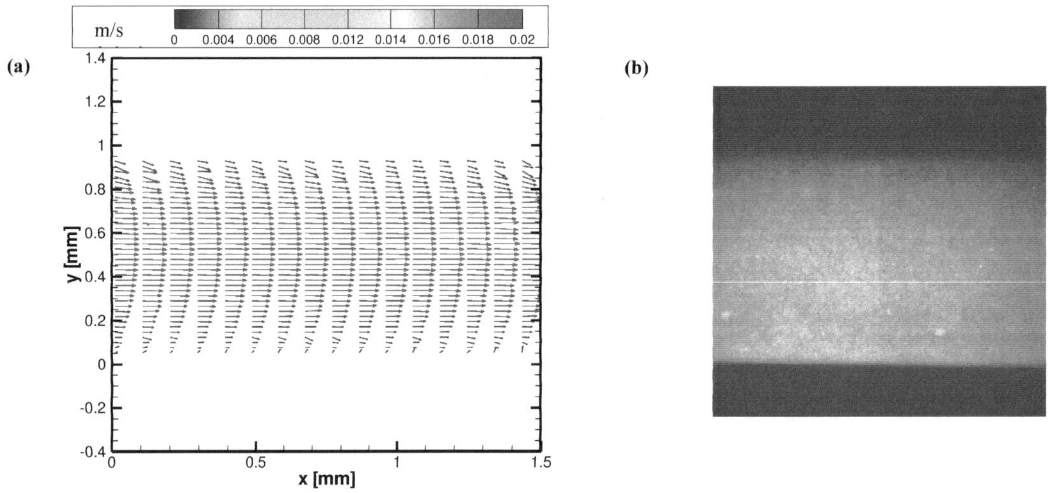

Figure 1. Circular channel: (a) velocity field and (b) frame of the flow.

IOP Conf. Series: Journal of Physics: Conf. Series **975** (2018) 012078 doi:10.1088/1742-6596/975/1/012078

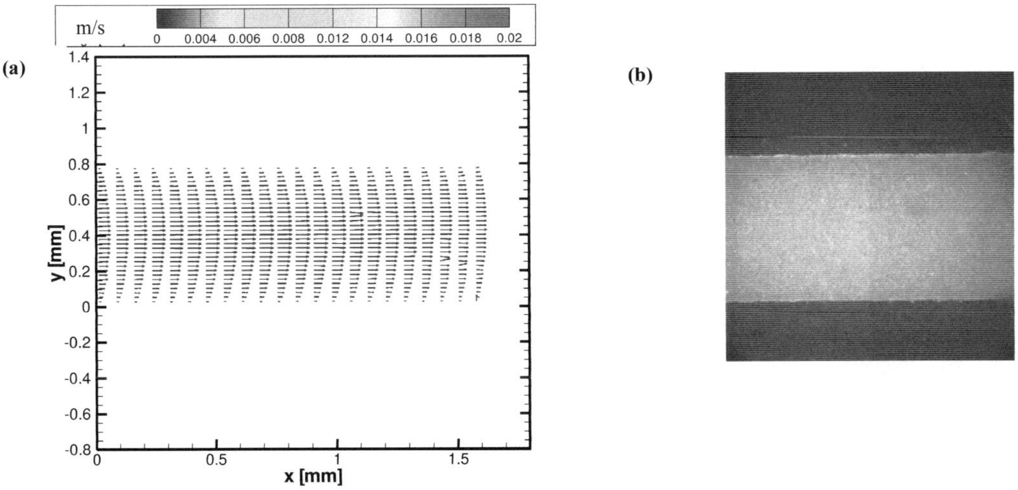

Figure 2. Rectangular channel (polished wall): (a) velocity field and (b) frame of the flow.

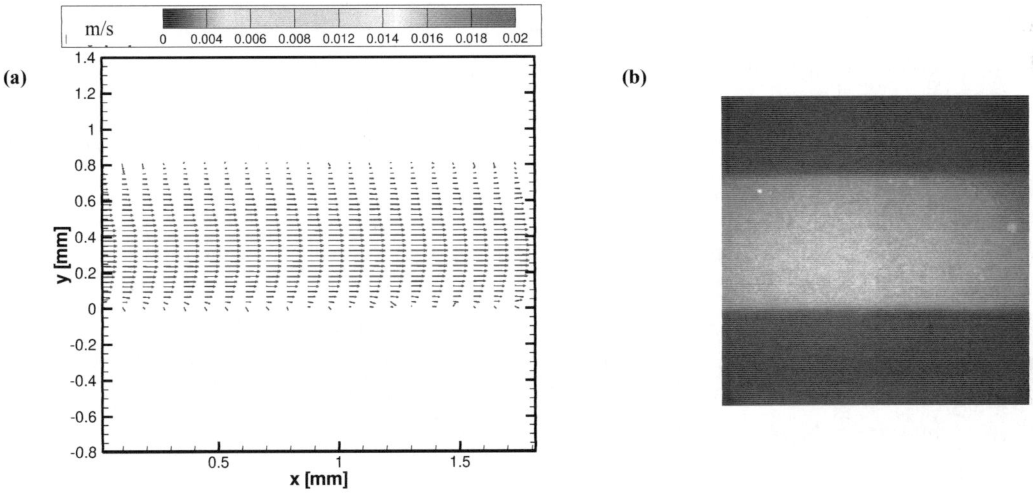

Figure 3. Rectangular channel (unpolished wall): (a) velocity field and (b) frame of the flow.

4.2. Water flowrate 500 μL/min

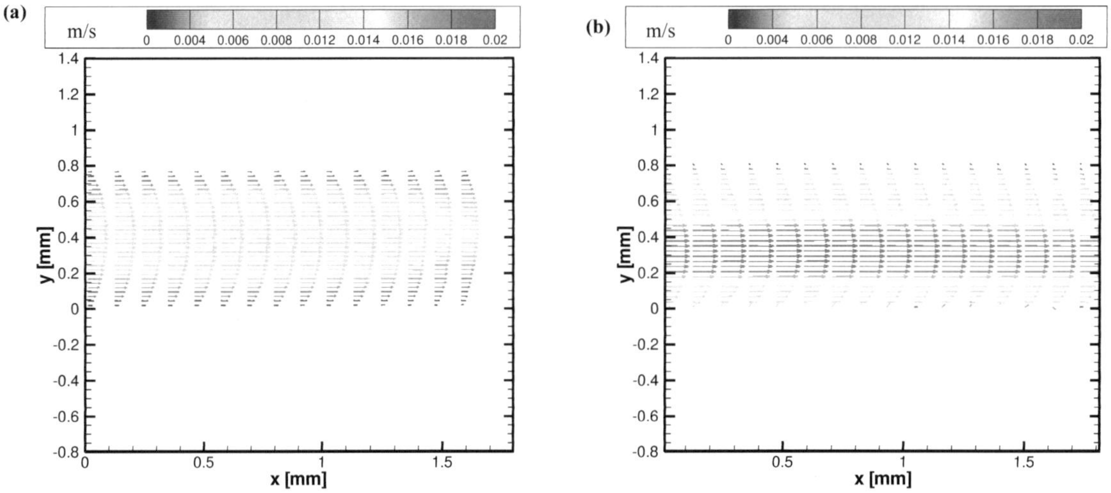

Figure 4. Velocity field in the rectangular channel (a) polished walls and (b) unpolished walls.

As can be seen in Figures 1 to 3, at very low flowrate the flow velocity field in the circular and also in the rectangular channels are within the range near zero. Also, a similarity between the flow velocity field in the rectangular channels shows that, although there are differences between the mean size of the wall roughness of the channels, as the flow is very slow, the differences between the resulting shear rate are not significant. However, when the flowrate is increased, differences are perceived. Since the width of the channel is in microscale, when the side wall is rough, the friction effects generated at the channel wall are propagated under considerable intensity and impact on all main flow, when compared to the case of smooth wall under a same flowrate. As consequence, the longitudinal components of the velocity vector around the centerline of the rough channel become less than those when the wall is smooth. Besides, an asymmetry of the flow field (Figures 3 and 4b) was observed, and it can be due to the unpolished walls be slightly wavy, or even due to some irregularities which could have arisen during the process of channel construction.

Then, the results aforementioned show that a metrological approach considering the surface condition of the geometries where a microflow occurs has to be accounted in microflow analysis, since this can impact on the results of microfluidic processes. Here were shown the first results assessed in this research line at Inmetro, and the investigations will be continued. Other fluids and discussions about influences due to microchannel fabrication techniques will be included in the next studies.

5. References

[1] Wolf H 2010 *Meas. Sci. Technol.* **21** p070101.

[2] Karniadakis G, Beskok A and Aluru N 2005 *Microflows and Nanoflows - Fundamentals and Simulation* Springer New York **25**.

[3] Dias B L, Aquino M H, Farias M H, Koiller J, Gabriel P C and Aibe V Y 2016 *J. Phys. Conf. Ser.* **733** p012007.

[4] Raffel M, Willert C, Wereley S and Kompenhans J 2007 *Particle Image Velocimetry* Springer Heidelberg.

Acknowledgements

The authors are grateful to CNPq, FAPERJ, FINEP, PETROBRAS and INMETRO for their support and sponsorship which have become possible the development of this research.

Institute of Physics
Dirac House, Temple Back
Bristol BS1 6BE UK

ISSN: 1742-6588
ISBN 978-1-5108-6043-8

CURRAN ASSOCIATES INC.
proceedings.com